Advanced Topics in Contemporary Physics for Engineering

This book highlights cutting-edge topics in contemporary physics, discussing exciting advances and new forms of thinking in evolving fields with emphases both on natural phenomena and applications to modern engineering. It provides material for thought and practice in nanophysics, plasma physics, and electrodynamics.

Nanophysics and plasmas are synergic physical areas where the whole is more than the sum of the parts (quantum, atomic and molecular, electrodynamics, photonics, condensed matter, thermodynamics, transport phenomena). The authors emphasize both fundamentals and more complex concepts, making the contents accessible as well challenging. Nanoscale properties and physical phenomena are explained under the umbrella of quantum physics. Advances made in the physical knowledge of the nanoworld, and its metrology are addressed, along with experimental achievements which have furthered studies of extreme weak forces present at nano- or sub-micron scales. The book does not focus in detail on the diversity of applications in nanotechnology and instrumentation, considering that the reader already has basic prior knowledge on that. It also covers an introduction to plasma universe phenomenology, the basics of advanced mathematics applied to the electromagnetic field, longitudinal forces in the vacuum, concepts of helicity and topological torsion, SU(2) representation of Maxwell equations, 2D representation of the electromagnetic field, the use of the fractional derivative, and ergontropic dynamics. The chapters include theory, applications, bibliographic references, and solved exercises.

The synergies of the book's topics demonstrate their potential in critical issues, such as relieving humans from barriers imposed by energetic and entropic dependencies and penetrating the realm of weak forces at the nanoscale. The book will boost both post-graduate students and mature scientists to implement new scientific and technological projects.

Rui F. M. Lobo is a professor of physics and physics engineering at the NOVA School of Science and Technology, NOVA University of Lisbon. He received his PhD in physics from the NOVA University of Lisbon and obtained a Habilitation in Nano-Engineering at this University. He also holds a degree in Chemical Engineering from Instituto Superior Técnico – University of Lisbon. He is among the pioneers working with single polyatomic molecule experiments in binary neutral collisions and is presently leading research in experimental nanophysics and nanotechnology for clean energy, hydrogen technology, and decarbonization. He has authored several scientific refereed publications in journals, books, book chapters, and conference proceedings. He has been a senior member of the Engineers National Order and has worked with international scientific committees in molecular beams and nanometer-scale technology. He was a Fellow Researcher at several renowned institutions including Max Planck Institute, Hahn Meitner Institute, Rice University, Osaka University, Ohio University, among others.

Mario J. Pinheiro is a professor of physics at the Instituto Superior Técnico, University of Lisbon. He received his PhD in physics and obtained the Habilitation in Physics from the Technical Institute of Lisbon and got his Diplome d'Études Approfondies in Plasma Physics at the University of Orsay Paris-XI. Mario's main interests are in the fundamentals of mechanics and the electromagnetic field, developing advanced propulsion systems for the next generation of spacecraft, and public awareness for the role of science in our societies. He was an invited researcher at the University of Knoxville-USA, working on plasma thrusters for drones, co-founder of the physics conferences for the Community of Portuguese Language Countries, and author of more than 100 papers, several books in advanced propulsion concepts for spaceflight, and developed a new theory of dynamics and electrodynamics, named ergontropic dynamics.

Advanced Topics in Contemporary Physics for Engineering

Nanophysics, Plasma Physics, and Electrodynamics

Rui F. M. Lobo and Mario J. Pinheiro

CRC Press is an imprint of the
Taylor & Francis Group, an **informa** business

First edition published 2023
by CRC Press
6000 Broken Sound Parkway NW, Suite 300, Boca Raton, FL 33487-2742

and by CRC Press
4 Park Square, Milton Park, Abingdon, Oxon, OX14 4RN

CRC Press is an imprint of Taylor & Francis Group, LLC

ISBN: 978-1-032-24763-2 (hbk)
ISBN: 978-1-032-25799-0 (pbk)
ISBN: 978-1-003-28508-3 (ebk)

DOI: 10.1201/9781003285083

Typeset in Palatino
by MPS Limited, Dehradun

To our families

Contents

Preface

When we decided to write this book on some relevant and current advanced topics in Physics, we were aware of the need to bring back to teaching topics of research of the 21st century. We felt that textbooks insisted too often on research conducted from 19th to the middle of the 20th century, and that stance could undermine scientific and technological progress. This book will be valuable for those who have previous training in classical and quantum physics, atomic and molecular physics, and the physics of continuous media. Students who desire to expand their background or carry out research in physics or physics engineering know that a significant number of diverse areas of interest are currently unavoidable. Among them, nanophysics, plasmas, and electrodynamics play a relevant role. Here both the motivated reader and the investigator's apprentice will find material for thought and practice in these areas together with bibliographic references, where they could deepen the knowledge exposed. Also, some specific topics are addressed, hoping to motivate students or general readers to search for more in-depth information in the technical literature. We recommend the reader to test ideas with accessible programs, converting theories and equations into numerical results. The software complements the hand and mind searching for a more understandable universe, unraveling its secrets.

This book intends to show that nanophysics and electrodynamics are synergetic physical areas where the whole is more than the sum of the parts (quantum, atomic and molecular, photonics, condensed matter, thermodynamics, transport phenomena). Both areas need to become part of the advanced training of students wishing to specialize in areas of physics and engineering. We expect they will contribute to the progress of society in a wise way, alleviating humans from traditional barriers imposed by the energetic and entropic dependencies.

23 May 2022, Lisbon, Portugal

1

Advanced Phenomena in Plasma

1.1 Introduction

The cosmic plasma theory is a distinctly scientific approach to the functioning of our universe that relies on a fundamentally different base than the orthodox cosmology, based on the dominance of gravitational forces, not duly recognizing the fundamental role of magnetic and electric fields in the formation of the galaxies and its cellular organization. It is an entirely different view of the nature and evolution of the universe that relies on data collected on the entire electromagnetic spectrum, starting from the observed fact that 99.999 percent of the universe is plasma, a universe dominated by electrical currents and electromagnetic forces that surpasses largely the gravitational force in magnitude. These facts do not exclude the validity of the general theory of relativity and related disciplines, which are extremely valuable to explain the behavior of matter in their advanced phase of existence, not in their origin. According to Tsiolkovsky's formula, a chemical-fuel rocket is not suitable for space flights and new solutions may be supplied by plasma physics. Hence, in this Chapter, we will introduce some new concepts and working mechanisms created in the framework of the cosmic plasma theory that may be useful to tackle problems in the terrestrial context as well.

1.2 The Plasma Universe Theory

According to the Plasma Universe Theory it is electromagnetic forces, and not gravity, that shapes the universe. Indeed, we live in a plasma dominated universe, and, therefore, it is natural to assume that, in the past, the universe was in a plasma state, which implies that this study also must be considered when dealing with cosmological issues [1, 2].

However, a growing number of experimental data confront scientists with the need to revise the field of theoretical plasma physics that was developed along the lines of thought provided by the kinetic theory of gases (e.g., Klimontovich's microscopic formalism and the Bogoliubov-Born-Green-Kirkwood-Yvon hierarchy equations for multiparticle distribution functions), with new concepts that may provide a new framework to understand more deeply important issues related to the origin of the cosmos, the solar system and planets, solar flares, or basic physics associated with plasma streamers.

The plasma state (see, e.g., Ref. [3] for the origin of the word) consists of electrically charged particles that behave collectively under the influence of electromagnetic fields. The charged particles are electrons, positive (protons H^+, and N_2^+, O^+, NO^+) and/or negative ions,

DOI: 10.1201/9781003285083-1

but they can also be formed by charged grains or dust particles. The density of the plasma is usually low (except inside stars) so that short-range binary collisions are negligible, while long-range electromagnetic forces, predominates.

The range of values of the plasma density spans from a few electrons per unit of volume up to electrons per cubic centimeter. The temperature range goes from a few milli-electronvolt (eV) registered in neon and fluorescent tubes to thousands eV in stars.

The plasma state can be created at the cost of a certain amount of energy, achieved by injecting energy to a gas or vapor in order to ionize the constitutive atoms and molecules. The source can be external, or placed at the heart of the medium through exothermal chemical reactions or nuclear reactions, for instance. After the energy source is disconnected, the natural process is the attraction of positive to negative particles, the plasma recombination process. The plasma state is characterized by certain properties, such as:

- the Debye length, λ_{De},
- neutrality,
- and the so called plasma parameter.

Besides the cellular structures that plasmas can form, it is also possible the formation of filamentary structures. The cause is the electrons that can flow in a background of positive ions, forming a ring of the magnetic field around it, subject to typical instability, known as the pinch effect.

The idea of the plasma universe model was principally advanced by Hannes Alfvén (1908–1995) and C. G. Fälthammar [1, 2]. The basic idea sustains that the same physical laws hold from the laboratory to the intergalactic space, where a network of electric currents fill the space, transferring momentum and energy over a large distance in the form of cellular structure, see Figure 1.1, illustrating the existence of a type of neuro-cosmos framework. The approximate magnitude of constitutive elements in the two different levels is remarkable: 100 billion brain cells, which seems close to the 125 billion galaxies estimated by the Hubble telescope of the Deep Field, a number that tends to increase with the astronomical observations with other instruments, such as radio telescopes, infrared cameras, X-ray and -ray cameras, opening the spectra well beyond.

This fact of nature is an argument to support those who believe that the human brain was created in the image of the cosmos, sustaining that the human being was capable of mapping the whole universe [4].

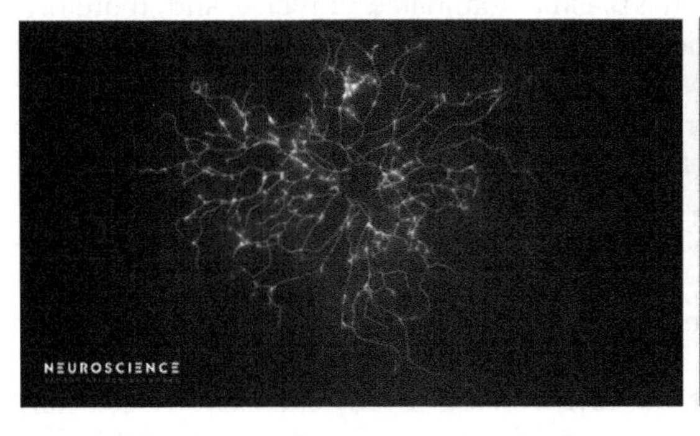

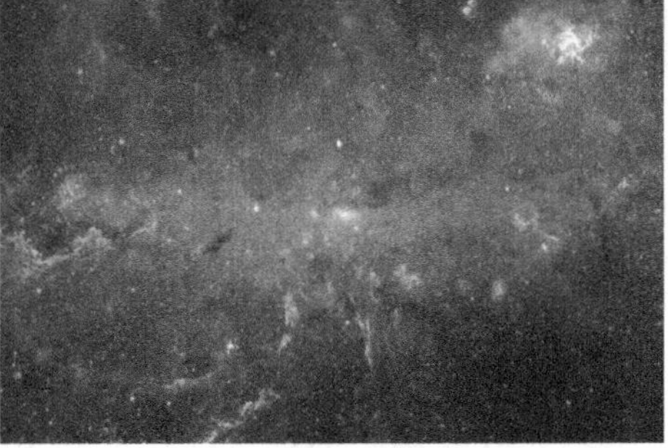

FIGURE 1.1
The structure of a brain cell resembles the entire universe.

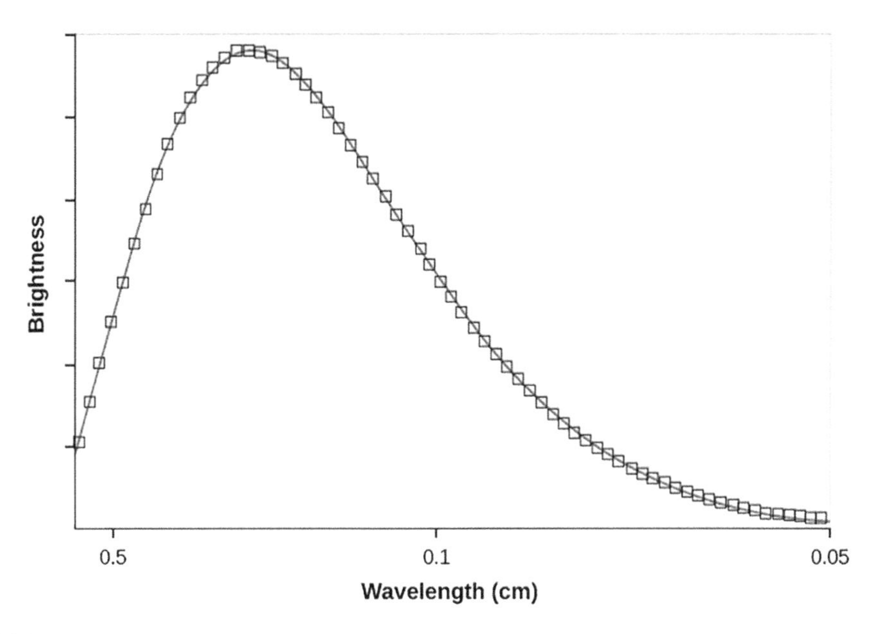

FIGURE 1.2
Spectrum of the cosmic background radiation. The curve through the data points is that of a thermal spectrum of 2.73 K. Image courtesy: Astronomy, Andrew Fraknoi et al. (OpenStax, Rice University, 2018). Download for free at https://openstax.org/details/books/astronomy.

Figure 1.2 shows the spectrum of the cosmic background radiation. Notice the peak at about the frequency $\nu = 2.79 \times 10^{11}$ Hz, corresponding to the temperature of the cosmic background radiation.

This frequency gives the temperature of the cosmic background radiation through Wien's displacement law $\lambda_m T = 2.898 \times 10^{-3}$ mK. The presence of this peak suggests conceivable means to extract energy from this astronomical "sea" of energy [5]. The potential extraction of energy from the "radiant energy," as termed by Nikola Tesla, attained now increasing value with the global energy crisis, since our civilization is ultimately condemned to collapse if no solution is found to an increasingly demanding need.

1.2.1 Parametric Resonance and Nonlinear Phenomenon

Michael Faraday (in 1831) was possibly the first to observe the phenomenon of **parametric resonance** when he noticed that surface waves excited by vertical excitation of a fluid-filled column had half the frequency of excitation. The theory of parametric resonance, established in large part by L. I. Mandelstam and N. D. Papaleksi (see, e.g., [6, 7]), suggests the possibility to control and amplify fluctuating external driven force acting on a physical system. The resonance phenomena hold singular relevance in the natural world, science and technology, medicine, civil engineering, aeronautics, rocket technics, and astronautics. Parametric resonance and nonlinear phenomena explain to some extent, the process of self-organization and adaptive grouping in stable formations. The most popular example of parametric resonance is a child swinging on a swing. The child's movements at the extreme points correspond to a periodic change of the physical pendulum length, and with this movement, the amplitude increases greatly. The phenomena of parametric resonance differ from the forced oscillations because, in this last one, what counts is the matching of the driving frequency with the oscillator's natural frequency, while the strongest parametric resonance is excited when the modulation's frequency ω is twice the system's natural frequency ω_0:

$$\omega = \frac{2\omega_0}{n} \tag{1.1}$$

where $n = 1, 2, \ldots$. In a linear system, the amplitude of parametrically excited oscillations can grow indefinitely. However, in real-life situations, nonlinearities restrict the growth of the amplitude. In addition, parametric resonance can only take place if the system is already characterized by, at least, some feeble natural oscillations, and with friction, being necessary that the parameter's amplitude of modulation exceeds a certain threshold.

Typical mechanical-electrical systems that exhibits parametric resonance can ve designed. For example, a circuit consists of an LC circuit with a constant inductor L connected in series with a capacitor whose value changes with time $C(t)$ due to its electrodes (with charge $q(t)$) subject to a prescribed motion $d(t)$, and hence $C(t) = \epsilon S/d(t)$, with the usual notation (and current $i = \dot{q}$). Kirchhoff's laws give immediately

$$\ddot{q} + \frac{d(t)}{\varepsilon SL} q(t) = 0 \tag{1.2}$$

The study of parametric resonance in a dynamical system, leads us frequently to the Mathieu's equation, a special case of second-order differential equation with a periodic coefficient

$$\frac{d^2u}{dt^2} + (a - 2q\cos 2t)u = 0 \tag{1.3}$$

or to the more standard form of the Mathieu-Hill equation

$$\frac{d^2u}{dt^2} + a(t)u = 0 \tag{1.4}$$

with $a(t + T) = a(t)$, T denoting the parametric resonance period of excitation. The general solution of **Mathieu's equation**, as a consequence of **Floquet's theorem**, is given by

$$u(t) = Ae^{\mu t}\phi(t) + Be^{-\mu t}\phi(-t) \tag{1.5}$$

where A and B are constants of integration, μ is the periodicity exponent and $\phi(t)$ has period π (see, e.g., [8]).

Parametric resonance is the basis of a new discipline called *Energy Harvesting*, and the usual devices used are micro-electromechanical system (MEMS) devices, made up of a micro-cantilever structure and a transducer. It is enough that an external force is applied to the cantilever perpendicularly to the length, instead of transversely, for the parametric resonance mechanism to generate more energy from the same amount of vibration, from either the natural vibrations of the environment or any other source [9].

In a plasma, parametric resonance can easily be achieved when a voltage of frequency $f = 2f_{pe}$ is applied to parallel plane electrodes, cathode and anode (with f_{pe}

denoting the electron plasma frequency, also known as the Langmuir frequency $f_{pe} = \frac{1}{2\pi}\sqrt{\frac{4\pi n_e q^2}{m_e}} = 5.7 \times 10^5 \sqrt{n_e}$ Hz). At this frequency, the plasma electron oscillation is amplified, and also the plasma ion oscillation by coupling between ions and electrons.

The dispersion relation $\omega = \omega(k)$ of an electron plasma wave is given by $\omega^2 = \omega_0^2(1 + 3\lambda_{De}k^2)$, ω denoting the frequency corresponding to the wave number k, λ_{De} the Debye length and ω_0 the (electron) plasma frequency. If we replace iku_x by the gradient, $iku_x \rightarrow \nabla$, assuming a one-dimensional plasma along direction u_x, the usual oscillatory harmonic equation for the electric potential ϕ, $\frac{\partial^2\phi}{\partial t^2} + \omega^2\phi = 0$, is replaced by

$$\frac{\partial^2\phi}{\partial t^2} + \omega_0^2\phi - 3\lambda_{De}\omega_0^2\nabla^2(\phi) = 0 \tag{1.6}$$

The solution of this equation in the case of a slab of length l_0, with $\phi(x = 0) = \phi(x = l_0) = 0$ is

$$\phi(x, t) = A[\cos(\Omega t) + B\sin(\Omega t)]\sin\left(\frac{n\pi x}{l_0}\right) \tag{1.7}$$

with $n = 1, 2, 3, \ldots$. It can be shown that, if we seek a solution of the form

$$\phi(x, t) = g(t)\sin\left(\frac{n\pi x}{l_0}\right) \tag{1.8}$$

we obtain Mathieu's equation.

Exercise: Show that:

$$\ddot{g} + \Omega^2[1 + 2\sigma\cos(2 + \epsilon)\Omega t]g = 0 \tag{1.9}$$

with $\epsilon \ll 1$, $\Omega = \omega_0\left[1 + \frac{3}{2}\left(\frac{n\pi\lambda_{De}}{l_0}\right)^2\right]$, and $\sigma = 3\left(\frac{n\pi\lambda_{De}}{l_0}h\right)$, with $h \ll 1$. [**hint:** see Ref. [10]]

1.2.2 Synergetics

The more comprehensive theory of resonance in nonlinear oscillating circuits is at the basis of Synergetics, or more generally, the theory of self-organizing systems, that aims to explain the formation of typical phenomena, as the intergalactic structure illustrated in Figure 1.1, or a dielectric barrier discharge, as shown, for example, in Ref. [11].

The study of complex systems requires tackling complexity, order parameters, and slaving principle (in the sense of Hermann Haken [12].[1] In systems composed of many parts it is adequate to represent the behavior of the system in the general form of an evolution equation for a state vector $\mathbf{q}(\mathbf{r}, t)$ to describe the system at a mesoscopic and/ or macroscopic level:

$$\boxed{\frac{\partial \mathbf{q}}{\partial t} = \mathbf{N}(\mathbf{q}\alpha) + \mathbf{F}(t)} \tag{1.10}$$

Here, **N** is a non-linear function of **q** which may also contain differential operators, such as the gradient of energy, α denoting a control parameter, e.g., the free energy. $\mathbf{F}(t)$ is a fluctuating force.

Which process brings galaxies together from apparently unrelated regions of the Cosmos to form systems with typical structures? Observational cosmology supports that the so-called large structure of our universe, i.e., clusters and superclusters, is most probably the outcome of small perturbations of matter and gravitational potential that occurred in the primordial gas in the isotropic Friedmann background. However, this view is not fully partaken by the plasma universe theory.

The cellular structure observed issues from the actuation of electric and magnetic fields associated with planets, stars, galaxies, and clusters of galaxies have been observed. Magnetic forces can be locally higher than gravitational forces and instruments carried on spacecraft measured EM fields locally. However, their role in the Earth's magnetic field requires the understanding of the collective properties of charged particles, at least in the fluid approximation, adding the action of current layers in space. Magnetized planets and cosmic bodies display a nearly magnetic dipole (axisymmetric) field, that concurrently with the interplanetary magnetic field (IMF), normally results in phenomena with higher complexity. The IMF has a solar origin and it was hinted at in sunspots, firstly by G. E. Hale, in 1908. It permeates the space between the Sun and planets. Near the Earth, the IMF has a week value of about 10^{-2} teslas (compared with its value at the surface of the Sun, which is around 10^{-4} teslas).

1.2.3 The Birkeland's Terrella Experiment

The inspirational first model of the Sun as the source of the solar corpuscular radiation was given by the Norwegian scientist Kristian Birkeland (1867–1917) when he did his terrella experiment (1896). It consisted of a 320-liter vacuum chamber with a hot cathode at the bottom and top surfaces simulating the Sun, and the uniformly magnetized sphere with 24 cm simulating the Earth (the terrella, Latin of "little earth") as the anode. An electrostatic potential between the cathode and the anode creates a stream of electrons (corona discharge) and its impact on a thin region of the polar caps gave a fluorescence that mimics the aurora lights. Birkeland was convinced that Earth and the Sun could be represented as a Crookes tube at a cosmic scale, with the sunspots acting as the cathodes, source of electrons that collide and stream along the geomagnetic field and giving rise to the aurora.

1.2.4 Celsius and the Aurora Borealis

Andres Celsius was the first to register observations of aurora borealis in Sweden and made a report to the Proceeding of the Royal Society entitled 'Observations of the Aurora Borealis in Sweden', in 1722. He made observations using a big compass needle, and thus referred for the first time, to electromagnetic phenomena as the source of the aurora.

Plasma pervades nearly the whole universe (the word expressing unity in diversity), it is the dominating state in the universe, except for some regions, for example, the Earth's surface and the upper atmosphere. The Earth's magnetic field is a shielding envelope that guards our planet, as illustrated in Figure 1.3.

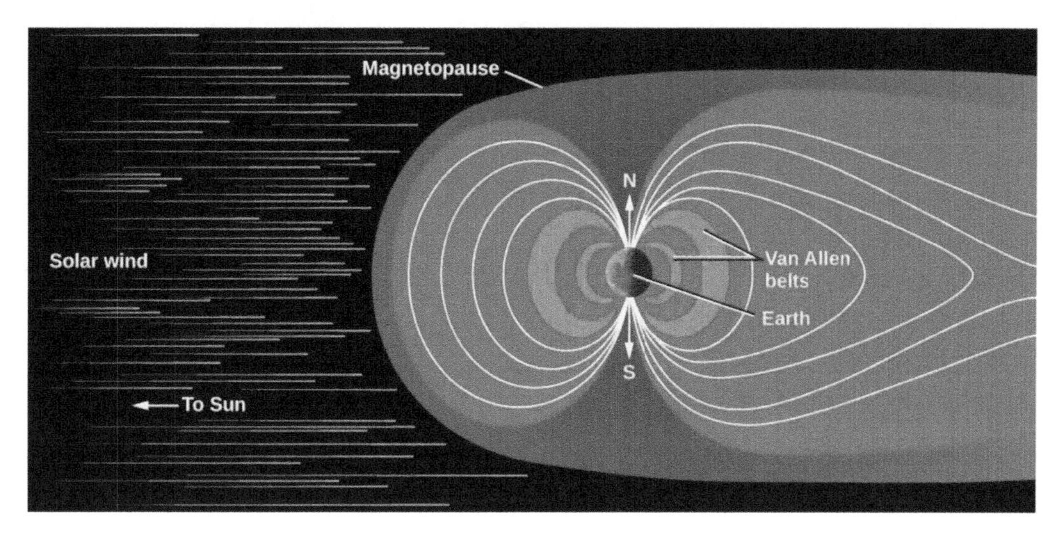

FIGURE 1.3
The earth magnetosphere. Image courtesy: Astronomy, Andrew Fraknoi et al (OpenStax, Rice University, 2018). Download for free at https://openstax.org/details/books/astronomy.

1.2.5 The Interplanetary Space

Before the 1950s it was commonly presumed that the interplanetary space was pure void, although it was known the Sun released powerful particles when solar flares occurred. Satellite observations have shown that the Sun continuously creates plasma in interplanetary space.

The source of the solar wind is the lower corona, the charged particles are accelerated radially from the Sun, attaining supersonic speeds at a few solar radii while the plasma becomes more rarefied (collisionless, meaning that the collisional mean-free-path exceeds the typical length for density variations, and resulting in an electric current flow with minor resistance). This situation favors the "frozen in" of the magnetic field lines to the solar wind, which is thus carried away from the Sun into space; this is the interplanetary magnetic field (IMF).

The Sun is in a plasma state, is not just a hot gaseous stellar object (expected to be cooling with distance), and the proof is that its temperature increase from its surface in the chromosphere to the corona, where its temperature increases abruptly. Our knowledge of the processes governing the abundance of elements, nuclear reactions, and the formation of the internal magnetic fields is still incomplete. However, the knowledge acquired about the Sun's surface structure is more complete. It is formed by a photosphere (a weekly ionized atmosphere with roughly 0.5 eV), chromosphere (about 1.7 eV, extending 5,000 km above the transition region to the inner corona), and corona (characterized by plasma temperatures of 170 eV and the place of explosively unstable magnetic field configurations, solar flares, and X-ray emission).

Besides the Sun, in our solar system stands another large plasma structure around Jupiter and Saturn, the Jupiter-Io plasma torus. The torus is filled with energetic sulfur and oxygen ions that have a temperature of about 100 thousand kelvin. This huge toroidal structure extends from an outer diameter 25 times the radius of the planet and an inner diameter of around 15 times the radius of Saturn (see Figure 1.4).

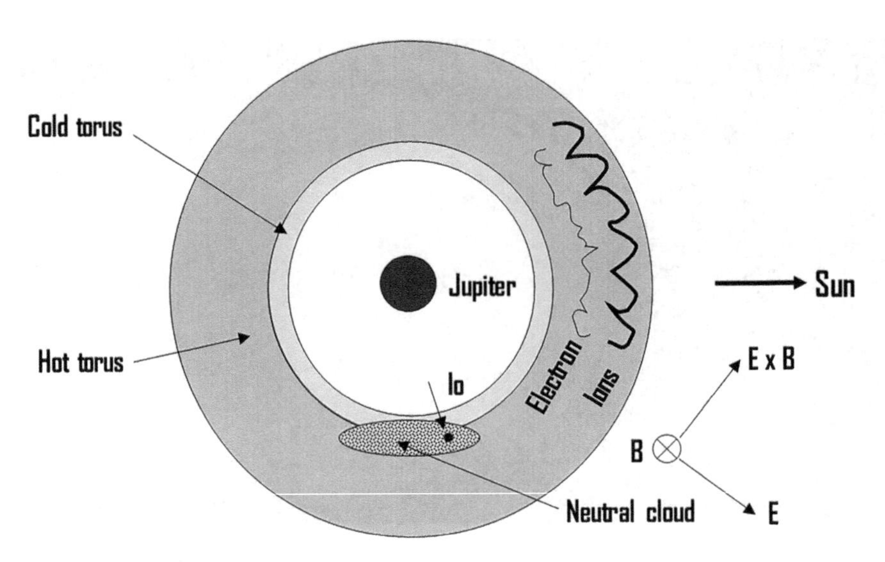

FIGURE 1.4
Jupiter-Io schematic of the plasma torus, projected onto Io's orbital plane. All dimensions are exaggerated.

1.2.6 The Birkeland Current

Connecting Jupiter and Io is a huge Birkeland current-carrying about seventy times the total electric power generated by all nations on our planet. No fusion experiment was capable of reproducing these configurations. This cosmic machine is an enormous electric generator. Birkeland[2] was the first initiating an experimental approach to connect experimental plasma physics and cosmic plasma physics.

Birkeland currents are field-aligned currents and can carry about 1 million amperes, heating the upper atmosphere and increasing drag on low-altitude satellites. Field-aligned current is a set of currents that flow in a direction parallel to the geomagnetic field lines connecting the Earth's magnetosphere to the Earth's high latitude ionosphere. They contribute to the Aurora, caused by the interaction of the plasma in the Solar Wind with the Earth's magnetosphere.

The field-aligned currents are closed by the discharge-current-carrying electrons arriving at the E region of the ionosphere,[3] where the electric conductivity perpendicular to the magnetic field is high. With the expected electron collisional processes, electrons lose energy, and hence they need a minimum of a few keV to reach the lower ionosphere. This current will create a perturbation $\delta\mathbf{B}$ in a direction perpendicular to the field:

$$\mu_0 J_{\|} = (\nabla \times \delta\mathbf{B})_{\|} \tag{1.11}$$

Birkeland currents emerge repeatedly in filamentary or twisted "rope-like" magnetic structures. Multi-terawatt pulsed power generators can generate kilometric filaments in the atmosphere and the laboratory allowing the study of those fundamental blocks of Nature, the Birkeland currents [13]. The propagation of electron beams along an axial magnetic field results in the beam breakup into discrete vortexlike current bundles when a threshold is surpassed [14]. The threshold depends on the beam current or a given distance of propagation is overcome. This breakup of the beam is analog to the Kelvin-Helmholz fluid dynamical shear instability occurring when there is velocity shear in a single continuous fluid or a velocity difference across the interface between two fluids. The hollow beam of electrons forms a circle of vortices in a formation called the diocotron instability that originates filamentation and the "auroral curls". Z-pinch may also occur with Birkeland currents, followed by twisted or braided rope. Adjacent Birkeland current filaments tend to be long-range attractive ($F \propto 1/r$), and short-range repulsive ($F \propto 1/r^3$).

1.2.7 The Jupiter-Io Generator

Comets also possess their magnetospheres, the formation beginning when they approach the Sun. Then, water molecules sublimate from the comet and become ionized by ultra-violet light from the Sun. The formed ions began to be accelerated by the solar wind electric field, and a sharp boundary assembled, shielding the comet atmosphere from direct interaction with the solar wind, a general mechanism actuating on comets and planets.

Charged particles accelerate in an electromagnetic field, according to $\mathbf{F} = q\mathbf{E}$. On Earth, particle accelerators can be operated through the potential gradient between a cathode and anode, like in a cathode ray tube (CRT), pulsed high-voltage, betatrons, resonant cavities, and waveguides, among additional mechanisms. In 1952, Alfvén and Wernholm suggested using an electron beam for an accelerating system [15, 16]. The main idea relies on the high fields associated with bunches of particles to drag other particles along with them, fields associated with one bunch of particles accelerating another bunch of particles. This simple idea is the basis of the class of collective accelerators.

The plasma accelerator is a very promising tool to obtain larger accelerating fields than in conventional devices, capable to generate high-energy particles (and as well, a neutral bunch of particles) with a relatively inexpensive apparatus, easily installed in any laboratory or university. Bostick [17] proposed that high energy ions could be accelerated by a collective acceleration to simulate astrophysical plasmas. The understanding of the accelerating mechanism that propels and heats the solar wind is a decade-old mystery.

One such cosmic accelerator is the Jupiter-Io system. Io's orbital velocity of 17 km/s is accompanied by an extended cloud of neutral particles (sodium, potassium, sulfur, and oxygen) probably as a result of its active volcanism. Due to collisional ionization with magnetospheric electrons, their lifetime is finite. The neutral cloud is ionized and co-rotates with Jupiter, forming a vast plasma torus, see Figure 1.4.

1.2.8 Electric Double Layers, Critical Velocity and Pinch Effect

According to Alfvén & Carlqvist [18], it is crucial to introduce in the study of the cosmic plasma physics, the concepts of electric double layers, critical velocity, and pinch effect. One potential source of instability in a flow of charged particles is its constriction, the Bennett pinch. This phenomenon might explain the formation and abundance of filamentary structures on a cosmic scale.

The Faraday disk dynamo has huge importance in our technological civilization, but this mechanism also occurs in astrophysics when the conductive plasma moves in the cosmic magnetic field. They are known to power aurorae on Uranus, sunspots, binary stars, the Jupiter-Io system, and the Earth, for example.

1.2.9 The Electric Charge of the Sun

The hypothesis that the Sun may have an electric charge given by (C), was proposed by Prof. Bailey in 1960 for the explanation of the maximum energy found for a primary cosmic ray particle and other astronomical phenomena [19]. Groundbreaking measurements of the Sun's electric field have shown that the Sun's electric field has some degree of influence over the solar wind but only plays a minor part in the acceleration provided to the solar wind. Then, other mechanisms might be giving the solar wind most of its kick [18].

Birkeland currents, Alfvé waves, electric double layers, pinch effect and the electric charge of the Sun should be considered the building blocks of a new theory of the plasma universe.

1.3 Transport Phenomena in Multicomponent Systems

In this section, we introduce the fundamental equations describing the flow of chemically reacting multicomponent mixtures in the presence of heat and mass transport. We begin with a microscopic approach and conclude with the basics of the *continuum approach*,[4] more useful for engineering fluid models.

1.3.1 Individual Trajectories

In conditions where collisional encounters between neutrals or charged particles are rare, such as in very low pressure or high-temperature plasmas,[5] the understanding of the individual trajectories in the self-consistent E and B fields leads to a reasonable understanding of the fundamental plasma phenomena.

The particle's trajectories must obey the Lorentz equation:

$$\mathbf{F} = e[\mathbf{E} + \mathbf{v} \times \mathbf{B}] \tag{1.12}$$

1.3.2 Microscopic Description of Plasma

1.3.2.1 Velocity Distribution Function

We designate the local velocity of species α with respect to fixed reference frame by the vector $\mathbf{v}_\alpha$. This is not the velocity of individual molecules of species α but rather the average velocity $\overline{\mathbf{v}}_\alpha$ with which a group of molecules are moving within a small volume dv, defined by

$$\overline{\mathbf{v}}_\alpha(\mathbf{r}, t) = \frac{1}{n_\alpha} \int \mathbf{v}_\alpha f_\alpha(\mathbf{r}, \mathbf{v}, t) dv_\alpha \tag{1.13}$$

Here, $f_\alpha(\mathbf{r}, \mathbf{v}, t)$ is the distribution function in coordinate-velocity space.

The description of the gas mixture of molecules in a non-equilibrium state can be done in terms of the distribution function $f_\alpha(\mathbf{r}, \mathbf{p}, t)$ (here now in phase space) such that the probable number of molecules of specie α dN_α with position coordinates in the range $d\mathbf{r}$ about $\mathbf{r}$ and with momentum coordinates in the range $d\mathbf{p}_\alpha$ about $\mathbf{p}_\alpha$ is given by:

$$dN_\alpha = f_\alpha(\mathbf{r}, \mathbf{p}_\alpha, t) d\mathbf{r} d\mathbf{v} \tag{1.14}$$

In the case a homogeneous thermodynamic equilibrium is onset, $f(v)$ is a Maxwellian function:

$$f(v) = \frac{n}{(a\sqrt{n})^3} e^{-\frac{v^2}{a^2}} \tag{1.15}$$

with $a = \sqrt{2k_B T/m}$. The particle density at a given point $\mathbf{r}$ is given by

$$n(\mathbf{r}) = \int f(\mathbf{r}, v, t) dv \tag{1.16}$$

The average speed is given by

$$\mathbf{u} = <\mathbf{v}> = \frac{1}{n} \int \mathbf{v} f(\mathbf{r}, \mathbf{v}, t) dv \tag{1.17}$$

We may define

$$\mathbf{w} = \mathbf{v} - \mathbf{u} \tag{1.18}$$

verifying

$$< \mathbf{w} > = 0; \quad < w^2 > \neq 0 \tag{1.19}$$

1.3.2.2 Evolution of the Distribution Function

The number of particles inside the element of volume $d\mathbf{r}d\mathbf{v}$ at instant t nearby the point $(\mathbf{r}, \mathbf{v})$ in phase space is given by $dN = f(\mathbf{r}, \mathbf{v}, t)d\mathbf{r}d\mathbf{v}$, but at later instant of time t' the same particles arrive at point $(\mathbf{r}', \mathbf{v}')$ occupying the same volume $d\mathbf{r}d\mathbf{v}$ of phase space. We can show that

$$d\mathbf{r}d\mathbf{v} = d\mathbf{r}'d\mathbf{v}' \tag{1.20}$$

This means that *the volume in phase space occupied by a given number of particles remains the same*. In the absence of collisions we have

$$dN = dN' \text{ No collisions} \tag{1.21}$$

and when in presence of collisions

$$dN - dN' = (\delta N)_{col} \text{ if collisions occur} \tag{1.22}$$

We may write

$$dN' = f(\mathbf{r}', \mathbf{v}', t + \delta t) d\mathbf{r}'d\mathbf{v}' \tag{1.23}$$

To obtain the Boltzmann kinetic equation, let us consider an **ideal gas**. The state of an individual particle of this gas at a given instant of time is described by its vector position $\mathbf{r}$, its velocity $\mathbf{v}$, and eventually its internal degrees of spin, which we shall denote by the generalized parameter I. The **ideality of a gas** is satisfied if the mean distance between the particles λ is much larger than their interaction radius R_i. We may write

$$n = \frac{N}{V} = \frac{3N}{4\pi\lambda^3} \Rightarrow \lambda = \left(\frac{3N}{4\pi n}\right)^{1/3} \tag{1.24}$$

$$\sigma = \pi R_i^2 \Rightarrow R_i = \sqrt{\frac{\sigma}{\pi}}$$

$$\lambda \gg R_i \Rightarrow n\sigma^{3/2} \ll 1$$

When the ideality condition 24 is verified, each particle moves as a free particle most of the time, except for a small fraction of the time of flight ($\sim n\sigma^{3/2}$) when it interacts with other particles and the state of the colliding particle changes. Within the range of validity of the ideality condition, the simultaneous collision of three particles is a more rare event than the collision between two particles, and we may neglect it, limiting ourselves to study only binary collisions.

All along the trajectory we have $\mathbf{r}' = \mathbf{r} + \mathbf{v}dt$, and so

$$f(\mathbf{r} + \mathbf{v}\partial t, \mathbf{v} + \mathbf{a}\partial t, t + \partial t) - f(\mathbf{r}, \mathbf{v}, t) = (\delta f)_{col} \tag{1.25}$$

and

$$f(\mathbf{r}, \mathbf{v}, t) + \frac{\partial f}{\partial t}\delta t + \mathbf{v}\cdot\frac{\partial f}{\partial \mathbf{r}}\delta t + \mathbf{a}\cdot\frac{\partial f}{\partial \mathbf{v}}\delta t - f(\mathbf{r}, \mathbf{v}, t) = (\delta f)_{col} \tag{1.26}$$

This last equation leads us immediately to the **Boltzmann kinetic equation**:

$$\frac{\partial f}{\partial t} + \mathbf{v}\cdot\frac{\partial f}{\partial \mathbf{r}} + \mathbf{a}\cdot\frac{\partial f}{\partial \mathbf{v}} = \left(\frac{\partial f}{\delta t}\right)_{col} \tag{1.27}$$

In the presence of binary collisions, the **Boltzmann collision term** is given by

$$\left(\frac{\partial f}{\partial t}\right)_{col} = \int \ldots \int (f'f_0' - ff_0)\,|\mathbf{w} - \mathbf{w}_0|\,bdbd\phi d\mathbf{w}_0 \tag{1.28}$$

Here, $f'f_0'$ are the product of the distribution functions *after* the binary collision, ff_0 is the product of the distribution functions *before* the collision; the subindex 0 refers to the *target* particles, and b is the *impact parameter*. The collision integral presented in 28 represents the number of particles in a given state with a given velocity in the unit of volume around a given point per unit of time.

Exercise: Using the Boltzmann kinetic equation, obtain the equation of diffusion in the sense of Einstein.

Solution: Let it be $f(x, t)$ at instant of time t the number of particles in the interval $(x, x + dx)$ and let us consider an instant of time $t + \tau$ ($\tau \ll 1$) at another point x' an interval of space $dx = dx'$ where the number of particles is $f(x', t + \tau)$. The particles that are within dx' come from nearby position of x' or were already there. Let us introduce a density of probability $\Phi_\tau(X)$ for a particle at position x has within the interval of time τ to undergo a displacement X from x:

$$\int_{-\infty}^{+\infty} \Phi_\tau(X)dX = 1 \tag{1.29}$$

$\Phi_\tau(X)$ is not known, but what is relevant here remains that this density of probability is expected to be non-null for small values of X and to verify $\Phi_\tau(-X) = \Phi_\tau(X)$. The introduction of $\Phi_\tau(X)$ allows us to determinate the number of particles from $(x, x + dx)$ that enter inside $(x', x' + dx')$ during the interval of time τ:

$$df(x', \tau) = f(x, 0)\Phi_\tau(x' - x) \tag{1.30}$$

Therefore, the total number of particles that come from every place around x' is given by

$$f(x', \tau) = \int_{x=-\infty}^{x=+\infty} f(x, 0)\Phi_\tau(x' - x)dx \tag{1.31}$$

We can rewrite the above 1.31 on a more suitable form by doing the change of variables, $X = x' - x$ and $dx = -dX$:

$$f(x', t + \tau) = \int_{-\infty}^{+\infty} f(x' - X, t)\Phi(X)dX \tag{1.32}$$

As $\tau \ll 1$, then

$$f(x', t + \tau) = f(x', t) + \tau\left(\frac{\partial f}{\partial t}\right) \tag{1.33}$$

We may develop $f(x' - X, t)$ in powers of X:

$$f(x' - X, t) = f(x', t) - X\frac{\partial f(x't)}{\partial x'} + \frac{X^2}{2!}\frac{\partial^2 f(x't)}{\partial x'^2} + \ldots \tag{1.34}$$

Now, we insert Eq. 1.34 into 1.33, putting $x' = x$, $f(x', t) = f(x, t) = f$, obtaining:

$$\begin{aligned} f + \tau\frac{\partial f}{\partial t} &= f\int_{-\infty}^{\infty} \Phi(X)dX - \frac{\partial f}{\partial x}\int_{-\infty}^{+\infty} X\Phi(X)dX \\ &+ \frac{\partial^2 f}{\partial x^2}\int_{\infty}^{+\infty} \frac{X^2}{2!}\Phi(X)dX \end{aligned} \tag{1.35}$$

Due to the symmetry condition (equation 1.30), we finally obtain from the above 1.35 the equation of diffusion:

$$\frac{\partial f}{\partial t} = D\frac{\partial^2 f}{\partial x^2} \tag{1.36}$$

Exercise: Using the Boltzmann kinetic equation, obtain the following equation:

$$\frac{\partial f}{\partial t} + \left[\mu E - \frac{\overline{x}}{\tau}\right] - \frac{1}{2}\frac{\overline{x}^2}{\tau}\frac{\partial^2 f}{\partial x^2} = 0 \tag{1.37}$$

Solution: Remark that

$$\tau\frac{df}{dt} = \tau\left(\frac{\partial f}{\partial t} + v_x\frac{\partial f}{\partial x}\right) = \int_{-\infty}^{\infty} [f(x_1)\phi(x - x_1) - f(x)\phi(x_1 - x)dx] \tag{1.38}$$

where $x' = x - x_1$. This gives

$$\frac{\partial f}{\partial t} + \mu E\frac{\partial f}{\partial x} = \frac{1}{\tau}\int_{-\infty}^{\infty} [f(x - x') - f(x')]\phi(x')dx' \tag{1.39}$$

But

$$f(x - x') \approx f(x) - x'\frac{\partial f}{\partial x} + \frac{x'^2}{2}\frac{\partial^2 f}{\partial x^2} + \ldots \tag{1.40}$$

Inserting into the integral 38,

$$\begin{aligned}\frac{\partial f}{\partial t} + \mu E\frac{\partial f}{\partial x} &= \frac{1}{\tau}\int_{-\infty}^{\infty} f(x)\phi(x')dx' - \frac{1}{\tau}\frac{\partial f}{\partial x}\int_{-\infty}^{\infty} \phi(x)x'dx' \\ &+ \frac{1}{2\tau}\frac{\partial^2 f}{\partial x^2}\int_{-\infty}^{\infty} x'^2\phi(x')dx'\end{aligned} \tag{1.41}$$

This finally gives the desired result

$$\frac{\partial f}{\partial t} + \mu E\frac{\partial f}{\partial x} = -\frac{1}{\tau}\overline{x}\frac{\partial f}{\partial x} + \frac{1}{2\tau}\overline{x^2}\frac{\partial^2 f}{\partial x^2} \tag{1.42}$$

At statistical equilibrium, $\partial f/\partial t = 0$, and the above equation reduces to

$$\mu E\frac{\partial f}{\partial x} - \frac{1}{2\tau}\overline{x^2}\frac{\partial^2 f}{\partial x^2} = 0 \tag{1.43}$$

We may search for solutions of the form as suggested by the Boltzmann theorem:

$$f = f_0 \exp\left(\frac{xE}{k_B T}\right) \tag{1.44}$$

We obtain immediately:

$$D = \frac{\overline{x^2}}{2\tau} = \mu k_B T \tag{1.45}$$

1.3.3 The Principle of Detailed Balance

The principle of detailed balance establishes a relationship between the forward and reverse directions of an elementary process. It is an outcome of the symmetry of the system concerning time reversal. Its application is important when we have, for example, to build a radiative collisional model with reactions of the type:

$$e^-(E) + A^{Z+} \rightleftarrows [(1)](2)e^-(E_1') + e^-(E_2') + A^{(Z+1)+} \tag{1.46}$$

In a plasma, particles can have a wide range of energies. The principle of micro-reversibility is concerned with interactions between particles for given collision energy and is an atomic property. The principle of detailed balance is concerned with the total effect of collisions and assuming that particles have a Maxwellian distribution.

Let us consider particles of specie (1) colliding against particles of specie (c) supposed at rest in the Laboratory referential. Let it be I_1 [atoms/cm^{2}s] the density of flux of specie (1) particles. The density of specie (c) must be low enough so that multiple scattering is not allowed. From the Laboratory viewpoint, the particles (1) are deflected, while (c) particles have a recoil.

The physical states of the system are characterized by the index i and j, respectively, before and after collision. We assume that a flux of particles I_1 with mass μ and momentum $\mathbf{p}_{\infty,i} = \mu\mathbf{v}_{\infty,i}$ impinge on N_c particles at rest. At great distance from the target they have momentum $\mathbf{p}_{\infty,j} = \mu\mathbf{v}_{\infty,j}$.

Let us denote by $dN_1(j)$ the number of particles deflected inside the solid angle $d\Omega_j$ during the interval of time t. This number $dN_1(j)$ can be measured with a device. $dN_1(j)$ is proportional to $I_1(i)$, to the number of target particles N_c, to the time interval t, and to the solid angle $d\Omega_j$:

$$dN_1(j) = I_1(i)N_c t\frac{d\sigma_{ij}}{d\Omega}d\Omega_j \tag{1.47}$$

The term $d\sigma_{ij}/d\Omega$ represents the differential cross-section (in units cm^2/steradians) for the collision $i \to j$, and depends on the relative kinetic energy $\mu v_{\infty,i}^2/2$ of the incident particles and the angle of diffusion Θ. Now, consider the flow of particles $I_{\bar{1}}$ coming with opposed velocity. After a collision with the target particles, one part of them $dI_{\bar{1}}$ is deflected with angle $d\Omega_i$:

$$dN_1(i) = I_{\bar{1}}(j)N_c\,|-t|\,\frac{d\sigma_{ji}}{d\Omega}d\Omega_i \tag{1.48}$$

with $d\sigma_{ji}/d\Omega$ denoting the differential cross-section for the reverse process $j \to i$. To guarantee that the probability is the same, two conditions must be verified:

$$\frac{dN_{\bar{1}}(j)}{N_c n_1(i)} = \frac{dN_1(i)}{N_c n_1(j)} \tag{1.49}$$

where n_1 represents the density of particles in the beam, and

$$d^3 r p_i^2 dp_i d\Omega = d^3 r p_j^2 dp_j d\Omega_j \tag{1.50}$$

Considering that $I_1(i) = n_1(i)v_i$ and $I_{\bar{1}}(j) = n_1(j)v_j$, we obtain

$$\frac{v_i d\sigma_{ij}/d\Omega}{p_j^2 dp_j} = \frac{v_j d\sigma_{ji}/d\Omega}{p_i^2 dp_i} \tag{1.51}$$

The conservation of energy gives

$$\frac{p_i^2}{2\mu} = \frac{p_j^2}{2\mu} + \Delta E \tag{1.52}$$

Differentiating the above expression we get $v_i dp_i = v_j dp_j$. Substituting into Eq. 1.49 we obtain an expression which traduces the **principle of microreversibility**:

$$p_i^2 \frac{d\sigma_{ij}(p_i\Theta)}{d\Omega} = p_j^2 \frac{d\sigma_{ji}}{d\Omega} \tag{1.53}$$

If both states $|i>$ and $|j>$ are degenerated, we must replace 1.52 by

$$g_i p_i^2 \frac{d\sigma_{ij}(p_i\Theta)}{d\Omega} = g_j p_j^2 \frac{d\sigma_{ji}}{d\Omega} \tag{1.54}$$

where g_i and g_j are the respective statistical weights. When multiplying Eq. 1.54 by 2μ and integrating in $d\Omega$, we obtain the **Klein-Rosseland relation**:

$$[E_\mu + \Delta E] g_i \sigma_{ij}(E_\mu + \Delta E) = E_\mu g_j \sigma_{ji}(E_\mu) \tag{1.55}$$

Here, $E_\mu = (\mu/2)v^2$, and $E_\mu + \Delta E$ and E_μ are, resp., the relative kinetic energy before and after collision. ΔE denotes the internal energy exchanged during the collision. 1.55 traduces the microreversibility for the total cross-sections.

Exercise: Express the collision integral in terms of the differential cross section of collisions between particles.

1.3.3.1 Momentum Transfert Cross Section

$$dN_1 = \Phi_1 d\sigma dt \tag{1.56}$$

where $d\sigma = \sigma(\chi, \phi)d\Omega$, and $\sigma(\chi, \phi)$ is the elastic cross section. Hence, we have

$$dN_1 = \Phi_1 N_2 \sigma(\chi, \phi) d\Omega dt \tag{1.57}$$

The total momentum transfer per unit of time is

$$\int \Delta p \frac{dN_1}{dt} = -\int \Phi_1 N_2 p(1 - \cos\chi)\sigma(\chi\phi)d\Omega = -Q\sigma_1 \tag{1.58}$$

where $Q = \Phi_1 N_2 p$ is the flux of momentum transferred to the target, and

$$\sigma_1 = \int_{\chi_m}^{\pi} (1 - \cos\chi)\sigma(\chi\phi)d\Omega \tag{1.59}$$

is the **momentum transfert cross section**. We should integrate from the minimum value χ_m since it is experimentally difficult to study diffusion at small angles.

1.3.4 Hydrodynamic Models

The equations of intraphase[6] transport are statements about the conservation of mass, momentum and energy applied to flow systems. The respective three equations are:

1. The equations of continuity for each chemical species (in terms of the mass flux with respect to fixed axes, $\mathbf{n}_\alpha \equiv \rho_\alpha \mathbf{v}_\alpha$, or mass flux with respect to the mass-average velocity $\mathbf{j}_\alpha = \rho_\alpha(\mathbf{v}_\alpha - \mathbf{v})$);
2. The equation of motion containing the momentum flux tensor **f** or the pressure tensor **p**;
3. The equation of energy in terms of energy flux *e* or *q*.

1.3.4.1 The Equation of Continuity

Applying the principle of mass conservation of species α within a *fixed* volume *V*, we may state

$$\begin{pmatrix}\text{Rate of increase}\\ \text{of mass of } \alpha\\ \text{crosses into V}\\ \text{over}\\ \text{surface S}\end{pmatrix} = \begin{pmatrix}\text{Rate}\\ \text{at which}\\ \text{mass of}\\ \text{species } \alpha\\ \text{crosses}\\ \text{into V}\\ \text{over}\\ \text{surface S}\end{pmatrix} + \begin{pmatrix}\text{Rate at which}\\ \text{mass of}\\ \alpha\\ \text{is produced}\\ \text{within V}\\ \text{due to}\\ \text{chemical}\\ \text{reactions}\end{pmatrix}$$

This statement can be written in mathematical terms:

$$\frac{d}{dt}\int_V \rho_\alpha dv = -\int_S (\mathbf{n}_\alpha \cdot \mathbf{n})dS + \int_V r_\alpha dv \tag{1.60}$$

If we suppose the volume *V* is fixed we can interchange d/dt with the integral sign, and convert the surface integral by using the Green-Gauss theorem into a volume integral:

$$\int_V \frac{\partial \rho_\alpha}{\partial t} dv = -\int_V (\nabla \cdot \mathbf{n}_\alpha)dv + \int_V r_\alpha dv \tag{1.61}$$

Considering that all integrals are extended over the same volume and that this volume is arbitrary, the integrals signs may be removed, and we obtain the **equation of continuity for species** α**:**

$$\frac{\partial \rho_\alpha}{\partial t} = -(\mathbf{n} \cdot \mathbf{n}_\alpha) + r_\alpha \tag{1.62}$$

where the index α run over all the ν species contained in flow system, $\alpha = 1, 2, \ldots \nu$. If we express the continuity equation in terms of moles of species α per unit volume, $c_\alpha \equiv \rho_\alpha / M_\alpha$, with M_α denoting the molecular weights, then we must write

$$\frac{\partial c_\alpha}{\partial t} = -(\mathbf{n} \cdot \mathbf{N}_\alpha) + R_\alpha \tag{1.63}$$

Here, $\mathbf{N}_\alpha \equiv c_\alpha \mathbf{v}_\alpha$, and

$$r_\alpha = M_\alpha R_\alpha \tag{1.64}$$

Both reaction rates r_α and R_α must contain all the contributions from chemical reactions where a given species A_α is a reactant or a product. This implies that if there is $R_{\alpha j}$ production rates for species A_α, we must add all contributions such as

$$R_\alpha = \sum_{j=1}^{K} R_{\alpha j} \tag{1.65}$$

where K is the number of reactions which are taking place in the flow system.

From the Boltzmamm equation (1.27) it is obtained

$$n_i m_i \left(\frac{\partial \mathbf{v}_i}{\partial t} + \mathbf{v}_i \cdot \nabla \mathbf{v}_i \right) = n_i Ze(\mathbf{E} + \mathbf{v}_i \times \mathbf{B}) - \nabla \cdot \overleftrightarrow{\Psi} - n_i m_i \nabla \phi + \mathbf{P}_{ei} \tag{1.66}$$

The **average speed** is given by:

$$\mathbf{v}_i = \frac{1}{n_i \Delta V} \sum \mathbf{w}_i \tag{1.67}$$

and with the dyadic defined by:

$$\overleftrightarrow{\Psi} = \frac{m_i}{\Delta V} \sum (\mathbf{w}_i - \mathbf{v}_i)(\mathbf{w}_i - \mathbf{v}_i) \tag{1.68}$$

It is verified

$$\nabla \cdot \overleftrightarrow{\Psi} = \nabla p \tag{1.69}$$

Now, define the following quantities:

$$\begin{aligned} \rho &= n_i m_i + n_e m_e \\ \mathbf{v} &= \frac{1}{\rho}(n_i m_i \mathbf{v}_i + n_e m_e \mathbf{v}_e) \\ \mathbf{j} &= e(n_i Z \mathbf{v}_i - n_e \mathbf{v}_e) \end{aligned} \tag{1.70}$$

Summation of the terms $n_i m_i \mathbf{v}_i$ with $n_e m_e \mathbf{v}_e$ from Eq. 1.66 leads to Table 1.1

TABLE 1.1

Components of the Stress Tensor in Spherical Coordinates for Newtonian Fluids

$\tau_{rr} = -\mu\left[2\frac{\partial v_r}{\partial r} - \frac{2}{3}(\nabla\cdot\mathbf{v})\right]$
$\tau_{\theta\theta} = -\mu\left[2\left(\frac{1}{r}\frac{\partial v_\theta}{\partial\theta} + \frac{v_r}{r}\right) - \frac{2}{3}(\nabla\cdot\mathbf{v})\right]$
$\tau_{\phi\phi} = -\mu\left[2\left(\frac{1}{r\sin\theta}\frac{\partial v_\phi}{\partial\phi} + \frac{v_r}{r} + \frac{v_\theta\cot\theta}{r}\right) - \frac{2}{3}(\nabla\cdot\mathbf{v})\right]$
$\tau_{r\theta} = \tau_{\theta r} = -\mu\left[r\frac{\partial}{\partial r}\left(\frac{v_\theta}{r} + \frac{1}{r}\frac{\partial v_r}{\partial\theta}\right)\right]$
$\tau_{\theta\phi} = \tau_{\phi\theta} = -\mu\left[\frac{\sin\theta}{r}\frac{\partial}{\partial\theta}\left(\frac{v_\phi}{\sin\theta}\right) + \frac{1}{r\sin\theta}\frac{\partial v_\theta}{\partial\phi}\right]$
$\tau_{\phi r} = \tau_{r\phi} = -\mu\left[\frac{1}{r\sin\theta}\frac{\partial v_r}{\partial\phi} + r\frac{\partial}{\partial r}\left(\frac{v_\phi}{r}\right)\right]$
$(\nabla\cdot\mathbf{v}) = \frac{1}{r^2}\frac{\partial}{\partial r}(r^2 v_r) + \frac{1}{r\sin\theta}\frac{\partial}{\partial\theta}(v_\theta\sin\theta) + \frac{1}{r\sin\theta}\frac{\partial v_\phi}{\partial\phi}$

$$\rho\frac{\partial\mathbf{v}}{\partial t} = [\mathbf{j}\times\mathbf{B}] - \nabla p - \rho\nabla\phi + \mathbf{P}_{ei} + \mathbf{P}_{ie} \tag{1.71}$$

Ignoring transient terms we have

$$\nabla p = [\mathbf{j}\times\mathbf{B}] - \rho\nabla\phi \tag{1.72}$$

since $\mathbf{P}_{ei} + \mathbf{P}_{ie} = 0$. We also have

$$\mathbf{E} + [\mathbf{v}\times\mathbf{B}] = \eta\mathbf{j} + \frac{1}{en_e}(\nabla p_i + \rho\nabla\phi) \tag{1.73}$$

The external product of $\mathbf{B}$ with 1.91 gives

$$\mathbf{B}\times\nabla p = \mathbf{B}\times[\mathbf{j}\times\mathbf{B}] - \rho\mathbf{B}\times\nabla\phi = \mathbf{j}B^2 - \mathbf{B}(\mathbf{B}\cdot\mathbf{j}) - \rho\mathbf{B}\times\nabla\phi$$

$$\Rightarrow \mathbf{j}_\perp = \frac{\mathbf{B}\times\nabla p}{B^2} \tag{1.74}$$

By other side, the external product of $\mathbf{B}$ with 1.92 gives

$$\mathbf{B}\times(\mathbf{E} + [\mathbf{v}\times\mathbf{B}]) = \eta[\mathbf{B}\times\mathbf{j}] + \frac{1}{en_e}[\mathbf{B}\times\nabla p_i]$$

$$\mathbf{B}\times\mathbf{E} + v_\perp B^2 - B(\mathbf{B}\cdot\mathbf{v}) = \eta[\mathbf{B}\times\mathbf{j}] + \frac{1}{en_e}[\mathbf{B}\times\nabla p_i]$$

$$\mathbf{v}_\perp = \frac{\mathbf{B}}{B^2}\times\left(-\mathbf{E} + \frac{1}{en_e}\nabla p_i\right) \tag{1.75}$$

1.3.5 The Stress Tensor in Newtonian Fluids

For **Newtonian fluids** the shear force per unit area is proportional to the negative of the local velocity gradient (*Newton's law of viscosity*). For example, a fluid with a velocity vector along Ox would be actuated by a force per unit area exerted in the x-direction equal to

TABLE 1.2

Components of the Stress Tensor in Cylindrical Coordinates for Newtonian Fluids

$$\tau_{rr} = -\mu\left[2\frac{\partial v_r}{\partial r} - \frac{2}{3}(\nabla\cdot\mathbf{v})\right]$$

$$\tau_{\theta\theta} = -\mu\left[2\left(\frac{1}{r}\frac{\partial v_\theta}{\partial \theta} + \frac{v_r}{r}\right) - \frac{2}{3}(\nabla\cdot\mathbf{v})\right]$$

$$\tau_{zz} = -\mu\left[2\frac{\partial v_z}{\partial z} - \frac{2}{3}(\nabla\cdot\mathbf{v})\right]$$

$$\tau_{r\theta} = \tau_{\theta r} = -\mu\left[r\frac{\partial}{\partial r}\left(\frac{v_\theta}{r}\right) + \frac{1}{r}\frac{\partial v_r}{\partial \theta}\right]$$

$$\tau_{\theta z} = \tau_{z\theta} = -\mu\left[\frac{\partial v_\theta}{\partial z} + \frac{1}{r}\frac{\partial v_z}{\partial \theta}\right]$$

$$\tau_{zr} = \tau_{rz} = -\mu\left[\frac{\partial v_z}{\partial r} + \frac{\partial v_r}{\partial z}\right]$$

$$(\nabla\cdot\mathbf{v}) = \frac{1}{r}\frac{\partial}{\partial r}(rv_r) + \frac{1}{r}\frac{\partial v_\theta}{\partial \theta} + \frac{\partial v_z}{\partial z}$$

$$\tau_{yx} = -\mu\frac{dv_x}{dy} \tag{1.76}$$

where μ is the viscosity of the fluid.

The components of the stress tensor for Newtonian fluids in spherical coordinates (r, θ, ϕ) is given by Table 1.2.

Exercise: Give the necessary conditions to initiation and sustaining a discharge plasma.

Exercise: Couette-Hatschek viscometers: Determine the tangential velocity and the shear stress distribution for the tangential laminar flow of an incompressible fluid between two vertical coaxial cylinders, with the inner one stationary and the outer one rotating with an angular velocity Ω_0. Neglect end effects.

Solution: The problem has cylindrical symmetry. The laminar flow is a reasonable assumption that leads us to zero velocity components v_r and v_z. We also assume there are no pressure gradients in θ-direction.

The equations of motion in cylindrical coordinates reduces here to the following set:

$$-\rho\frac{v_\theta^2}{r} = -\frac{\partial p}{\partial r} \tag{1.77}$$

$$0 = \frac{d}{dr}\left(\frac{1}{r}\frac{d}{dr}(rv_\theta)\right) \tag{1.78}$$

$$0 = -\frac{\partial p}{\partial z} + \rho g \tag{1.79}$$

1.79 can be integrated with respect to r with boundary conditions $v_\theta(R_1) = 0$ and $v_\theta(R_2) = \Omega_0 R_2$. It gives

$$\begin{aligned}
\frac{1}{r}\frac{d}{dr}(rv_\theta) &= K_1 \\
rv_\theta &= K_1\frac{r^2}{2} + K_2 \\
K_1 &= \frac{2\Omega_0 R_2}{1 - R_1^2 / R_2^2} \\
K_2 &= -\frac{K_1}{2}R_1^2 \\
v_\theta &= \frac{2\Omega_0 R_2}{1-\frac{R_1^2}{R_2^2}}\left(1 - \frac{R_1^2}{2r}\right)
\end{aligned} \tag{1.80}$$

Now, we can calculate the shear stress distribution $\tau_{r\theta}$:

$$\begin{aligned}
\tau_{r\theta} = \tau_{\theta r} &= -\mu\left[r\frac{\partial}{\partial r}\left(\frac{v_\theta}{r}\right) + \frac{1}{r}\frac{\partial v_r}{\partial \theta}\right] \\
&= -\mu r\frac{d}{dr}\frac{2\Omega_0 R_2}{1-\frac{R_1^2}{R_2^2}}\left(1 - \frac{R_1^2}{2r}\right) \\
&= -2\mu\Omega_0 R_2^2 \frac{1}{r^2}\frac{R_1^2}{R_2^2 - R_1^2}
\end{aligned} \tag{1.81}$$

The torque Γ can be calculated

$$\begin{aligned}
\Gamma &= |\mathbf{r}|\,|\mathbf{F}|\,S \\
S &= 2\pi R_2 L \\
\Gamma &= 2\pi R_2 L(-\tau_{r\theta})R
\end{aligned} \tag{1.82}$$

1.3.5.1 Homogeneous Reactions: Reaction Rates

A **homogeneous reaction** is a chemical change that occurs in the entire volume of the fluid. The rate of a homogeneous reaction is determined by the composition of the reaction mixture, the temperature, and the pressure.

In a chemical reaction

$$aA + bB \rightleftharpoons eE + fF \tag{1.83}$$

there are always *forward and reverse reactions* that can both occur to a certain extent, the process is described as a **reversible reaction.**[7] It has been found experimentally that after elapsing a sufficient interval of time all so-called reversible reactions attain a state of **chemical equilibrium**, meaning that no further change in the composition is detected.

According to the **law of mass action** (enunciated first by C. M. Guldberg and P. Waage), the driving force of a chemical reaction is proportional to the active masses of the reacting substances, that is, in modern language, the rate of a chemical reaction is proportional to the molar concentration of the reacting substances. Hence, in the case of reaction 83 we have

$$\text{Rate of forward reaction} = kc_A c_B \tag{1.84}$$

with k being a proportionality constant and c_A and c_B denoting the concentrations of both reactants. Correspondingly, it must be present the reverse reaction, such as

$$\text{Rate of reverse reaction} = k' c_E c_F \tag{1.85}$$

All the concentrations are referred to the equilibrium state and when chemical equilibrium is attained it must be verified the equality

$$k c_A c_B = k' c_E c_F \tag{1.86}$$

such that,

$$\frac{c_E c_F}{c_A c_B} = \frac{k}{k'} = K_c \tag{1.87}$$

The constant K_c is called the **equilibrium constant** of the reaction.

Neutral-neutral collisions involve elastic, inelastic and reactive processes. Due to the atom and molecules internal complexity, this class of process leads to larger number of processes than it happens with electronic collisions. The inelastic collisions change the internal energy of the molecules, they have a threshold thus depending of the incoming particle energy.

When equilibrium is onset, it is verified a simple relationship between the products, $[E]$ and $[F]$, and the reactants, $[A]$ and $[B]$. In fact,

$$\frac{[E]^e [F]^f}{[A]^a [B]^b} = K \tag{1.88}$$

where K is a constant at a given temperature, its ratio is constant and Eq. 1.88 is one of the most useful relationship in physical kinetics.

The rate equations and respective coefficients work as noted below:

- 1-Body ($A \rightarrow A'$): the rate of spontaneous emission from a radiating excited source

$$\frac{d[A]}{dt} = -\frac{[A]}{\tau} = -\nu [A] \tag{1.89}$$

  Here, $[A]$ denote particle density and τ is the natural lifetime of the state.

- 2-Body, binary collision and reaction of the kind ($A + B \rightarrow E + F$):

$$\frac{d[A]}{dt} = -\alpha [A][B] \tag{1.90}$$

  where $\alpha = \langle Qv \rangle$ and rate coefficients (in s^{-1} units) is given by $\nu = [B]/\langle Qv \rangle$. Usually, in 2-body electron-ion recombination we expect that $[A] \simeq [B]$.

- 3-Body ($A + B + C \rightarrow E + F + G$), e.g., 3-body attachment and recombination:

$$\frac{d[A]}{dt} = -K [A][B][C] \tag{1.91}$$

TABLE 1.3
List of Reactions for Exercice

Process	Rate coefficient
$O(3^P) + O_3 \rightarrow 2O_2(X^3\Sigma)$	$K_1 = 1.8 \times 10^{-11} \exp(-2300/T_g)$
$O(3^P) + wall \rightarrow \frac{1}{2}O_2(X^3\Sigma)$	$K_2 = 800 \times \sqrt{\frac{T_g}{300}}$ in units s^{-1}
$e + O_2(X^3\Sigma) \rightarrow O^- + O(^3P)$	$K_3 = f(E/N)$

For electron-ion recombination, usually $[A] \equiv [B] \simeq [C]$, in which case we obtain:

$$\frac{d[A]}{dt} = -Kn_e^2 n_+ \simeq -Kn_e^3 \tag{1.92}$$

Exercise: Let us suppose that the formation of the fundamental state of molecular oxygen $O_2(X^3\Sigma)$ occurs through the following set of reactions given in Table 1.3. This is an oversimplification; the entire set can include up to more than 50 reactions, see, for example, [20]. Write the governing equation for species $O_2(X^3\Sigma)$.

Solution: Note that $O_2(X^3\Sigma)$ is produced through reaction of atomic oxygen with ozone and atomic reassociation at the wall, and destroyed by electron impact. Hence we have

$$\frac{d[O_2(^3\Sigma)]}{dt} = 2K_1[O^3P)][O_3] + \frac{1}{2}K_2[O(^3P)] - K_3 n_e[O_2(^3\Sigma)] \tag{1.93}$$

1.3.5.2 Ponderomotive Forces Acting in a dielectric barrier discharge (DBD)

Siemens's experiments with the silent discharge for generating ozone in the 1850s (ozone synthesis is still until now the most important application of nonequilibrium plasma chemistry) was among the first industrial application of DBDs.

Let us consider a mixture of electrons and one kind of ion. The equations relevant are

$$\begin{aligned} n_e m_e \frac{\partial \mathbf{v}_e}{\partial t} &\cong n_e q_e \mathbf{E} + P_{en} \\ n_i m_i \frac{\partial \mathbf{v}_i}{\partial t} &\cong n_i q_i \mathbf{E} + P_{in} \end{aligned} \tag{1.94}$$

where

$$\begin{aligned} P_{en} &= -n_e m_e \nu_{en}(\mathbf{v}_e - \mathbf{v}_n) \\ P_{in} &= -n_i m_i \nu_{in}(\mathbf{v}_i - \mathbf{v}_n) \end{aligned} \tag{1.95}$$

The left hand side of 1.195 are the forces density acting on the electrons and ions in a given point of space. Its summation gives

$$\mathbf{f} = \mathbf{f}_{en} + \mathbf{f}_{in} \tag{1.96}$$

$$\mathbf{f} = e(Zn_i - n_e)\mathbf{E} - n_e m_e \nu_{en}(\mathbf{v}_e - \mathbf{v}_n) - n_i m_i \nu_{in}(\mathbf{v}_i - \mathbf{v}_n) \tag{1.97}$$

But note that

$$\mathbf{j}_\alpha = n_\alpha m_\alpha(\mathbf{v}_\alpha - \mathbf{v}_n) = -\rho \mathcal{D}_{\alpha n}\nabla\left(\frac{\rho_\alpha}{\rho}\right) \tag{1.98}$$

where $\rho_\alpha = n_\alpha m_\alpha$ and $\rho = \sum_\alpha n_\alpha m_\alpha$. If we assume $\rho \approx$ const, then

$$\mathbf{j}_\alpha \approx -\mathcal{D}_{\alpha n}\nabla\rho_\alpha = -m_\alpha \mathcal{D}_{\alpha n}\nabla n_\alpha \tag{1.99}$$

Hence, the term $n_e m_e \nu_{en}(\mathbf{v}_e - \mathbf{v}_n)$ becomes

$$n_e m_e \nu_{en}(\mathbf{v}_e - \mathbf{v}_n) = \nu_{en}\mathbf{j}_e = -m_e \nu_{en}\mathcal{D}_{en}\nabla n_e \tag{1.100}$$

But since $\mu_e \equiv -\frac{e m_e}{\nu_{en}}$, we obtain successively:

$$n_e m_e \nu_{en}(\mathbf{v}_e - \mathbf{v}_n) = -\nu_{en}\mathcal{D}_{en}\nabla n_e \tag{1.101}$$

$$= -\frac{n_e q_e}{m_e \mu_e}\mathcal{D}_{en}\nabla n_e \tag{1.102}$$

$$= -\frac{e\mathcal{D}_{en}}{\mu_e}\nabla n_e \tag{1.103}$$

$$= -k_B T_e \nabla n_e \tag{1.104}$$

since $k_B T \equiv \frac{eD}{\mu}$. Therefore, we finally obtain

$$\mathbf{f} = e(Zn_i - n_e)\mathbf{E} - k_B T_e \nabla n_e - k_B T_i \nabla n_i \tag{1.105}$$

$$= e(Zn_i - n_e)\mathbf{E} - \nabla(k_B T_e n_e + k_B T_i n_i) \tag{1.106}$$

= electric field force due to diffusion

This last result of Eq. 1.105 represent according to Boeuf *et al.* [21] the main forces acting in a plasma actuator. Notice that the force acting on the ionized gas can be written under the form of the stress tensor:

$$\mathbf{f} = \nabla\cdot\overleftrightarrow{T} \tag{1.107}$$

where $\overleftrightarrow{T}$ is such that $T_{ij} = \epsilon_0\left[E_i E_j - \frac{1}{2}E^2\delta_{ij}\right]$. This gives for one given component i the force density:

$$f_i = \partial_i T_{ij} = \epsilon_0 \partial_i\left[E_i E_j - \frac{1}{2}E^2\delta_{ij}\right] = \epsilon_0(\partial_i E_i)E_j - \frac{1}{2}2\epsilon_0 E_i \delta_{ij}\partial_j E_j = \rho E_i \tag{1.108}$$

since $\partial_i E_i = \rho/\varepsilon_0$, and $delta_j E_i = 0$ if the electric field is homogeneous.
The equation of motion is given by [22]:

$$\frac{\partial}{\partial t}\rho\mathbf{v} = -\nabla\cdot\rho\mathbf{v}\mathbf{v} - \nabla p - \nabla\cdot\overleftrightarrow{\tau} + \rho\mathbf{g} \tag{1.109}$$

Note that quantities as such $\rho\mathbf{v}\mathbf{v}$ and $\overleftrightarrow{\tau}$ are dyadics.

1.3.5.3 EHD Induced by Paraelectric Effects

The electric field lines begin and terminate on charges. Faraday suggests the heuristic vision of rubber bands in tension to pull charges of opposite signs together. In a glow discharge plasma, it generates a polarization field that acts on the charges, moving them toward increasing electric field gradients. As a result of frequent ion-neutral and electron-neutral collisions, the neutral gas is dragged with it.
The electrostatic body force on a plasma with a net charge ρ_c such as

$$\rho_c = e(Zn_i - n_e) \quad \mathrm{C/m^3} \tag{1.110}$$

is given by

$$\mathbf{F}_E = \rho_c\mathbf{E} \quad \mathrm{N/m^3} \tag{1.111}$$

This net charge creates a gas electric field given by the Poisson's equation

$$\nabla\cdot\mathbf{E} = \frac{\rho}{\varepsilon_0} \tag{1.112}$$

When inserting Eq. 1.112 into Eq. 1.111, we obtain

$$\mathbf{F}_E = \varepsilon_0\mathbf{E}\nabla\cdot\mathbf{E} = \frac{1}{2}\varepsilon_0\nabla^2 E \tag{1.113}$$

Remark that in 1-dim, we have

$$F_{Ex} = \frac{d}{dx}\left(\frac{1}{2}\varepsilon_0 E^2\right) \tag{1.114}$$

This means the electrostatic body force results from a pressure over ions and electrons, accelerating them and the momentum acquired are transmitted to the neutral gas by Lorentzian collisions. Neglecting other effects, like viscosity terms and centrifugal forces, the body forces due to gasdynamic and electrostatic forces, in an equilibrium state, is given by

$$\nabla p_g + \nabla p_E = 0 \tag{1.115}$$

that is, the gasdynamic ($p_g = nkT$) and the electrostatic pressures are approximately constant

$$p_g + p_E = \text{constant}, \tag{1.116}$$

or

$$nkT + \frac{\varepsilon_0}{2}E^2 = \text{constant} \tag{1.117}$$

This means that in regions of high electric field pressure, the gas-dynamic pressure as lower than in the surroundings, pulling the neutral gas to this region.

1.3.6 The Miller's Force

The acceleration of ions to any desired velocity by electrostatic or electromagnetic means are, in principle, possible.

In spatially varying high frequency electric fields it occurs an interesting phenomena of a single particle effect called the Miller force. Consider a particle subject to an oscillating field which varies smoothly with space, $E(x, t) = E_0(x)\cos(\omega t)$, with the field stronger to the right and weaker to the left. Suppose that at start the particle goes to the left region where the field is stronger. The particle receive a big punch to the right. But now the field is weaker at the right and so it receives a weaker punch to the left. Of course, the resulting motion is to the left, in fact, a net acceleration to the left that continues with successive cycles leading the particle away from the strong field region. The equation of motion is

$$m_i\ddot{x} = q_i E = q_i E_o(x)\cos(\omega t) \tag{1.118}$$

For calculations it is useful to decompose the motion in two terms, $x = x_o + x_i$, where x_o is the oscillating center, a time average over the short time $\frac{2\pi}{\omega}$. After a Taylor expansion about the oscillating center x_{oi} the above equation becomes

$$m_i(\ddot{x}_o + \ddot{x}_i) = q_i\left(E_i + x_i\frac{dE_i}{d\tau}\right)\cos(\omega t) \tag{1.119}$$

where $\frac{dE_i}{d\tau}$ must be evaluated at x_i. Averaging over time the above equation we obtain

$$m_i\overline{x}_o = q_i\overline{\frac{dE_i}{d\tau}x_i\cos(\omega t)} \tag{1.120}$$

So, Eq. 1.119 is reduced to

$$m_i x_i = q_i E_o\cos(\omega t) \tag{1.121}$$

whose solution is

$$x_i = -\frac{q_i E_o}{m_i\omega^2}\cos(\omega t) \tag{1.122}$$

Inserting into 1.120 and averaging over time, we get

$$x_o = -\frac{q_i^2 E_o}{2m_i^2\omega^2}\frac{dE_o}{d\tau} \tag{1.123}$$

The body force is $F_i = m_i\ddot{x}_o$ which gives the Miller force

$$F_i = -\frac{q_i^2}{4m_i\omega^2}\frac{d}{dx}(E_o^2) \tag{1.124}$$

This force plays an important role in laser fusion, electron-beam fusion, radio-frequency modification of the ionosphere, and solar radio bursts. This last phenomenon is associated with solar flares or another explosive event on the sun. A stream of energetic electrons that are injected from the sun into interplanetary space run roughly at 30% of the speed of light, reaching the Earth in about 20 mn. They follow the magnetic field lines with a given gyroradius and exciting the local plasma which radiates at its natural frequency or harmonics of it. At about 1 Earth orbit (1 AU) the electrons density is about 5–10 particles/cm^3 and the plasma radiates at 20–30 kHz.

1.3.6.1 A New Form of the Electromagnetic Energy Equation When Free Charges are Present

L. Tonks [19] proposed a new form of the electromagnetic energy equation when dealing with electron-beam tubes, relevant in regions in which both free charged particles and electromagnetic fields are present. The equation of balance of energy is, therefore, written as

$$\frac{1}{8\pi\epsilon_0}\frac{\partial}{\partial t}\int_V (E^2 + H^2)dV + \int_S [\mathbf{E}\times\mathbf{H}]\cdot d\mathbf{S} + \sum_i \int \rho_i \mathbf{v}_i\cdot\mathbf{E}dV = 0 \tag{1.125}$$

The summation Σ_i runs over all charged particles present in the volume V. Using

$$\frac{d\mathbf{v}}{dt} = \frac{e}{m}(\mathbf{E} + [\mathbf{v}\times\mathbf{H}]) \tag{1.126}$$

gives us

$$(\mathbf{v}\cdot\mathbf{E}) = \frac{m}{2e}\frac{d}{dt}(v^2) \tag{1.127}$$

Applying the convective derivative to v^2 yields:

$$\rho\frac{d}{dt}(v^2) = \rho\frac{\partial}{\partial t}(v^2) + \rho\mathbf{v}\cdot\nabla(v^2) \tag{1.128}$$

1.3.6.2 Equation of Conservation of Energy

In a given control volume at any instant of time we must have:

$$\dot{E}_{in} + (\dot{E}_g - \dot{E}_l) - \dot{E}_{out} = \frac{dE_{st}}{dt} \equiv \dot{E}_{st} \tag{1.129}$$

Here, $\dot{E}_{in}$ is the energy input per unit of time into the control volume; $\dot{E}_g$ is the energy gain per unit of time due to any process inside the control volume; $\dot{E}_l$ is the energy loss per unit of time due to any process inside the control volume; E_{st} is the energy stored inside the control volume. We may also write

$$\dot{E}_{st} = \frac{d}{dt}(\rho V c_v T) = \frac{d}{dt}(\text{internal energy of control volume}) \tag{1.130}$$

A common problem in transitory phenomena is the rapid temperature variation of a solid body in a medium. For example, a hot metal at temperature T_i is rapidly immersed in a liquid at temperature $T_\infty < T_i$. At the end of the interval of time, t the solid attains T_∞. The cooling is due to heat transfer at the interface solid-liquid. Frequently, it is assumed that the temperature gradients are negligible. Transient phenomena are determined through the general balance of energy. This energy balance must be related to the loss by energy convection:

$$-\dot{E}_{out} = \dot{E}_{st} \tag{1.131}$$

or

$$-hA_s(T - T_\infty) = \rho VC - v\frac{\partial T}{\partial t} \tag{1.132}$$

We define $\theta \equiv T - T - \infty$. Then it comes

$$\begin{aligned} \frac{\rho V c_v}{hA_s}\frac{d\theta}{dt} &= -\theta \\ \frac{\rho V c_v}{hA_s}\int_{\theta_i}^{\theta}\frac{d\theta}{\theta} &= -\int_0^t dt \end{aligned} \tag{1.133}$$

where $\theta_i \equiv T_i - T_\infty$. The solution is:

$$\begin{aligned} \frac{\rho V c_v}{hA_s}\ln\frac{\theta_i}{\theta} &= t \\ \frac{\theta}{\theta_i} = \frac{T - T_\infty}{T_i - T_\infty} &= \exp\left[-\left(\frac{hA_s}{\rho V c_v}\right)\right] \end{aligned} \tag{1.134}$$

This method is applicable (negligible gradients) whenever the adimensional parameter called **Biot number**:

$$B_i = \frac{hL}{K} = \frac{R_{\text{conduction}}}{R_{\text{convection}}} < 0.1 \tag{1.135}$$

In the case there is no external flux of heat neither heat sources by conduction or convection (vacuum) or these are negligible regarding the emitted radiation by the body, the equation becomes:

$$\rho V c_v \frac{\partial T}{\partial t} = -\epsilon A_{s,r} \sigma (T^4 - T_{sur}^4) \tag{1.136}$$

By integration this gives:

$$\frac{\epsilon A_{s,r} \sigma}{\rho V c_v} \int_0^t dt = \int_{T_i}^{T} \frac{dT}{T_{sur}^4 - T^4} \tag{1.137}$$

In the case the surface temperature is null, $T_{sur} = 0$, we obtain

$$t = \frac{\rho V c_v}{3\epsilon A_{s,r} \sigma} \left(\frac{1}{T^3} - \frac{1}{T_i^3} \right) \tag{1.138}$$

Exercise: Permanent space stations are increasing their dimension and this fact implies they augment their electric power dissipation. To maintain all the compartments at a safe temperature environment it is necessary to dissipate their temperature to the exterior. A new scheme for this purpose was proposed and it is called *Liquid droplet radiator*. The heat is transferred to a certain amount of oil at a recipient at high pressure and injected into external space in the form of a beam of small droplets. This beam cross a certain distance L along which it cools down by radiation to the external space at $T = 0K$. The droplets have the emissivity $\epsilon = 0.95$ and diameter $D = 0.5$ mm and are injected at $T_i = 500$ K at speed $v = 0.1$ m/s. The oil has the following properties: $\rho = 885$ kg/m^3, $c_v = 1900$ J/kg.K and $K = 0.14$ W/m.K.

a. Assume that each droplet radiate to the external space at $T_{sur} = 0$ K, determine the distance L necessary to collect the droplet at temperature final $T_f = 300$ K.
   Answer:

$$t = \frac{\rho V c_v}{3\epsilon A_s \sigma} \left(\frac{1}{T_f^3} - \frac{1}{T_i^3} \right) \Rightarrow t = 25.2\text{s} \Rightarrow D \simeq 2.5\text{cm} \tag{1.139}$$

b. What amount of thermal heat is rejected by each droplet?
   Answer:

$$\dot{E} = \epsilon\sigma (T_f^4 - T_i^4) A = 5.75 \times 10^{-4} W \Rightarrow E = \dot{E}t \simeq 58\text{mJ per droplet} \tag{1.140}$$

1.3.7 Effects of Space Charge

Plasmas created by laboratory devices are normally confined by walls constituting a physical boundary with huge importance. If the plasma boundary is an insulating wall, electrons and positive ions recombine into the surface in a time scale much shorter than in volume. Hence, the equality of fluxes of negative and positive charges follows:

$$\Gamma_e = \Gamma_i \tag{1.141}$$

The equality of densities does not follows because the electron and ion masses, and as well temperature and collision frequencies are not the same, otherwise, we should have, $n_e = n_i$ near and up to the wall. However, the temperatures of ions are nearly the same as the gas temperature and are mainly determined by thermal effects. Due to their smaller mass electron motion is mainly affected by thermal random motion near the wall with flux

$$\Gamma_{ew} = \frac{n_{ew}\bar{c}_e}{4} \tag{1.142}$$

where

$$\bar{c}_e = \sqrt{\frac{2k_B T_e}{\pi m_e}} \tag{1.143}$$

denotes the the mean random speed and n_{ew} is the electron density at the wall. The ion motion is more directed to the wall due to their inertia, and its speed is

$$v_i = \sqrt{\frac{k_B T_i}{m_i}} \tag{1.144}$$

These unbalanced tendencies result in the formation of a region where electrons and ions have significantly different densities, implying the existence of a space-charge field influencing the charged particle motion in this region.

1.4 Plasma Actuators Devices and Modeling

There has been a growing interest in the field of plasma aerodynamics related to its outstanding importance for active flow control, overriding the use of mechanical flaps [23–30]. Plasma actuators create a plasma above a blunt body that alters the laminar-turbulent transition inside the boundary layer [31–34], even at a high angle of attack [26, 35] they induce or reduce the fluid separation, diminishing drag [23] and increasing lift [36, 37]. They also allow sonic boom minimization schemes [38, 39], avoiding unwanted vibrations or noise [40, 41], sterilizing or decontaminating surfaces [42, 43], jet engines or wind turbines, via electromagnetic control. Its promising potential extends to flow control at hypersonic speeds [44].

The asymmetric dielectric barrier discharge (DBD) plasma actuator is a normal glow discharge that, like all normal glow discharges, operates at the Stoletow point [45, 46]. This guarantees that the generation of the ion-electron pairs in one atmosphere is done efficiently. In the air, the minimum energy cost is 81 electron volts per ion-electron pair formed in the plasma. In plasma torches, this energy cost can be of the order of one keV/ion-electron pair; in arcs, it can range from 10 to 50 keV/ion-electron pair.

In pure oxygen and pure nitrogen medium, particle and fluid simulations (e.g., Ref. [47]) done for a plasma actuator show the formation of an asymmetrical force that accelerates the ions dragging the neutral fluid in the direction of the buried electrode. A net force materializes because the plasma density and momentum transfer are more significant during the second half of the bias cycle. This effect is due partially to the ion density being higher during the second half-cycle. Thus, in each cycle, the induced total unidirectional force towards the buried electrode can create neutral fluid flow velocities on the order of 8 m/s [26].

Two-dimensional fluid models of a DBD plasma actuator (e.g., Ref. [48]) calculate the total force on ions and neutral particles. They have shown that the generated force retains identical nature to the electric wind in a corona discharge. The difference is that the force in the DBD is localized in the cathode sheath region of the discharge expanding along the dielectric surface. While the intensity of this force is much larger than the existing force of a dc corona discharge, it is active during less than a hundred nanoseconds for each discharge pulse and, consequently, the time-averaged forces are of the same magnitude in both cases [48]. The use of voltage pulses in plasma actuators, e.g., by modulation of the high-frequency excitation voltage carrier wave by a square wave, introduces mean and unsteady velocity components, and air momentum (including time-mean and oscillatory components) gains efficiency.

Actuators placed on the leading edge of an airfoil can control the boundary layer separation, while if located at the trailing edge, it can control lift. In particular, experiments found that thrust T and maximum induced speed u_{max} are proportional to the input power P, which depends nonlinearly on the voltage drop ΔV across the dielectric $T \propto u_{max} \propto P \propto \Delta V^{7/2}$.

Particle-in-cell and Monte Carlo calculations have shown that negative oxygen ions generated in the plasma modify the interplay between species diminishing the net ponderomotive force since negative ions impart momentum to the air in the opposite direction to positive ions. The atmospheric glow discharge has similar phenomenology as a low-pressure dc glow discharge. Research on high-speed jet control is progressing fast in a very competitive area (e.g., Ref. [49]).

1.5 Numerical Model

1.5.1 Description

To avoid numerical complexity neutral heavy species are not included in detailed plasma chemistry. At this stage, it is considered the kinetics involving electrically charged species supposedly playing a determinant role at atmospheric pressure: N_2^+, N_4^+, O_2^+, O_2^-, and electrons. Solving the set of governing equations for the charged species populations combined with the electric field that controls their dynamics (Poisson's equation), electrohydrodynamics (EHD) effects with interest to plasma actuators can be investigated. In particular, the body forces acting on the plasma horizontally (for neutral flow control) and perpendicularly (for boundary-layer control) to the energized electrode and the induced neutral particles' average speed using the Bernoulli equation.

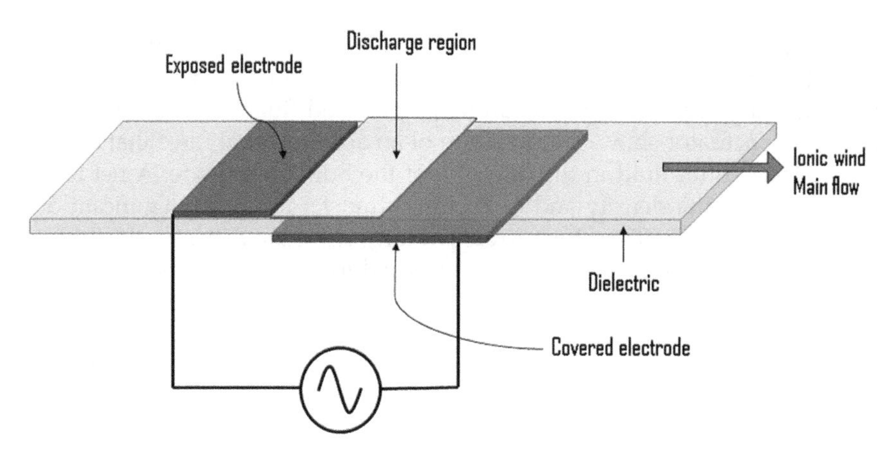

FIGURE 1.5
Schematic of the asymmetric plasma actuator.

The applied voltage has a sinusoidal wave form $V_a(t) = V_{dc} + V_{rms}\sin(\omega t)/\sqrt{2}$, where the root mean square voltage, V_{rms}, in this case study is 5 kV and the applied frequency is $f = 5$ kHz. Therefore, the dynamical time is $T = 200\ \mu$s.

The simulations were done for a two-dimensional flat staggered geometry while assuming the plasma was homogeneous along the OZ-axis (see Figure 1.5). The computational domain is a two-dimensional area with the total length along the Ox-axis $L_x = 4$ mm and height $L_y = 4$ mm. The grid has Cartesian coordinates and the *time stepping* was chosen typically of the order of one nanosecond. This is essentially a "surface discharge" arrangement with asymmetric electrodes. As shown in Figure 1.5, the simulated physical domain consists of conductive copper strips (with negligible thickness) of width $w = 1$ mm, separated by a d = 0.065 cm thick dielectric with a width equal to 3 mm and relative dielectric permittivity $\epsilon_r = 5$. The electrical capacity of the reactor is assumed to be given by the conventional formula $C = \epsilon_r \epsilon_0 S/d$.

1.5.2 Transport Parameters and Rate Coefficients

The working gas is an "airlike" mixture of a fixed fraction of nitrogen ($\delta_{N_2} = [N]/N = 0.78$) and oxygen ($\delta_{O_2} = [O_2]/N = 0.22$), as normally present at sea level at $p = 1$ atm. The electron homogeneous Boltzmann equation (EHBE) is solved with the 2-term expansion in spherical harmonics [50–52]. The gas temperature is assumed constant, both spatially and in a time frame, with $T_g = 300$ K, and the same applies to the vibrational temperature of nitrogen $T_v(N_2) = 2000$ K and oxygen $T_v(O_2) = 2000$ K, which are consistent with Ref. [53]. This assumption avoids the need to include a complex vibrational kinetic model. The result is the electrons energy distribution function (EEDF) which, for typical conditions, is independent of the electron concentrations. Electron-electron collisions make the EEDF Maxwellian, but they are not significant in molecular gases (like nitrogen) at a low degree of ionization (in our case, typically, the degree of ionization is $n_e/N \sim 10^{-5}$).

Using the set of cross-sections of excitation by electron impact taken from Ref. [54], rates coefficients and transport parameters needed for the electronic kinetics are obtained (see also Table 1.4). The species included in our model are the following: N_4^+, N_2^+, O_2^+, O_2^- and electrons. At atmospheric pressure N_4^+ ions need to be introduced since their

TABLE 1.4

List of Reactions taken into Account with CODEHD. Rate Coefficients for Chemical Processes were taken from Ref. [56]

Kind of reaction	Process	Rate coefficient
Ionization	$e + N_2 \rightarrow 2e + N_2^+$	$\nu_{ion}^{N_2\,a}$
Ionization	$e + O_2 \rightarrow 2e + O_2^+$	$\nu_{ion}^{O_2\,a}$
3-body electron attachment	$e + O_2 + O_2 \rightarrow O_2^- + O_2$	$K_{a1} = 1.4 \times 10^{-29}\left(\frac{300}{T_e}\right)\exp(-600/T_g)K_1(T_g, T_e)\ (\mathrm{cm^6/s})^b$
3-body electron attachment	$e + O_2 + N_2 \rightarrow O_2^- + N_2$	$K_{a2} = 1.07 \times 10^{-31}\left(\frac{300}{T_e}\right)^2 K_2(T_g, T_e)\ (\mathrm{cm^6/s})^c$
Collisional detachment	$O_2^- + O_2 \rightarrow e + 2O_2$	$K_{da} = 2.9 \times 10^{-10}\sqrt{\frac{T_g}{300}}\exp(-5590/T_g)\ (\mathrm{cm^3/s})$
e-ion dissociative recombination	$N_2^+ + e \rightarrow 2N$	$\beta = 2.8 \times 10^{-7}\sqrt{\frac{300}{T_g}}\ (\mathrm{cm^3/s})$
e-ion dissociative recombination	$O_2^+ + e \rightarrow 2O$	$\beta = 2.8 \times 10^{-7}\sqrt{\frac{300}{T_g}}\ (\mathrm{cm^3/s})$
2-body ion-ion recombination	$O_2^- + N_2^+ \rightarrow O_2 + N_2$	$\beta_{ii} = 2 \times 10^{-7}\sqrt{\frac{300}{T_g}}\left[1 + 10^{-19}N\left(\frac{300}{T_g}\right)^2\right]\ (\mathrm{cm^3/s})$
2-body ion-ion recombination	$O_2^- + O_2^+ \rightarrow 2O_2$	$\beta_{ii} = 2 \times 10^{-7}\sqrt{\frac{300}{T_g}}\left[1 + 10^{-19}N\left(\frac{300}{T_g}\right)^2\right]\ (\mathrm{cm^3/s})$
Ion-conversion	$N_2^+ + N_2 + N_2 \rightarrow N_4^+ + e$	$K_{ic1} = 5 \times 10^{-29}\ (\mathrm{cm^6/s})$
Recombination	$N_4^+ + e \rightarrow 2N_2$	$K_{r2} = 2.3 \times 10^{-6}/(Te/300)^{0.56}$
Ion-conversion	$N_4^+ + N_2 \rightarrow N_2^+ + 2N_2$	$K_{ic2} = 2.1 \times 10^{-16}\exp(T_g/121)\ (\mathrm{cm^3/s})$

Notes

a Numerical data obtained by solving the quasi-stationary, electron homogeneous Boltzmann equation. See Ref. [50] for details.

b With $K_1 = \exp(700(T_e - T_g)/(T_e T_g))$.

c With $K_2 = \exp(-70/T_g)\exp(1500(T_e - T_g)/(T_e T_g))$. In the set of governing equations, $\nu_{att} = K_{a1}\Delta_{O_2}^2 + K_{a2}\Delta_{O_2}\Delta_{N_2} + K_{da}\Delta_{O_2}$, with $\Delta_{O_2} = 0.22N$ and $\Delta_{N_2} = 0.78N$, denoting the fraction of oxygen and nitrogen molecules, respectively, and N denoting the total gas density.

concentrations can possibly be of the same order of magnitude or even higher than N_2^+ [55], partially due to the process of ion conversion [56]: $N_2^+ + N_2 + N_2 \rightarrow N_4^+ + N_2$, which occurs at a higher rate than the direct ionization, with constant $k_{ic1} = 5 \times 10^{-29}\ \mathrm{cm^6\ s^{-1}}$. Ion diffusion were obtained using Einstein-Smoluchowski relation and mobility coefficients were taken from [57, 58]: $\mu_{O_2^-}N = 6.85 \times 10^{21}\ \mathrm{V^{-1}\ m^{-1}\ s^{-1}}$ (on the range of E/N with interest here), $\mu_{O_2^+}N = 6.91 \times 10^{21}\ \mathrm{V^{-1}\ m^{-1}\ s^{-1}}$, and $\mu_{N_2^+}N = 5.37 \times 10^{21}\ \mathrm{V^{-1}\ m^{-1}s^{-1}}$. The gas density at $p = 1$ atm and assuming $T_g = 300$ K is $N = 2.447 \times 10^{25}\ m^{-3}$.

At atmospheric pressure the local equilibrium assumption holds and the transport coefficients ($\nu_{ion}^{N_2}, \nu_{ion}^{O_2}, \mu_e, \mu_p, D_e, D_p$) depend on space and time (r, t) only through the local value of the electric field $\mathbf{E}(\mathbf{r}, t)$; this is the so called *hydrodynamic regime*, thoroughly assumed in the present model.

The motion of the gas has an appreciable effect on the ions' mobility for gas flow velocity above $10^3 - 10^4$ m/s since this is comparable to the drift velocity of ions in the electric field. Although the gas flow has no direct effect on the motion of electrons, the coupling between electrons and ions through the ambipolar electric field does affect electrons' movement. To avoid the use of Navier-Stokes equations and to obtain a faster numerical solution to the present hydrodynamic problem, we can assume that the gas flow does not alter the plasma characteristics and is much smaller than the charged particle drift velocity.

With the above assumptions, charged species are represented by continuity and momentum transport equations, in the drift-diffusion approximation. This last approximation is valid if their drift energy is negligible concerning thermal energy. Furthermore, in the drift-diffusion equation, it is neglected the temperature gradient term (e.g., Ref. [48]).

The reactions included in this "effective" kinetic model are listed in Table 1.4. We assumed that all volume ionization is due to electron-impact ionization from the ground state and the kinetic set consists basically of ionization, attachment, and recombination processes. The kinetics of excited states and heavy neutral species are not considered, particularly the possible important role of metastable species (certainly improving plasma density through Penning ionization [59]) and nitrous oxide (poisoning gas) are omitted. The significant disadvantage associated with the present "effective" kinetic model is the shortage of consistency due to the neglect of atomic species but, as they are not directly involved in the plasma actuator performance [60], this simplification allows for investigation of the general trends of the plasma actuator operating characteristics. Furthermore, wind tunnel experiments have shown that relative humidity does not appreciably inhibits the performances of DBD actuators for relative humidity below 50% [61].

1.5.3 Numerical Model

The particle's governing equations are of convection-diffusion type, and they are solved using a well-known finite-element method proposed by Patankar [62]. Gauss's equation for the electric field can be solved, for example, by applying the successive over-relaxation method (SOR) for the boundary-value problem. The time step is restricted by the value of the dielectric relaxation time. For the present calculations, it was used (100x100) computational meshes.

The governing equation for N_4^+ (at atmospheric pressure the nitrogen ion predominant is N_4^+) is:

$$\frac{\partial n_{p2}}{\partial t} + \nabla\cdot(n_{p2}\mathbf{v}_{p2}) = \delta_{N_2}^2 N^2 n_{p1} K_{ic1} - K_{ic2}\delta_{N_2} N n_{p2} - K_{r2} n_{p2} n_e \tag{1.145}$$

For more concise notation, we put $n_{p2} \equiv [N_4^+]$; $n_{p1} \equiv [N_2^+]$; $n_p \equiv [O_2^+]$; $n_n \equiv [O_2^-]$, and $n_e \equiv [e]$. The governing equation for N_2^+ (one of the most mobile charged species in the plasma) is:

$$\begin{aligned} \frac{\partial n_{p1}}{\partial t} + \nabla\cdot(n_{p1}\mathbf{v}_{p2}) &= n_e \nu_{ion}^{N_2} + K_{ic2}\delta_{N_2} N n_{p2} - \beta_{ii} n_n n_{p1} \\ &\quad - \beta n_e n_{p1} - K_{ic1}\delta_{N_2}^2 N^2 n_{p1} \end{aligned} \tag{1.146}$$

The considered positive oxygen ion is O_2^+ and its resultant governing equation is given by:

$$\frac{\partial n_p}{\partial t} + \nabla\cdot(n_p \mathbf{v}_p) = n_e \nu_{ion}^{O_2} - \beta_{ii} n_n n_p - \beta n_e n_p \tag{1.147}$$

The negative ion O_2^- was introduced (due to high electronegativity of oxygen), and its governing equation was written as:

$$\frac{\partial n_n}{\partial t} + \nabla\cdot(n_n \mathbf{v}_n) = \nu_{att}^{O_2} n_e - \beta_{ii} n_{p1} n_n - K_{da}\delta_{O_2} N n_n \tag{1.148}$$

Finally, the governing equation for electrons can be written in the form:

$$\begin{aligned}\frac{\partial n_e}{\partial t} + \nabla\cdot(n_e \mathbf{v}_e) &= n_e(\nu_{ion}^{N_2} + \nu_{ion}^{O_2} - \nu_{att}^{O_2}) - \beta n_e(n_p + n_{p1}) \\ &\quad + K_{da}\delta_{O_2} N n_n - K_{r2} n_{p1} n_e\end{aligned} \tag{1.149}$$

Photo-ionization was not included as a non-local secondary effect. In order to avoid the use of Navier-Stokes equation, the *drift-diffusion approximation* for the charged particle mean velocities appearing in the continuity equations is used instead:

$$n_{n,p}\mathbf{v}_{n,p} = \pm n_{n,p}\mu_{n,p}\mathbf{E} - \nabla(n_{n,p}D_{n,p}) \tag{1.150}$$

Here, the subscript p corresponds to positive ions and n to electrons and negative ions. $n_{n,p}$, $\mu_{n,p}$ and $D_{n,p}$ represent their respective densities, charged particle mobility and diffusion coefficients. The minus sign is to be used for the negatively charged species. Also, in all the above 1.145–1.150, the different $\mathbf{v}$'s denote the averaged drift velocities of the differently charged particles. The applied voltage has a sinusoidal waveform:

$$V_a(t) = V_{dc} + V_0 \sin(\omega t) \tag{1.151}$$

where V_{dc} is the dc bias voltage (considered here as fixed to ground, $V_{dc} = 0$) and ω is the applied angular frequency. V_0 is the maximum amplitude, and the root mean square voltage in our case of study is $V_{rms} = 5$ kV, were the applied frequency is $f = 5$ kHz.

The total current (convective plus displacement current) was determined using the following equation given by Morrow and Sato [63]:

$$\begin{aligned}I_d(t) &= \frac{e}{V}\int_V \left(n_p\mathbf{v}_p - n_e\mathbf{v}_e - n_n\mathbf{v}_n - D_p\frac{\partial n_p}{\partial z} + D_e\frac{\partial n_e}{\partial z} + D_n\frac{\partial n_n}{\partial z}\right)\cdot\mathbf{E}_L dv \\ &\quad + \frac{\epsilon_0}{V}\int_V \left(\frac{\partial \mathbf{E}_L}{\partial t}\cdot\mathbf{E}_L\right) dv\end{aligned} \tag{1.152}$$

where $\int_V dv$ is the volume occupied by the discharge, $\mathbf{E}_L$ is the space-charge free component of the electric field. The last integral when applied to our geometry gives the displacement current component:

$$I_{disp}(t) = \frac{\varepsilon_0}{d^2}\frac{\partial V}{\partial t}\int_V dv \tag{1.153}$$

The flux density of secondary electrons onto the cathode is given by

$$\mathbf{j}_{se}(t) = \gamma \mathbf{j}_p(t) \tag{1.154}$$

with $\mathbf{j}_p$ denoting the flux density of positive ions. We assume throughout the calculations that Auger electrons are produced by impact of positive ions on the cathode with efficiency $\gamma = 5 \times 10^{-2}$.

Secondary electron emission plays a fundamental role in the working of the asymmetric DBD plasma actuator. The progressive accumulation of electric charges over the dielectric surface develops a so-called "memory voltage", whose expression is given by [59]:

$$V_m(t) = \frac{1}{C_{ds}} \int_{t_0}^{t} I_d(t')dt' + V_m(t_0) \tag{1.155}$$

Here, C_{ds} is the equivalent capacitance of the discharge.

As charged particles are generated in the plasma volume, the space-charge electric field is determined by solving the Poisson's equation coupled to the particle's governing equations:

$$\Delta V = -\frac{e}{\epsilon_0}(n_p - n_e - n_n) \tag{1.156}$$

Here, n_p, n_e and n_n denote, respectively, positive ion, negative ion and electron number densities. The following boundary conditions were assumed at the electrode and dielectric surfaces:

- over the electrode (Dirichlet boundary condition): $V(x, y = 0, t) = V - V_m$;
- over the insulator (Neumann boundary condition): $E_n = (\mathbf{E}\cdot\mathbf{n}) = \frac{\sigma}{2\epsilon_0}$.

The flux of electric charges impinging on the dielectric surface builds up a surface charge density σ which was calculated by balancing the flux to the dielectric and it is governed by

$$\frac{\partial \sigma}{\partial t} = e(|\Gamma_{p,n}| - |\Gamma_{e,n}|) \tag{1.157}$$

Here, $\mathbf{\Gamma}_{p,n}$ and $\mathbf{\Gamma}_{e,n}$ represent the normal component of the flux of positive and negative ions and electrons to the dielectric surface. It is assumed that ions and electrons recombine instantaneously on the perfectly absorbing surface. As will be discussed later, this simplified assumption constitutes a drawback of the present model, but to our knowledge, this important issue is not yet resolved in the literature. Figure 1.6 shows a typical evolution along a period of the calculated electric current, applied voltage, gas voltage, and memory voltage in a Dielectric Barrier Discharge. At about 740V, electron avalanches develop, replenishing the volume above the surface with charged particles. Hence, the charged particles flow to the dielectric, start accumulating on the surface, and build up an electric field that prevents the occurrence of a high current, and quenches the discharge development at an early stage.

1.5.4 The First Order Orbit Theory

To prepare the introduction of the topic of anomalous diffusion with our theoretical model we give here the basics of the motion of a charged particle with mass m, charge e in the presence of (uniform or time-dependents) electric $\mathbf{E}$ and magnetic fields $\mathbf{B}$.

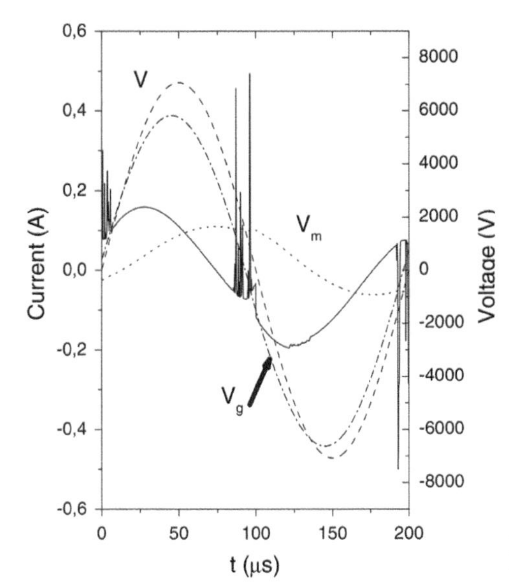

FIGURE 1.6
Electric current, applied voltage, gas voltage and memory-voltage as a function of time. Image courtesy: Ref: [28].

In the nonrelativistic approximation, the equation of motion is

$$m\frac{d\mathbf{v}}{dt} = \mathbf{F} + e(\mathbf{E} + [\mathbf{v} \times \mathbf{B}]) \tag{1.158}$$

We may distinguish the components parallel || and perpendicular ⊥ to the magnetic field. We see that

$$\frac{dv_{\|}}{dt} = \frac{e}{m}E_{\|} \tag{1.159}$$

It means charged particles uniformly accelerate along the magnetic field lines. It also means that for a plasma to be in equilibrium, $E_{\|}$ must be small or vanishingly small. The Lorentz force acts upon electrons and ions in the opposite direction, but the nonelectric force acts in the same path, disregarding the electric charge. Hence 1.158 can be written as

$$m\frac{d\mathbf{v}}{dt} = \mathbf{E}' + e(\mathbf{E} + [\mathbf{v} \times \mathbf{B}]) \tag{1.160}$$

where

$$\mathbf{E}' \equiv \mathbf{E} + \frac{\mathbf{F}}{e} \tag{1.161}$$

It signifies that $\mathbf{E}'$ is equivalent to an electric force per unit of charge. The work-energy theorem immediately gives us

$$\begin{aligned} W &= \Delta K \\ &\int_{\mathbf{r}_0}^{\mathbf{r}} (\mathbf{E}' \cdot d\mathbf{r}) \\ &= m\mathbf{v}^2 - m\mathbf{v}_0^2 \\ m\mathbf{v}^2 - m\mathbf{v}_0^2 &= 2e(\phi - \phi_0) \text{if E and F are both conservative} \end{aligned} \tag{1.162}$$

1.5.4.1 Particle Motion in the Presence of a Constant Magnetic Field

Assuming $\mathbf{E} = 0$ and $\mathbf{B} = B_z \mathbf{u}_z$, it can be shown that 1.158 has the solution:

$$\ddot{x} = \Omega \dot{y}; \quad \ddot{y} = -\Omega \dot{x}; \quad z\ddot{z} = 0 \tag{1.163}$$

Exercise: Obtain the solution of the set of 1.163.

Solution: Projecting the components, we have

$$\begin{aligned} \dot{v}_x &= \frac{eB}{m} v_y \\ \dot{v}_y &= -\frac{qB}{m} v_x \\ \dot{v}_z &= 0 \end{aligned} \tag{1.164}$$

We must give the initial conditions of the trajectory at $t = 0$:

$$\begin{aligned} x(0) = x_i; \quad & y(0) = y_i; \quad z(0) = z_i \\ v_x(0 =) = v_{xi}; \quad & v_{yi} = 0; \quad v_{zi} = 0 \end{aligned} \tag{1.165}$$

Integrating the first and second of 1.164, we have

$$\dot{x} = \Omega(y - y_0) + v_{xi} \quad \dot{y} = -\Omega(x - x_0) + v_{yi} \tag{1.166}$$

Taking the time derivative of both 1.166, we obtain

$$\begin{aligned} \ddot{x} &= -\Omega^2\left(x - x_0 - \frac{v_{yi}}{\Omega}\right) \\ \ddot{y} &= -\Omega^2\left(y - y_0 + \frac{v_{xi}}{\Omega^2}\right) \end{aligned} \tag{1.167}$$

where

$$\Omega \equiv \frac{eB}{m} \tag{1.168}$$

is the so called **cyclotron frequency**[8] and is directly proportional to **B**-field.

The general solution of both Eqs. 1.168 are of the form

$$x - x_0 - \frac{v_{yi}}{\Omega} = A \cos \Omega t + B \sin \Omega t \tag{1.169}$$

We need the two initial conditions x_0 and v_{xi} to obtain

$$x - x_0 - \frac{v_{yi}}{\Omega} = -\frac{v_{yi}}{\Omega}\cos\Omega t + \frac{v_{xi}}{\Omega}\sin\Omega t \tag{1.170}$$

and, by the same token,

$$y - y_0 + \frac{v_{xi}}{\Omega} = \frac{v_{xi}}{\Omega}\cos\Omega t + \frac{v_{yi}}{\Omega}\sin\Omega t \tag{1.171}$$

It easy to verify that

$$\left(x - x_0 - \frac{v_{yi}}{\Omega}\right)^2 + \left(y - y_0 + \frac{v_{xi}}{\Omega}\right)^2 = \frac{v_{xi}^2 + v_{yi}^2}{\Omega^2} = \frac{v_\perp^2}{\Omega^2} \tag{1.172}$$

where

$$v_\perp^2 = v_{xi}^2 + v_{yi}^2 \tag{1.173}$$

The projection of the trajectory in the plane transversal to **B** is a circle of radius

$$\rho = \frac{v_\perp}{\Omega} \tag{1.174}$$

and the center of gyration, also called the **guiding center**, has the coordinates $x_0 + v_{yi}/\Omega$, $y_0 - v_{xi}/\Omega$, and $z_0 + v_\parallel t$, at time t. Certainly the concept of guiding center is very helpful, since the gyration is much faster than the motion of the guiding center with vector position given by

$$\mathbf{r}_g = \dot{x}\mathbf{i} + \dot{y}\mathbf{j} + (z_0 + v_\parallel)\mathbf{k} \tag{1.175}$$

The guiding center has a linear motion along z at uniform speed $v_\parallel$, but shows an helical trajectory with **pitch angle** of the helix given by

$$\alpha = \arctan\left(\frac{v_\perp}{v_\parallel}\right) \tag{1.176}$$

1.5.5 Circuital Model of Anomalous Diffusion in a Cold Magnetized Plasma

All phenomena occurring with interfacial systems have a fundamental importance in science and technology (e.g., electrostatic charging of insulators, surface tension, forward conduction in p-n junctions). Specifically, the problem of the plasma-wall interactions are of primary importance in plasma physics.

Historically, the anomalously high diffusion of ions across magnetic field lines in Calutron ion sources (electromagnetic separator used by E. Lawrence for uranium isotopes) gave the firsts indications of the onset of a new mechanism [4]. It has been noticed that the plasma moves across the magnetic confining field at a much higher average velocity than is predicted by classical considerations. The classical diffusion coefficient is given by $D_{\perp} = \eta p/B^2$, while the anomalous diffusion coefficient by $D_{\perp} = \alpha kT/B$. Manifestly it is needed a better understanding of the physical laws governing the matter in the far nonequilibrium state and this is a challenging issue for the advancement of this frontier of physics.

General proposed explanations were advanced. The first one was Simon's "short-circuit" problem suggests that the observed losses could be explained by the highly anisotropic medium induced by the magnetic field lines, favoring electron current to the conducting walls [64]. Experiments were done by Geissler [65] in the 1960s have shown that diffusion in plasma across a magnetic field is near to classical (standard) diffusion type when insulating walls impose plasma ambipolarity, but in the presence of conducting walls charged particles diffused at a much higher rate.

This problem of plasma-wall interaction becomes more complex when looking at a complete description of a magnetized nonisothermal plasma transport in a conducting vessel. Beilinson *et al.* [66] have shown the possibility to control the discharge parameters by applying a potential difference to sectioned vessel conducting walls.

In the area of fusion reactors, there is a strong indication that for plasmas large but finite Bohm-like diffusion coefficient appears above a certain range of B [67]. Experiments give evidence of transport of particles and energy to the walls [68]. At the end of the 1960s, experimental results obtained in weakly ionized plasma [65] and in a hot electron plasma [69] (this one proposing a possible mechanism of flute instability) indicated a strong influence that conducting walls have on plasma losses across magnetic field lines. Geissler [65] suggested that the most probable explanation was the existence of diffusion-driven current flow through the plasma to the walls. Concerning fusion reactions, Taylor [70] provided a new interpretation of tokamak fluctuations as due to an inward particle flux resulting from the onset of filamentary currents.

Progress in the understanding of the generation of confinement states in plasma is fundamental [71] to pursue the dream of a fusion reactor [72, 73]. Anomalous diffusion is a cornerstone in this quest, as recent research with tokamaks suggest that the containment time is $\tau \approx 10^8 R^2/2D_B$, with R denoting the minor radius of a tokamak plasma and D_B is the Bohm diffusion coefficient [74]. Controlled nuclear fusion experiments have shown that the transport of energy and particles across magnetic field lines is anomalously large (i.e, not predicted by classical collision theory).

The conjecture made by Bohm is that the diffusion coefficient is $D_B = \alpha kT/eB$, where T is the plasma temperature and α is a numerical coefficient, empirically taken to be 1/16 [4]. Initially, the origin of the anomalous diffusion was assumed to be due to the turbulence of small-scale instabilities (see, for example, Refs. [67, 70, 75]). However, it is now clear that there is several different mechanisms that can lead to anomalous diffusion such as, coherent structures, avalanches type processes, and streamers, which have a different character than a purely diffusive transport process. Recent experimental results such as the scaling of the confinement time in L-mode plasmas and perturbative experiments undermine the previous paradigm built on the standard transport processes [76, 77] showing conclusively that there are many regimes where plasma diffusion does not scale as B^{-1}.

We introduce mow a mechanism of wall current drain set up together with the magnetic field "cutting" lines across the area traced by the charged particles trajectories. The proposed mechanism of anomalous diffusion is expected to be valid in purely diffusive regimes when plasma diffusion scales as Bohm diffusion, both in the edge and core of cold magnetized plasma. At his stage it was considered of secondary importance the role of collisions in randomizing the particle's distribution function. From collisional low-temperature plasmas to a burning fusion plasma subject the plasma confinement vessel to strong wall load, both in stellarator or tokamak operating modes, this explanation could be of considerable interest, particularly when diffusive transport process are dominant.

1.6 Simon's "Short-Circuit" Theory

The first attempt to explain why the plasma diffuses at a much higher average velocity than is predicted by classical theory has been advanced by A. Simon [64]. The magnetic field lines structure a highly anisotropic medium. Any fluctuation of the space charge builds up an electric field, which has a strong effect on the currents parallel to the magnetic field lines. The classical equation for conductivity across a magnetic field is given by

$$\sigma_{\perp} = \frac{\sigma_0}{1 + \frac{\Omega_c^2}{\nu_e^2}} \tag{1.177}$$

where $\Omega_c = eB/m$ is the electron cyclotron frequency, ν_e is the electron collision frequency, and $\sigma_0 = e^2 n_e / m\nu_e$ is the conductivity in the absence of a magnetic field. By the contrary, due to $\Omega_c/\nu_e \gg 1$, this electric field is too small to have any importance on the crossed-field conductivity. From this results that there is a strong current to the wall without a concomitant current to different regions of the plasma, making of this situation is a kind of circuital "short-circuit" problem. Although Simon attempted to explain the anomalously high rate of diffusion in Calutron ion sources in the frame of the classical diffusion theory calculating the coefficient $D_{\perp}$ as being approximately equal to the transverse diffusion coefficient of the ions. His proposal is not suitable, however, because the experimental determination of $D_{\perp}$ through a decaying plasma has shown that according to the magnetic field strength, the transverse diffusion coefficient can be much higher than the classical one or smaller than the transverse diffusion coefficient of the ions [65].

1.7 Plasma Turbulence and Transport

Purely diffusive transport models cannot give convincing explanations for a variety of experiments in magnetically confined plasmas in fusion engineering devices, particularly the scaling of the confinement time in L-mode plasmas. The assumed underlying instabilities are driven by either the pressure gradient or the ion temperature gradient. It is well-established fact that transport in high temperature confined plasmas is driven by turbulence and plasma profiles, and is subject to transition from L-mode to H-mode (characterized by a very steep gradient near the plasma surface) [71]. The non-linearity in the gradient-flux relation is the source of turbulence and turbulence-driven transport. The fluxes contain all the dynamic information on the transport process. Accordingly, changes in the gradient trigger local instabilities in the plasma. This local instability induces an increase in the nearby gradients, thus causing propagation of the instability all across the plasma. In particular, excessive pressure on the core propagates to the edge in a kind of avalanche. In weakly turbulent cold magnetized plasmas, besides the Calutron ion sources and the magnetron, the study of particle transport in crossed electric and magnetic fields result from applications to electromagnetic space propulsion (Hall thrusters). That plasma accelerators work with a radial magnetic field that prevents electron flow toward the anode and force the electrons closed-loop drift around the axis of the annular geometry. Neutrals coming from the anode are ionized in this rotating electrons cloud, while ions are accelerated by an axial electric field that freely accelerates them out from the device. This effect develops in the so-called extended acceleration zone (or electric-magnetic region plasma). In this acceleration zone the electron gyro-radius and the Debye shielding length are small relative to the apparatus dimension, while the ion gyroradius is larger than the apparatus typical length. From these spatial scales results that the electron motions are $[\mathbf{E} \times \mathbf{B}]$ drifts, but ions accelerate with the electric field that develops in the plasma. The first observations of a large amount of electron transport toward the anode have been noticed in the '60s 1960s (e.g., Ref. [78]) and they have been related to electric field fluctuations since they were correlated with the density variations to produce anomalous transport. Another possible mechanism that could explain the high electron transverse conductivity was advanced: collisions with the wall [79, 80]. Electrons moving freely along lines of forces of the magnetic field collide with the a wall more frequently than with ions and neutrals, while neutrals are reflected at the wall, enhancing emission of low-energy secondary electrons from the wall. As referred, the other strong candidate, which could possibly be the source of a higher axial electron current than the one predicted by the standard classical kinetic theory, is the turbulent plasma fluctuations. But there is no clear consensus on this issue [81–83].

The magnetron is a sputtering tool, used for reactive deposition and etching. The magnetron effect is applicable to different geometries and only need a closed-loop $[\mathbf{E} \times \mathbf{B}]$ drift to work. Rossnagel *et al.* [84] have shown that the Hall-to-discharge current ratio measured in those configurations could be explained if the high collision frequencies for electrons were associated to Bohm diffusion. In particular, Kaufman [85] argues that anomalous diffusion in closed-loop $[\mathbf{E} \times \mathbf{B}]$ thrusters could shift from core diffusion to edge diffusion (or wall effects) with increasing magnetic fields.

1.8 Circuital Model of Anomalous Diffusion

In a seminal paper [86] a conjecture was proposed based on the principle of minimum entropy-production rate, stating that a plasma will be more stable whenever the internal product of the current density $\mathbf{j}$ by an elementary conducting area $d\mathcal{A}$ at every point of the boundary - excluding the surface collecting the driving current - is null, $(\mathbf{j}\cdot\mathcal{A}) = 0$ at any point of the boundary (and excluding the surfaces collecting the discharge current), independently of the resistance R_i. The general idea proposed by Robertson [86] assumes that the plasma boundary is composed of small elements of area $\mathcal{A}_i$, each one isolated from the others, but each one connected to the exterior a common circuit through its resistor R_i and voltage V_i. The entropy production rate in the external circuits is given by:

$$\frac{dS}{dt} = \sum_i \frac{1}{T}(\mathbf{j}_i\cdot\mathcal{A}_i)^2 R_i \tag{1.178}$$

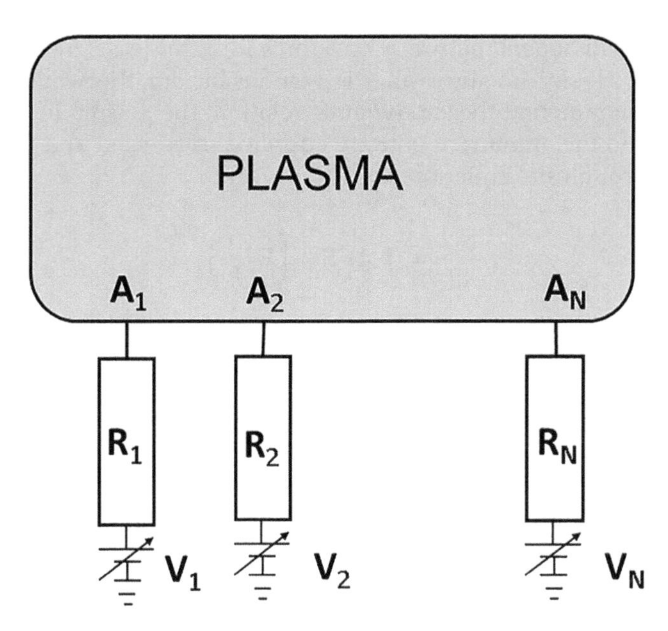

FIGURE 1.7
Schematic diagram of the plasma boundary connected to the common circuit through conducting walls.

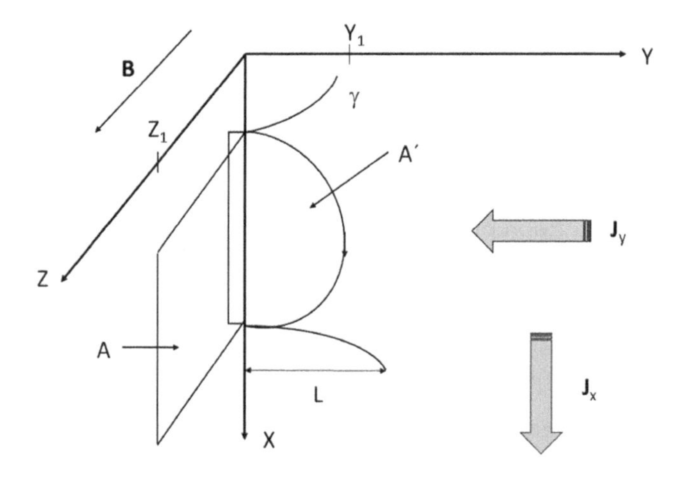

FIGURE 1.8
Schematic of the geometry for the plasma-wall current drain model. The uniform magnetic field points downward along Oz. Particles describe orbits in the plane xOy intersecting the wall (plan xOz). Orbits are represented by a semi-circular line for convenience. L is the maximum distance the trajectory attains from the wall. Image credit: Mario J Pinheiro 2007 J. Phys.: Conf. Ser. 71 012002. https://doi.org/10.1088/1742-6596/71/1/012002.

where T is the temperature of the resistors, supposed to be in thermal equilibrium with all the others. It is important to remark that the summation is over the different conducting areas eventually confining the plasma, *excluding* the areas of the electrodes. Figure 1.7 illustrates this concept.

We consider a simple axisymmetric magnetic configuration with magnetic field lines parallel to the z-axis with a plasma confined between two electrodes (see Figure 1.8). In general terms, a particle's motion in the plasma results in massive flux. As long as the flux is installed, the flux will depend naturally on a force **F** - in this case, the pressure gradient-driven process of diffusion to the wall - is responsible for the wall-driven current **j**. According to the fundamental thermodynamic relation, the plasma internal energy variation dU is related to the amount of entropy supplied or rejected and the work done by the driven force, through the equation:

$$\frac{dU}{dt} = (\mathbf{j}\cdot\mathcal{A})^2 R + \left(\mathbf{F}\cdot\frac{d\mathbf{r}}{dt}\right) \quad (1.179)$$

In the last term we identify with the macroscopic diffusion velocity $\mathbf{v}_d$ depicting the process of plasma expansion to the wall. To simplify somehow the calculations we assume a single plasma fluid under the action of a pressure gradient ($\mathbf{F} = \mathcal{A}Ldp/dy\vec{j}$, where $\vec{j}$ is a unit vector directed along the Oy axis).

In the presence of steady and uniform magnetic field lines (this simplifies the equations, but does not limit the applicability of the model), the particles stream freely along with them. From magnetohydrodynamics we have a kind of generalized Ohm's law (see, for example, Ref. [87]):

$$\nabla p = -en\mathbf{E} - en[\mathbf{v} \times \mathbf{B}] + [\mathbf{j} \times \mathbf{B}] - \frac{en\mathbf{j}}{\sigma} \quad (1.180)$$

where $\sigma = e^2 n\tau_e/m_e$ is the electric conductivity, with τ_e denoting the average collision time between electrons and ions. The force balance equation is given by:

$$\nabla p = [\mathbf{j} \times \mathbf{B}] \quad (1.181)$$

valid whenever the Larmor radius is smaller than the Debye radius. This assumption simplifies further the extension of our model to high enough magnetic fields. Therefore, after inserting Eq. 1.181 into Eq. 1.180 the y component of velocity is obtained:

$$v_y = -\frac{E_x}{B} - \frac{1}{\sigma B^2}\frac{dp}{dy} \quad (1.182)$$

From 1.182 we have the classical diffusion coefficient scaling with $1/B^2$ and thus implying a random walk of step length r_L (Larmor radius). To get the anomalous diffusion coefficient and as well as understand better its related physics, we must consider the process of diffusion to the wall - in the presence of an entropy source - with the combined action of the wall current drain, as already introduced in 1.179.

Therefore, using the guiding center plasma model the particle motion is made with velocity given by:

$$\mathbf{j} = en\mathbf{v}_d = -\frac{[\nabla p \times \mathbf{B}]}{B^2} \quad (1.183)$$

This equation forms the base of a simplified theory of magnetic confinement. The validity of 1.183 is restrained to the high magnetic field limit when the Larmor radius is shorter than the Debye radius.

Considering the motion along only one direction perpendicular to the wall (y-axis), it is clear that

$$(\mathbf{j}\cdot A)^2 = \frac{A^2}{B^2}\left(\frac{dp}{dy}\right)^2 \tag{1.184}$$

If we consider a quasi-steady state plasma operation, the plasma total energy should be sustained. Hence, $dU/dt = 0$, and the power associated with the driven pressure gradient is just maintaining the dissipative process of plasma losses on the wall. 1.179 governs the evolution of the diffusion velocity. Hence, we have

$$nv_d = -\frac{nR\mathcal{A}}{L}\frac{kT}{B^2}\frac{dn}{dy} = -D_T\frac{dn}{dy} \tag{1.185}$$

with D_T denoting the transverse (across the magnetic field) diffusion coefficient given by:

$$D_T = \frac{nR\mathcal{A}}{L}\frac{kT}{B^2} \tag{1.186}$$

This new result coincides with the classical diffusion coefficient [88] whenever $nR\mathcal{A}/L \equiv m\nu_{ei}/e^2$, containing a dependence on collision frequency and particle number density. Other theoretical approaches to this the problem was advanced by Bohm [4], who proposed an empirically-driven diffusion coefficient associating plasma oscillations as the source of the enhanced diffusion, while Tonks [89] have shown that the current density that is present in a magnetically immobilized plasma is only generated by the particle density gradient, not being associated with any drift of matter. Simon electron "short-circuit" [64] scheme attempt to explain the different rates of diffusion, electrons and ions experiment with the magnetic field. While the ion flux dominates the radial diffusion, the electron flux dominates axial losses, due to an unbalance of currents flowing to the wall.

In the absence of collisions, the guiding centers of charged particles behave as permanently attached to the same lines of force. On the contrary, as a result of collisions with other charged particles the guiding centers shift from one line of force to another resulting in a diffusion of plasma across the field lines. In our model, each orbit constitutes an elementary current I eventually crossing the wall.

However, the particle diffusion coefficient as shown in 1.186 gives evidence of an interplay with the resistance that the elementary circuit offers when in contact with the walls in the presence of the frozen-in effect. In fact, for sufficiently strong magnetic fields, apparently a hydrodynamic behavior of the plasma is installed [67, 90], with the appearance of "convective cells" and the $1/B$ behavior dominates, giving birth to the anomalous diffusion mechanism. The onset of freezing magnetic lines are valid whenever the Lundquist number $S \gg 1$ (convection of the magnetic field dominated medium). In this case the magnetic field lines are frozen-in in the medium (a consequence of a vortex type of character of the magnetic field $\mathbf{B}$) and the flux of them across a given surface is constant:

$$\Phi = B\mathcal{A}' = BL^2\alpha \tag{1.187}$$

Remark that $\mathcal{A}'$ is now the surface delimited by the elementary circuit γ (see Figure 1.8) and $\alpha \leq 1$ is just a geometrical factor (e.g., $\alpha = \pi/4$ at the limit of a circular orbit). This situation is fundamental to the onset of anomalous diffusion. Free electrons orbits are helical, but as Figure 1.8 shows, that their projections at right angles to the field are circular. Each particle orbit constitutes an elementary circuit with B-field cutting its surface being associated with it an elementary flux Φ. At the same time, we can envisage each orbit as constituting itself an elementary circuit, some of them intersecting the wall and thus the circuit is closed inside the wall. Therefore a resistance R drags the charged flow to the conducting wall. It is therefore plausible to associate to this elementary circuit a potential drop V and all the processes being equivalent to a current I flowing through the elementary circuit.

Assuming the plasma is a typical weakly coupled, hot diffuse plasma with a plasma parameter (number of particles in the Debye sphere) $\Lambda = n\lambda_{De}^3 \approx 1$, it is more likely to expect nearly equal average kinetic and potential energy. However, the typical plasma parameter encountered in glow discharges or nuclear fusion is $\Lambda \gg 1$. This means that the average kinetic energy is larger than the average potential energy. To contemplate all ranges of Λ we can relate them through the relationship

$$\rho V = (\mathbf{J} \cdot \mathbf{A})\delta \tag{1.188}$$

Here, ρ is the charge density, $\mathbf{A}$ is the vector potential, $\mathbf{J}$ is the current density, and $\delta \leq 1$ is just a parameter representing the ratio of potential to kinetic energy. Of course, when $\Lambda \geq 1$, then $\delta \leq 1$. This basic assumption is consistent with the hydrodynamic approximation taken in the development of equations. The limitations of the model are related to the unknowns Λ and δ that can be uncovered only through a self-consistent model of the plasma. However, our analysis of anomalous diffusion remains general and added new insight to the phenomena.

Now suppose that the diffusion current is along y-axis $\mathbf{J} = -J_y \mathbf{u}_y$ (see Figure 1.1). Consequently, $\mathbf{A} = -A_y \mathbf{u}_y$, and then the potential drop will depend on x-coordinate:

$$\rho[V(x_1) - V(x_0)] = J_y[A_y(x_1) - A_y(x_0)]\delta \tag{1.189}$$

Multiplying both members by the area $\mathcal{A}' = x_1 z_1$ and length $L = y_1$, we have

$$Q\Delta V = I y_1 [A_y(x_1) - A_y(x_0)]\delta = I\Phi\delta \tag{1.190}$$

$\Phi = \oint_\gamma (\mathbf{A} \cdot d\mathbf{x})$ is the flux of the magnetic field through the closed surface bounded by the line element $d\mathbf{x}$ (elementary circuit γ, see also Figure 1.8). By other side, naturally, the total charge present on the volume $\mathcal{V} = x_1 y_1 z_1$ is such as $Q = ie$, with i an integer. This integer must be related to ions charge number. From 1.190 we obtain

$$R = \frac{\Delta V}{I} = \delta \frac{\Phi}{Q} = \alpha\delta \frac{BL^2}{ie} \tag{1.191}$$

But, the particle density is given by $n = N/L\mathcal{A}$, with N being now the total number of charged particles present in volume $\mathcal{V} = \mathcal{A}L$. Since $i = N$, we retrieve finally the so-called Bohm-diffusion coefficient

$$D_B = \alpha\delta \frac{kT}{eB} \tag{1.192}$$

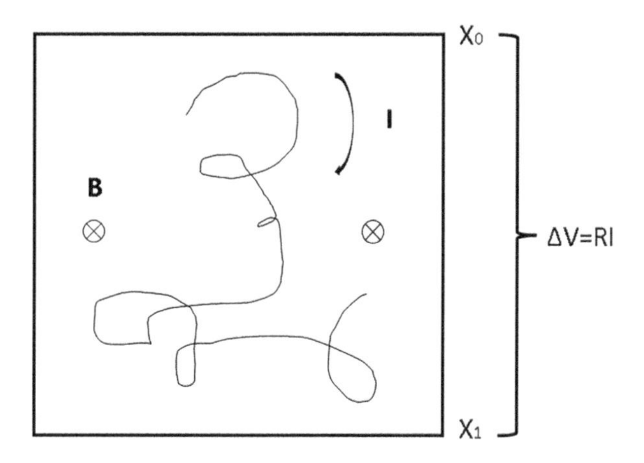

FIGURE 1.9
Volume control and particle's trajectory submitted to magnetic field. Magnetic field lines point downward. Image credit: Mario J Pinheiro 2007 J. Phys.: Conf. Ser. 71 012002. https://doi.org/10.1088/1742-6596/71/1/012002.

So far, our arguments were applied to edge anomalous diffusion. But they can be generalized to the core anomalous diffusion processes, provided that diffusive transport processes are dominant. For this purpose consider instead of a conducting surface a virtual surface delimiting a given volume, as shown in Figure 1.9.

Our coefficient is time-dependent and can be written under the form:

$$D_{\perp} = \delta \frac{kT(t)}{eB(t)^2} \frac{\Phi(t)}{L^2} \tag{1.193}$$

The nonrelativistic solutions of dynamical equation of a charged particle in time-dependent but homogeneous electric and magnetic field give the following approximative expressions for the width trajectories along the xOy plane (see, e.g., Ref. [91]) by:

$$\Delta x = \frac{v_{\perp}}{\Omega_c} \cos(\Omega_c t + \chi) \quad \Delta y = \mp \frac{v_{\perp}}{\Omega_c} \sin(\Omega_c t + \chi) \tag{1.194}$$

where χ is the initial phase, $v_{\perp}$ denotes the component of the velocity perpendicular to the magnetic field and the $\mp$ sign applies to electrons (-) or positive ions (+). From them we can retrieve the flux "cutting" area:

$$\mathcal{A} \approx \Delta x \Delta y = -\frac{1}{2}\left(\frac{v_{\perp}}{\Omega_c}\right)^2 \sin(2\Omega_c t + 2\chi) \tag{1.195}$$

Then the anomalous diffusion coefficient is just given by:

$$D_{\perp} \approx \mp \delta \frac{kT(t)}{eB(t_0)} \frac{1}{2} \frac{\sin(2\Omega_c t + 2\chi)}{|\cos(\omega t)|} \tag{1.196}$$

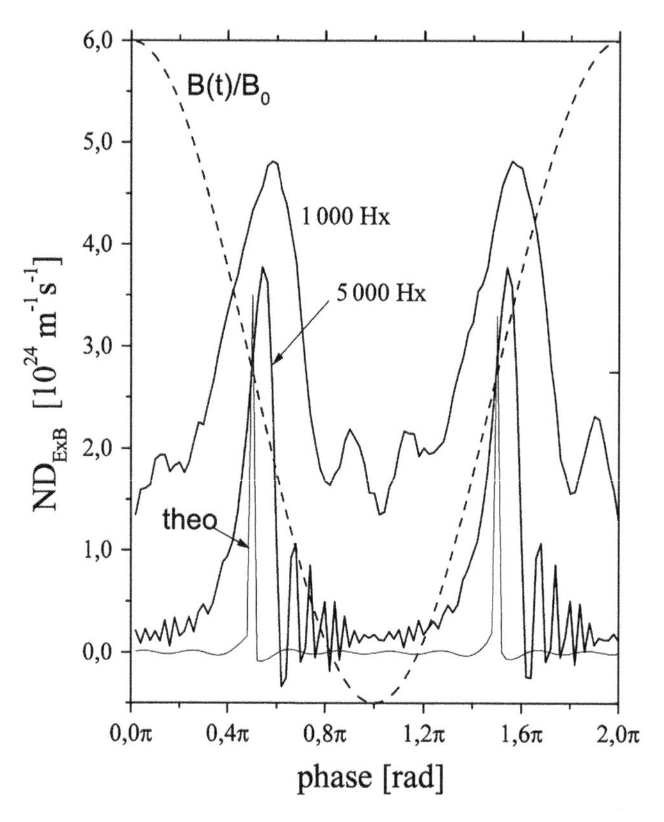

FIGURE 1.10
Comparison between numerical results for 5000 Hx obtained by Monte Carlo simulations Refs. [32, 33] of electron transport in crossed electric and magnetic RF fields in argon and 1.196 -theo. Dashed line: external time-dependent magnetic field. Parameters used: $\delta = \frac{1}{40}$; χ; applied frequency, f = 100 MHz; ciclotron frequency, $\Omega_c \approx 10^{10}$ Hz; $\frac{kT}{e} = 5.4$ eV; p = 1 Torr, $T_g = 300$ K, $\chi = 0$. Image credit: Mario J Pinheiro 2007 J. Phys.: Conf. Ser. 71 012002. https://doi.org/10.1088/1742-6596/71/1/012002.

As we can see in Figure 1.10 this last expression describes fairly well the diffusion process for high enough B/N values (the magnetic field to gas number density ratio) and explains the main processes building up such effects as i) the negative diffusion, which results from the contraction of the flux "cutting" area; ii) the ciclotronic modulation imprint on the transverse diffusion coefficient; iii) and the anomalous diffusion, due to the fast-flux rate of the magnetic field through the area $\mathcal{A}$). All this signs can be seen in Figure 1.10 where it is shown a comparison of numerical results (5000 Hx, 1 Hx = 10^{-27} T.m^3) obtained with Monte Carlo simulations of electron transport in crossed magnetic and electric fields by Petrović *et al.* [92, 93] with the theoretical prediction given by our 1.196. As long as only a self-consistent model could give us an exact value of the ratio of potential to kinetic energy δ, we assume here $\delta = 1/40$. The full agreement with the numerical calculations is not obtained due to neglecting effects related to the electric field variation in time and of the assumed collisionless approximation. This explains the big discrepancy is shown in Figure 1.10 when compared with the diffusion coefficient at 1000 Hx when collisions begin to be far more important to randomize individual trajectories and our approach is no more valid (at low enough B/N values).

1.9 Discussion and Summary

However, 1.192 suffers from the indetermination of the geometrical factor α. This factor is related to the ions charge number, it depends on the magnetic field magnitude and as well as on the external operating conditions (due to increased collisional processes, for ex.). The exact value of the product $\alpha\delta$ can only be determined through a self-consistent plasma model, but we should expect from the above discussion that $\alpha\delta < 1$. For a 100-eV plasma in a 1-T field, we obtain $D_B = 1.67$ m^2/s (using the thermal to magnetic energy ratio with particle's density $n = 10^{14}$ cm^{-3}). Furthermore, 1.191 can be used as a boundary condition (simulating an electrically floating surface) imposed when solving the Poisson equation.

Also, it is worth emphasizing that when inserting 1.191 into 1.186, and considering the usual definition of momentum transfer cross-section, then can be obtained a new expression for the classical diffusion coefficient as a function of the ratio of collisional ν and cyclotron frequency Ω, although (and in contrast with the standard expression), now also dependent on the geometrical factor α and energy ratio δ:

$$D_T = (\alpha\delta)\frac{\nu}{\Omega}\frac{kT}{m} \tag{1.197}$$

This explains the strong dependence of the classical diffusion the coefficient on ν/Ω showing signs of anomalous diffusion as discussed in Ref. [92] (obtained with a time-resolved Monte Carlo simulation in an infinite gas under uniform fields) and, in addition, the strong oscillations showed up in the calculations of the time dependence of the transverse component of the diffusion tensor for electrons in a low-temperature rf argon plasma. Those basic features result on one side from its dependence on R, which is proportional to the flux. Therefore, a flux variation can give an equivalent effect to the previously proposed mechanism: whenever a decrease (or increase) in the flux is onset through time dependence of electric and magnetic fields, it occurs a strong increase (or decrease) of the diffusion coefficient. On another side, when the resistance increases it occurs a related decrease of charged particles tangential velocity and its mean energy. So far, this the model gives a new insight into the results referred in Ref. [92] and also explains why the same effect is not obtained from the solution of the non-conservative Boltzmann equation as applied to an oxygen magnetron discharge with constant electric and magnetic fields [94].

A further application of Eq. 1.178 to a cold plasma can give a new insight into the "ambipolar-like" diffusion processes. Considering just one conducting surface (besides the electrodes driving the main current into the plasma) and the plasma build-up of electrons and one ion component to simplify matters, we obtain:

$$\frac{dS}{dt} = \frac{e^2}{T}(-n_e\mu_e\mathbf{E} + D_e\nabla n_e + n_i\mu_i\mathbf{E} - D_i\nabla n_i)^2\mathcal{A}^2R \tag{1.198}$$

Under the usual assumptions of quasi-neutrality and quasi-stationary plasma (see, for example, Ref. [88]), the following conditions must be verified:

$$\frac{n_i}{n_e} = \epsilon = \text{const}; \quad n_e\mathbf{v_e} = n_i\mathbf{v}_i \tag{1.199}$$

and hence, Eq. 1.178 becomes:

$$\frac{dS}{dt} = \frac{e^2}{T}[\mathbf{E}(\epsilon\mu_i - \mu_e)n_e + \nabla n_e(D_e - D_i\epsilon)]^2\mathcal{A}^2R \tag{1.200}$$

For a stable steady-state plasma with no entropy sources the condition $\dot{S} = 0$ prevails and then an "ambipolar-like" electric field is recovered [88]:

$$\mathbf{E} = \frac{D_e - \epsilon D_i}{\mu_e - \mu_i\epsilon}\frac{\nabla n_e}{n_e} \tag{1.201}$$

It means that the conducting surface must be at its floating potential. Such conceptual formulation provides new insight into "ambipolar-like" diffusion processes. In a thermal equilibrium state, a plasma confined by insulating walls will have an effective coefficient given by the above 1.201, a situation frequently encountered in industrial applications. This example by itself relates to ambipolar diffusion with no entropy production in the plasma. However, allowing plasma currents to the walls, entropy production is greatly enhanced generating altogether instabilities and plasma losses [86]. As long as confined plasmas are in a far-nonequilibrium state (with external surroundings) it is necessary to establish a generalized principle that rules matter and this circuital model for anomalous diffusion represents some progress in the physics of plasmas as nonequilibrium systems.

This simple circuital mechanism providing an interpretation of the anomalous diffusion in a magnetized confined plasma in a purely diffusive transport regime. The coupled action of the magnetic field "cutting" flux through the areas traced by the charge carriers' elementary orbits, together with the elementary electric circuit constituted by the charged particle trajectory itself is at the basis of the anomalous diffusion process. Whenever conducting walls bounding the plasma drain the current (edge diffusion) or, at the plasma core, the magnetic field flux through the areas traced by the charged particles varies, a Bohm-like behavior of the transverse diffusion coefficient can be expected. 1.196 can be used as an analytical formula when simulating plasma behavior at high B/N. In the near future we hope to generalize this model taking into account random collisions. The suggested mechanism could lead to a better understanding of the mechanism of plasma-wall interaction and help to develop full-scale numerical modeling of present fusion devices or collisional low-temperature plasmas.

Notes

1 But we would prefer to name it the "non-slaving principle", instead, for reasons discussed later in Section 2.3.

2 Kristian Olaf Bernhard Birkeland (13 December 1867 – 15 June 1917).

3 The E region is situated in the middle ionosphere; its lower boundary is at a level of about 100 km when the sun is vertical.

4 For example, we can find in S. Chapman and T. G. Cowling, *Mathematical Theory of non-uniform Gases* the *molecular viewpoint*.

5 This approach is good when the electromagnetic field is nearly constant along the particles mean free path.

6 It means: within a phase.
7 Not in the thermodynamical sense since a chemical reaction is an irreversible process.
8 Frequently and wrongly also called *Larmor frequency* or *gyrofrequency*. Unfortunately, it have been used incorrectly the terms Larmor frequency and the related Larmor radius, since the Larmor frequency for the precession of an orbital electron in a magnetic field is $eB/2m$, that is, half the value given by Eq. 1.168

References

1. H. O. G. Alfvén, "Cosmology in the plasma universe - an introductory exposition," *IEEE Trans. Plasma Sci.*, vol. 18, no. 1, pp. 5–10, 1990.
2. C.-G. Fälthammar, "Electrodynamics of cosmical plasmas-some basic aspects of cosmological importance," *IEEE Transactions on Plasma Science*, vol. 18, no. 1, pp. 11–17, 1990.
3. M. J. Pinheiro, "Plasma: the genesis of the word," *arXiv:physics/0703260v1 [physics.hist-ph]*, pp. 1–2, 2007.
4. D. Bohm, *Wholeness and the Implicate Order*. New York, NY: Routledge Classics, 1980.
5. S. Berkovich, *Obtaining Inexhaustible Clean Energy by Parametric Resonance Under Nonlocal Clocking*. Access online: www.chronos.msu.ru, 2010.
6. L. Mandeltam and N. Papaleksi, "Parametric excitation of electric oscillations," *ZhurnalTeknicheskoy Fiziki*, vol. 4, no. 1, pp. 5–29, 1934.
7. N. Papaleksi, A. Andronov, S. Chaikin, A. Witt and L. Mandelstam, "Exposé des recherches recentes sur les oscillations non-lineaires," *Technical Physics of the USSR*, vol. 2, no. 2–3, pp. 81–134, 1935.
8. H. Nijmeijer and T. I. Fossen, *Parametric Resonances in Dynamical Systems*. New York, NY: Springer, 2012.
9. D. J. Inman and A. Erturk, *Piezoelectric Energy Harvesting*. New York, NY, USA: John Wiley & Sons, Inc., 2011.
10. J.-I. Okutani, "Excitation of plasma oscillations by parametric resonance," *J. Phys. Soc. Japan*, vol. 23, no. 1, pp. 110–113, 1967.
11. K. Tachibana, T. Shirafuji, N. Sergey and S. N. Abolmasov, "Mechanisms of pattern formation in dielectric barrier discharges," *IEEE Trans. Plasma Sci.*, vol. 39, no. 11, pp. 2090–2091, 2011.
12. H. Haken, *Synergetics*. Proceedings of the International Workshop on Synergetics at Schloss Elmau, Bavaria, May 2–7, 1977.
13. K. Schuh, P. Panagiotopoulos, M. Kolesik, S. W. Koch and J. V. Moloney, "Multi-terawatt 10 μm pulse atmospheric delivery over multiple rayleigh ranges," *Opt. Lett.*, vol. 42, no. 19, pp. 3722–3725, Oct 2017.
14. A. L. Peratt and C. M. Snell, "Microwave generation from filamentation and vortex formation within magnetically confined electron beams," *Phys. Rev. Lett.*, vol. 54, pp. 1167–1170, Mar 1985.
15. P. Alfvén and H. Wernholm, "A new type of accelerator," *Arkiv Fysik*, vol. 5, pp. 2090–2091, 1952.
16. J. D. Lawson, "Collective and coherent methods of particle accelerator," *Particle Accelerators*, vol. 3, pp. 21–33, 1972.
17. W. H. Bostick, "What laboratory-produced plasma structures can contribute to the understanding of cosmic structures both large and small," *IEEE Trans. Plasma Sci.*, vol. 1, pp. 703–717, 1986.
18. P. Whittlesey, D. E. Larson, R. Livi, M. Berthomier, J. C. Kasper, A. W. Case, M. L. Stevens, S. D. Bale, R. J. MacDowall, J. S. Halekas, L. Berčič and M. P. Pulupa, "The sunward electron deficit: A telltale sign of the sun's electric potential," *ApJ*, vol. 916, no. 1, 916, pp. 16, 2021.

19. L. Tonks, "A new form of the electromagnetic energy equation when free charged particles are present," *Phys. Rev.*, vol. 54, pp. 863, 1938.
20. C. M. Ferreira, M. J. Pinheiro and G. Gousset, *Multicomponent Reactive Gas Dynamic Model for Low-Pressure Discharges in Flowing Oxygen*. Netherlands, Dordrecht: Springer, 1996.
21. J. P. Boeuf, Y. Lagmich, Th. Unfer, Th. Callegari and L. C. Pitchford, Electrohydrodynamic force in dielectric barrier discharge plasma actuators, *J. Phys. D: Appl. Phys.*, vol. 40, pp. 652, 2007.
22. W. E. Stewart, E. N. Lightfoot and R. B. Bird, *Transport Phenomena*. New York: John Wiley and Sons, Inc., 1960.
23. S. El-Khabiry and G. M. Colver, "Drag reduction by dc corona discharge along an electrically conductive flat plate for small reynolds number flow," *Phys. Fluids*, vol. 9, no. 3, pp. 587, 1997.
24. E. J. Jumper, M. L. Post, T. C. Corke and D. Orlov, "Application of weakly-ionized plasmas as wing flow-control devices," In *40th AIAA Aerospace Sciences Meeting and Exhibit, AIAA 2002-0350*. AIAA 2002-0350, 2002.
25. M. L. Post and T. C. Corke, "Separation control on high angle of attack airfoil using plasma actuators," AIAA-2003-1024, Reno, Nevada, 2003.
26. J. R. Roth, "Aerodynamic flow acceleration using paraelectric and peristaltic electrohydrodynamic effects of a one atmosphere uniform glow discharge plasma," *Phys. Plasmas*, vol. 10, no. 5, pp. 2117, 2003.
27. E. Moreau, G. Touchard, A. Labergue, L. Leger and J. P. Bonnet, "Experimental study of the detachment and the reattachment of an airow along an inclined wall controlled by a surface corona discharge – application to a plane turbulent mixing layer," *IEEE Trans. Ind. Appl.*, vol. 40, no. 5, pp. 1205, 2004.
28. M. J. Pinheiro, "Ehd ponderomotive forces and aerodynamic flow control using plasma actuators," *Plasma Process. Polym.*, vol. 3, pp. 135–141, 2006.
29. E. Moreau, "Airow control by non-thermal plasma actuators," *J. Phys. D: Appl. Phys.*, vol. 40, no. 3, pp. 605, 2007.
30. M. Frey, S. Grundmann and C. Tropea, "Unmanned aerial vehicle (uav) with plasma actuators for separation control," 47th AIAA Aerospace Sciences Meeting including The New Horizons Forum and Aerospace Exposition, AIAA-2009-698, Orlando, Florida, 2009.
31. L. S. Hultgren and D. E. Ashpis, "Demostration of separation delay with glow-discharge plasma actuators," 41st Aerospace Sciences Meeting and Exhibit, AIAA-2003-1025, Reno, Nevada, 2003.
32. T. E. McLaughlin, J. List, A. R. Byerley and R. D. Van Dyken, "Using plasma actuators flaps to control laminar separation on turbine blades in a linear cascade," 41st Aerospace Sciences Meeting and Exhibit, AIAA-2003-1026, Reno, Nevada, 2003.
33. A. Soldati and S. Banerjee, "Turbulence modification by large-scale organized electrohydrodynamic flows," *Phys.Fluids*, vol. 10, no. 7, pp. 1742, 1998.
34. A. Soldati, M. Fulgosi and S. Banerjee, "Turbulence modulation by an array of large-scale streamwise structures of ehd origin," Proc. of FEDSM'99 3rd ASME/JSME Joint Fluids Engineering Conference, 1999.
35. M. Yadav, J. Rahel, J. R. Roth, R. C. M. Madhan and S. P. Wilkinson, "Flow field measurements of paraelectric, peristaltic and combined plasma actuators based on the one atmospheric uniform glow discharge plasma (oaugdp^{TM})," 42nd AIAA Aerospace Sciences Meeting and Exhibit, AIAA-2004-0845, Reno, Nevada, 2004.
36. L. Leger, E. Depussay, V. Lago, E. Menier and G. Artana, "'Effect of a dc discharge on the supersonic rarefied air flow over a flat plate," *J. Phys. D: Appll. Phys.*, vol. 40, pp. 695, 2007.
37. S. Popovic and L. Vuskovic, "Aerodynamic effects in weakly ionized gas: phenomenology and applications," *The Physics of Ionized Gases, AIP Conf. Proc.*, vol. 876, pp. 272, 2006.
38. D. Siegelman, "Sonic boom minimization schemes," *J. Aircraft*, vol. 7, no. 3, pp. 280, 1970.
39. D. Boyd and T. Cain, "Electroaerodynamics and the effect of an electric discharge on cone/cylinder drag at mach 5," 37th Aerospace Sciences Meeting and Exhibit, AIAA-1999-0602, Reno, Nevada, 1999.

40. S. Chan, X. Huang and X. Zhang, "Atmospheric plasma actuators for aeroacoustic applications," *IEEE Trans. Plasma Sci.*, vol. 35, no. 3, pp. 693, 2007.
41. X. Zhang, S. Chan and S. Gabriel, "The attenuation of cavity tones using plasma actuators," 11th AIAA/CEAS Aeroacoustics Conference, AIAA-2005-2802, Monterey CA, 2005.
42. E. V. Shun'ko and V. S. Belkin, "Cleaning properties of atomic oxygen excited to metastable state $2s^22p^4(^1s_0$," *J. Appl. Phys.*, vol. 102, no. 8, pp. 083304, 2007.
43. J. R. Roth, "Method and apparatus for cleaning surfaces with a glow discharge plasma at one atmosphere of pressure," US Patent 5,938,854, 17 August 1999.
44. I. E. Ivanov, Y. Liao, I. V. Mursenkova and I. A. Kryukov, "Experimental and numerical investigation of a surface sliding discharge in a supersonic flow with an oblique shock wave," *Energies*, vol. 15, pp. 2189, 2022.
45. J. R. Roth, "Potential industrial applications of the one atmosphere uniform glow discharge plasma operating in ambient air," *Phys. Plasmas*, vol. 12, pp. 037103, 2005.
46. J. Rahel, X. Dai, J. R. Roth and D. M. Sherman, "The physics and phenomenology of one atmosphere uniform glow discharge plasma reactors for surface treatment applications," *J. Phys. D: Appl. Phys.*, vol. 38, pp. 555, 2005.
47. S. Jung, C. L. Enloe, T. E. McLaughlin, W. E. Morgan, G. I. Font and J. W. Baughn, "Simulation of the effects of force and heat produced by a plasma actuator on neutral flow evolution," 44th AIAA Aerospace Sciences Meeting and Exhibit, AIAA-2006-167, Reno, Nevada, 2006.
48. J. P. Boeuf and L. C. Pitchford, "Electrohydrodynamic force and aerodynamic flow acceleration in surface dielectric barrier discharge," *J. Appl. Phys.*, vol. 97, pp. 103307, 2005.
49. B. N. Ganguly, D. V. Wie, P. Bletzinger and A. Garscadden, "Plasmas in high speed aerodynamics," *J. Phys. D: Appl. Phys.*, vol. 38, pp. R33, 2005.
50. L. L. Alves, M. J. Pinheiro, C. M. Ferreira and A. B. Sá, "Modeling of low-pressure microwave discharges in ar, he and o2: Similarity laws for the maintenance field and mean powertransfer," *IEEE Trans. Plasma Sci.*, vol. 19, pp. 229, 1991.
51. C. M. Ferreira and J. Loureiro, "Electron kinetics in atomic and molecular plasmas," *Plasma Sources Sci. Technol.*, vol. 9, pp. 528, 2000.
52. M. J. Pinheiro and J. Loureiro, "Effective ionization coefficients and electron drift velocities in gas mixtures of sf_6 with he, xe, co_2 and n_2 from boltzmann analysis," *J. Phys. D: Appl. Phys.*, vol. 35, no. 23, pp. 3077, 2002.
53. J. Menart, A. S. Scott and C. DeJoseph Jr., "Vibrational temperatures and relative concentrations of $n_2(c^3\pi_u)$ and $n_2^+(b^2\sigma_u^+)$ for an asymmetric surface mode dielectric barrier discharge," *47th AIAA Aerospace Sciences Meeting including The New Horizons Forum and Aerospace Exposition*, 05 January 2009 - 08 January 2009, Orlando, Florida, https://doi.org/10.2514/6.2009-653, 2009.
54. J. P. Boeuf, L. Pitchford and L. Morgan, "Electron boltzmann equation solver, 2005," data retrieved from World Development Indicators, http://cpat.ups-tlse.fr/operations/operation03/POSTERS/BOLSIG.
55. A. W. Ali, "On electron beam ionization of air and chemical reactions for disturbed air deionization," *Naval Research Laboratory Memo Report 4619* Washington, DC, 1981.
56. A. Y. Kostinsky, A. A. Matveyev, I. A. Kossyi and V. P. Silakov, "Kinetic scheme of the non-equilibrium discharge in nitrogen-oxygen mixtures," *Plasma Sources Sci. Technol.*, vol. 1, pp. 207, 1992.
57. E. W. McDaniel, *Collision Processes in Ionizated Gases*. New York: John Wiley, 1964.
58. R. S. Sigmond, "Gas discharge data sets for dry oxygen and air," partly revised, *Gaseous Dielectrics V. Proceedings of the Fifth International Symposium on Gaseous Dielectrics*, Knoxville, Tennessee, U.S.A., May 3–7, 1983.
59. F. Gouda and G. Massines, "Role of excited species in dielectric barrier discharge mechanisms observed in helium at atmospheric pressure," *Conference on Electrical Insulation and Dielectric Phenomena*, vol. 2, pp. 496, 1999.

60. K. P. Singh and S. Roy, "Modeling plasma actuators with air chemistry for effective flow control," *J. Appl. Phys.*, vol. 101, pp. 123308, 2007.
61. R. Anderson and S. Roy, "Preliminary experiments of barrier discharge plasma actuators using dry and humid air," 44th AIAA Aerospace Sciences Meeting and Exhibit, AIAA-2006-369, Reno, Nevada, 2006.
62. S. V. Patankar, *Numerical Heat Transfer and Fluid Flow*. New York: Taylor & Francis, 1980.
63. R. Morrow and N. Sato, "The discharge current induced by the motion of charged particles in time-dependent electric fields; sato's equation extended," *J. Phys. D: Appl. Phys.*, vol. 32, pp. L20, 1999.
64. A. Simon, "Ambipolar diffusion in a magnetic field," *Phys. Rev.*, vol. 98, no. 2, pp. 317, 1955.
65. K. H. Geissler, "Classical and anomalous diffusion of an afterglow Plasma," *Phys. Rev.*, vol. 171, no. 1, pp. 179, 1968.
66. V. A. Rozhansky, L. L. Beilinson and L. D. Tsendin, "Two-dimensional nonuniformly heated magnetized plasma transport in a conducting vessel," *Phys. Rev. E*, vol. 50, no. 4, pp. 3033, 1994.
67. C.-S. Liu, D. Montgomery and G. Vahala, "Three-dimensional plasma diffusion in a very strong magnetic field," *Phys. Fluids*, vol. 15, no. 5, pp. 815, 1972.
68. C. C. Petty, T. C. Luce and J. C. M. de Haas, "Inward energy transport in tokamak plasmas," *Phys. Rev. Lett.*, vol. 68, no. 1, pp. 52, 1992.
69. L. A. Ferrari and A. F. Kuckes, "Instability in a hot electron plasma," *Phys. Fluids*, vol. 12, pp. 836, 1969.
70. J. B. Taylor and B. McNamara, "Plasma diffusion in two dimensions," *Phys. Fluids*, vol. 14, no. 7, pp. 1492, 1971.
71. A. Fukuyama, K. Itoh, S.-I. Itoh and M. Yagi, "Theory of plasma turbulence and structural formation– nonlinearity and statistical view –," *J. Plasma Fusion Res.*, vol. 79, no. 6, pp. 608, 2003.
72. R. J. Bickerton, "History of the approach to ignition," *Phil. Trans. R. Soc. Lond. A*, vol. 375, pp. 397, 1999.
73. V. D. Shafranov, "On the history of the research into controlled thermonuclear fusion," *Physics-Uspekhi*, vol. 44, no. 8, pp. 835, 2001.
74. H. J. Monkhorst, N. Rostoker and M. W. Binderbauer, "Colliding beam fusion reactor," *Science*, vol. 278, pp. 1419, 1997.
75. D. Montgomery and F. Tappert, "Conductivity of a two-dimensional guiding center plasma," *Phys. Fluids*, vol. 15, no. 4, pp. 683, 1972.
76. J. D. Callen and M. W. Kissick, "Evidence and concepts for non-local transport," *Plasma Phys. Control. Fusion*, vol. 39, pp. B173–B188, 1997.
77. B. N. Breizman, H. L. Berk and H. Ye, "Scenarios for the nonlinear evolution of alpha-particle-induced Alfvén wave instability," *Phys. Rev. Lett.*, vol. 68, no. 24, pp. 3563, 1992.
78. G. S. Janes and R. S. Lowder, "Anomalous electron diffusion and ion acceleration in a low-density plasma," *Phys.Fluids*, vol. 9, no. 6, pp. 1115, 1966.
79. G. N. Tilinin, A. V. Trofimov, Yu. A. Sharov, A. I. Morozov, Yu. V. Esinchuk and G. Ya. Shchepkin, "Plasma accelerator with closed electron drift and extended acceleration zone," *Sov. Phys.-Techn. Phys.*, vol. 17, no. 1, pp. 38, 1972.
80. A. I. Morozov, "Conditions of the current transfer performance by boundary conductivity," *Sov. Phys. Tech. Phys.*, vol. 32, no. 8, pp. 901, 1987.
81. J. P. Boeuf and L. Garrigues, "Low frequency oscillationsin a stationary plasma thruster," *J. Appl. Phys.*, vol. 84, no. 7, pp. 3541, 1998.
82. Y. Raitses, A. Smirnov and N. J. Fisch, "Electron cross-field transport in a low power cylindrical Hall thruster," *Phys. Plasmas*, vol. 11, no. 11, pp. 4922, 2004.
83. R. R. Hofer, I. Katz, I. Mikellides and M. Gamero-Castãno, "Heavy Particle Velocity and Electron Mobility Modeling in Hybrid-PIC Hall Thruster Simulations," *42nd AIAA/ASME/SAE/ASEE Joint Propulsion Conference & Exhibit, No. 2006-4658*, Sacramento, California, USA, July 2006.

84. S. M. Rossnagel and H. R. Kaufman, "Induced drift currents in circular planar magnetrons," *J. Vac. Sci. Technol. A*, vol. 5, no. 1, pp. 88, 1987.
85. H. R. Kaufman, "Technology of closed-drift thrusters," *AIAA J.*, vol. 23, pp. 78, 1985.
86. H. S. Robertson, "Ambipolarity and plasma stability," *Phys. Rev.*, vol. 118, no. 1, pp. 288, 1969.
87. B. B. Kadomtsev, "Particle transport, electric currents, and pressure balance in a magnetically immobilized plasma," *Phénomènes Collectifs dans les Plasmas*. Moscow: Mir Editions, 1979.
88. J. R. Roth, *Industrial Plasma Engineering, Vol 1 - Principles*. Bristol: Institute of Physics Publishing, 1995.
89. L. Tonks, *Phys. Rev.*, vol. 97, no. 6, pp. 1443, 1955.
90. P. B. Corkum, "Anomalous diffusion in a magnetized plasma," *Phys. Rev. Lett.*, vol. 31, no. 13, pp. 809, 1973.
91. S. Chandrasekhar, *Plasma Physics*. Chicago: Chicago Press, 1960.
92. T. Makabe, Z. M. Raspopović, S. Dujko and Z. Lj. Petrović, "Kinetic phenomena in charged particle transport in gases, swarm parameters and cross section data," *Plasma Sources Sci. Technol.*, vol. 14, pp. 293, 2005.
93. Z. Lj. Petrović, Z. Raspopović, S. Sakadžić and T. Makabe, "Monte Carlo studies of electron transport in crossed electric and magnetic fields in CF_4," *J. Phys. D: Appl. Phys.*, vol. 33, pp. 1298, 2000.
94. K. F. Ness, R. D. White, R. E. Robson and T. Makabe, "Electron transport coefficientsin O_2 magnetron discharges," *J. Phys. D: Appl. Phys.*, vol. 38, pp. 997, 2005.

2

Topics in Electromagnetism

2.1 Classical Electrodynamics

Classical electrodynamics (or classical electromagnetism) is a theory that investigates the action of forces of electromagnetic origin between electric charges and currents. In electromagnetism the interaction is described in terms of a potential or field permeating the whole space around the source charge or current – it is a classical example of **interaction at a distance**. Its limitations rely on the large length scales neglecting quantum mechanical effects, neglecting electron spin, and magnetic moment.[1]

It has been shown that $\mathbf{A}_\mu$ potentials are measurable (physical) in certain well-defined situations [1]. The set of situations when they are measurable possess a topology with transformation rules obeying SU(2) group, and the set of situations when the potential is not measurable transform according to the U(1) group. In this chapter, we will study the electromagnetic field within this last topology.[2]

In quantum theory, interactions are understood in terms of an *exchange interaction*, which means the exchange of a specific quantum (boson) is supposed to be associated with a particular type of interaction. The exchanged boson carries momentum from one charge to the other, and the rate of exchange of momentum gives the force.[3]

2.1.1 The Maxwell's Equations

One of the most fundamental principles of physics is the principle of charge conservation.[4] In order to construct it, we must use spatial distributions of charge $\rho(\mathbf{r}, t)$ and consider their velocity $\mathbf{u}(\mathbf{r}, t)$ at position $\mathbf{r}$ at time t. The density of current is

$$\mathbf{j}(\mathbf{r}, t) = \rho(\mathbf{r}, t)u(\mathbf{r}, t) \tag{2.1}$$

which is a measure of the quantity of charge crossing a transverse surface to $\mathbf{u}$ by unit of time. The principle of conservation relies on this idea that every decrease of charge inside volume V is because of (or equal) what is running out of its surface S:

$$-\Delta\left\{\int_V d^3\rho(\mathbf{r}, t)\right\} = \Delta t \oint_S (\rho(\mathbf{r}, t)\mathbf{u}(\mathbf{r}, t))\cdot d\mathbf{S} \tag{2.2}$$

The normal is directed toward the exterior of the surface. Stoke's theorem leads us to the result

$$\frac{\partial \rho}{\partial t} + div\mathbf{j} = 0 \tag{2.3}$$

DOI: 10.1201/9781003285083-2

Eq. 2.3 has a fundamental nature and it is experimentally firmly established being valid even for microscopically elementary processes, as for instance, the β-decay process $n \to p^+ + e^- + \bar{\nu}$, where n is the neutron, p is the proton, e is the electron, and $\bar{\nu}$ is the electron antineutrino.

Incidentally, charge-current conservation raises some doubts about the idea that photons are uncharged, or have no net charge, since electron-positron annihilation produces two photons

$$e^- + e^+ \to 2\gamma \tag{2.4}$$

However, if mass and charge are conserved in nature, they cannot be annihilated and it appears the possibility that electron and positron may combine to produce a new entity, a new kind of photon that moves at speed of light c and is made up of a combination of plus and minus e [2].

Example 2.1: *Inside a metallic conductor we may have* $\rho = \rho_+ + \rho_- = 0$, but $\mathbf{J} = \rho_+\mathbf{u}_+ + \rho_-\mathbf{u}_- \neq 0$.

Exercise: Obtain Eq. 2.2.

Maxwell's equations were obtained respecting this principle. In the case of a point charge at position $\mathbf{r}_q(t)$ we must consider instead

$$\begin{aligned} \rho(\mathbf{r}, t) &= q\delta(\mathbf{r} - \mathbf{r}_q(t)) \\ \mathbf{j}(\mathbf{r}, t) &= q\mathbf{r}_q(t)\delta(\mathbf{r} - \mathbf{r}_q(t)) \end{aligned} \tag{2.5}$$

with $\mathbf{u} = \dot{\mathbf{r}}_q \equiv \frac{d}{dt}\mathbf{r}_q$.

In the case of a distribution of charge, the force density (by unit of volume) is given by

$$\begin{aligned} d\mathbf{F} &= \rho d^3r(\mathbf{E} + [\mathbf{v} \times \mathbf{B}]) \\ \frac{d\mathbf{F}}{d^3r} &= \rho\mathbf{E} + [\mathbf{J} \times \mathbf{B}]. \end{aligned} \tag{2.6}$$

We may use one (discrete) description or a continuous description, depending on the scale of the phenomena being studied. At a macroscopic level charges are punctual since e.g., electrons and muons radius is $r < 10^{-13}$ Fm, or 10^{-18} m, and protons and quarks $r \lesssim 1$ Fm or 10^{-15} m. These dimensions are so small that we may consider them not describable by classical (nonquantic) physics.

Lorentz's law of force assumes charged particles are punctual and do not consider their electric dipole moment, nor magnetic moment, nor any supplementary term, like $\mathbf{F} = (\mu \cdot grad\mathbf{B})$.

The elementary sources terms that appear in the microscopic Maxwell's set of equations[5] are ρ, $\mathbf{J}$.

Denoting the microscopic fields with small letters, we may write the set of **microscopic Maxwell's equations** under the form

$$\begin{aligned} \nabla \times \mathbf{e} = -\dot{\mathbf{b}} \qquad & \nabla \cdot \mathbf{b} = 0 \\ \nabla \cdot \mathbf{e} = \frac{\rho}{\epsilon_0} \qquad & \nabla \times \mathbf{b} = \mu_0\mathbf{j} + \epsilon_0\mu_0\dot{\mathbf{e}} \end{aligned} \tag{2.7}$$

The constants appearing in the equations are the permittivity ϵ_0, the free-space permeability μ_0, and c is the velocity of light in vacuum

$$\begin{aligned} \epsilon_0 &= 8.854188 \times 10^{-12} \mathrm{As/Vm} \\ \mu_0 &= 4\pi \times 10^{-7} \mathrm{Vs/Am} \\ c^2 &= \frac{1}{\epsilon_0 \mu_0} \end{aligned}$$

The set of Eq. 2.7 is unsolvable because the matter is formed of an average number of particles[6] difficult to follow with variations in space typically of the order $1\text{Å} = 10^{-10}$ m and fluctuations in time of the order of 10^{-17} s in nucleons and 10^{-13} s in atoms. The theory must involve only average quantities in such a way that if $f(\mathbf{r}, t)$ is a microscopic field quantity, and $v(\mathbf{r})$ is a microscopically large but macroscopically small sphere with center at $\mathbf{r}$, then we obtain a *phenomenological average value*

$$\overline{f(\mathbf{r}, t)} = \frac{1}{v(\mathbf{r})} \int_{v(\mathbf{r})} \mathbf{f}(\mathbf{r}', \mathbf{t}) d^3 r' \tag{2.8}$$

We can assume that differentiation and averaging are interchangeable operations

$$\frac{\partial}{\partial t} \overline{f} = \overline{\frac{\partial f}{\partial t}} ; \nabla \overline{f} = \overline{\nabla f} \tag{2.9}$$

We can now define the **macroscopic fields** by averaging the microscopic fields

$$\mathbf{E}(\mathbf{r}, t) = \overline{\mathbf{e}(\mathbf{r}, t)} ; \mathbf{B}(\mathbf{r}, t) = \overline{\mathbf{b}(\mathbf{r}, t)} \tag{2.10}$$

The set of **macroscopic Maxwell's equations** is therefore given by

$$\begin{aligned} \nabla \times \mathbf{E} = -\dot{\mathbf{B}} \qquad & \nabla \cdot \mathbf{B} = 0 \\ \nabla \cdot \mathbf{E} = \frac{\rho}{\epsilon_0} \qquad & \nabla \times \mathbf{B} = \mu_0 \bar{\mathbf{j}} + \epsilon_0 \mu_0 \dot{\mathbf{E}} \end{aligned} \tag{2.11}$$

Exercise: Feynman's Programme According to Freeman Dyson [3], Feynman reconstructed Maxwell's equations only based on Newton's Second Law and the commutation relation between position and velocity for a single *nonrelativistic* particle. He *assumed* that a particle exists with position x_j ($j = 1,2,3$) and velocity $\dot{x}_j$ satisfying the following equations:

$$m\ddot{x}_j = F_j(x, \dot{x}, t) \tag{2.12}$$

$$[x_j, x_k] = 0 \tag{2.13}$$

$$m[x_j, \dot{x}_k] = i\hbar\delta_{jk} \tag{2.14}$$

Then, Feynman concluded that there exist fields $E(x, t)$ and $H(x, t)$ satisfying the Lorentz force equation:

$$F_j = E_j + \epsilon_{jkl} \dot{x}_k H_l \tag{2.15}$$

and the Maxwell Eq. 2.11. Give the proof.

We can verify that

$$[x_j, F_k] = m[x_j, \ddot{x}_k] \tag{2.16}$$

since

$$\frac{d}{dt}m[x_j\dot{x}_k - \dot{x}_k x_j] = 0 \tag{2.17}$$

$$\Rightarrow m[\dot{x}_j\dot{x}_k + x_j\ddot{x}_k - \ddot{x}_k x_j - \dot{x}_k\dot{x}_j] = 0 \tag{2.18}$$

$$[x_j, \ddot{x}_k] = [\dot{x}_k, \dot{x}_j] \tag{2.19}$$

$$\Rightarrow [x_j, F_k] = m[\dot{x}_k, \dot{x}_j] \tag{2.20}$$

The Jacobi identity is given by

$$[A, [B, C]] + [B, [C, A]] + [C, [A, B]] = 0, \tag{2.21}$$

and here it results in the form

$$[x_l[\dot{x}_j, \dot{x}_k]] + [\dot{x}_j, [\dot{x}_k, x_l]] + [\dot{x}_k, [x_l, \dot{x}_j]] = 0 \tag{2.22}$$

$$-\frac{1}{m}[x_l, [x_j, F_k]] - \frac{i\hbar}{m}[\dot{x}_j, \delta_{lk}] + \frac{i\hbar}{m}[\dot{x}_k, \delta_{lj}] = 0 \tag{2.23}$$

$$\Rightarrow [x_l, [x_j, F_k]] = 0 \tag{2.24}$$

But, $[x_j, F_k] = -[x_k, F_j]$ and, from aforementioned equations we have too

$$[x_j, F_k] = [x_j, E_k] + \epsilon_{klj}[x_j, \dot{x}_l H_j] \tag{2.25}$$

$$= \epsilon_{klj}H_j\frac{i\hbar}{m}\delta_{jl} = \frac{i\hbar}{m}\epsilon_{klj}H_l \tag{2.26}$$

$$= -\frac{i\hbar}{m}\epsilon_{jkl}H_l \tag{2.27}$$

The surprise that this result brings us resides in the assumptions that are Galilean invariant and end with Maxwell's equations that are Lorentz invariant.

2.1.2 Retarded Potentials and Fields

To calculate the scalar and vector potentials at a given point and a given time t, we must know the positions of the charges at a previous time $t - |\mathbf{r} - \mathbf{r}'|/c$ when the charged particles "launched" the fields from $\mathbf{r}'$ to arrive at $\mathbf{r}$ at time t propagating in space at speed of light c. The expressions for the scalar and vector potentials are given by

$$\Phi(\mathbf{r}, t) = \int_V \frac{\rho(\mathbf{r}', t - |\mathbf{r} - \mathbf{r}'|/c)}{|\mathbf{r} - \mathbf{r}'|} dv' \tag{2.28}$$

$$\mathbf{A}(\mathbf{r}, t) = \frac{1}{c} \int_V \frac{\mathbf{J}(\mathbf{r}', t - |\mathbf{r} - \mathbf{r}'|/c)}{|\mathbf{r} - \mathbf{r}'|} dv' \tag{2.29}$$

and are called **retarded potentials.**

Usually, the electric field is calculated from the potentials assuming that[7]

$$\mathbf{E} = -\nabla\Phi - \frac{1}{c}\frac{\partial \mathbf{A}}{\partial t} \tag{2.30}$$

Eqs. 2.28 and 2.29 can be rewritten in the more compact form

$$\Phi(\mathbf{r}, t) = \int_V \frac{[\rho(\mathbf{r}')]}{R} dv' \tag{2.31}$$

$$\mathbf{A}(\mathbf{r}, t) = \frac{1}{c} \int_V \frac{[\mathbf{J}(\mathbf{r}')]}{R} dv' \tag{2.32}$$

where $R \equiv |\mathbf{r} - \mathbf{r}'$ and the brackets mean evaluation at the retarded time $t_{red} \equiv t - R/c$:

$$\rho(\mathbf{r}', t - R/c) \equiv [\rho(\mathbf{r}')] \tag{2.33}$$

Now, remark that

$$\nabla\left(\frac{[\rho]}{R}\right) = \frac{1}{R}\nabla[\rho] + [\rho]\nabla\left(\frac{1}{R}\right) \tag{2.34}$$

and $\nabla(1/R) = -\mathbf{e}_R/R^2$. The first term on the RHS of Eq. 2.34 can be computed by remembering that $\nabla\Phi(u) = \partial\Phi/\partial u \nabla u$. It gives

$$\nabla[\rho(\mathbf{r}', t_{ret})] = \left[\frac{\partial\rho}{\partial t}\right]\nabla\left(t - \frac{1}{c}|\mathbf{r} - \mathbf{r}'|\right) = -\frac{1}{c}\left[\frac{\partial\rho}{\partial t}\right]\mathbf{e}_R \tag{2.35}$$

After inserting all the aforementioned terms into Eq. 2.30, we obtain the **generalized Coulomb–Faraday law**:

$$\mathbf{E}(\mathbf{r}, t) = \int_V \left(\frac{[\rho]\mathbf{e}_R}{R^2} + \frac{[\partial\rho/\partial t]\mathbf{e}_R}{cR} - \frac{[\partial\mathbf{J}/\partial t]}{c^2 R}\right) dv' \tag{2.36}$$

2.1.3 Vortices of an Electric Field

The electric field can have a vortical nature, a nonpotential character. A radio-frequency capacitive discharge, for example, creates a vortex electric field. We may consider a simple cylindrical plasma column of radius R formed between two plane electrodes distant by L from each other. The last equation from the set of Eq. 2.11, when used with Stocks' theorem, gives the magnetic field $B = 2E|\sigma'|R$. Repeating the same reasoning with $\nabla \times \mathbf{E} = \frac{\partial \mathbf{B}}{\partial t}$, the magnitude of the vortical field is obtained

$$E_{vort} = \frac{\omega HR}{2(R + L)} \tag{2.37}$$

Vortex electric current is known to induce a transition between spiral waves and other states, in particular, being able to abort the onset of turbulence in a fluid. A vertical electric field can be created by a long solenoid, and its value can be estimated by applying again Stokes's theorem

$$\oint (\mathbf{E}\cdot d\mathbf{l}) = -\frac{\partial}{\partial t} \oint\oint (\mathbf{B}\cdot d\mathbf{S}) \tag{2.38}$$

It is obtained

$$E = -\frac{R^2}{2r}\frac{dB}{dt} = -\frac{\mu_0 n R^2}{2r}\frac{dI}{dt} \tag{2.39}$$

Here, R is the radius of the solenoid, and I is the external electric current that feeds the solenoid. Additionally, we may notice that r is the radial distance to the axis, $r = \sqrt{(x - x_o)^2 + (y - y_0)^2}$, where (x_0, y_0) denotes the origin of the coordinates, and (x, y) defines the general position of the field. Assuming θ is the angle made by $\mathbf{E}$ relative to the Ox axis, the components of the electric field are: $E_x = E \cos\theta$, $E_y = E \sin\theta$.

The effect of the vertical electric field on excitable media can be described by a modified Fitzhugh–Nagumo model which is often used as a generic model for excitable media since it provides analytical solutions

$$\begin{cases} \frac{\partial u}{\partial t} = f(u, v) + \nabla^2 u + (\mathbf{E}\cdot\nabla u) \\ \frac{\partial v}{\partial t} = g(u, v) \end{cases} \tag{2.40}$$

The two-dimensional Laplace operator is $\nabla^2 = \frac{\partial^2}{\partial x^2} + \frac{\partial^2}{\partial y^2}$, and the variables u and v can be viewed as the "fast" and "slow" variables. The functions f and g will depend on the chosen model.[8] This method can be applied in several different contexts, such as to Winfree turbulence, one of the principal mechanisms underlying cardiac fibrillation. Winfree turbulence can be suppressed by locally injecting periodic signals in 3D excitable media.

2.1.4 Dependence of Electric Voltage on Integration Path

Let us consider an apparatus made by two resistors, R_1 and R_2, connected in a ring and placed over one extremity of an electromagnet, as shown in the schematic diagram of

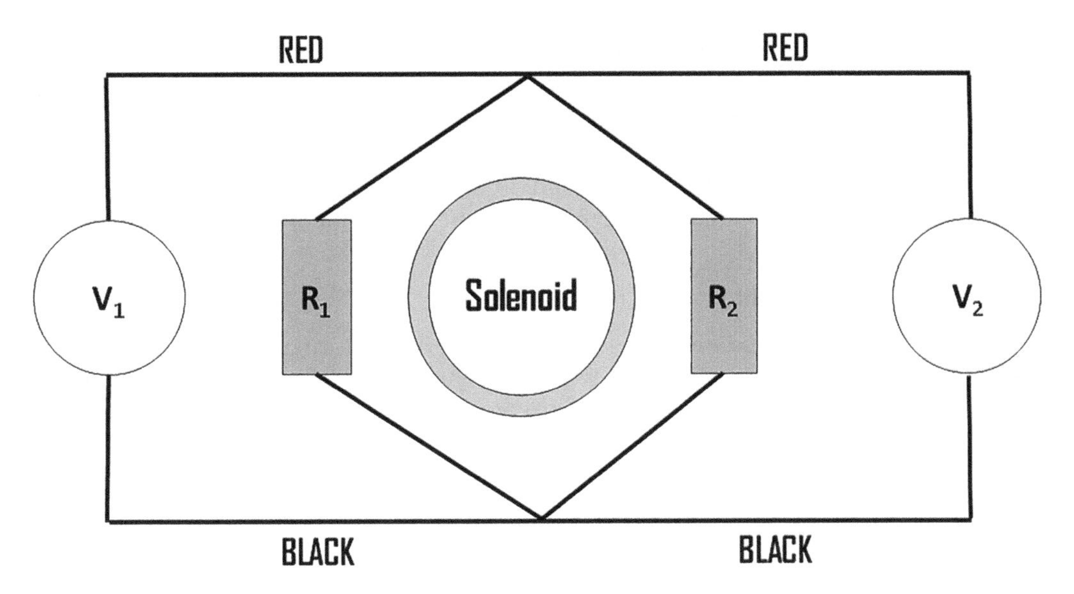

FIGURE 2.1
Voltage dependent on the integration path.

Figure 2.1. A long solenoid carrying a varying current produces a time-dependent magnetic field that induces an electric field. This electric field exists also outside of the solenoid where $\partial \mathbf{B}/\partial t$ is null and therefore where $\nabla \times \mathbf{E} = 0$. However, V_1 and V_2 are different. This problem is beautifully discussed in Ref. [4].

2.1.5 The Velocity Field and Radiation Fields

To obtain the fields at point r and instant of time t we must determine the retarded position r_{ret} at previous time t_{ret} when the particle had velocity $\mathbf{u} \equiv \dot{\mathbf{r}}(t_{ret})$ and acceleration $\dot{\mathbf{u}} \equiv \ddot{\mathbf{r}}_0(t_{ret})$. Let us introduce the following notation to make easier the representations of the fields

$$\beta \equiv \frac{\mathbf{u}}{c} \quad k \equiv 1 - \mathbf{n}\cdot\boldsymbol{\beta} \tag{2.41}$$

The fields are obtained by differentiation of the potentials and they are

$$\begin{aligned} \mathbf{E}(\mathbf{r}, t) &= q\left[\frac{(\mathbf{n}-\beta)(1-\beta^2)}{\kappa^3 R^2}\right] + \frac{q}{c}\left[\frac{\mathbf{n}}{\kappa^3 R} \times [(\mathbf{u} - \beta) \times \dot{\beta}]\right] \\ \mathbf{B}(\mathbf{r}, t) &= [\mathbf{n} \times \mathbf{E}(\mathbf{r}, t)] \end{aligned} \tag{2.42}$$

We may notice that the fields at time t are determined by the particle's position at a previous time. The magnetic field is perpendicular to the electric field $\mathbf{E}$ and $\mathbf{n}$. Also, remark that the electric field is composed of two terms

- The first term – the velocity field – falls off as $1/R^2$ and is a generalization of Coulomb's law. At constant velocity, this is the term that contributes to the field. In this case, the electric field always points along the line toward the particle's current position.
- The second term – the acceleration field – falls off as $1/R$, is proportional to the particle's acceleration, and is proportional to $\mathbf{n}$. This is the **radiation field**.

2.1.6 Schwinger's Variational Principle

Schwinger developed variational methods while working during the WWII on waveguides [5] that were later applied in nuclear and theoretical physics. At the end of the war, Schwinger understood that microwaves were capable to accelerate charged particles, and he later invented the microtron. The variational principle requires that some physical quantity be expressed in terms of a function of some other quantity, let us say, the current distribution. The function gives the required quantity if the true current distribution is inserted, otherwise, it gives a second-order error and the obtained (approximate) value is inferior to the true value. We exemplify the calculation of inductances. Suppose that a current I flows through a circuit C. The magnetic energy is given by

$$W = \frac{1}{2} I \int (\mathbf{B} \cdot d\mathbf{S}) = \frac{1}{2} L I^2 \tag{2.43}$$

is the self-inductance. We may calculate the magnetic energy by another equation

$$W = \frac{1}{2\mu} \int B^2 dv \tag{2.44}$$

According to the concept of energy by Maxwell, the integration is extended to infinity. Hence, we may write

$$W = \frac{\left(\frac{1}{2} I \int (\mathbf{B} \cdot d\mathbf{S})\right)^2}{\int \frac{B^2 dv}{2\mu}} \tag{2.45}$$

Then, it follows the variational expression of the self-inductance of the circuit C

$$L = \mu \frac{(\int \mathbf{B} \cdot d\mathbf{S})^2}{\int B^2 dv} \tag{2.46}$$

If we test the principle with another extra current δI in a circuit C' creating an extra induction field $\delta\mathbf{B}$. Looking at the variation of the self-inductance, δL, we have

$$0 = \delta L \int B^2 dv + 2L \int (\mathbf{B} \cdot \delta\mathbf{B}) dv - 2\mu \int (\mathbf{B} \cdot d\mathbf{S}) \int (\delta\mathbf{B} \cdot d\mathbf{S}) \tag{2.47}$$

But it can be shown that

$$\int \mathbf{B}\cdot\delta\mathbf{B}dv = \mu\frac{1}{\mu}\int \mathbf{B}\cdot\delta\mathbf{B}dv \quad (2.48)$$

$$= \mu \text{ mutual energy between I in C and } \delta I \text{ in } C' = \mu I\delta IL' \quad (2.49)$$

And it also can be shown that

$$2\mu\int(\mathbf{B}\cdot d\mathbf{S})\int(\delta\mathbf{B}\cdot d\mathbf{S} = 2\mu NN' \quad (2.50)$$

Inserting both equations, it can be shown that indeed $\delta L = 0$, i.e., L provides a variational expression when inserted an appropriated field to produce $\mathbf{B} = \nabla \times \mathbf{A}$.

2.1.7 Series of Fourier and Fourier Transform

Complicated functions can be represented as a power series. In particular, any given function can be represented as a sum of sine and cosine terms if they fulfill particular conditions called *Dirichlet conditions*. Such a representation is called a **Fourier series**.

Fourier transform is a generalization of the Fourier series and provides a representation of functions defined over an infinite interval with no particular periodicity. Frequently, Fourier transforms are used to represent time-varying functions, $f(t)$, requiring here that $\int_{\infty}^{\infty}|f(t)|dt$ be finite.

The transition from Fourier series to Fourier transform can be performed by using the complex Fourier series with period T

$$f(t) = \sum_{r=-\infty}^{\infty} c_r e^{2\pi irt/T} = \sum_{r=-\infty}^{\infty} c_r e^{i\omega_r t} \quad (2.51)$$

Here, $\omega_r \equiv 2\pi r/T$.

The **Fourier transform** of $f(t)$ is given by

$$\bar{f}(\omega) = \frac{1}{\sqrt{2\pi}}\int_{-\infty}^{\infty} f(t)e^{-i\omega t}dt \quad (2.52)$$

and its inverse is

$$\bar{f}(t) = \frac{1}{\sqrt{2\pi}}\int_{-\infty}^{\infty} \bar{f}(\omega)e^{i\omega t}d\omega \quad (2.53)$$

Example 2.2: *Evaluate the Fourier transform of the Coulomb potential of a point charge q, $V(r) = q/r$.*
As the Fourier transform of the Coulomb potential is not defined, we may first work with the **Yukawa potential**

$$V(r) = \frac{qe^{-\alpha r}}{r} \quad (2.54)$$

with $\alpha > 0$, since it will give us a well-defined function, and we may later take the limit $\alpha \to 0$ in order to get the Coulomb potential. In three dimensions, we may write

$$\overline{V}(\mathbf{k}) = \frac{1}{\sqrt[3]{2\pi}} \int\int\int \frac{qe^{-\alpha r}}{r} e^{-i\mathbf{k}\cdot\mathbf{r}} d^3x \tag{2.55}$$

To integrate this function we may preferentially choose spherical coordinates. We let $\mathbf{k} = k\mathbf{e}_z$, and $\mathbf{k}\cdot\mathbf{r} = kr\cos\theta$, where θ denotes the polar angle. Hence, we have

$$\overline{V} = \frac{q}{\sqrt[3]{2\pi}} \int_0^\infty r^2 dr \int_0^\pi \sin\theta d\theta \int_0^{2\pi} d\phi e^{-ikr\cos\theta} \frac{e^{-\alpha r}}{r} \tag{2.56}$$

We may start to integrate in variable θ

$$\int_0^\pi \sin\theta e^{-ikr\cos\theta} d\theta = \int_{-1}^1 e^{-ikru} du = \frac{1}{ikr}(e^{ikr} - e^{ikr}) \tag{2.57}$$

Inserting into Eq. 2.56, *we obtain successively*

$$\begin{aligned} \overline{V}(\mathbf{k}) &= \frac{q(2\pi)}{\sqrt[3]{2\pi}} \int_0^\infty dr r^2 \frac{e^{-\alpha r}}{r} \frac{1}{ikr}(e^{ikr} - e^{-ikr}) \\ &= \frac{q}{\sqrt{2\pi}} \frac{1}{ik} \int_0^\infty dr \left[e^{(-\alpha+ik)r} - e^{(-\alpha-ik)r}\right] \\ &= \frac{q}{\sqrt{2\pi}} \frac{1}{ik} \left(\frac{e^{(-\alpha+ik)r}}{-\alpha+ik} + \frac{e^{-(\alpha+ik)r}}{\alpha+ik} \right)_0^\infty \\ \overline{V}(\mathbf{k}) &= \frac{2q}{\sqrt{2\pi}} \frac{1}{k^2}. \end{aligned} \tag{2.58}$$

2.1.8 Lagrangian Formulation of Classical Electrodynamics

The lagrangian theory of fields has number of advantages:

- Invariances (or symmetries) are easily included, such as Lorentz, translations, rotations, Poicaré, gauge invariance.
- Constant of motion (momentum, energy) are easily obtained due to Noether's theorem.

2.1.8.1 Lagrangian of Free Classical Maxwell Fields

Instead of a scalar field $\phi(x)$ we have four fields $A_\mu(x)$ which are components of 4-vector.

2.1.8.2 Lagrangian of Topologically Massive Spinor Electrodynamics

Topologically massive spinor electrodynamics in three dimensions is governed by a lagrangian composed of gauge field, fermion, and interaction term

$$\mathcal{L} = \mathcal{L}_G + \mathcal{L}_F + \mathcal{L}_I \tag{2.59}$$

where the abelian gauge field is described by

$$\mathcal{L}_G = \frac{1}{4}F^{\mu\nu}F_{\mu\nu} + \frac{\mu}{4}\epsilon^{\mu\nu\alpha}F_{\mu\nu}A_\alpha \tag{2.60}$$

The antisymmetric field tensor $F_{\mu\nu} = \partial_\mu A_\nu - \partial_\nu A_\mu$. The Fermion field is given by

$$\mathcal{L}_F = i\overline{\psi}\eth\psi - m\overline{\psi}\psi \tag{2.61}$$

and the interacting field is

$$\mathcal{L}_I = -J^\mu A_\mu \tag{2.62}$$

with $J^\mu = -e\overline{\psi}\gamma^\mu\psi$. The coupling constant e has dimension $[mass]^{1/2}$.

Exercise: Find the equation describing a charged point particle minimally coupled to the vector potential, using the lagrangians Eqs. 2.60 and 2.62.

Solution: We recall here that the Euler–Lagrange equation reads

$$\partial_\mu \frac{\partial \mathcal{L}}{\partial(\partial_\mu\phi)} - \frac{\partial \mathcal{L}}{\partial\phi} = 0 \tag{2.63}$$

We may start by calculating step by step the terms in need. The first one is

$$\frac{\partial \mathcal{L}_G}{\partial(\partial_\nu A_\mu)} = \frac{1}{4}\frac{\partial}{\partial(\partial_\nu A_\mu)}[(\partial_\lambda A_\sigma - \partial_\sigma A_\lambda)(\partial_\lambda A_\sigma - \partial_\sigma A_\lambda)] \tag{2.64}$$

Continuing

$$\frac{\partial \mathcal{L}_G}{\partial(\partial_\nu A_\mu)} = \frac{1}{4}\frac{\partial}{\partial(\partial_\nu A_\mu)}[\partial_\lambda A_\sigma\partial_\lambda A_\sigma - \partial_\lambda A_\sigma\partial_\sigma A_\lambda - \partial_\sigma A_\lambda\partial_\lambda A_\sigma + \partial_\sigma A_\lambda\partial_\sigma A_\lambda] \tag{2.65}$$

Remark that terms like the first and fourth at the RHS of Eq. 2.65 are equal, and as well the second and third. Hence, we may write

$$\frac{\partial \mathcal{L}_G}{\partial(\partial_\nu A_\mu)} = \frac{1}{2}\frac{\partial}{\partial(\partial_\nu A_\mu)}[\partial_\lambda A_\sigma\partial_\lambda A_\sigma - \partial_\lambda A_\sigma\partial_\sigma A_\lambda] \tag{2.66}$$

Continuing one more line

$$\frac{\partial \mathcal{L}_G}{\partial(\partial_\nu A_\mu)} = \frac{1}{2}\left[\frac{\partial}{\partial(\partial_\nu A_\sigma)}(\partial_\lambda A_\sigma)\partial_\lambda A_\sigma - \frac{\partial}{\partial(\partial_\nu A_\mu)}(\partial_\sigma A_\lambda)\partial_\lambda A_\sigma\right] \tag{2.67}$$

and

$$\frac{\partial \mathcal{L}_G}{\partial(\partial_\nu A_\mu)} = \frac{1}{2}[2\partial_\nu A_\mu - 2\partial_\mu A_\nu] = [\partial_\nu A_\mu - \partial_\mu A_\nu] = F_{\mu\nu} \tag{2.68}$$

The next one is

$$\frac{\partial \mathcal{L}_G}{\partial(\partial_\nu A_\mu)} = \frac{\mu}{4}\frac{\partial}{\partial(\partial_\nu A_\mu)}[\epsilon^{\lambda\sigma\alpha}(\partial_\lambda A_\sigma - \partial_\sigma A_\lambda)A_\alpha] \tag{2.69}$$

whose derivative gives

$$\frac{\partial \mathcal{L}_G}{\partial(\partial_\nu A_\mu)} = \frac{\mu}{4}[\eta_{\lambda\nu}\eta_{\mu\sigma}\epsilon^{\lambda\sigma\alpha}A_\alpha - \eta_{\sigma\nu}\eta_{\mu\lambda}A_\alpha] \tag{2.70}$$

Of course, $\eta_{\lambda\nu}\eta_{\mu\sigma}=1$, when $\lambda = \nu$ and $\mu = \sigma$, and so on. Hence, we obtain

$$\frac{\partial \mathcal{L}_G}{\partial(\partial_\nu A_\mu)} = \frac{\mu}{4}[\epsilon^{\nu\mu\alpha} - \epsilon^{\mu\nu\alpha}]A_\alpha = \frac{\mu}{2}2\epsilon^{\nu\mu\alpha} = \frac{\mu}{2}\epsilon^{\nu\mu\alpha}A_\alpha \tag{2.71}$$

Exercise: Show that the Chern–Simons lagrangian is invariant by a gauge transformation, $A_\mu^{'} = A_\mu - \partial_\mu \Lambda$.

Solution: We have

$$\mathcal{L}'_{cs} = \kappa\epsilon^{\mu\nu\rho}A_\mu^{'}\partial_\mu A'_\rho \tag{2.72}$$

This gives

$$\delta L_{cs} = -\kappa\epsilon^{\mu\nu\rho}[A_\mu\partial_\nu\partial_\rho\Lambda + \partial_\mu\Lambda\partial_\nu A_\rho - \partial_\mu\Lambda\partial_\nu\partial^\nu\Lambda g_{\nu\rho}] \tag{2.73}$$

Remark that terms like $\partial_\nu\partial^\nu\Lambda = 0$ and **WA** = 0. Hence,

$$\delta L_{cs} = -\kappa\epsilon^{\mu\nu\rho}\partial_\mu(\Lambda\partial_\nu A_\rho) \tag{2.74}$$

The equations of motion can be obtained from the lagrangian

$$\begin{aligned} L_{cs} &= \tfrac{\kappa}{2}\epsilon^{\mu\nu\rho}A_\mu\partial_\nu A_\rho - A_\mu J^\mu \\ \frac{\partial L}{\partial A_\mu} - \partial_\nu\frac{\partial L}{\partial(\partial_\nu A_\mu)} &= 0 \\ \frac{\partial L}{\partial A_\mu} &= \tfrac{\kappa}{2}\epsilon^{\mu\nu\rho}\partial_\nu A_\rho - J^\mu \\ \frac{\partial L}{\partial(\partial_\nu A_\mu)} &= \tfrac{\kappa}{2}\epsilon^{\mu\nu\rho}A_\rho \end{aligned} \tag{2.75}$$

From Eq. 2.75, we deduce

$$-J^\mu + \frac{\kappa}{2}(\epsilon^{\mu\nu\rho}\partial_\nu A_\rho - \epsilon^{\rho\nu\mu}\partial_\nu A_\rho) = 0 \tag{2.76}$$

Eq. 2.76 can be presented as

$$J^\mu = \frac{\kappa}{2}\epsilon^{\mu\nu\rho}F_{\nu\rho} \tag{2.77}$$

Exercise: Show that $F_{\mu\nu} = \frac{1}{\kappa}\epsilon_{\mu\nu\rho}J^{\rho}$.
Solution:

$$\begin{aligned}
(\epsilon_{\mu\nu\sigma}\epsilon_{\mu\nu\rho})F_{\nu\rho} &= \tfrac{2}{\kappa}J^{\mu}\epsilon_{\mu\nu\sigma} \\
(2g_{\sigma\rho})F_{\nu\rho} &= \ldots \\
2F_{\nu\sigma} &= \tfrac{2}{\kappa}J^{\mu}\epsilon_{\mu\nu\sigma} \\
2F_{\nu\mu} &= \tfrac{2}{\kappa}J^{\rho}\epsilon_{\mu\rho\nu} \\
-2F_{muv} &= -\tfrac{2}{\kappa}J^{\rho}\epsilon_{\mu\nu\rho}
\end{aligned} \tag{2.78}$$

That is

$$F_{\mu\nu} = \frac{1}{\kappa}\epsilon_{\mu\nu\rho}J^{\rho} \tag{2.79}$$

Exercise: Show that the Bianchi identity $\epsilon^{\mu\nu\rho}\partial_{\mu}F_{\nu\rho} = 0$ is comparable with current conservation, $\partial_{\mu}J^{\mu} = 0$.

2.2 Chern–Simons Theory

The lagrangians we used before had a metric tensor $g_{\mu\nu}$ with a measure of space and time, and provided a set of instructions to calculate the scalar product of vectors by means of a contraction of indices. Chern–Simons theory (CST) is a three-dimensional topological field theory where "topological" means the theory does not depend on watches and measuring rods implicit in $g_{\mu\nu}$. CST depends on the topology of the manifold[9] chosen to describe the phenomena we are working on.

The generalized field strength is given by

$$F := dA + A^2 = dA + A \wedge A \tag{2.80}$$

According to Chern and Simons [6] the generalized Chern–Simons form is written as

$$\omega_{2n+1}(A, F) = kTr(F^{n+1}) = (n+1)\int_0^1 dtAF_t^n \tag{2.81}$$

where $F_t = tF + (t^2 - 1)A^2$. k is an operator defined through the equation $kP(A, F) = \int_0^1 dtk_tP(At, F_t)$. $P(A, F)$ is an arbitrary function of A and F, together with the set of equations

$$\begin{aligned}
A_t &= tA \\
F_t &= tdA + t^2A \wedge A \\
k_tA_t &= 0 \\
k_tF_t &= tA
\end{aligned} \tag{2.82}$$

It was shown that $Tr\,(F^{n+1}) = d\omega_{2n+1}(A, F)$ and so

$$\begin{aligned} Tr\,(F^2) &= d\omega_3(A, F) \\ &= d\,[Tr\,(AF) - \tfrac{A^3}{3} \\ &= d\Big[Tr\,(a\,(dA + A^2)) - \tfrac{A^3}{3}\Big] \\ &= d\Big[Tr\Big(AdA + \tfrac{2}{3}A^3\Big)\Big] \end{aligned} \tag{2.83}$$

Let us look at a simple example. Let the three-dimensional Minkovski space $M = \mathbb{R}^{2,1}$ represent the universe. The abelian Chern–Simons action is constructed from the $U(1)$ gauge field $A_\mu(x)$, and of the photon field in electrodynamics

$$S_{CS}[A] = \frac{k}{4\pi}\int_M d^3x\epsilon^{\mu\nu\rho}A_\mu\partial_\nu A_\rho \tag{2.84}$$

The equation of motion can be obtained from the condition $\delta S_{CS}[A']$=0.

$$\begin{aligned} \delta S_{CS}[A] &= \int d^2x\epsilon^{\alpha\beta\gamma}(\delta A_\alpha\partial_\beta A_\gamma + A_\alpha\partial_\beta\delta A_\gamma) \\ &= \int d^2x\epsilon^{\alpha\beta\gamma}(\delta A_\alpha\partial_\beta A_\gamma - (\partial_\beta A_\alpha)\delta A_\gamma) \end{aligned} \tag{2.85}$$

But $\delta\,(A_\alpha\partial_\beta A_\gamma) = 0 = A_\alpha\partial_\beta\delta A_\gamma + \delta A_\alpha\partial_\beta A_\gamma$. Hence,

$$\begin{aligned} \delta S_{CS}[A] &= \\ &= \int d^2x\epsilon^{\alpha\beta\gamma}(\partial_\beta A_\gamma\ \ \partial_\gamma A_\beta)\delta A_\alpha \\ &= \int d^2x\epsilon^{\alpha\beta\gamma}F_{\beta\gamma}\delta A_\alpha \end{aligned} \tag{2.86}$$

Remember that $A \wedge dA = \epsilon^{ijk}A_i\partial_j A_k$, and therefore, the aforementioned action only included the "kinetic term" $A \wedge dA$. The maximum of the action, $\delta S_{CS}[A] = 0$ thus implied $F_{\mu\nu} = 0$, showing also that $F_{\mu\nu}$ is gauge-invariant. But the Chern–Simons action $S_{CS}[A]$ is also gauge-invariant. Let us perform a gauge transformation $A_\alpha \to A'_\alpha = A_\alpha + \partial_\alpha\psi$. Then

$$\begin{aligned} S_{CS}[A'] &= S_{CS}[A] + \int d^2x\epsilon^{\alpha\beta\gamma}F_{\beta\gamma}\partial_\alpha\psi \\ &= S_{CS}[A] + \int d^2x\partial_\alpha(\epsilon^{\alpha\beta\gamma}F_{\beta\gamma}\psi) \end{aligned} \tag{2.87}$$

But $\partial_\alpha(\epsilon^{\alpha\beta\gamma}F_{\beta\gamma}\psi) = \epsilon^{\alpha\beta\gamma}\partial_\alpha F_{\beta\gamma}\psi + \epsilon^{\alpha\beta\gamma}F_{\beta\gamma}\partial_\alpha\psi$. As the lagrangian is invariant up to a total derivative, $S_{CS}[A'] = S_{CS}[A]$, and the Chern–Simons action is gauge invariant up to a boundary term.

Applying the aforementioned framework to the theory with the lagrangian

$$L = L_m + \kappa L_{CS} = -\frac{1}{4}F^{\alpha\beta}F_{\alpha\beta} + \frac{1}{2}\kappa\epsilon^{\alpha\beta\gamma}A_\alpha F_{\beta\gamma} \tag{2.88}$$

gives the equation of motion

$$\partial_\alpha F^{\alpha\beta} + \kappa\epsilon^{\beta\gamma\delta}F_{\gamma\delta}=0 \tag{2.89}$$

also known as *topologically massive Maxwell theory*. Therefore, the Chern–Simons term provides a topological gauge-invariant mass term for the photon. Note that the mass term $m^2 A_\alpha A^\alpha$ that appears in the Klein–Gordon equation is not gauge-invariant, the reason why the absence of mass of the photon is claimed to be due to gauge invariance.

We can see that if we introduce the dual of the field strength

$$G^\beta = \frac{1}{2}\epsilon^{\beta\gamma\delta}F_{\gamma\delta} \tag{2.90}$$

then

$$\begin{aligned} \partial_\alpha G_\beta - \partial_\beta G_\alpha &= 2\kappa\epsilon_{\alpha\beta\gamma}G^\gamma \\ \partial^\alpha\partial_\alpha G_\beta - \partial_\beta\partial^\alpha G_\alpha &= 2\kappa\epsilon_{\alpha\beta\gamma}\partial^\alpha G^\gamma \\ \partial_\beta G^\beta &= 0 \ (\textbf{Bianchi identity}) \end{aligned} \tag{2.91}$$

Using Bianchi identity, we obtain

$$\begin{aligned} WG_\beta &= \Sigma_a \partial^a\partial_a G_\beta \qquad = 2\kappa\epsilon_{\alpha\beta\gamma}partial^\alpha G^\gamma \\ &= \kappa(\epsilon_{\alpha\beta\gamma}\partial^\alpha G^\gamma - \epsilon_{\beta\gamma\alpha}\partial^\alpha G^\gamma \\ &= \kappa\epsilon_{\alpha\beta\gamma}(\partial^\alpha G^\gamma - \partial^\alpha G^\gamma) \\ &= 2\kappa^2\epsilon_{\alpha\beta\gamma}\epsilon^{\gamma\alpha\delta}G_\delta \end{aligned} \tag{2.92}$$

But $\epsilon_{\alpha\beta\gamma}\epsilon^{\gamma\alpha\delta} = 2\delta_{\delta\beta}$ and, therefore,

$$WG_\beta = 4\kappa^2 G_\beta = m^2 G_\beta \tag{2.93}$$

In analogy with Klein–Gordon equation, we may infer that this field possess mass, given by $m^2 = 4\kappa^2$. In conclusion, the Chern–simons theory describes excitations of mass $m^2 = 4\kappa^2$.

Electric and magnetic fields in two dimensions can be defined

$$\begin{aligned} E_i &= -\partial_i A_0 - \partial_0 A_i \\ B &= \epsilon^{ij}\partial_i A_j \end{aligned} \tag{2.94}$$

where the vector (gauge) potential is $A^\mu = (V, \mathbf{A})$, the matter current is $J^\mu = (\rho, \mathbf{J}$, and the fields constructed from the potentials are given by

$$F^\mu_\nu = \begin{bmatrix} 0 & E_1 & E_2 \\ -E_1 & 0 & B \\ -E_2 & -B & 0 \end{bmatrix} \tag{2.95}$$

We still use Lorentz-covariant notation with metric diag(1, − 1, − 1).[10]

Adding matter to the fields, the total lagrangian is

$$L = \frac{\kappa}{4\pi}\epsilon^{\mu\nu\rho}A_\mu\partial_\nu A_\rho + \overline{\Psi}(i\partial\!\!\!/ - m)\Psi - eA_\mu\overline{\Psi}\gamma^\mu\Psi \tag{2.96}$$

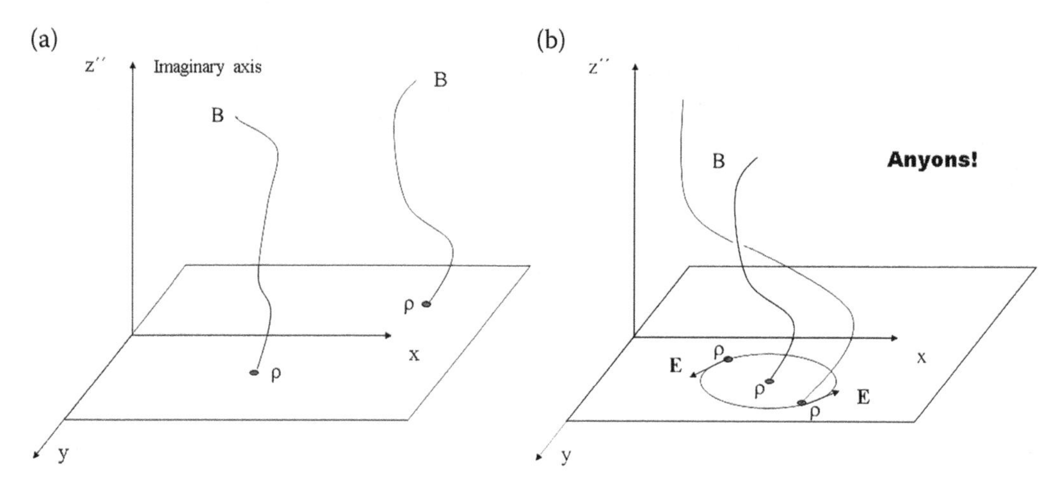

FIGURE 2.2
Anyons in a plane. (a) Two anyons, each with charge ρ and the corresponding electric field line (braid). (b) One anyon encircles the other and the braid corresponding to that process entangles.

The equation of motion follows

$$\frac{\kappa}{4\pi}\epsilon^{\mu\nu\rho}F_{\nu\rho} = \frac{\kappa}{2\pi}\epsilon^{\mu\nu\rho}\partial_\nu A_\rho = J^\mu \tag{2.97}$$

The $\mu = 0$ gives

$$J^0 = \frac{\kappa}{4\pi}\epsilon^{0\nu\rho}F_{\nu\rho} = \frac{\kappa}{4\pi}\epsilon^{012}F_{12} = \frac{\kappa}{4\pi}B \tag{2.98}$$

and, for $\mu = i$,

$$\begin{aligned} J^i &= \tfrac{\kappa}{4\pi}\epsilon^{i\nu\rho}F_{\nu\rho} \\ J^i &= \tfrac{\kappa}{2\pi}\epsilon^{ij}E_j \end{aligned} \tag{2.99}$$

To conclude, the scalar magnetic field B is sourced by electric charges, while $\mathbf{E}$ is resulting from the current. Just the opposite of electromagnetism! Figure 2.2 illustrates the concept.

2.3 Ergontropic Dynamics

In this section, we propose to introduce to the reader a summary of the methodology motivating our program to electromagnetic propulsion, as considered in our previous work. The classical mechanistic description presupposes the reversibility of time but didn't take into account the thermodynamic implications of the isotropy and homogeneity of Newtonian space-time. In previous work we introduced a novel approach differing in fundamental aspects from the standard treatments, particularly not deriving irreversible thermodynamics from the reversible microscopic dynamics, on the contrary, we relate dynamics and thermodynamics judiciously making use of the concepts of energy, work, and entropy. Incorporating the second law of thermodynamics which has a form of causality as a universal property, and working out the general method, at the end, a system of differential equations result able to describe a physical system in terms of dynamical quantities and free energy.

For this purpose we start with the total entropy of the system composed of N particles (or bodies), considering a simple dynamical system constituted by a set of N discrete interacting mass-points (although macroscopic) $m^{(\alpha)}$ ($\alpha = 1, 2, ..., N$) with x_i^α and v_i^α) ($i = 1, 2, 3; \alpha = 1, \dots, N$) denoting the coordinates and velocities of the mass point in a given inertial frame of reference. The inferior Latin index refers to the Cartesian components and the superior Greek index distinguishes the different mass points. The following step is the construction of the invariant entropy function $\bar{S}$, as a function of the so-far, *known* fundamental invariants in a closed system, hence, $\bar{S} = \bar{S}(E, \mathbf{p}, \mathbf{L})$:

$$\bar{S} = \sum_{\alpha=1}^{N} [S^{(\alpha)}(E^{(\alpha)} - \frac{(\mathbf{p}^{(\alpha)})^2}{2m^{(\alpha)}} - \frac{(\mathbf{J}^{(\alpha)})^2}{2I^{(\alpha)}} - q^{(\alpha)}V^{(\alpha)} + q^{(\alpha)}(\mathbf{A}^{(\alpha)}\cdot\mathbf{v}^{(\alpha)}) - U_{mec}^{(\alpha)}) + (\mathbf{a}\cdot\mathbf{p}^{(\alpha)} + \mathbf{b}\cdot([\mathbf{r}^{(\alpha)} \times \mathbf{p}^{(\alpha)}] + \mathbf{J}^{(\alpha)})] \tag{2.100}$$

or

$$\bar{S} = \sum_{\alpha=1}^{N} \mathfrak{S}^{(\alpha)} \tag{2.101}$$

Here, $\mathbf{a}$ and $\mathbf{b}$ are Lagrange multipliers vectors, related to the velocity of translation $\mathbf{v}_e = T\mathbf{a}$, and the velocity of rotation $\boldsymbol{\omega} = T\mathbf{b}$. The method of Lagrange multipliers is applied to find extremal values of a function of several variables subject to one or more constraints, namely the total linear momentum and the total angular momentum about the origin, both with respect to a definite inertial reference frame

$$\sum_{\alpha} \mathbf{p}_\alpha = \mathbf{P} = \text{const} \tag{2.102a}$$

$$\sum_{\alpha} ([\mathbf{r}_\alpha \times \mathbf{v}_\alpha] + \mathbf{J}_\alpha) = \mathbf{J} = \text{const} \tag{2.102b}$$

with $\mathbf{J}$ denoting the total sum of the orbital and spin angular momentum, both equations, due to the principles of energy and angular momentum conservation in an isolated body, constituted by all $\alpha = 1, 2, \dots N$ particles, or bodies, all at the same temperature to avoid, at this level of description, mathematical complexity. Eqs. 2.102a,b are the constraints equations referred above.

The meaning of the absolute temperature T plays here the role of kinetic temperature, the equipartition expression in momenta but its true essence might not be entirely captured yet. Finally, the total entropy is inserted on the fundamental equation of entropy written in the differential form

$$\nabla_k U = T\nabla_k \bar{S} - F_k \tag{2.103}$$

summing over the all ensemble of N particles ($k = 1, \ldots N$). We may notice now that from the kinetic energy definition the definition of force as defined by Newton is obtained, and this finding led us to associate the gravitational forces with the distribution of particles in space [7]. The underlying mathematical formalism leads us to the well-known equations of (electro)dynamics (if electromagnetic entities enter into the system), and through this methodology a powerful new approach to mechanics and electrodynamics is provided.

Following the mathematical procedure proposed in Ref. [8] the total entropy of the system $\bar{S}$ (where the symbols have the usual meaning) is thus given by

$$\bar{S} = \sum_{\alpha=1}^{N} [S^{(\alpha)} (E^{(\alpha)} - \frac{(\mathbf{p}^{(\alpha)})^2}{2m^{(\alpha)}} - \frac{(\mathbf{J}^{(\alpha)})^2}{2I^{(\alpha)}} - q^{(\alpha)} V^{(\alpha)} + q^{(\alpha)} (\mathbf{A}^{(\alpha)} \cdot \mathbf{v}^{(\alpha)})$$

$$- U_{mec}^{(\alpha)}) + (\mathbf{a} \cdot \mathbf{p}^{(\alpha)} + \mathbf{b} \cdot ([\mathbf{r}^{(\alpha)} \times \mathbf{p}^{(\alpha)}] + \mathbf{J}^{(\alpha)})] = \sum_{\alpha=1}^{N} \mathfrak{S}^{(\alpha)} \tag{2.104}$$

where $\mathbf{a}$ and $\mathbf{b}$ are Lagrange multipliers (as vectors). It can be shown that $\mathbf{v}_{rel} = \mathbf{a}T$ and $\omega = \mathbf{b}T$. The conditional extremum points form the dynamical equations of motion of a general physical system (the equality holds whenever the physical system is in thermodynamic and mechanical equilibrium), which is defined by two first-order differential equations

$$\frac{\partial \bar{S}}{\partial \mathbf{p}^{(\alpha)}} = 0, \text{ canonical momentum} \tag{2.105}$$

$$\frac{\partial \bar{S}}{\partial \mathbf{r}^{(\alpha)}} = -\frac{1}{T} \nabla_{r^{(\alpha)}} U^{(\alpha)} - \frac{1}{T} m^{(\alpha)} \frac{\partial \mathbf{v}^{(\alpha)}}{\partial t} = 0, \text{ fundamental equation of dynamics} \tag{2.106}$$

At thermodynamic equilibrium, the total entropy of the body has a maximum value, constrained through the supplementary conditions referred before (on total energy, linear momentum, and angular momentum), which can be obtained by means of the minimization techniques associated with Lagrange multipliers. Extremum conditions imposed on entropy or internal energy not only provide criteria for the evolution of the system but determine as well the stability of thermodynamic systems at equilibrium, onset when the entropy increase with distance from equilibrium

$$\frac{\partial \bar{S}}{\partial \mathbf{r}} > 0 \tag{2.107}$$

Also, the gradient of the total entropy in momentum space multiplied by factor T gives (from now on, we drop the superscript α and retain the i-index (i = x,y,z) for the spatial coordinates)

$$T\frac{\partial \bar{S}}{\partial \mathbf{p}} = \left\{ -\frac{\mathbf{p}}{m} + \frac{q}{m}\mathbf{A} + \mathbf{v}_{rot} + [\boldsymbol{\omega} \times \mathbf{r}] \right\} \tag{2.108}$$

The maximization of Eq. 2.108 leads to the total (canonical) momentum, given by

$$\mathbf{p} = m\mathbf{v}_{rot} + m_i [\boldsymbol{\omega} \times \mathbf{r}] + q\mathbf{A} \tag{2.109}$$

However, in nonequilibrium conditions, the exact canonical momentum for each particle should be read

$$\boxed{\mathbf{p} = m\mathbf{v}_{rot} + m[\boldsymbol{\omega} \times \mathbf{r}] + q\mathbf{A} - mT\nabla_p \bar{S}} \tag{2.110}$$

2.3.1 Dynamics

Taking the gradient of $\bar{S}$ in Eq. 2.100, we obtain

$$\nabla\bar{S}_{r_i} = \frac{\partial S_i}{\partial U_i}\frac{\partial U_i}{\partial \mathbf{r}_i} + m\mathbf{a}_i\cdot\frac{\partial \mathbf{v}_i}{\partial \mathbf{r}_i} + \mathbf{b}\cdot\frac{\partial}{\partial \mathbf{r}_i}([\mathbf{r}_i \times \mathbf{p}_i] + \mathbf{J}_i) \tag{2.111}$$

Here, we recall that by definition of temperature, $\frac{1}{T} = \frac{\partial S_i}{\partial U_i}$, and $\mathbf{v}_e = \mathbf{a}T, \boldsymbol{\omega} = \mathbf{b}T$. Then (now dropping the i-index to simplify), the following is obtained:

$$T\nabla\bar{S}_r = \frac{\partial U}{\partial \mathbf{r}} + m\mathbf{v}_e\cdot\frac{\partial \mathbf{v}}{\partial \mathbf{r}} + \frac{\partial}{\partial \mathbf{r}}(\boldsymbol{\omega}\cdot[\mathbf{r} \times \mathbf{p}]) + \boldsymbol{\omega}\cdot\frac{\partial \mathbf{J}}{\partial \mathbf{r}} \tag{2.112}$$

The internal energy is the summing up of the kinetic energy with the potential energy (e.g., gravitational, electromagnetic) and, therefore,

$$\frac{\partial U}{\partial \mathbf{r}} = -m\mathbf{v}\cdot\frac{\partial \mathbf{v}}{\partial \mathbf{r}} - m\frac{\partial \Phi}{\partial \mathbf{r}} \tag{2.113}$$

Hence,

$$T\nabla\bar{S}_r = -m\mathbf{v}\cdot\frac{\partial \mathbf{v}}{\partial \mathbf{r}} - m\frac{\partial \Phi}{\partial \mathbf{r}} + m\frac{\partial}{\partial \mathbf{r}}(\mathbf{v}\cdot(\mathbf{v}_e + [\boldsymbol{\omega} \times \mathbf{r}])) + \frac{\partial}{\partial \mathbf{r}}(\boldsymbol{\omega}\cdot\mathbf{J}) \tag{2.114}$$

or,

$$T\nabla\bar{S}_r = -m\mathbf{v}\cdot\frac{\partial \mathbf{v}}{\partial \mathbf{r}} - m\frac{\partial \Phi}{\partial \mathbf{r}} + m\frac{\partial}{\partial \mathbf{r}}(\mathbf{v}\cdot\mathbf{v}) + \frac{\partial}{\partial \mathbf{r}}(\boldsymbol{\omega}\cdot\mathbf{J}) \tag{2.115}$$

Finally, getting them together, we get

$$T\nabla\bar{S}_r = m\mathbf{v}\cdot\frac{\partial \mathbf{v}}{\partial \mathbf{r}} - m\frac{\partial \Phi}{\partial \mathbf{r}} - \frac{\partial}{\partial \mathbf{r}}\left(\frac{J^2}{2I} - \boldsymbol{\omega}\cdot\mathbf{J}\right) \tag{2.116}$$

The first term on the RHS of Eq. 2.116 is the convective term and we know that

$$\frac{d\mathbf{v}}{dt} = \frac{\partial \mathbf{v}}{\partial t} + \mathbf{v}\cdot\frac{\partial \mathbf{r}}{\partial \mathbf{v}} \tag{2.117}$$

However, the net force acting on the volume element of the body causes acceleration $\frac{\partial \mathbf{v}}{\partial t}$, in which case we should replace

$$- m\mathbf{v}\cdot\frac{\partial \mathbf{v}}{\partial \mathbf{r}} \rightarrow m\frac{\partial \mathbf{v}}{\partial t}$$

This amount to the recognition of the principle of equivalence, and for an observer sitting on the extended particle center of mass, he is at rest. Then it results

$$T\nabla\bar{S}_r = -m\frac{\partial \mathbf{v}}{\partial t} - m\frac{\partial \Phi}{\partial \mathbf{r}} - \frac{\partial}{\partial \mathbf{r}}\left(\frac{J^2}{2I} - \boldsymbol{\omega}\cdot\mathbf{J}\right) \tag{2.118}$$

We can rewrite Eq. 2.118 in a more suitable form

$$m\frac{\partial \mathbf{v}}{\partial t} = -m\frac{\partial \Phi}{\partial \mathbf{r}} - \frac{\partial}{\partial \mathbf{r}}\left(\frac{J^2}{2I} - \boldsymbol{\omega}\cdot\mathbf{J} + TS\right) \tag{2.119}$$

assuming, to simplify, that the process has no thermal gradients. Finally, and for more practical uses, we introduce instead the free energy term, in which case the final form of the Newtonian dynamics becomes more consistently represented in the complete form

$$\boxed{m\frac{\partial \mathbf{v}}{\partial t} = -m\frac{\partial \Phi}{\partial \mathbf{r}} - \frac{\partial}{\partial \mathbf{r}}\left(\frac{J^2}{2I} - \boldsymbol{\omega}\cdot\mathbf{J} - F\right)} \tag{2.120}$$

This equation describes the combined motion of rotation and translation, with a rotation about an *instantaneous center* with angular velocity $\boldsymbol{\omega}(t)$. The free energy term inside Eq. 2.120 ensures the tendency of the free energy to the minimum, tending to the smallest possible value, connected to the term $-TS$ that actuates to give the maximum possible value, that is, ensuring that all possible states can be realized equally.

Example 2.3: *Rolling body on an inclined plane*
A rigid body of mass M rolling down an inclined plane making an angle θ with the horizontal. Eq. 2.119 *can be applied to solve the problem, with $\omega = 0$ (there is no rotation of the frame of reference) and considering that only the gravitational force acts on the rolling body, with inertial moment relative to its own center of mass given by $I_c = \beta MR^2$. Hence, we obtain*

$$M\ddot{x} = Mg\sin\theta - \partial_x w \tag{2.121}$$

Here, $w \equiv \frac{(J_c)^2}{2I_c}$. Assuming that the x-axis is directed along the inclined plane, and considering that the angular momentum relative to the rigid body center of mass is given by $J_c = I_c\omega'$, with $\omega' = d\theta/dt$, and noticing that $dx = v_x dt$ (holonomic constraint), it is readily obtained

$$M\ddot{x} = Ma_x = Mg\sin\theta - I_c\omega'\frac{d\omega'}{v_x dt} = Mg\sin\theta - \beta MR^2\frac{\omega'}{\omega' R}\alpha \tag{2.122}$$

Since $\alpha = a/R$, then it results in the well-known equation

$$a_x = \frac{g \sin \theta}{(1 + \beta)} \tag{2.123}$$

This example shows that translation and rotation (or accelerated motion) are self-consistently integrated into Eq. 2.119, *with clear advantages.*

2.3.2 Electrodynamics

The energetics of the charged particles is represented by Maxwell–Lorentz's theory of the electromagnetic field. Therefore, the internal energy U_α of each subsystem moving with momentum $\mathbf{p}_\alpha$ in a Galilean frame of reference is given by

$$E_\alpha = U_\alpha + \frac{p_\alpha^2}{2m_\alpha} + \frac{J_\alpha^2}{2I_\alpha} - q_\alpha V_\alpha + q_\alpha (\mathbf{A}_\alpha \cdot \mathbf{v}_\alpha) \tag{2.124}$$

where, other than the mechanical energy, the electrostatic and magnetic energy was re-instated with the appropriate energy term: $U_{m,\alpha} \to -q_\alpha (\mathbf{A}_\alpha \cdot \mathbf{v}_\alpha)$. Returning to Eq. 2.100, when considering the flux of the internal energy, by Eq. 2.124 it is straightforward to get the following complete expression:

$$\begin{aligned} \frac{T}{2}\frac{\partial \bar{S}}{\partial \mathbf{r}_i} &= -q_i \nabla_{r_i} V_i + q_i \nabla_{r_i} (\mathbf{v}_i \cdot \mathbf{A}_i) \\ &\quad + m_i \frac{d\mathbf{v}_i}{dt} - \frac{1}{2} \nabla_{r_i} \left(\frac{J_i^2}{I_i} \right) + \frac{1}{2} \nabla_{r_i} (\omega \cdot \mathbf{J}_i) \end{aligned} \tag{2.125}$$

It is worth rearranging some terms to give a more precise physical meaning to whole expressions. Accordingly, utilizing the definition of $\mathbf{B}$ in terms of the vector potential, we have the mathematical identity

$$\nabla_{r_i} (\mathbf{A}_i \cdot \mathbf{v}_i) = (\mathbf{v}_i \cdot \nabla_{r_i}) \mathbf{A}_i + (\mathbf{A}_i \cdot \nabla_{r_i}) \mathbf{v}_i + [\mathbf{v}_i \times \mathbf{B}_i] + [\mathbf{A}_i \times [\nabla_{r_i} \times \mathbf{v}_i]] \tag{2.126}$$

The cross product $(\mathbf{v}_i \cdot \nabla) \mathbf{A}_i$ demands an interpretation inside physical grounds. Consider a charge q moving with velocity v along a given direction x, such that it is displaced by $x = vt$. The Maxwell equations determine the electromagnetic field in terms of the position of field sources relative to the position of the charge, $A(x - vt, y, z)$. It follows, for a planar wave, a time dependence of the form $\frac{\partial A}{\partial t} = -v \frac{\partial A}{\partial x}$. Taking this into description, the extremum condition $\frac{\partial \bar{S}}{\partial \mathbf{r}_i} = 0$, for example, points us to the association

$$E_i + [\mathbf{v}_i \times \mathbf{B}_i] = \frac{1}{2q_i} \nabla_{r_i} \left(\frac{J_i^2}{I_i} - \omega \cdot \mathbf{J}_i \right) - [\mathbf{A}_i \times \omega] \tag{2.127}$$

which is a condition for equilibrium of charges, as in plasma rotation in tokamaks. The fields are defined as usually by $\mathbf{E} = -\nabla V - \frac{\partial \mathbf{A}}{\partial t}$ and $\mathbf{B} = [\nabla \times \mathbf{A}]$. Inside the latter expression, the electromagnetic field was calculated explicitly and hence the equilibrium condition received a new form. In special, we remark that a gyroscopic term serves a vital part of plasma equilibrium.

Accordingly, executing the same procedure as recommended in the section for the dynamics, and regarding some vectorial algebra, the force operating on electrically charged particles is quickly obtained

$$\begin{aligned}\mathbf{f}_i &= q_i\mathbf{E}_i + q_i[\mathbf{v}_i \times \mathbf{B}_i] + q_i(\mathbf{A}_i\cdot\nabla_{r_i})\mathbf{v}_i + 2q_i[\mathbf{A}_i \times \omega] + \frac{1}{2}\nabla_{r_i}(\omega\cdot\mathbf{J}_i) \\ &- \frac{1}{2}\nabla_{r_i}\left(\frac{J_i^2}{I_i}\right) - \nabla_{r_i}U_i\end{aligned} \tag{2.128}$$

However, at this stage, we require the physical meaning of the mathematical entities associated, especially the last term, the gradient of energy in space. It is presumed that, not far away from equilibrium, we might state that

$$-\nabla_{r_i}U_i = -\nabla_{r_i}U_i^{eq} + \mathbf{f}^S \tag{2.129}$$

We express by $\mathbf{f}_i^S$ a force term depicting a small disturbance of a constant equilibrium state discriminating *ith* system, enabling the former expression to be reworked into the form

$$\begin{aligned}\mathbf{f}_i &= \mathbf{f}_i^M + \mathbf{f}_i^L + q_i(\mathbf{A}_i\cdot\nabla_{r_i})\mathbf{v}_i - 2q_i[\omega \times \mathbf{A}_i] \\ &- \frac{1}{2}\nabla_{r_i}\left(\frac{J_i^2}{I_i}\right) + \nabla_{r_i}(\omega\cdot\mathbf{J}_i) + \mathbf{f}_i^S\end{aligned} \tag{2.130}$$

The expression $\mathbf{f}_i^M$ holds for the mechanical force pulling over the *ith* particle, $\mathbf{f}_i^L = q_i\mathbf{E}_i + q_i[\mathbf{v}_\times\mathbf{B}_i]$ signifies the corresponding Lorentz's force. The third term on the right-hand side of the preceding equation is the rate of variation of the particle's velocity along the potential vector acting on it. Remark that $\mathbf{A}_i = \sum_j \frac{q_j\mathbf{v}_j}{cr_{ij}}$ is the potential vector (here, in Gaussian units) at the point where the charge q_i is, as obtained by a development to first order in $\mathbf{v}_i$ of the Liénard–Wiechert potentials. This solution corresponds to a retarded potential and is accurate exclusive for $\frac{v}{c} \ll 1$, oppositely, we should combine a relativistic correction. It can be easily demonstrated that the resulting relationship exists

$$q_i(\mathbf{A}_i\cdot\nabla_{r_i})\mathbf{v}_i = -q_i[\omega \times \mathbf{A}_i] \tag{2.131}$$

To reveal its complete physical meaning after a better arrangement, it is quickly seen to match a kind of gyroscopic force

$$q_i(\mathbf{A}_i\cdot\nabla_{r_i})\mathbf{v}_i = \sum_j \frac{q_iq_j}{cr_{ij}}[\omega \times \mathbf{v}_j] \tag{2.132}$$

Inserting Eq. 2.132 into Eq. 2.130 the total force acting over the *ith* particle can be written in the final form

$$\begin{aligned}\mathbf{f}_i &= \mathbf{f}_i^M + \mathbf{f}_i^L - 2\sum_j m_{ij}^{em}[\omega \times \mathbf{v}_j] \\ &- \frac{1}{2}\nabla_{r_i}\left(\frac{J_i^2}{I_i}\right) + \nabla_{r_i}(\omega\cdot\mathbf{J}_i) + \mathbf{f}_i^S\end{aligned} \tag{2.133}$$

We have included in the latter equation the electromagnetic mass of the system emerging from the interaction among charges q_i and q_j distant r_{ij} apart

$$m_{ij}^{em} = \frac{1}{2}\frac{q_i}{c}\frac{q_j}{r_{ij}} \tag{2.134}$$

The third term on the RHS – a variety of gyroscopic forces – is unexpected and demands additional investigation. It can point to uncovering new phenomena, such as levitation, because it suggests that the interaction between a system of charges with different signs leads to a mass increase, whereas equal sign charged particles when interacting lead to a mass decrease. An effect similar was introduced by Boyer [9] when examining the change in weight connected with the electrostatic potential energy for a system of two-point charges supported side by side against a weak downwards gravitation field.

Finally, the fourth term on the RHS is closely related to the Einstein-de Haas effect. There is a natural relationship between $\mathbf{J}_i$ and the Ampèrian current constituting part of the material medium and the magnetization vector (in special, $\mu = g\frac{e\mathbf{J}}{2mc}$), and this result in a body force. The fifth term on the RHS is Stern-Gerlach's force when applied to a magnetic moment.

The last term on the RHS, $\mathbf{f}_i^S$, was introduced *mutatis mutandis* to represent nonequilibrium processes, such as bremsstrahlung radiation, line radiation, turbulence, *inter alia*. Conjointly, to replicate fully nonequilibrium processes, other terms should be added, such as the pressure gradients ∇p as well the dissipative terms related to the momentum gain of the ion fluid induced by collisions with electrons – as admitted in [10], they are of the form $\mathbf{P}_{ei} = \eta e^2 n^2 (\mathbf{v}_e - \mathbf{v}_i)$, where η is the specific resistivity (in general, a tensor), but in particular to an isotropic plasma medium it remains a scalar. When seeking a more complete description, and transposing the equations from the discrete to the continuum level, the magnetohydrodynamics fluid equations of motion should be recovered.

We don't attempt to derive Maxwell equations from the actual framework for they are inherently relativistic equations responding to Lorentz transformations and a truly accurate definition of time and space should be addressed.

Example 2.4: *Electromotive force and cell potential*
With the introduction of the concepts of anode, cathode, electrode, electrolyte, and ion, in 1835 by Michael Faraday, it was possible to describe phenomena and give explanations in electrochemistry.

The natural inclusion of entropy (or free energy) inside the fundamental equation of dynamics opens the door to treating a more general class of problems, particularly those related to electrochemistry. Taking the equation of force along a closed curve γ, *with* $\oint_\gamma (\mathbf{F}^{ext} \cdot d\mathbf{s}) = q\mathcal{F}\mathcal{E}$, *and taking additionally,* $\Delta G = -q\mathcal{F}\mathcal{E}$, *according to the laws of electrolysis, and from Maxwell's equations* $\Delta S = -\left(\frac{\partial \Delta G}{\partial T}\right)_p$, *then, we obtain the Helmholtz formula for the electromotive force of a battery (or cell potential* $\mathcal{E}$*)*

$$\mathcal{E} = U + T\frac{\partial \mathcal{E}}{\partial T} \tag{2.135}$$

The consistent integration of dynamics with thermodynamics permits addressing electrochemical processes via the modified Newton equation.

Such a formulation suggests that the external forces do not dictate the fate of a physical system *per se*, as in fact, due to entropy (and information entropy, via Shannon's theorem) related phenomena, the physical system can organize itself and can tend toward novel behavioral patterns. Here I also wish to emphasize that the action-to-reaction law needs a reinterpretation, different from the deprecated form as appears in standard textbooks. The reaction follows the action in a time lag that depends on the entropy (or information) gradients; the release from constraints follows the tendency toward a new arrangement of the parts of the system.

2.4 An Overview of Other Electromagnetic Theories

> *That one body may set upon another at a distance, through vacuum, without mediation of anything else by and through which their action may be conveyed from one to another, is to me so great an absurdity that I believe no man, who has in philosophical matters a competent faculty of thinking, can ever fall into it.*
>
> –*Newton*

We intend to give here a brief description of the mainstream ideas to lead to the foundations of electrodynamics. As electromagnetism is at the base of the new forms of propulsion it is worth having an embracing view of this vast subject. Nevertheless, due to the complexity of the subject and the large contributions given by outstanding people to this area, we do not intend to be complete.

We start by distinguishing between action-at-a-distance and field theories, as they were generally understood in the 19th century. Action at a distance means that force emanated from one particle of matter acting on another particle at a distance. In the 19th century, there were two kinds of force fields: force fields and ether fields. Depending on whether the force was distributed in space or whether the force was carried by a medium or ether. An example of a force field is the theory of lines of force due to Faraday.

Laplace and Poisson formalized action-at-a-distance theories after followed by Gauss, Wilhelm Weber, and Helmholtz.

In 1600 William Gilbert, from Colchester, England, published the book *De Magnete*. Gilbert was the court physician and physicist to Queen Elizabeth I and sustained that the Earth was a big magnet, wrote about magnetic declination[11] and magnetic inclination.[12] He was firmly convinced of the power of scientific method and experimentation:

> In the discovery of secret things and in the investigation of hidden causes, stronger reasons are obtained from sure experiments and demonstrated arguments than from probable conjectures and the opinions of philosophical speculators of the common sort; therefore to the end that the noble substance of that great loadstone, our common mother (the Earth) still quite unknown, and also the forces extraordinary and exalted of this globe may the better be understood, we have decided first to begin with the common stony and ferruginous matter, and magnetic bodies, and the parts of the earth that we may handle and may perceive with the senses; then to penetrate to the inner parts of the earth.
>
> –*De Magnete*

Robert Boyle, an Irish amateur scientist, was the first to publish a book about electricity. He was the first to notice that amber (a fossilized resin), through friction with silk, attracted small fragments of paper (producing some light), similar to what a magnet does with pieces of iron. At the end of the 16th century, the study of magnetism acquired a scientific character.

In 1752, Benjamin Franklin completed the famous experiment with a paper kite, a wire, and a key, proving that lightning was an electric spark.

2.4.1 The Experiment of Oersted

In 1820, Hans Christian Oersted demonstrated the effect of an electrical current on a magnet.[13] This experiment represents one of the most significant scientific discoveries of all time.

With this new phenomenon discovered, Biot,[14] Savart, Ampère,[15] encouraged by Laplace, started experimentally investigating to quantify the effect of a conducting wire on a magnet. Ampère saw in Oersted's experiment evidence of the existence of thin electrical currents turning around molecules in the interior of matter. With this idea in mind, he started a series of experiments to check out if two electrical conductors affected each other in the same way as it does a single wire on a magnet. At first, he succeeded to show that two straight parallel wires attract or repel each other whenever the current in both wires has the same or opposing directions. Subsequently, Ampère succeeded to show that whenever a helicoidal wire (which he later called a solenoid) is crossed by a current, it develops a north and south pole in a similar way as a magnet.

Ampère succeeded in a short time to discover the first empirical laws of a new science he named *electrodynamics*.

To find the fundamental laws he imagines the action of each other of *current elements* (small portions of current flowed wire) of two nearby wires crossed by the current. His results evidenced not only an inverse square dependence on the distance between the current elements but as well on the direction of these current elements.

2.4.2 The Method of Hypothesis – Introduced by Ampère

Ampère avoids empiricism, the idea that through experiments all fundamental laws could be retrieved. However, the thought is that hypothesis of the existence of very small electrical currents circling (still) invisible molecules could only be formulated by theoretical thinking.

2.4.3 Ampère's Law

The concept of current elements proved to be very fruitful. Ampère questioned which magnitude would attain the force between two nearby current elements, whenever they are positioned not only in parallel side-by-side, but also in whatever position, in particular, along the line of the other current element. Let lambda be the ratio of the force exerted when both current elements are parallel to the force, when they are longitudinal. It is not simple to measure the longitudinal force between adjacent elements in a wire, that is why Ampère used his method of hypothesis. In his second equilibrium experiment, Ampère adduces as an experimental fact [11]:

> ... that an infinitely small portion of electric element exerts no action on another infinitely small portion of current situated in a plane that passes through its midpoint, and which is perpendicular to its direction.

2.4.4 The Experiment of Ampère

One of the experimental pieces of evidence found by Ampère was that a current element of a line perpendicular to a plane exerts no action on any current element sitting in the plane.

The second experiment conducted by Ampère with two parallel wires showed that the wires attract each other when the current flows in the same direction, but repel each other when the current flow in the opposite direction. From this experiment Ampère defines *de force* between current elements as dependent on the current elements' length, their respective intensities, and their relative position, that is, the distance r connecting their midpoints and the angles θ and θ' which they form with the line r. This led him to assume that the force between two different elements should be of the form

$$\frac{ii'dsds'}{r^n}\phi(\theta, \theta') \tag{2.136}$$

$\phi(\theta, \theta')$ denotes one unknown function to be determined from the experiment. Thus, two unknowns are to be determined: n and ϕ. The experiment so far indicates that the current elements draw into a plan, that is, the longitudinal $ds \cos\theta$, $ds' \cos\theta'$ components and the parallel components, $ds \sin\theta$, $ds' \sin\theta'$ are such that, from the second equilibrium experiment, the force between one longitudinal and parallel pair is zero. Therefore, the force is exerted between two pairs, either parallel or longitudinal

$$\frac{ii'dsds' \sin\theta \sin\theta'}{r^n} \tag{2.137}$$

and

$$\frac{ii'dsds'k \cos\theta \cos\theta'}{r^n} \tag{2.138}$$

k is the ratio of the parallel to the longitudinal force, taking the parallel force as the unity. The total force is obtained by summing up the two terms

$$\frac{ii'dsds'}{r^n}(\sin\theta \sin\theta' + k \cos\theta \cos\theta') \tag{2.139}$$

We deduced this formula by assuming the current elements to be in a plane; in the general case of two pairs lying in two planes whose angle with each other is ω was deduced by Ampère

$$\frac{ii'dsds'}{r^n}(\sin\theta \sin\theta' \cos\omega + k \cos\theta \cos\theta') \tag{2.140}$$

Through two additional equilibrium experiments, Ampère determined that $n = 2$ and $k = -1/2$.

2.4.5 Weber's Law

In 1846, Weber enunciated a law for the force acting at-a-distance between electrically charged particles e and e' that depend not only on the distance r between them but also on their relative velocity v and relative acceleration a

$$F = \frac{ee'}{r^2}\left(1 - \frac{1}{c^2}v^2 + \frac{2a}{cs}\right) \tag{2.141}$$

The constant introduced was in 1857 shown by Weber and Kohlraush[16] to be approximately $\sqrt{2}$ times the velocity of light.

Weber nurtured the idea that Eq. 2.141 was the core of an incomplete theory that would unify all-natural phenomena under the single law of force. In 1848 he succeeded to demonstrate that this force could be derived from a potential. The existence of this potential guaranteed conservation of *vis viva.*

It is worth mentioning that Ampère builds up the ponderomotive force between two abstract elements of current that would conserve *vis viva*. Weber and Fechner, following the French conception, build up the force between point atoms of the two electrical fluids which were supposed to constitute the electrical current. In the end, the net force would result from four interactions between those atoms. Weber expected his formula to describe all major phenomena of electricity and magnetism through the three terms that constitute his law of force:

1. The first term describes Coulomb inverse square the electrostatic force between two electrically charged particles: attractive between unlike charges, and repulsive between like particles.
2. Electromagnetic force between constant velocity currents, a manifestation of Ampère law.
3. Electromotive force induced between accelerating currents, a manifestation of Faraday's law.

Ampère's electrodynamic law is valid only in moving currents in fixed conductors. Besides that, the electrostatic force stood separately from it. Weber believed that the phenomena of induction should be integrated into the electrostatic and electrodynamic laws.

Gustav Fechner[17] suggested an interesting conception. The flow of electricity could consist of oppositely charged particles moving through the conductor with the same velocity but in opposite directions. That is, any short segment of wire could have a negative and positive charged particles speeding past one another. The modern view of the charge transport consists of a positive charged particle practically immobile and negative particle moving slowly. The interaction between two current elements involved four interactions:

1. Between +q and +q′
2. Between +q and −q′
3. Between −q and −q′
4. Between −q and +q′

Weber established the following:

1. Two current elements lying on the same straight line either repel or attract each other, depending on whether currents flow in the same or opposite directions.
2. Two parallel current elements perpendicular to the line joining them either attract or repel each other, depending on whether their currents flow in the same or opposite directions.

3. A current element, which lies on a straight line with a wire element induces a similar directed or opposed current, depending on whether its own current intensity decreases or increases.

From these assumptions, Weber established a theorem: the electrostatic force must be reduced when the electrical particles are in relative motion. The relative velocity between two particles is represented by dr/dt if r denotes the relative position between them. As this force should act when they are approaching or receding, Weber uses $(dr/dt)^2$. His theorem is expressed in the form

$$F = \frac{ee'}{r^2}\left(1 - a^2\frac{dr^2}{dt^2}\right) \tag{2.142}$$

Two parallel elements exert a force, as it was experimentally shown by Ampère. Using the Fechner hypothesis it can be noticed that when two parallel elements cross each other their relative velocity changes from negative (when they are approaching) to zero (when they are directly opposite to each other) and finally, becomes positive (when they are receding from each other). As we see, when they cross each other, their relative velocity goes from negative to positive and this means that they have a relative acceleration. The electrostatic force must be increased by a relative acceleration and Weber added a new term to the electrostatic force

$$F = \frac{ee'}{r^2}\left(1 - a^2\frac{dr^2}{dt^2} + b\frac{d^2r}{dt^2}\right) \tag{2.143}$$

where b is a constant to be determined. In fact, Weber showed that

$$\frac{d^2r}{dt^2} = \frac{1}{r}\frac{dr^2}{dt^2} \tag{2.144}$$

Eq. 2.141 then becomes

$$F = \frac{ee'}{r^2}\left(1 - a^2\frac{dr^2}{dt^2} + \frac{b}{r}\frac{dr^2}{dt^2}\right) \tag{2.145}$$

To find the relationship between a and b, Weber returned to Ampère's previous results, namely, that the ratio between the second and third elements is nothing other than the ratio between the parallel and longitudinal elements, that is, $1/2$. Hence,

$$b = 2a^2r \tag{2.146}$$

and the force is rewritten as

$$F = \frac{ee'}{r^2}\left(1 - a^2\frac{dr^2}{dt^2} + 2a^2r\frac{d^2r}{dt^2}\right) \tag{2.147}$$

By considering the mutual interaction between each of the four pairs of current elements, Weber divided the constant a by 4, determining the following general expression for the force:

$$F = \frac{ee'}{r^2}\left(1 - \frac{a^2}{16}\frac{dr^2}{dt^2} + \frac{a^2}{8}r\frac{d^2r}{dt^2}\right) \tag{2.148}$$

Weber and Kohlrausch in 1855 determined experimentally the value of a and found that $4/a = 4.395 \times 10^{11}$ mm/s, a value known as the Weber constant. This value will be denoted by c. It is an interesting explanation of its meaning by Weber: "that relative velocity which electrical masses e and e' have and must retain, if they are not to act on each other at all." Weber fundamental law is now written as

$$F = \frac{ee'}{r^2}\left(1 - \frac{1}{c^2}\frac{dr^2}{dt^2} + \frac{2r}{c^2}\frac{d^2r}{dt^2}\right) \tag{2.149}$$

2.4.5.1 "Catalytic Forces" and the Fundamental Length

In 1846 Weber conceived that the two electrical fluids formed a neutral ether in which ponderable molecules were immersed. His project consisted in explaining the wave theory of light on the basis of his force law and on the oscillations of this ether. The former discovery of diamagnetism by Faraday in 1845 was perfectly explained by his theory as induction of Amperian double currents in the neutral ether surrounding molecules. Diamagnetism is a phenomenon consisting of the induction in a nonmagnetic body, when subjected to a strong magnetic field, of a magnetic polarity opposite to the one normally induced in normal magnetic substances called paramagnetic.

By 1880, Weber had developed an advanced electrical theory of matter, which included gravitation. The central notion was the pairwise relation. In 1846, Weber thought that the acceleration-dependent term in the law of force might indicate two atoms interact through the ether or even, through Newton's second law of motion, it could point to the action of a third particle on the first two. Berzelius[18] named such forces catalytic forces, meaning that Weber's force law might contain the reduction of chemical to electrical action.

The potential, which by taking the gradient gives Weber force law, is

$$V = \frac{ee'}{r}\left(\frac{v^2}{c^2} - 1\right) \tag{2.150}$$

Weber soon realized that potential energy has an extra physical meaning. After him:

> In many respects one can speak with more justification of the physical existence of work, expressed through the potential, then of the physical existence of force, of which one can only say that it seeks to change physical relations of bodies.

In 1852, Weber advanced a model for resistance to electrical conduction that was a premonitory sign of the modern atomic concept. If free ether consisted of neutral pairs of positive and negative particles, as Weber thought, then the particles of each pair should turn around each other, constituting what Weber called the *atom pairs*. With this heuristic model in mind, he could explain how electric current could flow inside a wire when submitted to an applied electromotive force: the orbiting pairs would undergo successive dissociation and recombination into new pairs, thereby resulting in the opposite motion of positive and negative particles. Difficulties with this model are evident since it did not explain the different densities of ether in different materials.

In 1862, Weber worked on the problem of converting work done to produce electrical motion in a current to heat. Instead of the previous explanation of two particles turning around in smaller and larger circles, Weber thought that molecules should have one kind of electricity (e.g., negative) in their mass center, and the positive charge should move around. To move one particle from one atom pair to another should need an applied electromotive force. This motion should appear as the heating of ponderable matter.

2.4.5.2 Nuclear Forces Dependent on Velocity and Weber's Electrodynamics

Velocity-dependent forces have widespread use in nuclear physics. The field formulation has many advantages, especially when it is considered a system with a large number of particles. But, when only two particles are interacting, instead of the Lorentz force (which is manifestly noncentral and not obeying Newton's third law) more insight can be obtained when use is made of the force postulated by Weber

$$\mathbf{F}_{12} = q_1 q_2 \left\{ \frac{1}{r^2} + \frac{1}{c^2} \left[\frac{1}{r} \frac{d^2 r}{dt^2} - \frac{1}{2r^2} \left(\frac{dr}{dt} \right)^2 \right] \right\} \mathbf{e}_r \tag{2.151}$$

2.4.6 Neumann's Principle of Electrodynamics

In 1868, Carl Neumann published his work entitled *Die Principien der Elektrodynamik*, addressing particularly the relationship between Weber's law and the conservation of *vis viva* [12]. To Neumann, there was a hindrance with the usual formulation of the *vis viva*

$$\frac{1}{2} \sum_i^N m_i v_i^2 = U + k \tag{2.152}$$

where it is assumed for the conservative forces $\mathbf{F} = -\nabla U$ (k is a constant), but no clear method existed for velocity-dependent forces (kind of entropic forces). So, Neumann hypothesized that potentials propagate with a large finite velocity and he distinguished two kinds of potentials: the emitted and the received potential. According to him, it was the received potential that acts on point electric charges. The emitted potential by two electric charges q and q_1 at distance r and at time $t - \Delta t$ was assumed similar to the Newtonian potential but with retardation, which we may write in the form

$$\phi(r - \Delta r) = \frac{q q_1}{r - \Delta r} \tag{2.153}$$

This emitted potential is received at time t when the distance between charged particles is r and the received potential is U after the interval of time $\Delta t = \frac{r}{c}$. As r is a function of time, $r = f(t)$, then $r - \Delta r = f(t - \Delta t)$, and $\dot{r} = f'(t)$. Expanding in a Taylor series

$$\begin{aligned} r - \Delta r &= r - \Delta t \dot{r} + \left[\frac{(\Delta t)^2}{2} \ddot{r} \right] + \ldots \\ &\approx r - \frac{r}{c} \dot{r} + \frac{r^2}{2c^2} \ddot{r} \end{aligned} \tag{2.154}$$

Hence, the *received potential* at time t is given by

$$U(r,\dot{r}) = \phi(r-\Delta r) = \frac{qq_1}{r - \frac{r}{c}\dot{r} + \frac{r^2}{2c^2}\ddot{r}} \tag{2.155}$$

Repeating the Taylor series for the denominator, the following is obtained:

$$U = \phi(r) - \frac{r}{c}\dot{r}\phi'(r) + \frac{r^2}{2c^2}\ddot{r}\phi'(r) + \frac{r^2}{2c^2}(\dot{r})^2\phi''(r) \tag{2.156}$$

From Eq. 2.156, the U potential can be written as the sum of two functions, one of them, an exact differential with respect to t:

$$U = V_1 + V_2 \tag{2.157}$$

where

$$V_1 = qq_1\left[\phi + \frac{r^2\phi''}{2c^2}\left(\frac{dr}{dt}\right)^2 - \frac{(r^2\phi')'}{2c^2}\left(\frac{dr}{dt}\right)^2\right] \tag{2.158}$$

and

$$V_2 = qq_1\frac{d}{dt}\left(\frac{r^2\phi'}{2c^2}\frac{dr}{dt} - \frac{\int r\phi' dr}{c}\right) \tag{2.159}$$

V_1 is called by Neumann the effective potential and V_2 the ineffective potential (so named because when performing Hamilton's principle, the variations of V_2 disappears). Substituting $\phi = \frac{1}{r}$ into the aforementioned potentials, it can be obtained the new forms of the propagated Newtonian potential

$$V_1 = \frac{qq_1}{r}\left[1 + \frac{1}{c^2}\left(\frac{dr}{dt}\right)^2\right] \tag{2.160}$$

and

$$V_2 = qq_1\left(\frac{\log r}{c} - \frac{1}{2c^2}\frac{dr}{dt}\right) \tag{2.161}$$

Applying Hamilton's principle

$$\delta\int_{t_1}^{t_2}(T-U)dt = 0 \tag{2.162}$$

where T is the kinetic energy, U is the received potential, it is found that the V_2 term vanishes, leaving us with

$$\delta \int_{t_1}^{t_2} (T - V_1)dt = 0 \tag{2.163}$$

and, therefore, with the Euler–Lagrange equation for the r variable

$$\frac{d}{dt}\left(\frac{\partial V_1}{\partial \dot{r}}\right) - \frac{\partial V_1}{\partial r} = F_r \tag{2.164}$$

giving the force

$$F_r = \frac{qq_1}{r^2}\left[1 - \frac{1}{c^2}\left(\frac{r}{dt}\right)^2 + \frac{2r}{c^2}\left(\frac{d^2r}{dt^2}\right)\right] \tag{2.165}$$

similar to Weber's law. Then, Neumann separates two components of the effective potential V_1, the static potential $V_{11} = \frac{qq_1}{r}$, and the moor potential $V_{12} = qq_1\left(\frac{d\psi}{dt}\right)^2$, where $\psi = \frac{2}{c}\sqrt{r}$. In doing so, Neumann was able to obtain a conservation law for velocity-dependent interactions

$$T + V_{11} - V_{12} = \text{constant} \tag{2.166}$$

Weber did not use the concepts of electric and magnetic fields to describe electromagnetic phenomena. He postulated, instead, an (instantaneous) action-at-a-distance formulation for the (central) force between charges. Weber's electrodynamics can be utilized to tackle nuclear forces interactions. Curiously, Weber coined the term unipolar induction for Faraday's homopolar dynamo in 1839.

2.4.7 Field Theories and Action-at-a-Distance Theories

The understanding of natural phenomena in the 19th century was dominated by the notion of a pervasive ether. The source of this notion recedes Descartes, Leibniz, and Kant[19] since they introduced the idea of primitive matter. Many ethereal media existed depending on the subject, such as electric, magnetic, caloric, luminiferous and gravitational. The wave theory of light gave a strong impetus to unify them in a single medium. In fact, by postulating the ether heat and light could be incorporated into mechanics verifying the theory of conservation of *vis viva*.

2.4.8 Maxwell's Electrodynamics

The wave theory of light was widely accepted in the middle of the 19th century and with it the existence of a luminiferous ether. James Clerk Maxwell showed that the optical ether was the seat of electrical and magnetic effects. William Thomson (Lord Kelvin) showed how atoms could be nothing else but vortex motions in a medium that is called "the Universal Plenum." These primordial ideas of Maxwell and Thomson were later developed by a pleiad of scientists, among them Fitzgerald, Oliver Lodge, and Joseph Larmor.

Thomson and Maxwell saw in Faraday's lines of forces a mechanical imprint of a material medium that could provide the background for a unified theory of electromagnetism and optical phenomena. Faraday did not support the view of electric and magnetic fluids and the related action-at-a-distance related to their action. Instead, he characterized electric and magnetic phenomena by the geometrical pattern in space imprinted by them, represented by lines of forces whose directions followed the directions of forces and their spacing was an indication of their magnitudes: close spacing meaning strong force and less dense spacing meaning weak forces. Faraday succeeded in developing a rudimentary calculus of them and this image had a strong and fruitful influence on the discovery of new phenomena.

2.4.9 Helmholtz's Electrodynamics

The electrodynamics developed by James Clerk Maxwell is based on the concept of fields. During the 1870s Hermann von Helmholtz tried to develop electrodynamics based on the notion of fields (and with it the ether) and particles subject to distant sources transmitted instantaneously across space. Those were times when philosophy played an open role in the development of science.

2.4.10 The Biefeld–Brown Effect

Maxwell's theory is the basis of our civilization but some experimental observations continue to be puzzling and need an intellectual breakthrough. In the years the 1920s as Thomas Townsend Brown was doing X-rays experiments with a Coolidge tube (invented in 1913 by William D: Coolidge), Brown found that a net force was exerted on the Coolidge tube when the current was turned on. Brown applied for a British patent entitled "Method of Producing Force or Motion," issued on November 15, 1928. Later on, this discovery was named the Biefeld–Brown effect in honor of his mentor, Dr. Paul Alfred Biefeld, professor of physics and astronomy at Denison University in Granville, Ohio. In the 1920s Biefeld and Brown experimented on capacitors and later on Brown issued another patent entitled "Electrokinetics Apparatus" which made the following observations: the greatest force on the capacitor is produced when the small electrode is connected to a positive pole; the effect occurs in a dielectric medium. Brown explained the effect in terms of ionic wind but did not give a thorough explanation of the effect. In 1965, Brown issued another patent entitled "Electrokinetic Apparatus" where he reposts the existence of a net force even in a vacuum. Probably the effect which is playing any role is the electrostatic pressure. High electric field gradients can give the same effect. If only ionic wind is the main mechanism, we can expect a very small effect due to corona discharge between electrodes.

2.4.11 Vacuum and Casimir Forces

"The word force is understood by many to mean simply the tendency of a body to pass from one place to another," stated Faraday, and it was his concept that prevailed until modern times, where it was replaced by the concept of field. The concept of "engineering the vacuum" was introduced by T. D. Lee in his textbook *Particle Physics and Introduction to Field Theory* [13]. He discusses the possible properties vacuum is endowed and concludes that

> The experimental method to alter the properties of the vacuum may be called vacuum engineering ... . If indeed we are able to alter the vacuum, then we may encounter some new phenomena, totally unexpected.

Einstein's theory of relativity has reduced the vacuum or ether to a useless concept, but with the advent of quantum theory and especially quantum electrodynamics, the vacuum is the seat of energetic particles and field fluctuations. Later on, in building up the general theory of relativity, Einstein recognizes that the vacuum should have a structure, being the seat of a spacetime structure that encodes the distribution of matter and energy.

Notes

1. A deficiency corrected by SU(2) group formulation of electromagnetic field, see Ch. 7.
2. Topology studies the spatial properties of objects that are preserved under continuous deformation of objects. One of the first treatises on modern topology is Euler's 1736 paper entitled "Seven Bridges of Königsberg" dealing with a famous problem solved by Euler. "The master of us all," a title by which Euler is known among mathematicians, succeeded in creating this area of mathematics that extracts the essence of a phenomenon without taking into account the details. After all, rarely the world can be well modeled by mathematics. Euler defined topology as *geometria situs*, because it is concerned with the properties of position without taking into account their magnitude, a task undertaken by geometry that requires the notion of angle and distance, that is, a metric.
3. A good image is given by the example of two ice skaters traveling side by side. If they are initially moving parallel to each other, they will start to "repel" each other by throwing a ball back and forth.
4. Historical note about mathematical symbols: The symbols dx, dy, or dy/dx were introduced in the literature by Gottfried Wilhelm Leibniz (1646–1716), which also introduced the symbol of integral $\int$. The first derivative $f'(x)$ and so on, were introduced by Joseph Louis Lagrange (1736–1813). The partial derivative was introduced by Antoine-Nicolas Caritat, Marquis de Condorcet (1743–1794), but in the form ∂z. It was Adrien Marie Legendre in 1786 the first to propose the use of the form $\partial u/\partial x$. The limits of integration were proposed by Jean Baptiste Joseph Fourier (1768–1830). The symbol of integration around a closed curve $\oint$ was apparently first introduced by Arnold Sommerfeld (1868–1951).
5. These equations were established by James Clerk Maxwell around 1861–1864 from experiments on electric and magnetic phenomena investigated by Michael Faraday.
6. On average 10^{23} particles (electrons, atoms, ...) per cm^3 in constant motion generating fields rapidly oscillating in space and time $\mathbf{e}$ and $\mathbf{b}$ difficult to evaluate.
7. We will use here the Gaussian system of units. It is of great help to succeed easily SI (needed in engineering) and Gaussian systems of units.
8. See, e.g., Yuan Xiao-Ping, et al. "Spiral wave generation in a vortex electric field," *Chinese Physics Letters*, vol. 28, no. 10, p. 100505, 2011.
9. A manifold is a topological space that resembles locally an euclidean space.
10. The theory has now lost its Lorentz invariance while still is conform to Galilean invariance.
11. The angle between magnetic north and true north
12. The angle between a compass needle and the horizontal plane at a given point on the Earth's surface.
13. Actually it was a compass, an instrument containing a freely suspended magnetic bar, which points along the horizontal component direction of the Earth's magnetic field. It was probably invented by the Chinese, during the Qin dynasty (221–206 BC)
14. Jean-Baptiste Biot (1774–1862), was a French physicist, astronomer, and mathematician. He used first polarized light to study solutions.
15. André-Marie Ampère, (1775–1836) was the founding father of electrodynamics, in particular the discoverer of a fundamental force of electrodynamics.

16 Friedrich Wilhelm Georg Kohlrausch, an important German experimental physicist investigated the electro-conductive properties of electrolytes. He made electrical and magnetic precision measurements and contributed to extending the absolute system of Gauss and Weber to include electric and magnetic measuring units.
17 Gustav Theodor Fechner (1801–1887) was a German philosopher and physicist. He invented "psychophysics," the study of the relationship between stimulus intensity and the subjective experience of the stimulus. With Weber, he proposed the law the intensity of a sensation increases as the logarithm of the stimulus.
18 Jöns Berzelius (1779–1848) was a Swedish chemist and disciple of Dalton. Lavoisier, he is considered the father of chemistry.
19 Kant used the word Uestoff or Wehlther.

References

1. T. W. Barrett, "Topological foundations of electromagnetism," *Annales de la Fondation Louis de Broglie*, vol. 26, no. 55, pp. 55–79, 2001.
2. M. W. Evans and J. P. Vigier, *The Enigmatic Photon: Theory and Practice of the B(3) field* (in the series Fundamental Theories of Physics, 1994).
3. J. Dyson, "Freeman. Feynman's proof of the maxwell equations," *American Journal of Physics*, vol. 58, pp. 209–211, 1990.
4. R. H. Romer, *American Journal of Physics*, vol. 50, pp. 1089–1093, 1982.
5. J. Schwinger and D. S. Saxon, *Discontinuities in Waveguides - Notes on Lectures by Julian Schwinger*, 1968.
6. S. S. Chern & J. Simons, "Characteristic forms and geometric invariants," *Annals of Mathematics*, vol. 99, pp. 48–69, 1974.
7. M. J. Pinheiro and J. Loureiro. "Effective ionization coefficients and electron drift velocities in gas mixtures of sf_6 with he, xe, co_2 and n_2 from Boltzmann analysis," *Journal of Physics D: Applied Physics*, vol. 35, no. 23, p. 3077, 2002.
8. M. J. Pinheiro, "A variational method in out-of-equilibrium physical systems," *Scientific Reports*, vol. 3, pp. 1–9, 2013.
9. T. H. Boyer, "Electrostatic potential energy leading to an inertial mass change for a system of two point charges," *American Journal of Physics*, vol. 46, no. 383, pp. 383–385, 1978.
10. L. Spitzer, *Physics of Fully Ionized Gases*. New York: Interscience Publishers, 1967.
11. A.-M. Ampère, *Mémoire Sur la Théorie Mathématique des Phénomènes Electrodynamiques Uniquement Deduites de L'experience*. Paris: A. Hermann, 1883.
12. T. Archibald, "Carl Neumann versus Rudolf Clausius on the propagation of electrodynamic potentials," *American Journal of Physics*, vol. 54, pp. 786–790, 1986.
13. T. D. Lee, *Particle Physics and Introduction to Field Theory*. Boca Raton: CRC Press-Taylor and Francis Group, 2004.

3

Tensors, Spinors, and Higher Representations of the Electromagnetic Field

NOTATIONS

4-dimensional quantities
Tensoriels indices are 0,1,2,3
The metric has the signature (+−−−)
The 4-vectors are noted $A^i = (A^0, \mathbf{A})$

3.1 Tensorial Calculus, Differential Forms, and Spinorial Calculus

3.1.1 Introduction to Tensorial Calculus

3.1.1.1 Historical Introduction

In 1693 Gottfried Wilhelm Leibniz provided the first account on this subject in his correspondence with Marquis De L'Hôspital, giving awkward statements on how to eliminate the unknowns from systems of linear equations. In 1750, Gabriel Cramer provided for the first time a comprehensible formulation and general definitions of the functions. The name "determinant" was given by Gauss, with Vandermonde, Lagrange, Cauchy, and Jacobi, among others, among those who laid the foundations. Considering the pervasive presence of determinants in fundamental aspects of mathematics, we may say that mathematical physics is a kind of chess board game.

3.1.2 Origin of Determinants

We denote determinants with qth column and nth row by the notation

$$|d_{qr}| \equiv \begin{bmatrix} d_{11} & d_{21} & \dots & d_{m1} \\ d_{12} & d_{22} & \dots & d_{m2} \\ \vdots & \vdots & \vdots & \vdots \\ d_{1m} & d_{2m} & \dots & d_{mm} \end{bmatrix} \tag{3.1}$$

where it is verified $|d_{qr}| = |d_{rq}|$.

The main properties which obey this class of functions are set out in the next section.

DOI: 10.1201/9781003285083-3

3.1.2.1 Matrix Characteristics

As we have seen, a column matrix is a vector and a square matrix defines transformations over vectors. Essentially a matrix is a table.

Let us consider a linear transformation, that changes point M to N in such a way that vector $\vec{U} \equiv \overrightarrow{OM}$ transform to vector $\vec{V} \equiv \overrightarrow{ON}$

$$\begin{aligned} a_{11}u_1 + a_{12}u_2 &= v_1 \\ a_{21}u_1 + a_{22}u_2 &= v_2 \end{aligned} \tag{3.2}$$

We may write in matrix notation

$$\begin{pmatrix} a_{11} & a_{12} \\ a_{21} & a_{22} \end{pmatrix} \begin{pmatrix} u_1 \\ u_2 \end{pmatrix} = \begin{pmatrix} v_1 \\ v_2 \end{pmatrix} \tag{3.3}$$

that can be written in a more compact form

$$au = v$$

or, in higher dimensions,

$$\begin{pmatrix} a_{11} & a_{12} & \dots & a_{1n} \\ a_{21} & a_{22} & \dots & a_{2n} \\ \cdot & \cdot & \dots & \cdot \\ a_{n1} & a_{n2} & \dots & a_{nn} \end{pmatrix} \begin{pmatrix} u_1 \\ u_2 \\ \cdot \\ u_n \end{pmatrix} = \begin{pmatrix} v_1 \\ v_2 \\ \\ v_n \end{pmatrix} \tag{3.4}$$

Example 3.1: *If we have*

$$\begin{pmatrix} \cos\theta & -\sin\theta \\ \sin\theta & \cos\theta \end{pmatrix} \begin{pmatrix} u_1 \\ u_2 \end{pmatrix} = \begin{pmatrix} v_1 \\ v_2 \end{pmatrix} \tag{3.5}$$

This represents a **rotation**.

Example 3.2: *If we have*

$$\begin{pmatrix} K & 0 \\ 0 & K \end{pmatrix} \begin{pmatrix} u_1 \\ u_2 \end{pmatrix} = \begin{pmatrix} v_1 \\ v_2 \end{pmatrix} \tag{3.6}$$

In this case, this transformation is a **homothety**.

3.1.2.2 Eigenvectors and Eigenvalues of a Transformation Matrix

Let us consider a vector $[U] \neq 0$ to which we apply a linear transformation represented by a matrix $[a]$. If the result of this transformation is another matrix $[V]$, such as

$$[V] = \lambda [U] \tag{3.7}$$

where λ is a scalar, we say that $[U]$ is an **eigenvector**[1] of $[a]$ and λ is an **eigenvalue** of $[a]$.

Example 3.3:

$$\begin{pmatrix} v_1 \\ v_2 \end{pmatrix} = \begin{pmatrix} 3 & 0 \\ 0 & -2 \end{pmatrix} \begin{pmatrix} u_1 \\ u_2 \end{pmatrix} = \begin{pmatrix} 3u_1 \\ -2u_2 \end{pmatrix} \tag{3.8}$$

We see that $0x_1$ *is an eigenvector of* $\begin{pmatrix} 3 & 0 \\ 0 & -2 \end{pmatrix}$ *and 3 its eigenvalue; also,* $0x_2$ *is an eigenvector of the transformation matrix with eigenvalue* -2.

Example 3.4:

$$\begin{pmatrix} v_1 \\ v_2 \end{pmatrix} = \begin{pmatrix} 5 & -9 \\ 1 & -5 \end{pmatrix} \begin{pmatrix} u_1 \\ u_2 \end{pmatrix} \tag{3.9}$$

If λ *is an eigenvalue, we may write*

$$\begin{pmatrix} 5 & -9 \\ 1 & -5 \end{pmatrix} \begin{pmatrix} u_1 \\ u_2 \end{pmatrix} = \lambda \begin{pmatrix} u_1 \\ u_2 \end{pmatrix} \Rightarrow \begin{cases} 5u_1 - 9u_2 &= \lambda u_1 \\ u_2 - 5u_2 &= \lambda u_2 \end{cases} \tag{3.10}$$

or

$$\begin{cases} (5-\lambda)u_1 - 9u_2 &= 0 \\ u_1 - (5+\lambda)u_2 &= 0 \end{cases}$$

This system of algebraic equations has solutions $\neq 0$ *if*

$$\begin{vmatrix} 5-\lambda & -9 \\ 1 & -(5+\lambda) \end{vmatrix} = 0$$

with solutions

$$\lambda^2 - 16 = 0 \Rightarrow \lambda_1 = 4, \quad \lambda_2 = -4$$

Let us first search for the eigenvector associated to $\lambda_1 = 4$. *We obtain*

$$\begin{cases} u_1 - 9u_2 &= 0 \\ u_1 - 9u_2 &= 0 \end{cases}$$

We can choose arbitrarily $u_2 = 0 \Rightarrow u_1 = 9a$ *and hence, the vector* $[A] = \begin{pmatrix} u_1 \\ u_2 \end{pmatrix} = a\begin{matrix} 9 \\ 1 \end{matrix}$ *is an eigenvector with eigenvalue* $\lambda_1 = 4$.

For the other eigenvalue $\lambda_2 = -4$ *we must solve*

$$\begin{cases} 9u_1 - 9u_2 &= 0 \\ u_1 - u_2 &= 0 \end{cases}$$

Using the same procedure as before, let us put arbitrarily $u_2 = b \Rightarrow u_1 = b$ *and then* $[B] = \begin{pmatrix} b \\ b \end{pmatrix} = b\begin{matrix} 1 \\ 1 \end{matrix}$ *is the eigenvector that corresponds to the eigenvalue* $\lambda_2 = -4$.

3.1.2.3 Application to the Special Theory of Relativity

The theory of linear transformations on vector spaces plays a fundamental role in most areas of mathematical physics.

Consider a vector x in the plane $\mathbf{R}^2$ written in terms of its x^1 and x^2 components

$$x = x^1 e_1 + x^2 e_2 = x^i e_i \tag{3.11}$$

The vectors e_1 and e_2 form the **basis** of the linear vector space $\mathbf{R}^2$, and naturally the components of the vector x are determined by the choice we make of the basis.[2]

The set of coordinates of an event is (t, x, y, z) and it is useful to consider them as four coordinates (in a 4-dimensional space) $x^i = (x^0, x^1, x^2, x^3)$, with $i = 0, 1, 2, 3$

$$x^0 = ct \quad x^1 = x \quad x^2 = y \quad x^3 = z \tag{3.12}$$

Time appears in a purely formal manner as the fourth coordinate. We have

$$(x^0)^2 - (x^1)^2 - (x^2)^2 - (x^3)^2 \tag{3.13}$$

It is generally called **4-vector** a collection of four quantities A^0, A^1, A^2, and A^3 which transform themselves in a transformation of coordinates as the coordinates of the 4-position; (A^0, A^1, A^2, A^3) transforms like (x^0, x^1, x^2, x^3). In particular, in a Lorentz transformation

$$A^0 = \frac{A^{0'} + \frac{v}{c}A^{1'}}{\sqrt{1 - \frac{v^2}{c^2}}} \quad A^1 = \frac{A^{1'} + \frac{v}{c}A^{0'}}{\sqrt{1 - \frac{v^2}{c^2}}} \quad A^2 = A^{2'} \quad A^3 = A^{3'} \tag{3.14}$$

We thus have

$$(A^0)^2 - (A^1)^2 - (A^2)^2 - (A^3)^2 \tag{3.15}$$

We also have the following relationship between the contravariant components of a 4-vector A^i and the covariant components A_i

$$A_0 = A^0 \quad A_1 = -A^1 \quad A_2 = -A^2 \quad A_3 = -A^3 \tag{3.16}$$

The square of the 4-vector is

$$\sum_{i=0}^{3} A^i A_i = A^i A_i = A^0 A_0 + A^1 A_1 + A^2 A_2 + A^3 A_3 \tag{3.17}$$

In a Lorentz transformation we have

$$\begin{aligned} x^{\alpha} \rightarrow x^{\alpha} &= \Lambda^{\alpha}_{\beta} x^{\beta} \\ A^{\alpha} &= \Lambda^{\alpha}_{\beta} A^{\beta} \end{aligned} \tag{3.18}$$

3.1.3 Tangent Vectors and Mappings

3.1.4 Contravariant Tensors

An example of a contravariant tensor is a vector formed by taking the total differential of a variable in one coordinate system concerning the variables in another coordinate system. A tangent vector, or contravariant vector (or tensor) at the point $p_0 \in M^n$, denoted by $\mathbf{X}$, assigns to each coordinate patch (U, x) holding p_0, an n-tuple of real numbers[3]

$$(X^i_U) = (X^1_U, \ldots, X^n_U) \tag{3.19}$$

such that if $p_0 \in U \cap V$, then

$$X^i_V = \sum_j \left[\frac{\partial x^i_V}{\partial x^j_U}(p_0) \right] X^j_U \tag{3.20}$$

The prototype of a (contravariant) vector is the flow velocity of a particle.

Example 3.5: *Let us consider the relationship between the cartesian coordinates of a point with its polar coordinates*

$$x = r \cos\theta \quad y = r \sin\theta \tag{3.21}$$

The differential dx in terms of r and θ is a contravariant tensor of rank-1.

$$dx = \frac{\partial x}{\partial r} dr + \frac{\partial x}{\partial \theta} d\theta \tag{3.22}$$

where

$$\begin{aligned} \frac{\partial x}{\partial r} &= \cos\theta \\ \frac{\partial x}{\partial \theta} &= -r\sin\theta \end{aligned} \tag{3.23}$$

If we divide Eq. 3.22 *by the time differential dt, we obtain the x component of the velocity vector*

$$\frac{dx}{dt} = \frac{\partial x}{\partial r}\frac{dr}{dt} + \frac{\partial \theta}{\partial \theta}\frac{\theta}{dt} \tag{3.24}$$

and, accordingly

$$\frac{dr}{dt} = \frac{\partial r}{\partial x}\frac{dx}{dt} + \frac{\partial r}{\partial y}\frac{dy}{dt} \tag{3.25}$$

We can condensate the writing of the aforementioned equation in tensor notation of a **contravariant tensor**, *rank 1, under the form*

$$B^i = \frac{\partial y^i}{\partial x^\alpha} A^\alpha \tag{3.26}$$

Hence, using the example above, we may put

$$B^i = \frac{dx}{dt} \tag{3.27}$$

and $x^1 = x$, $x^2 = y$ *and* $A^1 = dx/dt$, $A^2 = dy/dt$, *then, the contravariant law for tensors, Rank 1,* Eq. 3.26 *is written in a detailed form as*

$$\begin{aligned} B^1 = \frac{dr}{dt} &= \frac{\partial y^1}{\partial x^1} A^1 + \frac{\partial y^1}{\partial x^2} A^2 = \cos\theta \frac{dx}{dt} + \sin\theta \frac{dy}{dt} \\ B^2 = \frac{d\theta}{dt} &= \frac{\partial \theta}{\partial x} A^1 + \frac{\partial \theta}{\partial y} A^2 = -\frac{1}{r\sin\theta}\frac{dx}{dt} + \frac{1}{r\cos\theta}\frac{dy}{dt} \end{aligned} \tag{3.28}$$

3.1.5 Covariant Tensors

The prototype of a covariant vector is the gradient of a function. We will give next the example of the gravitational potential Φ in cartesian coordinates and polar coordinates. The gravitational potential is given by

$$\Phi = -\frac{GMm}{r} \tag{3.29}$$

with $r = \sqrt{x^2 + y^2}$ and $\theta = \arctan\frac{y}{x}$. The force along the Ox direction is given by

$$\begin{aligned} F_x &= -\frac{\partial \Phi}{\partial x} \\ \frac{\partial \Phi}{\partial x} &= \frac{\partial \Phi}{\partial r}\frac{\partial r}{\partial x} + \frac{\partial \Phi}{\partial \theta}\frac{\partial \theta}{\partial x} \end{aligned} \tag{3.30}$$

Hence, putting

$$\begin{aligned} x^1 = r \quad y^1 = x \\ x^2 = \theta \quad y^2 = y \end{aligned} \tag{3.31}$$

We notice that we can write

$$B_1 = \frac{\partial \Phi}{\partial x} = \frac{\partial r}{\partial x}\frac{\partial \Phi}{\partial r} + \frac{\partial \theta}{\partial x}\frac{\partial \Phi}{\partial \theta} = \frac{\partial x^1}{\partial y^1} A_1 + \frac{\partial x^2}{\partial x^1} A_2 \tag{3.32}$$

and

$$B_2 = \frac{\partial \Phi}{\partial y} = \frac{\partial r}{\partial y}\frac{\partial \Phi}{\partial r} + \frac{\partial \theta}{\partial y}\frac{\partial \Phi}{\partial \theta} = \frac{\partial x^1}{\partial y^2}A_1 + \frac{\partial x^2}{\partial y^2}A_2 \tag{3.33}$$

With the above example, we see that in tensor notation a covariant tensor of Rank 1 can be expressed under the form

$$B_i = \frac{\partial x^{(\alpha)}}{\partial y^i}A_{(\alpha)} \tag{3.34}$$

3.1.6 Higher Ranks and Mixed Tensors

3.1.7 Space Curves

A curve C in space can be described by the vector $\mathbf{r}(u)$ joining the origin O of a given coordinate system to a point on the curve, see Figure 3.1. The parameter u varies and, in cartesian coordinates,

$$\mathbf{r}(u) = x(u)\mathbf{i} + y(u)\mathbf{j} + z(u)\mathbf{k} \tag{3.35}$$

where $x = x(u)$, $y = y(u)$, and $z = z(u)$ are called the **parametric equations** of the curve.

A curve can be described in a parametric form by a vector $\mathbf{r}(s)$, with s representing the arc length along the curve measured since a given initial point. Let us see how we can do it. For this purpose, let us take the infinitesimal vector displacement

$$d\mathbf{r} = dx\mathbf{i} + dy\mathbf{j} + dz\mathbf{k} \tag{3.36}$$

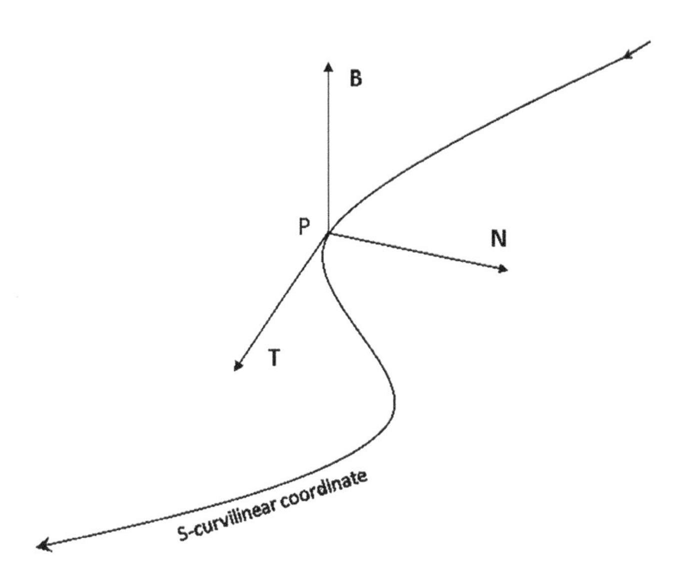

FIGURE 3.1
Curved trajectory in space C described by the unit tangent $\hat{t}$, normal $\hat{n}$, and binormal $\hat{b}$ at a particular point P.

The square of the vector displacement gives

$$(ds)^2 = d\mathbf{r}\cdot d\mathbf{r} = (dx)^2 + (dy)^2 + (dz)^2 \tag{3.37}$$

and, from it we also have

$$\left(\frac{ds}{du}\right)^2 = \frac{d\mathbf{r}}{du}\frac{d\mathbf{r}}{du} \tag{3.38}$$

The arc length between two points u_1 and u_2 on the curve is given by

$$s = \int_{u_1}^{u_2} \sqrt{\frac{d\mathbf{r}}{du}\cdot\frac{d\mathbf{r}}{du}}\,du \tag{3.39}$$

Exercise: Consider a curve lying in the xy-plane, defined by a given function $y = y(x)$, and $z = 0$. Obtain the integral of the arc length.

Solution: We have $ds^2 = dx^2 + dy^2$. If we identify u with x (since the curve is a function of x), we have

$$s = \int_a^b \sqrt{\frac{ds^2}{dx^2}}\,dx = \int_a^b \sqrt{\left(\frac{dx}{dx}\right)^2 + \left(\frac{dy}{dx}\right)^2}\,dx = \int_a^b \sqrt{1 + y'^2}\,dx \tag{3.40}$$

with $y' \equiv dy/dx$.

Exercise: The brachistochrone curve,[4] or curve of fastest descent: given two points a and b, with a higher than b, give the curve of fastest descent for a body submitted to the gravitational field.[5]

Solution: The energy conservation gives

$$\begin{aligned} mgy_0 &= \tfrac{1}{2}mv^2 + mgy \\ g(y_0 - y) &= \tfrac{1}{2}\left[\left(\tfrac{dx}{dt}\right)^2 + \left(\tfrac{dy}{dt}\right)^2\right] \\ (dt)^2 &= \frac{ds^2}{2g(y_0 - y)} \end{aligned} \tag{3.41}$$

The given body spends the total time

$$T = \int_0^T dt = \int_a^b \sqrt{\frac{1 + (dy/dx)^2}{2g(y_0 - y)}}\,dx \tag{3.42}$$

It is clear that to get the minimum time we have to solve a variational problem of the form

$$\delta \int_a^b F(x, y, y')dx = 0 \tag{3.43}$$

subject to the constraints

$$y(a) = y_0; \quad y(b) = 0 \tag{3.44}$$

Eq. 3.43 can be developed to yield

$$0 = \int_a^b \left(\frac{\partial F}{\partial y}\delta y + \frac{\partial F}{\partial y'}\delta y' \right) dx = \int_a^b \left(\frac{\partial F}{\partial y} - \frac{d}{dx}\frac{\partial F}{\partial y'} \right) dx \tag{3.45}$$

From the above equality, the Euler–Lagrange equation is obtained

$$\frac{\partial F}{\partial y} - \frac{d}{dx}\frac{\partial F}{\partial y'} = 0 \tag{3.46}$$

Taking the derivative of F with respect to x

$$\frac{df}{dx} = \frac{\partial F}{\partial x} + \frac{\partial F}{\partial y}y' + \frac{\partial F}{\partial y'}y'' \tag{3.47}$$

Now, remark that if we multiply the Euler–Lagrange equation by y'

$$y'\frac{\partial F}{\partial y} - y'\frac{d}{dx}\left(\frac{\partial F}{\partial y'} \right) = 0 \tag{3.48}$$

Inserting Eq. 3.47 into Eq. 3.48, we obtain

$$\begin{aligned} \frac{dF}{dx} - \frac{\partial F}{\partial y'}y'' - \frac{\partial F}{\partial x} - \frac{d}{dx}\left(\frac{\partial F}{\partial y'} \right) &= 0 \\ -\frac{\partial F}{\partial x} + \frac{d}{dx}\left(F - y'\frac{\partial F}{\partial y'} \right) &= 0 \end{aligned} \tag{3.49}$$

In particular, if $\partial F/\partial x = 0$, we have

$$\begin{aligned} \frac{d}{dx}\left(F - y'\frac{\partial F}{\partial y'} \right) &= 0 \\ \Rightarrow F - y'\frac{\partial F}{\partial y'} &= C \end{aligned} \tag{3.50}$$

The last equation is the **Beltrami identity**.[6]

We then have the Beltrami identity as a new point of departure

$$F - y'\frac{\partial F}{\partial y'} = C \tag{3.51}$$

But $F = \sqrt{\frac{1 + y'^2}{2g(y_0 - y)}}$, so then

$$y'\frac{\partial F}{\partial y'} - F = \frac{y'^2}{\sqrt{2g(y_0 - y)(1 + y'^2)}} - \sqrt{\frac{1 + y'^2}{2g(y_0 - y)}} = C \tag{3.52}$$

$$\Rightarrow \frac{y'^2}{\sqrt{1 + y'^2}} - \sqrt{1 + y'^2} = C\sqrt{2g(y_0 - y)} \tag{3.53}$$

and

$$\Rightarrow 1 = C^2 2g(y_0 - y)(1 + y'^2) \tag{3.54}$$

Now, put $y' = -\cot\Theta/2$. Then, we have

$$\begin{aligned} \tfrac{1}{C^2} = 2g(y_0 - y)(1 + \cot\Theta/2) &= 2g(y_0 - y)\tfrac{1}{\sin^2\Theta/2} \\ \Rightarrow y &= y_0 - \tfrac{1}{4C^2 g}(1 - \cos\Theta) \end{aligned} \tag{3.55}$$

On another side, we know that $dy/dx = -\cot\Theta/2$. Then, by integration, we obtain successively

$$\begin{aligned} dx &= -\frac{dy}{\cot\Theta/2} \\ dx &= \frac{1}{4C^2 g}\frac{2\sin\Theta/2\cos\Theta/2\sin\Theta/2}{\cos\Theta/2}d\Theta \\ dx &= \frac{1}{2C^2 g}\sin^2\frac{\Theta}{2}d\Theta \\ x &= x_0 + \frac{1}{2C^2 g}\int_0^{\Theta}\sin^2\frac{\Theta}{2}d\Theta' \\ x &= x_0 + \frac{1}{4C^2 g}(\Theta - \sin\Theta) \end{aligned} \tag{3.56}$$

The two obtained equations are the parametric representation of the **cycloid**

$$x = \tfrac{1}{4C^2 g}(\Theta - \sin\Theta); \quad y = y_0 - \tfrac{1}{4C^2 g}(1 - \cos\Theta) \tag{3.57}$$

3.1.8 The Tensors δ_{ij} and ϵ_{ijk}

The Kronecker symbol, with components of a second-order Cartesian tensor δ_{ij}, is defined by

$$\delta_{ij} = \begin{cases} 1 & \text{if i = j} \\ 0 & \text{otherwise} \end{cases} \tag{3.58}$$

A *second-order Cartesian tensor* transforms as

$$T'_{ij} = L_{ik}L_{jl}T_{kl} \tag{3.59}$$

Here, we have

$$\delta'_{kl} = L_{ki} L_{lj} \delta_{ij} = L_{ki} L_{li} = \delta_{kl} \tag{3.60}$$

where we have made use of the orthogonality relation $L_{ik} L_{jk} = \delta_{ij}$. The transformation gives indeed the same expression, and it transforms according to the second-order tensor transformation law. Prove that δ_{ij} is a second-order tensor.

Another third-order Cartesian tensor has great importance, the **Levi-Civita symbol** ϵ_{ijk}, the value of which is

$$\epsilon_{ijk} = \begin{cases} +1 & \text{if i, j, k is an even permutation of 1, 2, 3} \\ -1 & \text{if i, j, k is an odd permutation of 1, 2, 3} \\ 0 & \text{otherwise} \end{cases} \tag{3.61}$$

ϵ_{ijk} is totally anti-symmetric since it changes sign under the interchange of any pair of subscripts.

δ_{ij} and ϵ_{ijk} can be used to write well-known expressions from vector algebra. The vector product $\mathbf{a} = [\mathbf{b} \times \mathbf{c}]$, has an i-component $a_i = \epsilon_{ijk} b_j c_k$.

Exercise: Write the contracted Cartesian tensors $\mathbf{a}\cdot\mathbf{b}$; $\nabla^2\phi$; $\nabla \times \mathbf{v}$; $\nabla \times [\nabla \times \mathbf{v}]$; $[\mathbf{a} \times \mathbf{b}]\cdot\mathbf{c}$.

Solution: We can easily transpose what we talked about before and write

$$\mathbf{a}\cdot\mathbf{b} = a_i b_i = \delta_{ij} a_i b_j \tag{3.62}$$

$$\nabla^2\phi = \frac{\partial^2\phi}{\partial x_i \partial x_i} = \delta_{ij} \frac{\partial^2\phi}{\partial x_i \partial x_j} \tag{3.63}$$

$$[\nabla \times \mathbf{v}]_i = \epsilon_{ijk} \frac{\partial v_k}{\partial x_j} \tag{3.64}$$

$$[\nabla(\nabla\cdot\mathbf{v})]_i = \frac{\partial}{\partial x_i}\left(\frac{\partial v_i}{\partial x_j}\right) = \delta_{jk} \frac{\partial^2 v_j}{\partial x_i \partial x_k} \tag{3.65}$$

$$[\nabla \times (\nabla \times \mathbf{v})]_i = \epsilon_{ijk} \frac{\partial}{\partial x_j}\left(\epsilon_{klm} \frac{\partial v_m}{\partial x_l}\right) = \epsilon_{ijk}\epsilon_{klm} \frac{\partial^2 v_m}{\partial x_j \partial x_l} \tag{3.66}$$

$$(\mathbf{a} \times \mathbf{b})\cdot\mathbf{c} = \delta_{ij} c_i \epsilon_{jkl} a_k b_l = \epsilon_{ikl} c_i a_k b_l \tag{3.67}$$

We notice the important relationship between ϵ and δ tensors

$$\epsilon_{ijk}\epsilon_{klm} = \delta_{il}\delta_{jm} - \delta_{im}\delta_{jl} \tag{3.68}$$

We can use the Levi–Civita symbol to write down the expression for the determinant of any matrix A, for example, for a 3 × 3 matrix

$$A\epsilon_{lmn} = a_{li} a_{mj} a_{nk} \epsilon_{ijk} \tag{3.69}$$

The determinant of A is given by

$$\det A = |A| = \begin{bmatrix} a_{11} & a_{21} & a_{31} \\ d_{21} & a_{22} & a_{32} \\ a_{3m} & a_{32} & a_{33} \end{bmatrix} \tag{3.70}$$

Exercise: Apply Eq. 3.70 to the matrix

$$A = \begin{bmatrix} 2 & 1 & -3 \\ 3 & 4 & 0 \\ 1 & -2 & 1 \end{bmatrix}$$

Solution: Set $l = 1$, $m = 2$, $n = 3$. Then we write

$$\begin{aligned} |A| &= a_{1i}a_{2j}a_{3k}\epsilon_{ijk} \\ &= a_{11}a_{22}a_{33}\epsilon_{123} + a_{12}a_{23}a_{31}\epsilon_{231} + a_{13}a_{21}a_{32}\epsilon_{312} \\ &\quad + a_{11}a_{23}a_{32}\epsilon_{132} + a_{12}a_{21}a_{33}\epsilon_{213} + a_{13}a_{22}a_{31}\epsilon_{321} \\ &= 2(4)1 + 1(0)1 + (-3)3(-2) \\ &\quad - 2(0)(-2) - (1(3)1 - (-3)4(1)) \\ &= 35 \end{aligned} \tag{3.71}$$

Notice that we only have the following combinations (123), (231), (312), (132), (213), and (321).

3.1.8.1 Metric Tensors and the Line Element

Extending the concepts acquired by experience in common Euclidean space, we can build a mathematical n-dimensional space E_n. The magnitude of a given vector is given by $\mathbf{r} = \sqrt{x^2 + y^2 + z^2} = \sqrt{x^i x^i}$. If we consider the displacement vector dx^i ($i = 1, 2, \ldots, n$) determined by two points of space $P(x)$ and $P(x + dx)$, with x^i normal cartesian coordinates, then, using again the Pythagoras theorem, the distance squared between the two points is given by

$$ds^2 = dx^i dx^i; \ (i = 1, 2, \ldots, n) \tag{3.72}$$

called **arc element** in space E_n.

If we change of coordinates, $x^i = x^i(y^1, \ldots, y^n)$, we may write Eq. 3.72

$$ds^2 = \frac{\partial x^i}{\partial y^\alpha}\frac{\partial x^i}{\partial y^\beta} dy^\alpha dy^\beta \tag{3.73}$$

since $dx^i = \frac{\partial x^i}{\partial y^\alpha} dy^\alpha$. We may write the above equation as

$$ds^2 = g_{\alpha\beta} dy^\alpha dy^\beta \tag{3.74}$$

introducing the **metric tensor**

$$g_{\alpha\beta} = \frac{\partial x^i}{\partial y^\alpha}\frac{\partial x^i}{\partial y^\beta} \tag{3.75}$$

which are symmetric coefficients with respect to the indexes α and β, and function of the variables y^i.

Exercise: (a) Write the Lagrangian of a particle in a central potential $U(r)$ (an arbitrary function of all three coordinates); (b) write the Lagrangian in matrix-vector notation; (c) convert the Lagrangian to spherical coordinates.

Solution:

a. The Lagrangian for the problem is

$$L = K - U = \frac{1}{2}m(\dot{x}^2 + \dot{y}^2 + \dot{z}^2) - U(r) \tag{3.76}$$

b. In **matrix-vector notation**:
An algebraic form[7] of the type

$$A(x, y) = a_{11}y_1x_1 + a_{12}y_1x_2 + a_{21}y_2x_1 + a_{22}y_2x_2 \tag{3.77}$$

is named a bilinear form[8] of degree 2. We can write Eq. 3.99 as

$$A(x, y) = (y_1 y_2)\begin{pmatrix} a_{11} & a_{12} \\ a_{21} & a_{22} \end{pmatrix}\begin{pmatrix} x_1 \\ x_2 \end{pmatrix} \tag{3.78}$$

This representation allows us to write instead

$$L = \frac{1}{2}(\dot{x} \quad \dot{y} \quad \dot{z})\begin{bmatrix} 1 & 0 & 0 \\ 0 & 1 & 0 \\ 0 & 0 & 1 \end{bmatrix}\begin{bmatrix} \dot{x} \\ \dot{y} \\ \dot{z} \end{bmatrix} - U(x, y, z) \tag{3.79}$$

c. The definition of (x, y, z) in terms of (r, θ, ϕ) is

$$\begin{aligned} x &= r \sin\theta \cos\phi \\ y &= r \sin\theta \sin\phi \\ z &= r \cos\theta \end{aligned} \tag{3.80}$$

The Pythagorean theorem is expressed like

$$ds^2 = dx^2 + dy^2 + dz^2 \tag{3.81}$$

and, considering that

$$\begin{aligned} dx &= dr \sin\theta \cos\phi + r\cos\theta\cos\phi d\theta - r\sin\theta\sin\phi d\phi \\ dy &= dr\sin\theta\sin\phi + r\cos\theta\sin\phi d\theta + r\sin\theta\sin\phi d\phi \\ dz &= dr\cos\theta - r\sin\theta d\theta \end{aligned} \tag{3.82}$$

Squaring all the three terms we obtain

$$ds^2 = dr^2 + r^2(d\theta^2 + \sin^2\theta d\phi^2)) \tag{3.83}$$

Considering that

$$ds^2 = g_{rr}dr^2 + g_{\theta\theta}d\theta^2 + g_{\phi\phi}d\phi^2 \tag{3.84}$$

we can write the Lagrangian in a matrix-form

$$L = \frac{1}{2}(\dot{r}\ \ \dot{\theta}\ \ \dot{\phi})\begin{bmatrix} 1 & 0 & 0 \\ 0 & r^2 & 0 \\ 0 & 0 & r^2\sin^2\theta \end{bmatrix}\begin{bmatrix} \dot{r} \\ \dot{\theta} \\ \dot{\phi} \end{bmatrix} - U(r, \theta, \phi) \tag{3.85}$$

The metric tensor calculates the square length of a single vector or the scalar product of two vectors. This means that, if we have a vector expanded in terms of basis $\vec{e_i}$

$$\vec{u} = u^\alpha \vec{e_\alpha} \tag{3.86}$$

the metric tensor may be represented in the form

$$\mathbf{g}(\vec{u}, \vec{v}) = \vec{u}\cdot\vec{v} \tag{3.87}$$

with the property

$$\mathbf{g}(\vec{u}, \vec{v}) = \mathbf{g}(\vec{v}, \vec{u}) \tag{3.88}$$

The metric coefficients g_α may be defined in this way

$$g_{\alpha,\beta} \equiv \mathbf{g}(\vec{e_\alpha}, \vec{e_\beta}) = \vec{e_\alpha}\cdot\vec{e_\beta} \tag{3.89}$$

Exercise: Calculate the square length of separation vector $\vec{r} = \Delta x^\alpha \vec{e_\alpha}$ between two events.
Solution:

$$(\Delta s)^2 = \mathbf{g}(\Delta x^\alpha \vec{e_\alpha}, \Delta x^\beta \vec{e_\beta}) = \Delta x^\alpha \Delta x^\beta \mathbf{g}(\vec{e_\alpha}, \vec{e_\beta}) = g_{\alpha\beta}\Delta x^\alpha \Delta x^\beta \tag{3.90}$$

which results in

$$(\Delta x)^2 = -(\Delta x)^2 + (\Delta x)^2 + (\Delta y)^2 + (\Delta z)^2 \tag{3.91}$$

provided we define metric in the Lorentz frame

$$g_{\alpha,\beta} = \begin{bmatrix} -1 & 0 & 0 & 0 \\ 0 & 1 & 0 & 0 \\ 0 & 0 & 1 & 0 \\ 0 & 0 & 0 & 1 \end{bmatrix} \tag{3.92}$$

Exercise: Show that the rest mass of a particle is related to its energy and momentum by the equation $(mc^2)^2 = E^2 - (\vec{p}c)^2$.

Solution: The 4-momentum is defined by the expression $\vec{p} = m\vec{u}$, where $\vec{u}$ denotes the 4-velocity and m is its rest mass. We have

$$\vec{u} = \frac{d\mathcal{P}}{d\tau} = \frac{dx^{\mu}}{d\tau}\vec{e_{\mu}} = u^0\vec{e_0} + u_1\vec{e_1} + u^2\vec{e_2} + u^3\vec{e_3} \tag{3.93}$$

$\mathcal{P}(\tau)$ represents the world-line, and the time and space components of the velocity in special relativity are given by

$$u^0 = \frac{dt}{d\tau} = \frac{1}{\sqrt{1 - \vec{v}^2}} \tag{3.94}$$

$$u^i = \frac{dx^i}{d\tau} = \frac{v^i}{\sqrt{\vec{v}^2}} \tag{3.95}$$

(we used here geometrical units). Hence,

$$\vec{p}^2 = \vec{p}\cdot\vec{p} = m^2(\vec{u}\cdot\vec{u}) = m^2\mathbf{g}(\vec{u}, \vec{u}) = m^2u^{\alpha}\vec{e_{\alpha}}\cdot u^{\beta}\vec{e_{\beta}} \tag{3.96}$$

$$= m^2u^{\alpha}u^{\beta}g_{\alpha\beta} = -m^2(u^0)^2 + m^2(u^i)^2 \tag{3.97}$$

$$= -\frac{m^2}{1 - v^2} + \frac{m^2\vec{v}^2}{1 - v^2} = -\frac{m^2}{1 - v^2}(1 - v^2) = -m^2 \tag{3.98}$$

3.1.8.2 Christoffel Symbols

While in the special theory of relativity Lorentz transformation relates the observations made by different observers in two inertial frames of reference, in general relativity the description of relative motion is more complex because measurements made by different observers are made in a gravitational field.

Christoffel symbols are combinations of partial derivatives of fundamental tensors g_{ij} and g^{ij}. They are used to form tensor equations (covariant derivatives, and intrinsic derivatives), but they are not tensors.

We will introduce now the Christoffel symbol of first species

$$[ij, k] \equiv \frac{1}{2}\left(\frac{\partial g_{ik}}{\partial x^j} + \frac{\partial g_{jk}}{\partial x^i} - \frac{\partial g_{ij}}{\partial x^k}\right) \tag{3.99}$$

The Christoffel symbols of second species are defined by the formula

$$\Gamma^k_{ij} \equiv g^{k\alpha}[ij, \alpha] \tag{3.100}$$

Also, note that

$$\Gamma^k_{ij} = \Gamma^k_{ji} \tag{3.101}$$

The symbol Γ^i_{kj} behaves like a vector only for linear or affine transformations of coordinates $x^i = x^i(x^1, \ldots, x^n)$, with $\frac{\partial^2 x^i}{\partial x^k \partial x^j} \equiv 0$, for any values of i, k, j.

The alternate expression

$$T^i_{kj} = \Gamma^i_{kj} - \Gamma^i_{jk} = \Gamma^i_{[kj]} \tag{3.102}$$

is a tensor called the **torsion tensor**.

3.1.8.3 Covariant Derivation

The covariant derivative of a vector is defined by

$$\nabla_\alpha A^\beta \equiv A^{\beta;\alpha} \equiv \frac{\partial A^\beta}{\partial x^\alpha} + A^\gamma \Gamma^\beta_{\gamma\alpha} \tag{3.103}$$

whilst the covariant derivative of a covector is given by the formula

$$\nabla_\alpha A_\beta \equiv A_{\beta;\alpha} \equiv \frac{\partial A_\beta}{\partial x^\alpha} - A_\gamma \Gamma^\gamma_{\beta\alpha} \tag{3.104}$$

We also have

$$A_{i,jk} = \frac{\partial A_{i,j}}{\partial x^k} - \Gamma^a_{ij} A_{a,j} - \Gamma^a_{jk} A_{i,a} \tag{3.105}$$

Let us use the notation

$$\nabla_k T^{(i)}_{(j)} \equiv T^{(i)}_{(j),k} \tag{3.106}$$

We have

$$\nabla_k \nabla_l T^i = \nabla_k \left(\frac{\partial T^i}{\partial x^l} + \Gamma^i_{ql} T^q \right) \tag{3.107}$$

and we can obtain successively

$$\frac{\partial}{\partial x^k} = \left(\frac{\partial T^i}{\partial x^l} + \Gamma^i_{ql} T^q \right) + \Gamma^i_{pk} \left(\frac{\partial T^p}{\partial x^l} + \Gamma^p_{ql} T^q \right) - \Gamma^p_{lk} \left(\frac{\partial T^i}{\partial x^p} + \Gamma^i_{qp} T^q \right) \tag{3.108}$$

that is

$$\frac{\partial}{\partial x^k} = \frac{\partial^2 T^i}{\partial x^k \partial x^l} + \frac{\partial T^q}{\partial x^k}\Gamma^i_{ql} + T^q \frac{\partial \Gamma^i_{ql}}{\partial x^k} + \Gamma^i_{pk}\frac{T^p}{\partial x^l} - \Gamma^p_{lk}\frac{\partial T^i}{\partial x^p} + \Gamma^i_{pk}\Gamma^p_{ql} - \Gamma^p_{lk}\Gamma^i_{qp}T^q \tag{3.109}$$

From Eq. 3.109 we obtain

$$(\nabla_k \nabla_l - \nabla_l \nabla_k)T^i = \left(\frac{\partial \Gamma^i_{ql}}{\partial x^k} - \frac{\partial \Gamma^i_{qk}}{\partial x^l}\right)T^q + \left(\Gamma^i_{pq}\Gamma^p_{ql} - \Gamma^i_{pl}\Gamma^p_{qk}\right)T^q - (\Gamma^p_{lk} - \Gamma^p_{kl})\frac{\partial T^i}{\partial x^p} \tag{3.110}$$

We may define

$$-R^i_{qkl} = \frac{\partial \Gamma^i_{ql}}{\partial x^k} - \frac{\partial \Gamma^i_{qk}}{\partial x^l} + \Gamma^i_{pk}\Gamma^p_{ql} - \Gamma^i_{pl}\Gamma^q_{ql} \tag{3.111}$$

Then, the above Eq. 3.110 converts to

$$(\nabla_k \nabla_l - \nabla_l \nabla_k)T^i = -R^i_{qkl}T^q + T^p_{kl}\frac{\partial T^i}{\partial x^p} \tag{3.112}$$

Here, R^i_{qkl} denotes the **Riemann tensor** and T^p_{kl} the **torsion tensor**.

3.1.9 Curvature: The Riemann, Torsion, and Weyl Tensors

The quantity R^i_{qkl} is a tensor called the Riemann tensor

$$-R^i_{qkl} = \frac{\partial \Gamma^i_{ql}}{\partial x^k} - \frac{\partial \Gamma^i_{qk}}{\partial x^l} + \Gamma^i_{pk}\Gamma^p_{ql} - \Gamma^i_{pl}\Gamma^q_{ql}$$

while the torsion tensor is defined by

$$T^i_{kj} = \Gamma^i_{kj} - \Gamma^i_{jk} = \Gamma^i_{[kj]} \tag{3.113}$$

Exercise: Determine the divergence of the covariant vector $\Phi_{,i}$ in curvilinear coordinates, that is, the Laplacian, which may be written in tensorial notation in the following form:

$$\nabla^2 \Phi = g^{ij}(\Phi_{,i})_j = g^{ij}\left[\frac{\partial^2 \Phi}{\partial x^i x^j} - \Gamma^k_{ij}\frac{\partial \Phi}{\partial x^k}\right] \tag{3.114}$$

Solution: The contravariant metric tensor is $g_{k\alpha} = e_k e_\alpha$, where e_k are the base vectors, such as for a radius vector

$$\vec{r} = x\vec{i} + y\vec{j} + z\vec{k}$$

the base vectors are given by

$$\frac{\partial \vec{r}}{\partial x} = e_1 = \vec{i}$$
$$\frac{\partial \vec{r}}{\partial y} = e_2 = \vec{j}$$
$$\frac{\partial \vec{r}}{\partial z} = e_3 = \vec{k} \tag{3.115}$$

In curvilinear coordinates r and θ, we have $x^1 = r$, and $x^2 = \theta$, and hence, $g_{11} = 1$, $g_{22} = r^2$, $g^{11} = 1$, $g^{22} = 1/r^2$. From here, we obtain

$$\nabla^2\Phi = 1\left[\frac{\partial^2\Phi}{\partial r^2} - \Gamma^1_{11}\frac{\partial\Phi}{\partial r} - \Gamma^2_{11}\frac{\partial\Phi}{\partial\theta}\right] + \frac{1}{r^2}\left[\frac{\partial^2\Phi}{\partial\theta^2} - \Gamma^1_{21}\frac{\partial\Phi}{\partial r} - \Gamma^2_{22}\frac{\partial\Phi}{\partial\theta}\right] \tag{3.116}$$

We know that

$$\Gamma^\alpha_{ij} = g^{k\alpha}[ij, k] = \frac{g^{k\alpha}}{2}\left[\frac{\partial g_{ik}}{\partial x^j} + \frac{\partial g_{jk}}{\partial x^i} - \frac{\partial g_{ij}}{\partial x^k}\right] \tag{3.117}$$

For polar coordinates the Christoffel symbols are the following:

$$\begin{aligned} \Gamma^1_{11} &= 0; \quad \Gamma^2_{22} = 0 \\ \Gamma^1_{21} &= 0; \quad \Gamma^2_{12} = \tfrac{1}{r} \\ \Gamma^1_{12} &= 0; \quad \Gamma^2_{21} = \tfrac{1}{r} \\ \Gamma^1_{22} &= -r; \quad \Gamma^2_{11} = 0 \end{aligned} \tag{3.118}$$

Therefore, we obtain the Laplacian in polar coordinates

$$\nabla^2\Phi = \frac{\partial^2\Phi}{\partial r^2} + \frac{1}{r}\frac{\partial\Phi}{\partial r} + \frac{1}{r^2}\frac{\partial^2\Phi}{\partial\theta^2} \tag{3.119}$$

3.1.10 Dragging in the Kerr Field

This metric is valid for slow speeds like $(V/c)^2 \ll 1$. Imagine we are freely moving in a Lense–Thirring field at a constant radial coordinate like $dr/dt = 0$. The radial component of the geodesic equation is the Lense–Thirring metric given by

$$\begin{aligned} ds^2 &= \left(1 - \tfrac{2Gm}{c^2 r}\right)c^2dt^2 - \left(1 - \tfrac{2Gm}{c^2 r}\right)^{-1}dr^2 - r^2(d\theta^2 + \sin^2\theta d\phi^2) \\ &\quad + 2\tfrac{2Gm}{c^2 r}\sin^2\theta dt d\phi \end{aligned} \tag{3.120}$$

$$\left(\frac{d}{ds}g_{11}\frac{dr}{ds}\right) = \frac{1}{2}\frac{\partial g_{\mu\nu}}{\partial r}\frac{dx^\mu}{ds}\frac{dx^\nu}{ds} \tag{3.121}$$

The second derivative relatively to r is

$$\frac{d^2r}{ds^2} = \frac{1}{2g_{11}} \frac{\partial g_{\mu\nu}}{\partial r} \frac{dx^\mu}{ds} \frac{dx^\nu}{ds} \tag{3.122}$$

which gives

$$\begin{aligned} \frac{d^2r}{ds^2} = \frac{1}{2g_{rr}} \Bigg[& \frac{\partial g_{tt}}{\partial r}\left(c\frac{dt}{ds}\right)^2 + 2c\frac{\partial g_{t\phi}}{\partial r}\frac{dt}{ds}\frac{d\phi}{ds} \\ & + \frac{\partial g_{\phi\phi}}{\partial r}\left(\frac{d\phi}{ds}\right)^2 + \frac{\partial g_{\theta\theta}}{\partial r}\left(\frac{d\theta}{ds}\right)^2 \Bigg] \end{aligned} \tag{3.123}$$

The only term that has interest now is the specific part of the radial acceleration, the second derivative of r with respect to s, which gives

$$\left(\frac{d^2r}{ds^2}\right)_{rot} = \frac{c}{g_{rr}} \frac{\partial g_{t\phi}}{\partial r} \frac{dt}{ds} \frac{d\phi}{ds} \tag{3.124}$$

Retaining only terms with the same order of magnitude as Gm/c^2r, we obtain

$$\left(\frac{d^2r}{ds^2}\right)_{rot} = \frac{2Gma \sin^2\theta}{c^2r^2} \frac{dt}{ds} \frac{d\phi}{ds} \tag{3.125}$$

for an observer moving slowly, $ds \approx cdt$, and then, it results

$$\left(\frac{d^2r}{dt^2}\right)_{rot} = \frac{2Gma}{c^2r^2} \sin^2\theta\omega \tag{3.126}$$

with $\omega = \frac{d\phi}{dt}$ the instantaneous angular velocity of the observer about the z axis. We realize that due to the rotation of the gravitational field source (dragging force), it appears as an additional force. If the observer moves in the same direction as the field source rotates, it appears as an outward dragging force, while if the observer moves counter to the field source's sense of rotation, the resulting dragging force is inward, analog to the Coriolis force in a river flowing along the equator (Baer's law).

> There is no choice to genius. A great man does not wake up on some fine morning, and say, I am full of life, I will go to sea, and find an antarctic continent: today, I will square the circle: I will ransack botany, and find a new food for man: I have a new architecture in my mind: I foresee a new mechanic power: No; but he finds himself in the river of thoughts and events, forced onward by the ideas and necessities of his contemporaries.
>
> *–Ralph Waldo Emerson (1803-1882)*

3.1.11 Spinorial Calculus

Any physical entity V (e.g., mass, velocity, stress, ...) can be described by a finite number of "components" V_1, V_2, ..., which can be put into a column matrix

$$\begin{bmatrix} V_1 \\ V_2 \\ \cdots \\ V_n \end{bmatrix} \tag{3.127}$$

The general term V_n is the "carrier" of the representation. We can represent a transformation of the matrix V in the form

$$V' = T(V) \tag{3.128}$$

or, in the form

$$\begin{aligned} T_M T_L &= T_{ML} \\ T_{s's''} T_{ss'} &= T_{ss''} \end{aligned} \tag{3.129}$$

where $T_{ss'}$ is the matrix of the transformation from the frame s to the frame s'. First, apply $T_{ss'}$ then apply $T_{s's''}$. In a nonrelativistic theory, one particle of spin s is described by $(2s + 1)$ quantities which is a **spinor** symmetric of rank-2.

From the mathematical viewpoint, spinors are irreducible representations of the group of spatial rotations.

In relativistic theory, we need 4-dimensional spinors.

A spinor ξ^α is one quantity with two quantities ($\alpha = 1, 2$). For example, it is considered that the two components of the wave function of a spin-1/2 particle are ξ^1 and ξ^2 corresponding to the eigenvalues of the projection of spin 1/2 in z-axis, giving the values +1/2 or −1/2.

Under an arbitrary transformation of the Lorentz group, these two quantities transform like

$$\begin{aligned} \xi'^1 &= \alpha\xi^1 + \beta\xi^2 \\ \xi'^2 &= \gamma\xi^1 + \delta\xi^2 \end{aligned} \tag{3.130}$$

Here, $(\alpha, \beta, \gamma, \delta)$ are functions of the angles in 4-dimensions and verify the relation

$$\alpha\delta - \beta\gamma = 1 \tag{3.131}$$

Let ξ^α and Ξ^α be two given spinors. Then, the bilinear form $\xi^1\Xi^2 - \xi^2\Xi^2$ is invariant under transformation 130.

It is useful to introduce covariant and contravariant components

$$\xi_\alpha = g_{\alpha,\beta}\xi^\beta \tag{3.132}$$

with

$$g_{\alpha,\beta} = \begin{bmatrix} 0 & 1 \\ -1 & 0 \end{bmatrix} \tag{3.133}$$

Then

$$\xi_1 = \xi^2; \;\; \xi_2 = -\xi^1 \tag{3.134}$$

and

$$\begin{aligned} \eta^{\dot{1}'} &= \alpha^* \eta^{\dot{1}} + \beta^* \eta^{\dot{2}} \\ \eta^{\dot{2}'} &= \gamma^* \eta^{\dot{1}} + \delta^* \eta^{\dot{2}} \end{aligned} \tag{3.135}$$

3.1.12 Maxwell's Equations in Spinorial Representation

Two of the four Maxwell's equations can be written as

$$\begin{aligned} &\nabla \times \mathbf{E} + \frac{\partial \mathbf{B}}{\partial t} = 0 \quad \nabla \times \mathbf{B} - \frac{\partial \mathbf{E}}{\partial t} = 0 \\ &i\frac{\partial (i\mathbf{B})}{\partial t} = \frac{1}{i}\mathbf{S}\cdot\nabla(\mathbf{E}) \quad i\frac{\partial \mathbf{E}}{\partial t} = \frac{1}{i}\mathbf{S}\cdot\nabla(\mathbf{E}) \end{aligned} \tag{3.136}$$

where $i = \sqrt{-1}$ and $(S^i)_{jk} \equiv (1/i)\varepsilon_{ijk}$.

As the Klein–Gordon equation has difficulties at a physical level, it appeared the need to construct a new wave equation as

$$i\frac{\partial \Psi}{\partial t} = \left(\frac{1}{i}\mathbf{a}\cdot\nabla + \beta m\right) \equiv H\Psi \tag{3.137}$$

Here, Ψ is a vector wave function and $\mathbf{a}$, β are hermitian matrices in order to attribute to the Hamiltonian H hermitian properties and reassure that positive conserved probabilities density exists. An explicit representation in terms of the 2×2 unit matrix I and Pauli σ^i matrices (see Section 7) is provided by

$$\beta = \begin{bmatrix} I & 0 \\ 0 & I \end{bmatrix} \quad \alpha^i = \begin{bmatrix} 0 & \sigma^i \\ \sigma^i & 0 \end{bmatrix} \tag{3.138}$$

In the 4D representation of 3.137, the wave function Ψ may be written as a bispinor $\Psi = \begin{bmatrix} \phi \\ \chi \end{bmatrix}$ in terms of two-component spinors ϕ and χ, which are known in quantum mechanics as the large and small components, respectively. The resulting equation is

$$i\frac{\partial \phi}{\partial t} = m\phi + \frac{1}{i}\mathbf{s}\cdot\nabla\chi \tag{3.139}$$

$$i\frac{\partial \chi}{\partial t} = -m\chi + \frac{1}{i}\mathbf{s}\cdot\nabla\phi \tag{3.140}$$

The aforementioned equations can be put in a more compact formulation introducing the operator

$$P = i\hbar\sigma_\mu\delta^\mu \equiv i\hbar\left(\sigma_0\frac{\partial}{c\partial t} - \sigma_x\frac{\partial}{\partial x} - \sigma_y\frac{\partial}{\partial y} - \sigma_z\frac{\partial}{\partial z}\right) \tag{3.141}$$

We may define F

$$F = -E^k\sigma_k + icB^k\sigma_k \tag{3.142}$$

and

$$J = J^\mu\sigma_\mu = \rho c\sigma_0 + j^k\sigma_k \tag{3.143}$$

where $\mu = 0, 1, 2, 3$ and $k = 1, 2, 3$. We must use here the Pauli matrix (quaternion) multiplication table

$$\sigma_x\sigma_y = i\sigma_z = -\sigma_y\sigma_x \tag{3.144}$$

$$\sigma_y\sigma_z = i\sigma_x = -\sigma_z\sigma_y \tag{3.145}$$

$$\sigma_y\sigma_z = i\sigma_y = -\sigma_x\sigma_z \tag{3.146}$$

$$\sigma_x\sigma_x = \sigma_y\sigma_y = \sigma_z\sigma_z = \sigma_0 \tag{3.147}$$

Maxwell's equations can be written in terms of the Pauli matrices

$$PF = \epsilon' J \tag{3.148}$$

where the constant $\epsilon' = i\hbar/c\epsilon_0$. The matrix F is

$$\begin{aligned} F = & -E^0\begin{bmatrix}1 & 0\\ 0 & 1\end{bmatrix} - E^x\begin{bmatrix}0 & 1\\ 1 & 0\end{bmatrix} - E^y\begin{bmatrix}0 & -i\\ i & 0\end{bmatrix} - E^z\begin{bmatrix}1 & 0\\ 0 & -1\end{bmatrix} \\ & + icB^0\begin{bmatrix}1 & 0\\ 0 & 1\end{bmatrix} + icB^x\begin{bmatrix}0 & 1\\ 1 & 0\end{bmatrix} + icB^y\begin{bmatrix}0 & -i\\ i & 0\end{bmatrix} + icB^z\begin{bmatrix}1 & 0\\ 0 & -1\end{bmatrix} \end{aligned} \tag{3.149}$$

This gives

$$F = \begin{bmatrix} -E^0 + icB^z - E^z + icB^0 & -E^x + iE^y + cB^y + icB^x \\ -E^x - iE^y - cB^y + icB^x & -E^0 - icB^z + E^z + icB^0 \end{bmatrix} \tag{3.150}$$

We also have

$$P = i\hbar \begin{bmatrix} \partial_t - \partial_z & -\partial_x + i\partial_y \\ -\partial_x - i\partial_y & \partial_t - \partial_z \end{bmatrix} \tag{3.151}$$

It is convenient to introduce the spin wave function components

$$\begin{matrix} \phi_{11} = E_x + iE^y & \phi_{12} = E^z + icB^0 & cB^y + icB^x & \phi_{14} = -E^0 + icB^z \\ \phi_{21} = E^x + iE^y & \phi_{22} = E^z + icB^0 & \phi_{23} = cB^y - icB^x & \phi_{24} = -E^0 - icB^z \end{matrix} \tag{3.152}$$

We can write

$$PF = i\hbar \begin{bmatrix} \frac{1}{c}\partial_t - \partial_z & -\partial_x + i\partial_y \\ -\partial_x - i\partial_y & \frac{1}{c}\partial_z \end{bmatrix} \begin{bmatrix} \Phi_1 & \Phi_2 \\ \Phi_3 & \Phi_4 \end{bmatrix} = \frac{i\hbar}{c\epsilon_0} \begin{bmatrix} \rho c + j_z & j_x - ij_y \\ j_x + ij_y & \rho c - j_z \end{bmatrix} \tag{3.153}$$

Solving Eq. 3.153 we can obtain several equations, among them are the two following ones:

$$-\frac{1}{c}\frac{\partial E_0}{\partial t} + \nabla\cdot\mathbf{E} = \frac{\rho}{\epsilon_0} \tag{3.154}$$

$$-\frac{1}{c^2}\frac{\partial \mathbf{E}}{\partial t} + \frac{1}{c}\nabla E_0 + \nabla \times \mathbf{B} = \mu_0 \mathbf{j} \tag{3.155}$$

From Eq. 3.154 we obtain

$$-\frac{1}{c}\frac{\partial}{\partial t}\left(\frac{1}{c}\partial_t\Phi + \nabla\cdot\mathbf{A}\right) - \nabla\cdot\left(-\frac{1}{c}\partial_t\mathbf{A} - \nabla \times \mathbf{M} - \nabla\Phi\right) = \frac{\rho}{\epsilon_0} \tag{3.156}$$

We can rewrite as

$$\frac{1}{c^2}\frac{\partial \Phi}{\partial t^2} + \nabla\cdot\nabla \times \mathbf{M} + \nabla^2\Phi = \frac{\rho}{\epsilon_0} \tag{3.157}$$

A kind of Umov–Poynting theory can lead us to the continuity equation

$$\frac{1}{c}\frac{\partial W}{\partial t} + \nabla\cdot\mathbf{S} = 0, \tag{3.158}$$

where

$$W = \frac{1}{2}(\mathbf{E}^2 + \mathbf{H}^2 + E_0^2 + H_0^2) \tag{3.159}$$

and

$$\mathbf{S} = \mathbf{E} \times \mathbf{H} + E_0\mathbf{E} + H_0\mathbf{H} \tag{3.160}$$

It is convenient to use the following relationships:

$$\mathbf{E} = -\frac{\partial \mathbf{A}}{\partial t} - \nabla \times \mathbf{M} - \nabla A_0 \quad E_0 = \frac{\partial A_0}{\partial t} + \nabla \cdot \mathbf{A} \tag{3.161}$$

$$\mathbf{H} = -\frac{\partial \mathbf{M}}{\partial t} + \nabla \times \mathbf{A} - \nabla M_0 \quad H_0 = \frac{\partial M_0}{\partial t} + \nabla \cdot \mathbf{M} \tag{3.162}$$

Here,

$$M_0 = \frac{(\mathfrak{m} \cdot \mathbf{r})}{r^3} \tag{3.163}$$

$$\mathbf{A} = \frac{\mu_0}{4\pi}\frac{\mathfrak{m} \times \mathbf{r}}{r^3} \text{ in SI units} \tag{3.164}$$

$$\mathbf{M} = \chi_m \mathbf{H} \tag{3.165}$$

where χ_m is the magnetic susceptibility. Recall that [1]

$$\mathbf{H} = -\frac{\mathfrak{m}}{r^3} + \frac{3(\mathfrak{m} \cdot \mathbf{r})}{r^5} \tag{3.166}$$

Exercise: Using Eq. 3.157 predict the existence of longitudinal scalar waves (see Ref. [2]).

Solution: In fact, neglecting magnetization, the equation reduces to

$$\nabla^2 \Phi - \frac{1}{c^2}\frac{\partial^2 \Phi}{\partial t^2} = -\frac{\rho}{\epsilon_0} \tag{3.167}$$

Here, Φ is the scalar potential and ρ is the source charge density. After Eqs 3.159 and 3.160, the energy flux $\mathbf{S}$ is

$$\mathbf{S} = E_0\mathbf{E} = \left(\frac{1}{c}\partial_t \Phi + \operatorname{div} \mathbf{A}\right)\left(-\frac{1}{c}\partial_t \mathbf{A} - \nabla \Phi\right) = -\frac{1}{c^2}\nabla \Phi \frac{\partial \Phi}{\partial t} \tag{3.168}$$

and the energy density is given by

$$W = \frac{1}{2}\left[(\nabla \Phi)^2 + \left(\frac{1}{c}\partial_t \Phi\right)^2\right] \tag{3.169}$$

Exercise: What kind of source can you suggest to generate longitudinal waves?

Solution: The most simple form is a spherical surface with a uniform periodic change of the net charge q pulsating in time

$$\rho = q\delta(\mathbf{r} - \mathbf{r}')\sin(\omega t) \tag{3.170}$$

We can verify that the solution of Eq. 3.167, putting the charge at the origin, $\mathbf{r}' = 0$, is given by a spherical wave

$$\Phi = \frac{q}{r}\sin(kr - \omega t) \tag{3.171}$$

with $k = 2\pi/\lambda$. Monstein and Wesley [2] detected such types of waves in an experiment conducted in 2002.

> Trayaṁ vā idam, nāma rūpaṁ karma: Name, form and action are the three categories into which everything can be brought together.
>
> –*The Brihadaranyaka Upanishad*

3.1.13 Differential Forms

Let us consider a tensor of the kind (0,k), rank $0 + k$, called a **covector**, for example, the gradient of a function $f(x)$: $T_i = \frac{\partial f}{\partial x}$ is a covector.

We recall from analysis that the differential of a function f is given by

$$df = \frac{\partial f}{\partial x^i}dx^i \tag{3.172}$$

Given a coordinate transformation $x^i = x^i(x^{1'}, \ldots, x^{n'})$, we have

$$dx^i = \frac{\partial x^i}{\partial x^{i'}}dx^{i'} \tag{3.173}$$

and so

$$df = \frac{\partial f}{\partial x^i}\frac{\partial x^i}{\partial x^{i'}}dx^{i'} = \frac{\partial f}{\partial x^{i'}}dx^{i'} \tag{3.174}$$

This result means that df is invariant by coordinate transformation. Therefore, if we associate to each covector T_i a differential form $T_i dx^i$, this one will be invariant by coordinate transformations.

But, what is dx^i? Covectors e^i transform like

$$e^i = \frac{\partial x^i}{\partial x^{i'}}de^{i'} \Leftrightarrow dx^i = \frac{\partial x^i}{\partial x^{i'}}dx^{i'} \tag{3.175}$$

We may say that dx^i are the covectors of base e^i, and $t_i dx^i$ is equivalent to the decomposition of $T_i e^i$ of the covector relatively to the given basis.

A differential 1-form[9] on an open subset[10] of R^2 is an expression of the form

$$F(x, y)dx + G(x, y)dy \tag{3.176}$$

where F, G are both R-valued functions on the open set.

If $f(x, y)$ is a C^1 R-valued function on an open set,[11] then its exterior derivative (or total differential) is given by

$$df = \frac{\partial f}{\partial x}dx + \frac{\partial f}{\partial y}dy \tag{3.177}$$

A 1-form can be converted from a vector field (and back) by the following operation:

$$F_1\vec{i} + F_2\vec{j} + F_3\vec{k} \leftrightarrow F_1dx + F_2dy + F_3dz \tag{3.178}$$

and we can convert vector fields to two forms and back by the same procedure

$$F_1\vec{i} + F_2\vec{j} + F_3\vec{k} \leftrightarrow F_1dy \wedge dz + F_2dz \wedge dx + F_3dx \wedge dy \tag{3.179}$$

Exterior differential forms are mathematical objects that appear under integral signs. For example, a line integral

$$\int Adx + Bdy + Cdz \tag{3.180}$$

gives the first 1-form

$$\omega = Adx + Bdy + Cdz \tag{3.181}$$

The surface integral

$$\iint Pdydz + Qdzdx + Rdxdy \tag{3.182}$$

gives us the 2-form

$$\alpha = Pdydz + Qdzdx + Rdxdy \tag{3.183}$$

A volume integral

$$\iiint Hdxdydz \tag{3.184}$$

gives us the 3-form

$$\lambda = Hdxdydz \tag{3.185}$$

Those are examples of **differential forms** in space variables, in space E^3, but we can build r-form in n variables, as well.

In general, an exterior r-form in n variables $x^1, \ldots, x^n$ is expressed as

$$\omega = \frac{1}{r!}\sum A_{i_1,\ldots,i_r}dx^{i_1}\ldots dx^{i_r} \tag{3.186}$$

where the coefficients A represent smooth functions of the variables and skew-symmetry in the indices.

It is associated with each r-form ω an $(r + 1)$-form $d\omega$ which is called the **exterior derivative** of ω. the exterior derivative d: $\Omega^p(M) \to \Omega^{p+1}(M)$ (p = 0,..., dim M) is defined through the formula

$$d\omega = d\left(\eta(x)dx^{i1} \wedge \ldots \wedge dx^{i_p}\right) = \frac{\partial \eta(x)}{\partial x^j}dx^j \wedge dx^{i_1} \wedge \ldots \wedge dx^{i_p} \tag{3.187}$$

and extended by linearity to all of $\Omega^p(M)$.

The exterior derivative is given in such a way to validate general Stokes' formula

$$\int_{\partial\Sigma} \omega = \int_{\Sigma} d\omega \tag{3.188}$$

In this integral Σ is an $(r + 1)$-dimensional oriented variety and $\partial\Sigma$ is its boundary.

We can introduce the **Poincaré Lemma**

$$d(d\omega) = 0 \tag{3.189}$$

Exercise: Prove Poincaré Lemma:

Solution: We start from the general expression of a p-form

$$\begin{aligned} \omega(x) &= f(x)dx^{i_1} \wedge \ldots \wedge dx^{i_p} \\ dd(\omega(x)) &= d\left(\tfrac{\partial f}{\partial x^j}dx^j \wedge dx^{i_1} \wedge \ldots \wedge dx^{i_p}\right) \\ &= \tfrac{\partial^2 f}{\partial x^j \partial x^k}dx^k \wedge dx^j \wedge dx^{i_1} \wedge \ldots \wedge dx^{i_p}=0 \end{aligned} \tag{3.190}$$

because it is verified $\frac{\partial^2 f}{\partial x^j \partial x^k} = \frac{\partial^2 f}{\partial x^k \partial x^j}$, and $dx^j \wedge dx^k = -dx^k \wedge dx^j$. You can start to prove it by starting with a 2-form. In mathematical jargon, we say that the operator d is *nilpotent*[12] and in 3D Euclidean geometry, the Poincaré Lemma is equivalent to the identities $\nabla \times \nabla\phi = 0$ and $\nabla\cdot(\nabla \times \mathbf{v}) = 0$, where ϕ is a scalar field and $\mathbf{v}$ a vector field.

3.1.13.1 Lorentz Force and Electromagnetic Field Tensor

The Lorentz force law is written in 3D notation in the form

$$\frac{d\mathbf{p}}{dt} = e(\mathbf{E} + \mathbf{v} \times \mathbf{B}) \tag{3.191}$$

In relativistic notation we write

$$\frac{d\mathbf{p}}{d\tau} = e(u^0\mathbf{E} + \mathbf{u} \times \mathbf{B}) \tag{3.192}$$

or

$$\frac{dp}{d\tau} = eF(u) \tag{3.193}$$

We can also write this in components

$$\frac{dp^\alpha}{d\tau} = eF^\alpha_\beta u^\beta \tag{3.194}$$

provided we define the components of the electromagnetic field tensor

$$F^\alpha_\beta = \begin{pmatrix} 0 & -E_x & -E_y & -E_z \\ E_x & 0 & -B_z & B_y \\ E_y & B_z & 0 & -B_x \\ E_z & -B_y & B_x & 0 \end{pmatrix} \tag{3.195}$$

and

$$F_{\alpha\beta} = \begin{pmatrix} 0 & E_x & E_y & E_z \\ -E_x & 0 & -B_z & B_y \\ -E_y & B_z & 0 & -B_x \\ -E_z & -B_y & B_x & 0 \end{pmatrix} \tag{3.196}$$

It is more usual for the presentation under the covariant form that can be obtained by lowering an index with the metric components

$$F_{\alpha\beta} = \eta_{\alpha\gamma} F^\gamma_\beta \tag{3.197}$$

where the "metric coefficients" are

$$\eta_{\alpha\gamma} = \begin{bmatrix} -1 & 0 & 0 & 0 \\ 0 & 1 & 0 & 0 \\ 0 & 0 & 1 & 0 \\ 0 & 0 & 0 & 1 \end{bmatrix} \tag{3.198}$$

$$F_{\alpha\beta} = \begin{pmatrix} 0 & -E_x & -E_y & -E_z \\ E_x & 0 & B_z & -B_y \\ E_y & -B_z & 0 & B_x \\ E_z & B_y & -B_x & 0 \end{pmatrix} \tag{3.199}$$

The **dual** of the antisymmetric tensor F can be constructed using the **Hodge (star) operator** $*$

$$^*F_{\alpha\beta} = \frac{1}{2} F_{\mu\nu} \varepsilon_{\mu\nu\alpha\beta} \tag{3.200}$$

This operation can be done for any vector **J**

$$^*\mathbf{J}_{\alpha\beta\gamma} = J^\mu \varepsilon_{\mu\alpha\beta\gamma} \tag{3.201}$$

or any third-rank antisymmetric tensor B ($B_{\alpha\beta\gamma} = B_{[\alpha\beta\gamma]}$)

$$^{*}B_{\alpha} = \frac{1}{3!} B^{\lambda\mu\nu} \varepsilon_{\lambda\mu\nu\alpha} \tag{3.202}$$

The electromagnetic 4 force is given by

$$F_{\mu} = F_{\mu\nu} J^{\nu} \tag{3.203}$$

with J^{μ} denoting the 4 current density, given by

$$J = (\rho, \mathbf{J}) \tag{3.204}$$

Here, ρ is the charge density and $\mathbf{j}$ is the 3 current density. Also, we have

$$F_{\mu} = T^{\lambda}_{\mu,\lambda} \tag{3.205}$$

where T^{λ}_{μ} is the energy tensor of electromagnetic field

$$T^{\lambda}_{\mu} = -F_{\mu\rho} F^{\lambda\rho} + \frac{1}{4} \delta^{\lambda}_{\mu} F^{\rho\sigma}_{\mu\sigma} \tag{3.206}$$

Exercise: Obtain the $\bigstar F$ of the electromagnetic field F.

Solution: The dual electromagnetic field tensor $\bigstar F$ is an antisymmetric pseudo-4-tensor, defined by

$$\bigstar F_{\mu\nu} = \frac{1}{2} \epsilon^{\mu\nu\alpha\beta} F_{\alpha\beta} \tag{3.207}$$

We can calculate, step by step, each term. For example

$$\begin{array}{cc}
F^{12} & = \\
\frac{1}{2}\epsilon^{12\alpha\beta}F_{\alpha\beta} & = \\
\frac{1}{2}\epsilon^{12\alpha 4}F_{\alpha 4} + \epsilon^{014\alpha}F_{3\alpha} & \\
F^{12} & = \\
\frac{1}{2}[\epsilon^{12\alpha 4}F_{\alpha 4} + (-\epsilon^{12\alpha 4})(-F_{\alpha 4})]F^{12} & = \\
\epsilon^{12\alpha 4}F_{\alpha 4} & = \\
B_{x} &
\end{array}$$

and, following the same procedure

$$F_{21} = \frac{1}{2} \epsilon^{21\alpha\beta} F_{\alpha\beta} \tag{3.208}$$

$$F_{21} = \frac{1}{2} [\epsilon^{2143} F_{43} + \epsilon^{2134} F_{34}] \tag{3.209}$$

$$F_{21} = \epsilon^{4123} F_{43} \tag{3.210}$$

$$F_{21} = -E_z \tag{3.211}$$

The continued calculation of each term, at the end gives

$$(\bigstar F)_{\alpha\beta} = \begin{pmatrix} 0 & B_x & B_y & B_z \\ -B_x & 0 & E_z & -E_y \\ -B_y & E_z & 0 & E_x \\ -B_z & E_y & -E_x & 0 \end{pmatrix} \tag{3.212}$$

We may conclude that by taking the dual of F it amounts to doing the replacements

$$E_i \mapsto -B_i, \quad B_i \mapsto E_i \tag{3.213}$$

This matrix equation shows the unity of the electric and magnetic fields, which merge into a single entity, denoted by a letter F, from Faraday. Hence, in Eq. 3.230, **E** or **B** acquires a physical meaning that is independent of the chosen coordinates or the reference frames.

3.1.13.2 Maxwell's Equations in Differential Form

3.1.13.3 Dyadic Notation

Vector notation is not appropriate to handle second-order or higher-order tensor quantities, such as the stress-energy momentum tensor. A **dyadic** is a second-order tensor quantity, third-order tensors are called **triadic**,and fourth-order tensors are called **tetradic**. Dyadic symbols represent the second-order tensor quantities and vector symbols represent the first-order tensor quantities. For example, dyadic notation is very appropriate to represent elasticity equations but, Hooke's law which contains fourth-order tensor quantity cannot be written in dyadic notation.

Dyadic may be expressed in terms of three vectors

$$\overleftrightarrow{T} = \mathbf{i}\mathbf{T}_x + \mathbf{j}\mathbf{T}_y + \mathbf{k}\mathbf{T}_z \tag{3.214}$$

where

$$\begin{aligned} \mathbf{T}_x &= T_{xx}\mathbf{i} + \mathbf{T}_{xy}\mathbf{j} + \mathbf{T}_{xz}\mathbf{k} \\ \mathbf{T}_y &= T_{yx}\mathbf{i} + T_{yy}\mathbf{j} + T_{yz}\mathbf{K} \\ \mathbf{T}_z &= T_{zx}\mathbf{i} + T_{zy}\mathbf{j} + T_{zz}\mathbf{k} \end{aligned} \tag{3.215}$$

We see that each component of the dyadic T_j can be expressed as

$$\mathbf{T}_j = \sum_i T_{ij}\mathbf{u}_i \tag{3.216}$$

and the dyadic function

$$\overleftrightarrow{T} = \sum_j \sum_i T_{ij} \mathbf{u}_i \mathbf{u}_j \tag{3.217}$$

where $\mathbf{u}_i$ is unitary vector in a given system of coordinates. The components T_{ij} may also appear in a matrix notation

$$[T_{ij}] = \begin{bmatrix} T_{xx} & T_{xy} & T_{xz} \\ T_{yx} & T_{yy} & T_{yz} \\ T_{zx} & T_{zy} & T_{zz} \end{bmatrix} \tag{3.218}$$

The conjugate of the dyadic $\overleftrightarrow{T}$ is denoted by $\overleftrightarrow{T_c}$ and takes the form

$$\overleftrightarrow{T_c} = \mathbf{T}_x\mathbf{i} + \mathbf{T}_y\mathbf{j} + \mathbf{T}_z\mathbf{k} \tag{3.219}$$

We must keep in mind that most of the operations involving dyadic are logical extensions of the corresponding vector operations. Thus, we have the inner product of a vector and a dyadic

$$\mathbf{A} \cdot \overleftrightarrow{T} = A_x \mathbf{T}_x + A_y \mathbf{T}_y + A_z \mathbf{T}_z \tag{3.220}$$

and the inner product between a vector and the conjugate dyadic

$$\mathbf{A} \cdot \overleftrightarrow{T_c} = (A_x \mathbf{T}_x)\mathbf{i} + (A_y \mathbf{T}_y)\mathbf{j} + (A_z \mathbf{T}_z)\mathbf{k} = \overleftrightarrow{T} \cdot \mathbf{A} \tag{3.221}$$

The external product between a vector and a dyadic is given by

$$\begin{aligned} \overleftrightarrow{B} = [\mathbf{A} \times \overleftrightarrow{T}] &= \mathbf{A} \times \sum_{ij} T_{ij} \mathbf{u}_i \mathbf{u}_j \\ &= \sum_{ijk} T_{ij} A_k [\mathbf{u}_k \times \mathbf{u}_i] \mathbf{u}_j \end{aligned} \tag{3.222}$$

with ijk taken in a cyclic order. In matrix notation and cartesian unit vectors, we have

$$\mathbf{A} \times \overleftrightarrow{T} = \begin{bmatrix} \mathbf{i} & \mathbf{j} & \mathbf{k} \\ A_x & A_y & A_z \\ \mathbf{T}_x & \mathbf{T}_y & \mathbf{T}_z \end{bmatrix} \tag{3.223}$$

The operator divergence is given by

$$\begin{aligned} \operatorname{div} \overleftrightarrow{T} = \nabla \cdot \overleftrightarrow{T} &= \left(\mathbf{i}\frac{\partial}{\partial x} + \mathbf{j}\frac{\partial}{\partial y} + \mathbf{k}\frac{\partial}{\partial z} \right) \cdot (\mathbf{i}\mathbf{T}_x + \mathbf{j}\mathbf{T}_y + \mathbf{k}\mathbf{T}_z) \\ &= \frac{\partial}{\partial x}\mathbf{T}_x + \frac{\partial}{\partial y}\mathbf{T}_y + \frac{\partial}{\partial z}\mathbf{T}_z \end{aligned} \tag{3.224}$$

$$\text{div}\,\overleftrightarrow{T_c} = \nabla\cdot\overleftrightarrow{T_c} = (\nabla\cdot\mathbf{T}_x)\mathbf{i} + (\nabla\cdot\mathbf{T}_y)\mathbf{j} + (\nabla\cdot\mathbf{T}_z)\mathbf{k} \tag{3.225}$$

$$\text{curl}\,\overleftrightarrow{T} = \nabla\times\overleftrightarrow{T} = \begin{bmatrix} \mathbf{i} & \mathbf{j} & \mathbf{k} \\ \frac{\partial}{\partial x} & \frac{\partial}{\partial y} & \frac{\partial}{\partial z} \\ \mathbf{T}_x & \mathbf{T}_y & \mathbf{T}_z \end{bmatrix} = \mathbf{i}\times\frac{\partial\overleftrightarrow{T}}{\partial x} + \mathbf{j}\times\frac{\partial\overleftrightarrow{T}}{\partial y} + \mathbf{k}\times\frac{\partial\overleftrightarrow{T}}{\partial z} \tag{3.226}$$

Exercise: Give the time component of F_μ.

Solution: If we use Eq. 3.204 we obtain

$$\begin{aligned} F_0 &= F_{0\nu}J^\nu = F_{00}J^0 + F_{01}J^1 + F_{02}J^2 + F_{03}J^3 \\ F_0 &= (-E_x)J^x + (E_y)J^y + (-E_z)J^z = -(\mathbf{E}\cdot\mathbf{j}) \end{aligned} \tag{3.227}$$

We can also use Eq. 3.205, but it is more lengthy in the calculation.

It leads to the result

$$F_0 = \frac{\partial}{\partial t}\left(\frac{E^2}{2} + \frac{H^2}{2}\right) + \nabla\cdot\mathbf{E}\times\mathbf{H} \tag{3.228}$$

Consider the form

$$\omega = (E_1dx^1 + E_2dx^2 + E_3dx^3)dt + (H_1dx^2dx^3 + H_2dx^3dx^1 + H_3dx^1dx^2) \tag{3.229}$$

Then

$$^*\omega = -(B_1dx^1 + B_2dx^2 + B_3dx^3)dt + (E_1dx^2dx^3 + E_2dx^3dx^1 + E_3dx^1dx^2) \tag{3.230}$$

Defining the **Faraday** 2-form with Eq. 3.229, $\omega \equiv F$, and introducing the 1-form of Action on a 4D space-time of independent variables (x, y, z, t)

$$A = \sum_{k=1}^{3} A_k(x, y, z, t)dx^k - \phi(x, y, z, t)dt \equiv \mathbf{A}\cdot d\mathbf{r} - \phi dt \tag{3.231}$$

Maxwell's equations can be obtained subject to the constraint of the exterior differential system, the 2 form of field intensities, F

$$F = dA \tag{3.232}$$

Exercise: Show that

$$F = dA = \left(\frac{\partial A_k}{\partial x^j} - \frac{\partial A_j}{\partial x^k}\right)dx^j \wedge dx^k \tag{3.233}$$

Solution: It is more appropriate here to write instead

$$F = dA = \sum_{k=1}^{4} A_k dx^k \tag{3.234}$$

$$= \frac{\partial A_x}{\partial y} dydx + \frac{\partial A_x}{\partial z} dzdx \tag{3.235}$$

$$+ \frac{\partial A_y}{\partial x} dxdy + \frac{\partial A_y}{\partial z} dzdy \tag{3.236}$$

$$\frac{\partial A_z}{\partial x} dxdz + \frac{\partial A_z}{\partial y} dydz \tag{3.237}$$

$$\frac{\partial A_t}{\partial x} dxdt + \frac{\partial A_t}{\partial y} dydt + \frac{\partial A_t}{\partial z} dzdt \tag{3.238}$$

It is relatively easy to see that we can write the last expression under the condensed form

$$F = dA = \sum_{j=1}^{4} \sum_{k=1;k\neq j}^{4} \left(\frac{\partial A_k}{\partial x^j} - \frac{\partial A_j}{\partial x^k} \right) dx^j \wedge dx^k \tag{3.239}$$

We can rewrite it still as

$$F = dA = \left(\frac{\partial A_k}{\partial x^j} - \frac{\partial A_j}{\partial x^k} \right) dx^j dx^k \tag{3.240}$$

$$F = dA = F_{jk} dx^j \wedge dx^k \tag{3.241}$$

Eq. 3.240 can be developed

$$F = F_{01} dx^0 dx^1 + F_{02} dx^0 dx^2 + F_{03} dx^0 dx^3 + F_{10} dx^1 dx^0 + F_{12} dx^1 dx^2 + \cdots \tag{3.242}$$

The electromagnetic Faraday tensor $F_{\mu\nu}$ is a 4×4 spacetime matrix having rows and columns of time t, x, y, and z cartesian coordinates.[13] The first index μ refers to the row, and the second index ν refers to the column. Substituting with the components of the electromagnetic field tensor, see Eq. 3.199, we obtain

$$\begin{aligned} F &= \tfrac{1}{2}(B_z dxdy + E_x dxdt - B_y dxdz - E_x dtdx - E_y dtdy - E_z dtdz \\ &\quad + E_y dydt - B_z dydx + B_x dydz + E_z dtdz + B_y dzdx - B_x dzdy) \\ F &= B_z dxdy - Bydxdz - E_x dtdx - E_y dtdy - E_z dtdz + B_x dydz \end{aligned} \tag{3.243}$$

We just insert each term of the electromagnetic tensor one by one multiplying by the wedge product $dx^i \wedge dx^j = dx^i dx^j$. Now, comparing 3.242 with Eq. 3.240, that we should read like

$$F = \left(\frac{\partial A_y}{\partial x} - \frac{\partial A_x}{\partial y}\right)dxdy \tag{3.244}$$

$$+\left(\frac{\partial A_z}{\partial x} - \frac{\partial A_x}{\partial z}\right)dxdz \tag{3.245}$$

$$+\left(\frac{\partial A_x}{\partial t} - \frac{\partial A_t}{\partial x}\right)dtdy \tag{3.246}$$

$$+\left(\frac{\partial A_y}{\partial t} - \frac{\partial A_t}{\partial y}\right)dtdy \tag{3.247}$$

$$+\left(\frac{\partial A_z}{\partial t} - \frac{\partial A_t}{\partial z}\right)dtdz \tag{3.248}$$

We may define the 2-form

$$F: =\frac{1}{2}F_{\mu\nu}dx^{\mu} \wedge dx^{\nu} = \frac{1}{2}F_{\mu\nu}dx^{\mu}dx^{\nu} \tag{3.249}$$

and try to link it to the 1-form A. But first, let us remind you of a very interesting idea proposed by C. N. Yang and R. Mills in 1954. A complex scalar field with N-components, $\phi(x) = \phi_1(x), \phi_2(x), \ldots, \phi_N(x)$, transforms according to $\phi(x) \to U\phi(x)$, with U denoting an element of SU(N). The transposed matrix transforms as $\phi^{\dagger} \to \phi^{\dagger}U^{\dagger}$, with $U^{\dagger}U = 1$. The product $\phi^{\dagger}\phi$ is invariant, but $\partial\phi^{\dagger}\partial\phi$ is not. In fact,

$$\partial_{\mu}\phi \to \partial_{\mu}(U\phi) = U\partial_{\mu}\phi + (\partial_{\mu}U)\phi = U[\partial_{\mu}\phi + (U^{\dagger}\partial_{\mu}U)\phi)] \tag{3.250}$$

The last term in Eq. 3.250 breaks the invariance, and one solution to restore it is to generalize the ordinary derivative ∂_{μ} to a covariant derivative

$$D_{\mu}\phi(x) = \partial_{\mu}\phi(x) - iA_{\mu}(x)\phi(x) \tag{3.251}$$

where the field A_{μ} is the gauge potential. We can verify that invariance is restored, $D_{\mu}\phi(x) \to U(x)D_{\mu}\phi(x)$ provided

$$A_{\mu} \to UA_{\mu}U^{\dagger} - i(\partial_{\mu}U)U^{\dagger} = UA_{\mu}U^{\dagger} + iU\partial_{\mu}U^{\dagger} \tag{3.252}$$

Exercise: Show it.

$$\begin{aligned} D_{\mu}\phi &= \partial_{\mu}\phi - iA_{\mu}\phi \\ &= \partial_{\mu}(U\phi) - i[UA_{\mu}U^{\dagger} + iU\partial_{\mu}U^{\dagger}]U\phi \\ &= U\partial_{\mu}\phi + \phi\partial_{\mu}U - iUA_{\mu}\phi + U(\partial_{\mu}U^{\dagger})U\phi \\ &= U\partial_{\mu}\phi + \phi\partial_{\mu}U - iUA_{\mu}\phi + U(\partial_{\mu}U^{\dagger})U\phi \\ &= U\partial_{\mu}\phi - U\partial_{\mu}U^{\dagger}U - iUA_{\mu}\phi + U(\partial_{\mu}U^{\dagger})U\phi \end{aligned} \tag{3.253}$$

Solution: Notice that the second term on the RHS changes the sign because $U^{+}U = 1$, and so $\partial_\mu U^{+}U + U^{+}\partial_\mu U = 0$. Finally, the second and fourth terms cancel each other. Then, we have

$$D_\mu\phi = U\left[\partial_\mu\phi - iA_\mu\phi\right] = UD_\mu\phi \tag{3.254}$$

We remind here that U is a unitary matrix of order n and has n^2 independent elements. If H is a hermitian matrix,[14] $exp(iH)$ is a unitary matrix, and we can write

$$U = exp(iH) \tag{3.255}$$

We conclude that, considering that $\phi = -A_t$

$$E_x = -\frac{\partial A_x}{\partial t} - \frac{\partial\phi}{\partial x} \tag{3.256}$$

$$\ldots B_x = (\mathrm{curl}A)_x \ldots \tag{3.257}$$

that is

$$\mathbf{E} = -\frac{\partial\mathbf{A}}{\partial t} - \nabla\phi \quad \mathbf{B} = \nabla\times\mathbf{A} \tag{3.258}$$

We also can show that

$$\begin{aligned}
dF = &\left(\frac{\partial E_x}{\partial t}dt + \frac{\partial E_x}{\partial x}dx + \frac{\partial E_x}{\partial y} + \frac{\partial E_x}{\partial z}\right)\wedge dx\wedge dt\\
&+\left(\frac{\partial E_y}{\partial t}dt + \frac{\partial E_y}{\partial x}dx + \frac{\partial E_y}{\partial y} + \frac{\partial E_y}{\partial z}\right)\wedge dy\wedge dt\\
&+\left(\frac{\partial E_z}{\partial t}dt + \frac{\partial E_z}{\partial x}dx + \frac{\partial E_z}{\partial y} + \frac{\partial E_z}{\partial z}\right)\wedge dz\wedge dt\\
&+\left(\frac{\partial B_x}{\partial t}dt + \frac{\partial B_x}{\partial x}dx + \frac{\partial B_x}{\partial y} + \frac{\partial B_x}{\partial z}\right)\wedge dy\wedge dz\\
&+\left(\frac{\partial B_y}{\partial t}dt + \frac{\partial B_y}{\partial x}dx + \frac{\partial B_y}{\partial y} + \frac{\partial B_y}{\partial z}\right)\wedge dz\wedge dx\\
&+\left(\frac{\partial B_z}{\partial t}dt + \frac{\partial B_z}{\partial x}dx + \frac{\partial B_z}{\partial y} + \frac{\partial B_z}{\partial z}\right)\wedge dx\wedge dy
\end{aligned} \tag{3.259}$$

We can rearrange the terms as

$$\begin{aligned}
dF = &\left(\frac{\partial B_x}{\partial x} + \frac{\partial B_y}{\partial y}\frac{\partial B_z}{\partial z}\right)dx\wedge dy\wedge dz\\
&+\left(\frac{\partial B_x}{\partial t} + \frac{\partial E_z}{\partial y} - \frac{\partial E_y}{\partial z}\right)dt\wedge dy\wedge dz\\
&+\left(\frac{\partial B_y}{\partial t} + \frac{\partial E_x}{\partial z} - \frac{\partial E_z}{\partial x}\right)dt\wedge dz\wedge dx\\
&+\left(\frac{\partial B_z}{\partial t} + \frac{\partial E_y}{\partial x} - \frac{\partial E_x}{\partial y}\right)dt\wedge dx\wedge dy
\end{aligned} \tag{3.260}$$

The first term of Eq. 3.230 gives null value because $\nabla \cdot \mathbf{B} = 0$, and the second term is also null because curl$\mathbf{E} = -\frac{\partial \mathbf{B}}{\partial t}$. Hence,

$$dF = 0 \tag{3.261}$$

which means that the general 2-form F is a closed 2-form; the tube of F nowhere comes to an end.

Exercise: Show that $d\star F = \star J$.

Solution: The Navier–Stokes equation can be obtained with a similar procedure. We start from the 1-form

$$A = \sum_{i=1}^{3} v_i dx^i - Hdt \tag{3.262}$$

with H given by

$$H = \frac{v^2}{2} + \int \frac{dp}{\rho} - \lambda \mathrm{div}\vec{v} + k_B T \tag{3.263}$$

We have therefore,

$$F = dA = d\left(v_x dx + v_y dy + v_z dz\right) - dHdt \tag{3.264}$$

or

$$\begin{aligned} F = dA \;=\; & \left(\frac{\partial v_x}{\partial x}dx + \frac{\partial v_x}{\partial y}dy + \frac{\partial v_x}{\partial z}dz + \frac{\partial v_x}{\partial t}dt\right)dx \\ & + \left(\frac{\partial v_y}{\partial x}dx + \frac{\partial v_y}{\partial y}dy + \frac{\partial v_y}{\partial z}dz + \frac{\partial v_y}{\partial t}dt\right)dy \\ & + \left(\frac{\partial v_z}{\partial x}dx + \frac{\partial v_z}{\partial y}dy + \frac{\partial v_z}{\partial z}dz + \frac{\partial v_z}{\partial t}dt\right)dz \\ & - \left(\frac{\partial H}{\partial x}dx + \frac{\partial H}{\partial y}dy + \frac{\partial H}{\partial z}dz + \frac{\partial H}{\partial t}dt\right)dt \end{aligned} \tag{3.265}$$

Of course, this leads us to the equation (recall that $dxdy = dx \wedge dy$ and so forth)

$$\begin{aligned} F = dA \;=\; & \left(-\frac{\partial v_x}{\partial y} + \frac{\partial v_y}{\partial x}\right)dx \wedge dy + \left(\frac{\partial v_z}{\partial y} - \frac{\partial v_y}{\partial z}\right)dy \wedge dz + \left(\frac{\partial v_x}{\partial z} - \frac{\partial v_z}{\partial x}\right)dz \wedge dx \\ & + \frac{\partial v_x}{\partial t}dt \wedge dx + \frac{\partial v_y}{\partial t}dt \wedge dy + \frac{\partial v_z}{\partial t}dt \wedge dz \\ & - \frac{\partial H}{\partial x}dx \wedge dt - \frac{\partial H}{\partial y}dy \wedge dt - \frac{\partial H}{\partial z}dz \wedge dt \end{aligned} \tag{3.266}$$

It is easy to verify that

$$\begin{aligned}\frac{\partial H}{\partial x} &= \vec{\nabla}\left(\frac{v^2}{2}\right)\\ &+ \vec{\nabla}\left(\frac{dp}{\rho}\right)\\ &- \lambda\vec{\nabla}(\vec{\nabla}\cdot\vec{v}) + \vec{\nabla}(k_B T)\end{aligned} \tag{3.267}$$

Taking into account that

$$c\vec{\omega} = \vec{\nabla} \times \vec{v} \tag{3.268}$$

$$\vec{a} = -\frac{\partial\vec{v}}{\partial t} - \vec{\nabla}H \tag{3.269}$$

our Eq. 3.266 becomes

$$\frac{\partial\vec{v}}{\partial t} + \vec{\nabla}\left(\frac{v^2}{2}\right) = -\vec{\nabla}\left(\frac{dp}{\rho}\right) + \lambda\vec{\nabla}(\vec{\nabla}\cdot\vec{v}) - \vec{\nabla}(k_B T) \tag{3.270}$$

We can obtain the 3-form of **Helicity** or **Topological Torsion**

$$\begin{aligned} H &= A \wedge dA\\ H &= H_{ijk}dx^i \wedge dx^j \wedge dx^k\\ &(v_x dx + v_y dy + v_z dz - Hdt)\wedge\\ &(\omega_z dxdy + \omega_x dydz + \omega_y dzdx + a_x dxdt + a_y dydt + a_z dzdt)\end{aligned} \tag{3.271}$$

or

$$\begin{aligned} H = &(v_x\omega_x + v_y\omega_y + v_z\omega_z)dxdydz\\ &(v_x a_y - v_y a_x - H\omega_z)dxdydt\\ &(v_z a_x - v_x a_z - H\omega_y)dzdydt\\ &(v_y\omega_z - v_z\omega_y - H\omega_x)dydzdt\end{aligned} \tag{3.272}$$

or,

$$H = -T_x dydzdt - T_y dzdxdt - T_z dxdydt + hdxdydz \tag{3.273}$$

where

$$h = (\vec{\omega}\cdot\vec{v}) \tag{3.274}$$

$$\vec{T} = [\vec{a} \times \vec{v}] + H\vec{\omega} \tag{3.275}$$

The terms $[\vec{a} \times \vec{v}]$ and $H\vec{\omega}$ are, respectively, called the *shear of transversal acceleration* and *shear of rotational acceleration*.

The **Topological Parity** is given by

$$K = dH = dA \wedge dA \tag{3.276}$$

$$= (\partial_x T_x + \partial_y T_y + \partial_z T_z + \partial_t h)dxdydzdt \tag{3.277}$$

and it can be demonstrated along the same lines of procedure to be given by the expression

$$K = dH = -(\vec{a}\cdot\vec{\omega})dxdydzdt \tag{3.278}$$

Hence, in x, y, z, t representation, we have the equation as

$$\text{div}\mathbf{T} + \frac{\partial h}{\partial t} = -2(\vec{a}\cdot\vec{\omega}) \tag{3.279}$$

Hence, there is an anomaly with the helicity-torsion conservation law, since the term $-(\vec{a}\cdot\vec{\omega})$ is not null.

Notice that we can write the acceleration in the form

$$\vec{a} = -\frac{\partial\vec{v}}{\partial t} - \text{grad}H = -[\vec{v}\times\vec{\omega}] + \nu\text{curl}\vec{\omega} \tag{3.280}$$

Considering that the Lagrangian is

$$L = (\vec{v}\cdot\vec{v}) - H \tag{3.281}$$

the torsion axial vector may be written as

$$T = (h\vec{v} - L\vec{\omega}) - \nu[\vec{v}\times\text{curl}\vec{\omega}] \tag{3.282}$$

Hence, even for Euler flows (when $\nu = \lambda = 0$), the torsion axial vector current is not null.

Topological torsion is necessary for understanding turbulent and chaotic flow.

Exercise: Knowing that the topological torsion of the electromagnetic field is given by $T = (\mathbf{E}\times\mathbf{A} + \phi\mathbf{B}, \mathbf{A}\cdot\mathbf{B})$, determine the governing equation of the helicity in a plasma.

Solution: Considering the analogy

$$\begin{array}{ll} \nu & \to \eta \\ \mathbf{v} & \to \mathbf{A} \\ \text{curl}\,\mathbf{v} & \to \mathbf{B} \\ \text{curl curl}\,\mathbf{v} & \to \mathbf{J} \end{array} \tag{3.283}$$

we may write

$$\frac{\partial}{\partial t}(\mathbf{A}\cdot\mathbf{B}) + \text{div}(\phi\mathbf{B} + \mathbf{E}\times\mathbf{A}) = -2\eta\mathbf{B}\cdot\mathbf{J} \tag{3.284}$$

We can integrate into a given region of space

$$\iiint_V \frac{\partial}{\partial t}(\mathbf{A}\cdot\mathbf{B})d^3r + \iiint_V \operatorname{div}(\phi\mathbf{B} + \mathbf{E}\times\mathbf{A})d^3r = -2\iiint_V (\mathbf{B}\cdot\mathbf{J})d^3r \tag{3.285}$$

or, using Green–Gauss theorem

$$\iiint_V \frac{\partial}{\partial t}(\mathbf{A}\cdot\mathbf{B})d^3r + \oint_S (\phi\mathbf{B} + \mathbf{E}\times\mathbf{A})\cdot\mathbf{n}dS = -2\iiint_V (\mathbf{B}\cdot\mathbf{J})d^3r \tag{3.286}$$

where $\iiint (\mathbf{A}\cdot\mathbf{B})d^3r$ is the helicity density. Its density, $\mathbf{A}\cdot\mathbf{B}$, is called flux linkage. η is the electrical resistivity, given by

$$\eta = \frac{Ze^2 m_e^{1/2}}{3\sqrt{\sqrt{\pi}\pi\varepsilon_0^2}}\frac{L}{(2k_B T_e)^{3/2}} \tag{3.287}$$

in units Ohm-m, so that $\mathbf{E} = \eta\mathbf{J}$ (Ref. [3]).

3.1.14 Topology and Turbulence

There is no clear definition of a turbulent state, although it is considered in general terms, to be a time-dependent, dissipative, intermittent, irreversible, 3D flow. We can refer to at least four different well-known theories: (i) the Kolmogorov theory of turbulence (1941) which proposes that the turbulent state is formed by vortices of all scales with random intensities; (ii) the wavelet theory of Zimin; (iii) The Hopf–Landau theory, proposing that the transition to turbulence is a "cascade" of successive instabilities; (iv) Ruelle–Takens theory with a strange attractor defining the turbulent state.

3.1.15 The Chern–Simon Lagrangian

We have defined the helicity with Eq. 3.271. We may define a **magnetic tubular link** (or more briefly, a **magnetic link** as a smooth immersion into $\mathbb{R}^3$ of finitely many disjoint standard solid tori $\bigsqcup_{i=1}^{n}\mathbb{T}_i$, such as

$$L: \bigsqcup_{i=1}^{n}\mathbb{T}_i \to \mathbb{R}^3$$

and a smooth magnetic field $\mathbf{B}$ on $\mathbb{R}^3$. For each solid torus $\mathbb{T}_i$ and a meridional disk $\mathcal{D}_i$, we define the **magnetic flux** $\Phi_i = \Phi(L\mathbb{T}_i)$ through the ith component given by the surface integral

$$\Phi_i = \iint_{L\mathcal{D}_i} (\mathbf{B}\cdot\mathbf{n})dS \tag{3.288}$$

where $\mathbf{n}$ denotes the normal to the surface $L\mathcal{D}_i$, directed along the $\mathbf{B}$-field.

We may note that the **helicity** $H(L)$ is identical to the Chern–Simon action

$$H(L) = \int \mathbf{A} \wedge \mathbf{A} = \int \mathrm{Tr}\left(\mathbf{A} \wedge d\mathbf{A} + \frac{2}{3}\mathbf{A} \wedge \mathbf{A} \wedge \mathbf{A}\right) \tag{3.289}$$

where **A** is the 1-form vector potential.

3.1.16 The Lie Derivative

In the language of differential forms, the displacement of a mathematical object inducing a change of scale can be done using the **Lie differential** with respect to a direction field, $X = [x^k]$ acting on p-forms, $\omega(x^k, dx^k)$, and recreating the same form multiplied by a scale factor D

$$L_X\omega = i(X)d\omega + d(i(X)\omega) \tag{3.290}$$

This definition of the Lie derivative is not easy to apply.

Example 3.6: *Biot–Savart's law is given as*

$$d\mathbf{B}(r) = \frac{\mu_0}{4\pi}\mathbf{J}(\mathbf{r}') \times \frac{\mathbf{r} - \mathbf{r}'}{|\mathbf{r} - \mathbf{r}'|}d^3r' \tag{3.291}$$

Integrating, we have

$$\mathbf{B}(\mathbf{r}) = \frac{\mu_0}{4\pi}\int_{\mathbb{R}^3} \mathbf{J}(\mathbf{r}') \times \frac{\mathbf{r} - \mathbf{r}'}{|\mathbf{r} - \mathbf{r}'|}d^3r' \tag{3.292}$$

We may remark first that

$$\frac{r}{|r|^3} = -\nabla\left(\frac{1}{|r|}\right) \tag{3.293}$$

is a conservative vector field. We may proceed with the calculation to obtain

$$\mathbf{B}(r) = \frac{\mu_0}{4\pi}\int_{\mathbb{R}^3} \mathbf{J}(r') \times \left(-\nabla\left(\frac{1}{|\mathbf{r} - \mathbf{r}'|}\right)\right)d^3r' \tag{3.294}$$

$$= \frac{\mu_0}{4\pi}\int_{\mathbb{R}^3}\left[\nabla\left(\frac{1}{|\mathbf{r} - \mathbf{r}'}\right. \times \mathbf{J}(r') + \frac{1}{|\mathbf{r} - \mathbf{r}'|}(\nabla_r \times \mathbf{J}(\mathbf{r}))\right]d^3r' \tag{3.295}$$

$$= \frac{\mu_0}{4\pi}\int_{\mathbb{R}^3} \nabla \times \left(\frac{1}{|\mathbf{r} - \mathbf{r}'|}\right.\mathbf{J}(r')d^3r' \tag{3.296}$$

$$= \nabla_r \times \left(\frac{\mu_0}{4\pi}\int_{\mathbb{R}^3} \frac{\mathbf{J}(r')}{|\mathbf{r} - \mathbf{r}'|}d^3r'\right) \tag{3.297}$$

The operator ∇_r acts only on the r variables and the term inside it is the potential vector **A**. Eq. 3.296 *could be written in the form of an exterior derivative using*

$$d(\alpha_1 \wedge \alpha_2 \wedge \ldots \wedge \alpha_k) \sum_{i=1}^{k} (-1)^k \alpha_1 \wedge \ldots \wedge (d\alpha_i) \wedge \ldots \wedge \alpha_k \tag{3.298}$$

In the case of the product of a function with a 1-form, we have

$$d(f\alpha) = df \wedge \alpha + fd\alpha \tag{3.299}$$

If instead of a 1-form we consider a vector field $\vec{F}$, we may write

$$\vec{\nabla} \times (f\vec{F}) = f\vec{\nabla} \times \vec{F} + [\vec{\nabla} \times \vec{F}] \tag{3.300}$$

For any vector fields the Lie derivative $L_X Y$ is the same as the bracket $[X, Y]$

$$[X, Y](fg) = XY(fg) - YX(fg) = [X, Y](f)g + f[X, Y](g) \tag{3.301}$$

Example 3.7: *Consider a stationary fluid flow described by a 3-velocity* $\mathbf{v}(\mathbf{r})$.

Example 3.8: *Obtain the Lie derivative, also known as the Cartan magic formula.*
Consider the 1-form $\omega = \omega_i dx^i$. Then take the inner product $i_X\omega$ of ω and X, $i_x\omega(X_1, \ldots, X_{p-1})$. The definition of inner product is

$$i_X\omega = \frac{1}{(p-1)!}\omega_{ki_2\ldots i_p} X^k dx^{i_2} \wedge \ldots \wedge dx^{i_p} \tag{3.302}$$

It gives

$$i_x\omega = i_x(\omega_j dx^j) = \frac{1}{0!}\omega_h X^h = <\omega, X> \tag{3.303}$$

So,

$$\begin{aligned} di_x\omega &= d\omega_i X^i + \omega_i dX^i \\ &= \frac{\partial\omega_i}{\partial x_j} X^i dx^j + \omega_i \frac{\partial X^i}{\partial x^j} dx^j \\ &= \partial_j\omega_i X^i dx^j + \omega_i \partial_j X^i dx^j \end{aligned} \tag{3.304}$$

The last line was intended to simplify the writing.
By other side

$$d\omega = \partial_j\omega_i dx^j \wedge dx^i \tag{3.305}$$

Hence,

$$i_X\omega = i_x(\partial_j\omega_i dx^j \wedge dx^i) \tag{3.306}$$

$$= (\partial_j\omega_i)X^j dx^i - (\partial_j\omega_i)X^i dx^j \tag{3.307}$$

Now, we can sum up

$$\Rightarrow (di_X + i_X d)\omega = (\partial_j\omega_i)X^j dx^i + \omega_i(\partial_j X^i)dx^j \tag{3.308}$$

$$= [X^k\partial_k\omega_i + \omega_k(\partial_i X^k)]dx^i \tag{3.309}$$

$$= L_X\omega \tag{3.310}$$

According to R. M. Kiehn [4], Maxwell's electromagnetic theory is a set of topological statements, that is, independent of metric, connection, dimension, and ideas that were proposed in 1934 by D. Van Dantzig [5].

Hence, we may start to study any kind of system such as physical, biological, and economic on this ground. The methodology is the following:

- Start with an ordered array: 1, 2, 3, 4, ...
- Assert the existence of an ordered set of independent variables: $t, x, y, z, \ldots$
- Construct a set of C^2 functions of $t, x, y, z, \ldots$, such as $\phi, A_x, A_y, A_z, \ldots$
- Then build up the 1-form of action: $A = A_m(\chi^k)d\chi^k = -\phi dt + A_k dx^k$.

The electromagnetic field can be constructed on these grounds. In a space-time frame t, x, y, z, the action is the 1-form

$$A = A_k(t, x, y, z)dx^k = \mathbf{A}\cdot d\mathbf{r} - \phi dt \tag{3.311}$$

and the field intensities are given by the 2-form $F = dA$

$$\begin{aligned} F = dA &= (\partial_j A_k - \partial_k A_j)dx^j \wedge dx^k \\ &= F_{jk}dx^j \wedge dx^k = +B_z dx \wedge dy + \ldots + E_x dx \wedge dt \end{aligned} \tag{3.312}$$

The work done by an electromagnetic field can be calculated using the formula

$$W = i_{\rho \mathbf{V}_4} dA = i_J F \tag{3.313}$$

Note that this method gives us the Lorentz force as a derivative, without any *a priori* assumptions. In addition, the thermodynamic 1-form W may be path dependent, since it is not necessarily a perfect differential.

We can calculate the internal energy term U along the same lines

$$U = i_J A = (\mathbf{J}\cdot\mathbf{A}) - \rho\phi = \rho[(\mathbf{v}\cdot\mathbf{A}) - \phi] \tag{3.314}$$

Exercise: Obtain the result of Eq. 3.313.
Solution: We have

$$W = \frac{1}{(2-1)!} F_{ki_2} J dx^{i_2} \tag{3.315}$$

since F is a 2-form field intensity and note that $J = [\rho, \mathbf{J}]$. Then, it follows

$$\begin{aligned} W &= J^1[F_{12}dx^2 + F_{13}dx^3 + F_{14}dx^4] \\ &\quad + J^2[F_{21}dx^1 + F_{23}dx^3 + F_{24}dx^4] \\ &\quad + J^3[F_{31}dx^1 + F_{32}dx^2 + F_{34}dx^4] \\ &\quad + J^4[F_{41}dx^1 + F_{42}dx^2 + F_{43}dx^3] \end{aligned} \tag{3.316}$$

Replacing the components, we have

$$\begin{aligned} W &= \rho[E_x dx + E_y dy + E_z dz] \\ &\quad + \rho V_x[-E_x dt - B_z dy + B_y dz] \\ &\quad + \rho V_y[-E_y dt + B_z dx - B_x dz] \\ &\quad + \rho V_z[-E_z dt - B_y dx + B_x dy] \end{aligned} \tag{3.317}$$

that we may put under the more compact form

$$W = \rho\mathbf{E}\cdot d\mathbf{r} + [\mathbf{J} \times \mathbf{B}]\cdot d\mathbf{r} - (\mathbf{J}\cdot\mathbf{E})dt \tag{3.318}$$

In Eq. 3.318 we recognize two well-known terms, the Lorentz force

$$d\mathbf{F}_L \equiv [\rho\mathbf{E} + \mathbf{J} \times \mathbf{B}]\cdot d\mathbf{r} \tag{3.319}$$

and the dissipative power

$$(\mathbf{J}\cdot\mathbf{E})dt \tag{3.320}$$

component.

3.1.17 Elements of Set Theory

$A = \{a\}$ set formed by a single element
$B = \{a, b\}$ set formed by two elements a,b
$C = \{1, 2, 3, 4\}$ set of natural numbers lesser than 5
$D = \{\ldots, -15, -5, 0, 5, 10, 15, \ldots\}$ set of all integers divisible by 5

We can define the above sets C and D in the following manner

$$C = \{x \colon x \in N, x < 5\}$$

$$D = \{x \colon x \in I, \text{x is divisible by 5}\}$$

In 1879 another great German mathematician published a small book entitled *Begriffsschrift*.[15] This work is considered probably the most important work on logic.

According to Frege, the phrase

All horses are mammals

can be written using the logic relation *if…, then*

If x is a horse, then x is a mammal.

3.1.18 The Yang–Mills Field

In 1954, C. N. Yang and R. Mills published a paper entitled "Conservation of Isotopic Spin and Isotopic Gauge Invariance" [6] proposing a gauge theory based on the SU(N) group. This gauge[16] theory is a local non-Abelian theory built with the purpose to explain strong interactions.

3.1.19 Elements of Group Theory

The notion of group appeared from the investigations of number theory at the end of the 18th century, the theory of algebraic equations at the end of the 18th century, and geometry at the beginning of the 19th century with the study of symmetry transformations in plane and solid geometry.

A *group* G is a nonempty ensemble constituted by a set (of symmetry operations, numbers, …) of elements (a, b, c, …) in which a binary operation $\circ$ is defined[17] having the following properties

1. *Closure*: If a and b are in G, $c = ab$ is also in G;
2. *Associative*: $(a \circ b) \circ c = a \circ (b \circ c)$;
3. *Identity*: There exists an element e such that $a \circ e = e \circ a = a$;
4. *Inverse*: For every a in G, there exists an element a^{-1} such that $a \circ a^{-1} = a^{-1} \circ a$.

The element identity has essentially nothing to do the inverse element is to undo what was done.

A group is Abelian if the binary operation (e.g., multiplication) is commutative: $a \circ b = b \circ a$; if not, the group is called a non-Abelian group.

Loosely speaking, a group can be a set of elements, numbers, objects, and so on, but they work together in an *algebraic structure*, which gives to them emergent properties. The forthcoming examples clarify this idea.

Many physical systems underlying dynamics have some kind of symmetry.

The most important applications of group theory can be found in quantum mechanics.

Example 3.9: *show that there is only one group of order three, employing a step-by-step procedure to find the multiplication table.*

Let $G = e, a, b$ be *the referred group, and let us assume further that* $e \neq a \neq b$ *and let us establish one operation that will give* G *the group properties.*

TABLE 3.1

To be Filled According to the Rules of the Group of Transformations

e	a	b
a		
b		

TABLE 3.2

Example of an Abelian Group

e	a	b
a	b	e
b	e	a

TABLE 3.3

Cayley Table for the Dihedral Group

	u	ρ	ρ^2	ρ^3	σ^2	τ^2	b	e
u	u	ρ	ρ^2	ρ^3	σ^2	τ^2	b	e
ρ	ρ	ρ^2	ρ^3	u	b	e	τ^2	σ^2
ρ^2	ρ^2	ρ^3	u	ρ	τ^2	σ^2	e	b
ρ^3	ρ^3	u	ρ	ρ^2	e	b	σ^2	τ^2
σ^2	σ^2	e	τ^2	b	u	ρ^2	ρ^3	ρ
τ^2	τ^2	b	σ^2	e	ρ^2	u	ρ	ρ^3
b	b	σ^2	e	τ^2	ρ	ρ^3	u	ρ^2
e	e	τ^2	b	σ^2	ρ^3	ρ	ρ^2	u

When the domain of a binary operation is finite we can build a table, for example, Table 3.1, *where the result of the binary operation for each pair of elements is shown.*

The Rearrangement Lemma obliges us to write the rows and columns inside the multiplication table in such a way that all are distinct: a given element of the group can not appear twice in a given row or column. Thus, we can only have $a^2 = b$. And the table must be as similar to Table 3.2.

3.1.20 Cyclic Groups

Example 3.10: The *cyclic group of order n, Z/nZ.*
A cyclic group $\mathcal{G}$ is a group that can be generated by a single element, g, called a "generator" of the group, such that, for some $g \in \mathcal{G}$, all $x \in \mathcal{G}$ has the form g^m when written multiplicatively, where $m \in I$. Every element of the group is a power of g (a multiple of g when the notation is additive, see Table 3.3).

3.1.21 Permutation Groups

Let $S = 1, 2, \ldots, n$ and let us consider the set S_n of the $n!$ permutations of these n symbols. It is irrelevant if they are natural numbers. We can define a permutation operation $\circ$

among the elements of S_n. It is usual to call S_n the symmetric group of n symbols and any subgroup of S_n a permutation group of n symbols.

Example 3.11: *All permutations of an ordered set of objects.*
A permutation is a procedure to reorder a set.

Example 3.12: *The diedral group (Klein).*
Show that the subset $u = (1), \rho, \rho^2, \rho^3, \sigma^2, \tau^2, b = (1, 3), e = (2, 4)$ *of* S_4 *is a group called the dihedral group.*

We can use the properties of symmetry of a square to show these properties. We advise you to draw on a cardboard the letters and numbers as shown in Figures 3.2 and 3.3. *Hence, consider the square with vertex denoted by* 1, 2, 3, 4 *with center in O, diagonals* 103 *and* 204 *and the AOB and COD. We shall all have possible motions that the square can be submitted to.*

ρ *denotes a rotation of the square by* 90° *around O in the counterclockwise.* $\rho^2 = \rho \circ \rho$ *is the rotation by* 180° *around O,* ρ^3 *is a rotation of* 270° *and* $\rho^4 = u$ *is the rotation of* 360° *around O.* σ^2 *is the rotation of* 180° *around AOB and* τ^2 *is the rotation of* 180° *around COD. e and b are, respectively, the rotation of* 180° *around diagonals* 103 *and* 204. *Remark that the group is non-Abelian.*

The importance of group theory is that it provides arithmetic representations of geometrical operations.

Example 3.13: *The symmetric group, or permutation group,* S_n.

Example 3.14: *The unitary group,* $U(n)$.

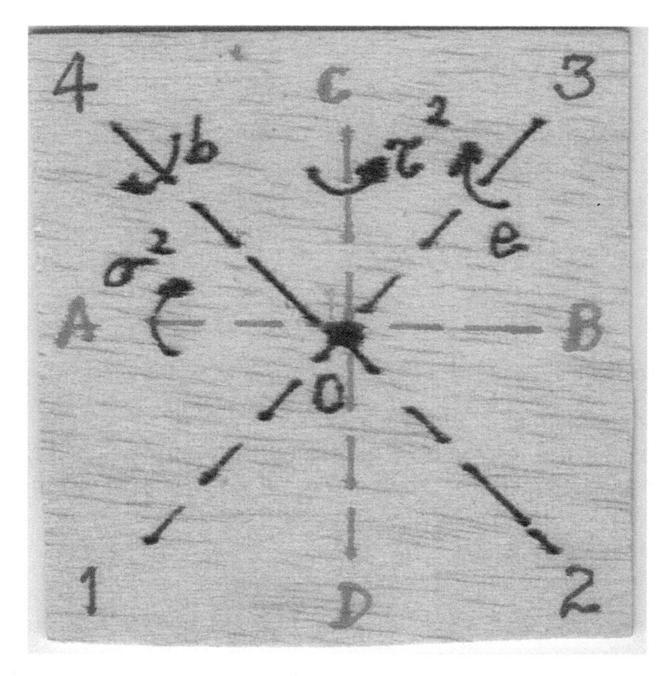

FIGURE 3.2
Diedral group cartoon: verse.

FIGURE 3.3
Diedral group cartoon: reverse.

A *group representation* describes abstract groups in terms of linear transformations of vector spaces. It can be used to represent group elements as matrices in such a way that the group operation can be represented by matrix multiplication. Group representations allow many group-theoretic problems to be reduced to problems in linear algebra.

It is a mapping of the abstract group elements to a set of matrices $a \to D(a)$, such that, if $ab = c$, then $D(a)D(b) = D(c)$.

A *representation* of a finite group G on a finite-dimensional complex vector space V is a homomorphism ρ: $G \to GL(V)$ of G to the group of automorphisms of V.

3.1.22 Classical Groups

The elements of the groups defined in this course are matrices with elements belonging to one of the four fields:

- **R**, the field of real numbers;
- **C**, the field of complex numbers;
- **H**, the field of quaternions;
- **O**, the field of octonions (or Cayley numbers).

The algebras of $n \times n$ matrices with entries (or elements) in **R**, **C**, **H**, and **O** are denoted by $M_n(\mathbf{R})$, $M_n(\mathbf{C})$, $M_n(\mathbf{H})$, $M_n(\mathbf{O})$, respectively.

We denote by $\mathrm{End}_F(V)$ the F-linear maps from a vector space V, with scalar field F, into itself. Then, we have

- $M_n(\mathbf{R}) \cong \text{End}_R(\mathbf{R}^n)$;
- $M_n(\mathbf{C}) \cong \text{End}_C(\mathbf{C}^n)$;
- $M_n(\mathbf{H}) \cong \text{End}_H(\mathbf{H}^n)$;
- $M_n(\mathbf{O}) \cong \text{End}_O(\mathbf{O}^n)$.

We will deal with representations in the next section. The symbol $\cong$ holds for **congruence**, that is, an equivalence relation on an algebraic structure (e.g., group) that is compatible with the structure.

3.1.23 Group Representations

A representation of a symmetry group is a set of square matrices, all of the same dimension.

3.1.23.1 Group of Rotations in Three Dimensions

Let us consider two transformations A and B, and let be $\overrightarrow{A}$ and $\overrightarrow{B}$ the two vectors associated with the transformations. In order that $\overrightarrow{A}$ and $\overrightarrow{B}$ play the role of vectors, they must satisfy the commutative law: $\overrightarrow{A} + \overrightarrow{B} = \overrightarrow{B} + \overrightarrow{A}$.

But the addition of two rotations, that is the product AB of two matrices is not commutative: $AB \neq BA$.

As we can not represent a finite rotation by one vector, we can however do it for an infinitesimal rotation. Then, the component x_1' of a vector $\overrightarrow{r}$ is practically equal to x_1

$$x_i' = x_i + \sum_j \epsilon_{ij} x_j = \sum_j (\delta_{ij} + \epsilon_{ij}) x_j \tag{3.321}$$

or, written in matricial notation

$$x_i' = (1 + \epsilon) x \tag{3.322}$$

The order of successive operations has no importance. If $(1 + \epsilon_1)$ and $(1 + \epsilon_2)$ are two infinitesimal transformations, then

$$(1 + \epsilon_1)(1 + \epsilon_2) = 1^2 + \epsilon_1 1 + 1\epsilon_2 + \epsilon_1\epsilon_2 \sim 1 + \epsilon_1 + \epsilon_2 + \mathcal{O}(\epsilon^2) \tag{3.323}$$

These simple considerations show that the commutative property allows the use of infinitesimal transformations to be represented by vectors.

Hence, if

$$\begin{aligned} A &= 1 + \epsilon \\ A^{-1} &= 1 - \epsilon \\ \Rightarrow AA^{-1} &= (1 + \epsilon)(1 - \epsilon) = 1 \end{aligned} \tag{3.324}$$

Example 3.15: *Infinitesimal rotation around Oz axis.*

$$A = \begin{pmatrix} \cos\phi & \sin\phi & 0 \\ -\sin\phi & \cos\phi & 0 \\ 0 & 0 & 1 \end{pmatrix} \tag{3.325}$$

When $\phi \to d\phi \ll 1$, *then*

$$A = \begin{pmatrix} 1 & d\phi & 0 \\ -d\phi & 1 & 0 \\ 0 & 0 & 1 \end{pmatrix} = 1 + \epsilon \tag{3.326}$$

This means that

$$\epsilon = \begin{pmatrix} 0 & d\phi & 0 \\ -d\phi & 0 & 0 \\ 0 & 0 & 0 \end{pmatrix} = d\phi \begin{pmatrix} 0 & 1 & 0 \\ -1 & 0 & 0 \\ 0 & 0 & 0 \end{pmatrix} \tag{3.327}$$

Now, we may recall that the transpose of a matrix is the object obtained by replacing all elements a_{ij} by a_{ji}

$$(A^{\top})^{-1} = (A^{-1})^{\top} \tag{3.328}$$

An antisymmetric matrix is a square matrix that satisfies the identity $A = -A^{\top}$. This condition is verified here, since

$$\epsilon^{\top} = d\phi = \begin{pmatrix} 0 & -1 & 0 \\ 1 & 0 & 0 \\ 0 & 0 & 0 \end{pmatrix} \Rightarrow A^{\top} = 1 + \epsilon^{\top} = 1 - \epsilon \tag{3.329}$$

All the rotations in space around a fixed point O with orthogonal axis Ox_1, Ox_2, Ox_3 transform in the following manner. If we consider the rotation around axis Ox_1 with angle θ_1, the transformation takes place according to the relationship

$$\begin{aligned} x_1' &= x_1 + 0x_2 + 0x_3 \\ x_2' &= 0x_1 + \cos\theta_1 x_2 + \sin\theta_1 x_3 \\ x_3' &= 0x_1 - \sin\theta_1 x_2 + \cos\theta_1 x_3 \end{aligned} \tag{3.330}$$

This rotation can be defined by means of the matrix

$$g_{\theta_1} = \begin{pmatrix} 1 & 0 & 0 \\ 0 & \cos\theta_1 & \sin\theta_1 \\ 0 & -\sin\theta_1 & \cos\theta_1 \end{pmatrix} \tag{3.331}$$

In a similar way, we can build the representative matrix for rotations around axis Ox_2 and Ox_3

$$g_{\theta_2} = \begin{pmatrix} \cos\theta_2 & 0 & \sin\theta_2 \\ 0 & 1 & 0 \\ -\sin\theta_2 & 0 & \cos\theta_2 \end{pmatrix}; \; g_{\theta_3} = \begin{pmatrix} \cos\theta_3 & \sin\theta_3 & 0 \\ -\sin\theta_3 & \cos\theta_3 & 0 \\ 0 & 0 & 1 \end{pmatrix} \tag{3.332}$$

The set of all rotations around an arbitrary axis forms a group. This group is infinite and continues.[18] The number of independent parameters that form the group is called their order. The group of rotations in three dimensions has three (θ_1, θ_2, θ_3).

The order of the transformation matrix gives the dimension of the group, in this case, the order of the group of rotations in their dimensions is three.

The groups continuous of finite order are called **Lie groups**. The group of rotations in three dimensions is a Lie group.

3.1.23.2 Lie Algebra

It is more appropriate to make Taylor developments of the trigonometric functions in order to study infinitesimal rotations. Therefore, we have

$$A_1 = i\left(\frac{\partial}{\partial \theta_1} g_{\theta_1}\right)_{\theta_1=0} = \begin{pmatrix} 0 & 0 & 0 \\ 0 & 0 & -i \\ 0 & i & 0 \end{pmatrix} \tag{3.333}$$

We also have

$$A_2 = \begin{pmatrix} 0 & 0 & i \\ 0 & 0 & 0 \\ -i & 0 & 0 \end{pmatrix}; A_3 = \begin{pmatrix} 0 & i & 0 \\ -i & 0 & 0 \\ 0 & 0 & 0 \end{pmatrix} \tag{3.334}$$

A_1, A_2, and A_3 are called the group generators. The number of group generators is equal to the group order. We can verify that the group generators obey to the following commutation relations

$$\begin{aligned} [A_1, A_2] \equiv A_1A_2 - A_2A_1 &= iA_3 \\ [A_2, A_3] &= iA_1 \\ [A_3, A_1] &= iA_2 \end{aligned} \tag{3.335}$$

The set of generators A_1, A_2, and A_3 generates the Lie group algebra of the group of rotations in three dimensions.

We call **Lie algebra of Lie group** the set of elements N that observe the following conditions:

1. If X and Y are elements of the N set, then the sum $X + Y$ and the product αX (with α any number) still belong to the N set.
2. The commutator of two elements X and Y of the set is expressible with elements of the N set.
3. The commutators of elements of the N set verify the following identities

$$\begin{aligned} [XY] + [YX] &= 0 \\ [X(Y + Z)] &= [XY] + [XZ] \\ [X[YZ]] + [Y[ZX]] + [Z[XY]] &= 0 \text{ Jacobi relationship} \end{aligned} \tag{3.336}$$

The existence of a unit element and its inverse in a Lie algebra is not necessary.[19]

Lie algebra of an arbitrary group G is given by the relationship

$$[T_i, T_k] = f_{ikl} T_l \tag{3.337}$$

where T_i is a group of generators of G. The constants f_{ikl} are called **structure constants**. According to Jacobi relationships, they have the following properties

$$\begin{aligned} f_{ikl}f_{lmn} + f_{kml}f_{lin} + f_{mil}f_{lkn} &= 0 \\ f_{ikl} &= -f_{kil} \end{aligned} \tag{3.338}$$

Example 3.16: *Consider the reaction*

$$\pi^+ + p \to \pi^+ + p \tag{3.339}$$

The scattering amplitude has spin dependence. We can follow J. Sakurai treatment and write the incident plane wave as

$$\Psi_{inc} = e^{ikz}\chi \tag{3.340}$$

Here, χ is the spin eigenfunction for a spin $\frac{1}{2}$ particle, $\chi_+ = \begin{pmatrix}1\\0\end{pmatrix}$ and $\chi_- = \begin{pmatrix}0\\1\end{pmatrix}$. The scattered wave function is given by

$$\Psi_{sca} = \frac{e^{ikr}}{r}M\chi \tag{3.341}$$

where M is a 2×2 matrix defined by Eq. 3.341 *and is called the scattering matrix. In* Eq. 3.339 *the incident particle π^+ is a spinless particle with angular momentum l being scattered by a spin $\frac{1}{2}$ target. Angular momentum and parity conservation require that the matrix elements of M be scalars and invariant under rotation and reflection.*

We lead to write

$$M = f(\theta) + ig(\theta)\vec{\sigma}\cdot\vec{n} \tag{3.342}$$

with $f(\theta)$ and $g(\theta)$ being dependent on the energy through the angle θ. Why this important matrix has this form? This is a 2×2 matrix and it can be formed in terms of the unit matrix and the three Pauli matrix σ_x, σ_y, and σ_z.

If we choose the Oz axis along the direction of the incident proton and noting that

$$\vec{n} = \frac{\vec{p_i} \times \vec{p_f}}{\left|\vec{p_i} \times \vec{p_f}\right|} \tag{3.343}$$

resulting in

$$\vec{n} = (\sin\varphi, -\cos\varphi, 0) \tag{3.344}$$

then

$$\vec{\sigma}\cdot\vec{n} = \begin{pmatrix} 0 & 1 \\ 1 & 0 \end{pmatrix} \sin\varphi + \begin{pmatrix} 0 & -i \\ i & 0 \end{pmatrix} (-\cos\varphi) = \begin{pmatrix} 0 & ie^{-i\varphi} \\ -ie^{i\varphi} & 0 \end{pmatrix} \tag{3.345}$$

and hence

$$M = f(\theta)\begin{pmatrix} 1 & 0 \\ 0 & 1 \end{pmatrix} + ig(\theta)\begin{pmatrix} 0 & ie^{-i\varphi} \\ -ie^{i\varphi} & 0 \end{pmatrix} = \begin{pmatrix} f(\theta) & -g(\theta)e^{-i\varphi} \\ g(\theta)e^{i\varphi} & f(\theta) \end{pmatrix} \tag{3.346}$$

The angle φ can be defined at once if the scattering occurs in the x-z plane, then $\varphi = \pi$.

3.1.23.3 Lie Algebras

Lie algebras[20]

From the group generators, we can form new operators that commute with all the generators. These generators are called **Casimir operators**.

Example 3.17: *The rotation group in three dimensions has one Casimir operator*

$$C = A_1^2 + A_2^2 + A_3^2 \tag{3.347}$$

This operator can be associated with the angular momentum squared.

3.1.24 SU(2) Representation of Maxwell's Equations

According to Barrett [7] the SU(2) representation of the electromagnetic field represents situations in which the vector potential $\mathbf{A}_\mu$ ($\mu = 0, 1, 2, 3$) has physical significance. See Section 3.1.23 about general linear groups.

In this theoretical frame, electrons possess charge, mass, spin– 1/2, and magnetic moment. The SU(2) representation predicts the existence of magnetic monopoles, that is, elementary particles that carry units of magnetic charge. On another side, spin has important theoretical implications and practical applications, among them are nuclear magnetic resonance spectroscopy, electron spin resonance spectroscopy, and magnetic resonance imaging (MRI) in medicine, which relies on proton spin density; giant magnetoresistive (GMR) sandwich structure consisting of alternating ferromagnetic and nonmagnetic metal layers on the base of drive head technology in modern hard disks. It is envisioned direct application of spin as a binary information carrier in spin transistors. This concept was proposed in 1990 and is known as Datta and Das spin transistor [8].[21] Electronics based on spin transistors are called **spintronics**. Figure 3.4 illustrates the concept of the Datta and Das spin transistor.

There is an essential difference between electromagnetism and Yang–Mills theory in the sense that the concept of photon is replaced by a set of local non-Abelian massless potential fields with internal-space noncommuting operators.

The Yang–Mills field equation is

$$D^\mu F_{\mu\nu} = \partial^\mu F_{\mu\nu} - iq[A^\mu, F_{\mu\nu}] = j_\nu \tag{3.348}$$

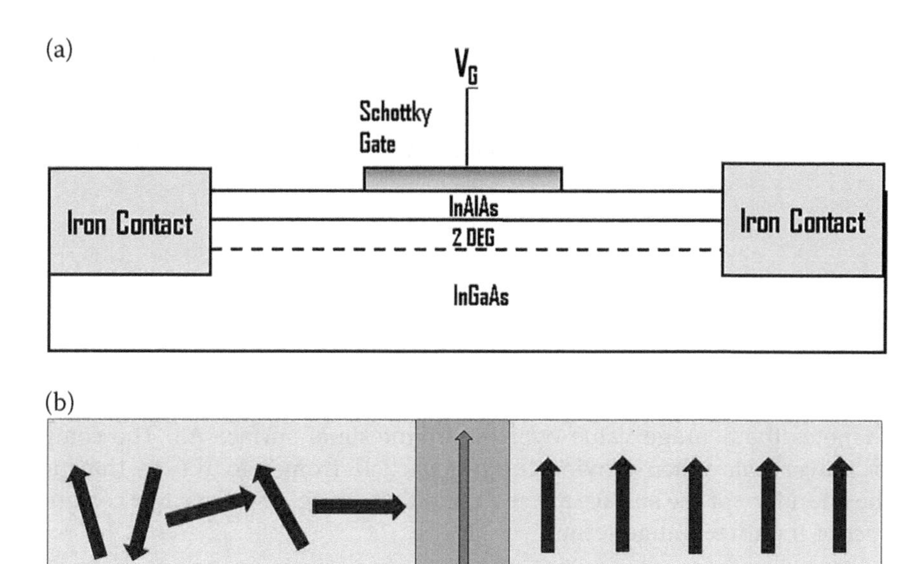

FIGURE 3.4
(a) Building structure of the Datta and Das spin transistor. (b) With the ferromagnetic source acting as a spin polarizer.

In electrodynamics, the U(1) representation of the homogeneous Maxwell equation is written as

$$\partial_\mu F_{\nu\lambda} + \partial_\lambda F_{\mu\nu} + \partial_\nu F_{\lambda\mu} = 0 \tag{3.349}$$

We remark here that Eq. 3.350 is not derivable from a Lagrangian and only depends on the space-time properties of the Maxwell tensor. Hence, we can replace Eq. 3.350 with the gauge covariant derivative, such as

$$D_\mu F_{\nu\lambda} + D_\lambda F_{\mu\nu} + D_\nu F_{\lambda\mu} = 0 \tag{3.350}$$

This equation means that the total flux out of the test volume vanishes [9].

To understand the origin of the new term on the Yang–Mills field equation we need to calculate the value of the field $F_{\mu\nu}$ at two nearby positions x and $x + dx$ in order to obtain its rate of change. Considering that gauge fields do not commute, we need to calculate the magnitude and direction (in what is called the particle internal space) of $F_{\mu\nu}$. So, its value depends on the particular path taken from x to $x + dx$. For the purpose of the calculation, we shall consider a test particle moving around the boundary of a surface: the internal space direction will rotate and the phase of the test charge wavefunction will variate. Now, we know that Stoke's theorem relates the net phase change to the total flux through the surface. The infinitesimal gauge transformation acting on the charged particle wavefunction for the spatial displacement $x \to x + dx$ is

$$U_x(dx) = 1 - iqA{\cdot}dx \tag{3.351}$$

The derivative of the flux can be calculated by measuring the flux at x and $x + dx$. So, we move the charged particle from x to $x + dx$ and move back again the charged particle to its initial position.

Along the considered path the net change of the wavefunction is given by

$$d\psi = -iq \sum A \cdot dx\psi \tag{3.352}$$

The summation over the vector potential can be related to the Maxwell tensor $F_{\mu\nu}$ using Stoke's theorem

$$\sum A \cdot dx = \iint D \cdot dS \simeq \widetilde{F} \cdot \Delta S \tag{3.353}$$

Here, $\widetilde{F}$ denotes the average value over the infinitesimal surface ΔS. The complete sequence of movements when moving the test particle from x to $x + dx$, transporting it around the boundary of the surface at $x + dx$ and then bringing it back to x corresponds to the product of the three gauge terms

$$[1 - iaA \cdot dx][1 - iq\widetilde{F}(x + dx) \cdot \Delta S][1 + iqA \cdot dx] \tag{3.354}$$

But, developing the $\widetilde{F}(x + dx)$ to first order in x, we obtain

$$\widetilde{F}(x + dx) \cdot \Delta S \simeq \widetilde{F}(x) \cdot \Delta S + \partial_\mu \widetilde{F} \cdot \Delta S dx^\mu \tag{3.355}$$

Inserting Eq. 3.354 into Eq. 3.355, it gives

$$\widetilde{F} \cdot \Delta S + \partial_\mu \widetilde{F} \cdot \Delta S dx^\mu - iq[A_\mu, \widetilde{F} \cdot \Delta S] dx^\mu \tag{3.356}$$

After subtracting the flux at x, $\widetilde{F}(x) \cdot \Delta S$, we obtain the net change of flux between x and $x + dx$

$$\{\partial_\mu \widetilde{F} \cdot \Delta S - iq[A_\mu, \widetilde{F} \cdot \Delta S]\} dx^\mu \tag{3.357}$$

In our notation $\widetilde{F} \cdot \Delta S = F_{\mu\nu} dS^{\mu\nu}$, and hence we arrive at our gauge covariant derivative

$$D^\lambda F_{\mu\nu} = \partial^\lambda F_{\mu\nu} - iq[A^\lambda, F_{\mu\nu}] \tag{3.358}$$

Next, we will define the Yang–Mills equation as

$$\partial^\mu F_{\mu\nu} = J_\nu \tag{3.359}$$

defining a new total current density

$$J_\nu = j_\nu + iq[A^\mu, F_{\mu\nu}] \tag{3.360}$$

We will write now the non-Abelian equations in terms of the electric and magnetic fields. The electric and magnetic fields components are defined through the relations

$$F_{0i} = E_i, \quad F_{ij} = \epsilon_{ijk} B_k \tag{3.361}$$

Let us write Eq. 3.359 for the component $\nu = 1$. We have

$$\begin{aligned}\partial^0 F_{01} + \partial^1 F_{11} + \partial^2 F_{21} + \partial^3 F_{31} = J_1 = j_1 + iq[A^0, F_{01}] + iq[A^1, F_{11}] + iq[A^2, F_{21}] \\ + iq[A^3, F_{31} - \partial_t E_x + \partial_y H_z - \partial_z H_y = j_x + iq[A^t, E_x] \\ + iq[A^{1y}, H_z] - iq[A^z, H_y]\partial_t E_x - [\nabla \times H]_x + iq[A_0, E_x] \\ + iq\{[A_y, H_z]_x + [A_z, H_y]_x\} = j_x \Rightarrow \partial_t \mathbf{E} - \nabla \times \mathbf{H} \\ + iq[A_0, \mathbf{H}] + iq(\mathbf{A} \times \mathbf{E} - \mathbf{E} \times \mathbf{A}) = 0 \qquad (3.362)\end{aligned}$$

We can obtain the non-Abelian version of Gauss's law putting $\nu = 0$. It gives

$$\nabla \cdot \mathbf{E} + iq(\mathbf{A} \cdot \mathbf{E} - \mathbf{E} \cdot \mathbf{A}) = j_0 \qquad (3.363)$$

Exercise: Obtain Eq. 3.363.
Solution: We have, putting $\nu = 0$ in Eq. 3.359

$$\partial^0 F_{00} + \partial^1 F_{10} + \partial^2 F_{20} + \partial^3 F_{30} = J_0 \qquad (3.364)$$

or,

$$\begin{aligned}-\partial^x E_x - \partial^y E_y - \partial^z E_z = j_0 + iq[A^0, F_{00}] + iq[A^1, F_{10}] + iq[A^2, F_{20}] + iq[A^3, F_{30}]\partial_x E_x \\ + \partial_y E_y + \partial_z E_z = j_0 - iq[A^x, E_x] - iq[A^y, E_y] - iq[A^z, E_z]\nabla \cdot \mathbf{E} \\ + iq\{A^x E_x - E_x A^x + A^y E_y - E^y A_y + A^z E_z - E^z A_z\} = j_0 \qquad (3.365)\end{aligned}$$

The last equation can be reset under the compact form

$$\nabla \cdot \mathbf{E} + iq(\mathbf{A} \cdot \mathbf{E} - \mathbf{E} \cdot \mathbf{A}) = j_0 \qquad (3.366)$$

The conservation of electric charge in QED is expressed by means of the equation

$$\partial^\nu (\partial^\mu F_{\mu\nu}) = \partial^\nu j_\nu = 0 \qquad (3.367)$$

Once more, we generalize the above equation to the conservation of non-Abelian charge by simply replacing the partial derivatives with the gauge covariant derivative

$$D^\nu (D^\mu F_{\mu\nu}) = D^\nu j_\nu = 0 \qquad (3.368)$$

3.1.25 SO(3,1) Representation of Maxwell's Equations

For a massless particle, any $|\psi\rangle$ physical state can be expanded in terms of either the helicity basis states $|k; \xi\rangle$ or the direct product states $|k\rangle \left| \begin{matrix} j & j' \\ \mu & \mu' \end{matrix} \right\rangle$

$$|\psi\rangle = \sum_{k,\xi} |k;\xi\rangle\langle k;\xi|\psi\rangle \tag{3.369}$$

and

$$|\psi\rangle = \sum_{k,\mu,\mu'} |k\rangle \left| \begin{matrix} j & j' \\ \mu & \mu' \end{matrix} \right\rangle \left\langle k; \begin{matrix} j & j' \\ \mu & \mu' \end{matrix} \middle| \psi \right\rangle \tag{3.370}$$

Here, k is a 4-vector obeying for free-fields (no sources), with $\mathbf{k}$ representing the 3-momentum vector and $k_4 = \hbar\omega/c$. In the presence of sources, the 4-vector k does not verify $k{\cdot}k = 0$. For photons the helicity states ξ ($\xi = \pm$), the helicity states are given by the angular momentum aligned along the direction of motion. when the helicity is +1 the photon has right-hand polarization; when the helicity is –1 the polarization is on left-hand. Also, note that Eq. 3.370 is the sum for indexes μ, μ': $-j \le \mu \le +j$, $-j' \le \mu' \le +j'$.

It can be shown that Maxwell's equations can be written for massless particles with positive helicity under the form [10, 11]

$$\left\{ J_i^j k_i^0 - j k_4^0 I_{2j+1} \right\} \left\langle k^0; \begin{matrix} j & 0 \\ m & 0 \end{matrix} \middle| \psi \right\rangle = 0 \tag{3.371}$$

For the negative helicities $\xi = -j$, the valid equation is

$$\left\{ J_i^j k_i^0 + j k_4^0 I_{2j+1} \right\} \left\langle k^0; \begin{matrix} 0 & j \\ 0 & m \end{matrix} \middle| \psi \right\rangle = 0 \tag{3.372}$$

with the notation $J_i k_i \equiv \mathbf{J}{\cdot}\mathbf{k}$.

Complex fields can be defined by the identity

$$\psi_{jm}(x) = \langle x|k\rangle \left\langle k; \begin{matrix} j & 0 \\ m & 0 \end{matrix} \middle| \psi \right\rangle \tag{3.373}$$

with $j = 1$, and $m = +1, 0, -1$, or x, y, z, or $1, 2, 3$. We can replace $k_4 = -\frac{1}{i}\frac{1}{c}\frac{\partial}{\partial t}$ and $\mathbf{k} = \frac{1}{i}\nabla$. Here, the angular matrices are obtained from

$$\mathbf{J} = \frac{i}{2}\boldsymbol{\alpha} \tag{3.374}$$

where the $\boldsymbol{\alpha}$ are given by

$$\alpha_1 = \begin{bmatrix} 0 & 0 & 0 \\ 0 & 0 & i \\ 0 & -i & 0 \end{bmatrix}, \alpha_2 = \begin{bmatrix} 0 & 0 & -i \\ 0 & 0 & 0 \\ i & 0 & 0 \end{bmatrix}, \alpha_3 = \begin{bmatrix} 0 & i & 0 \\ -i & & 0 \\ 0 & 0 & 0 \end{bmatrix} \tag{3.375}$$

The matrices I are the following [11]

$$I_1 = \begin{bmatrix} 0 & -\alpha_1 \\ \alpha_1 & 0 \end{bmatrix}, \; I_2 = \begin{bmatrix} 0 & -\alpha_2 \\ \alpha_2 & 0 \end{bmatrix}, \; I_3 = \begin{bmatrix} 0 & -\alpha_3 \\ \alpha_3 & 0 \end{bmatrix} \tag{3.376}$$

Hence, we can write Maxwell's equations for positive helicities as

$$\langle k | x \rangle \left\{ \mathbf{J} \cdot \frac{1}{i} \nabla + \frac{1}{i} \frac{\partial}{\partial (ict)} I_3 \right\} \langle x | k \rangle \left\langle k; \begin{matrix} 1 & 0 \\ m & 0 \end{matrix} \middle| \psi \right\rangle = 0 \tag{3.377}$$

We can write under a standard form

$$\begin{bmatrix} -\frac{i}{c}\frac{\partial}{\partial t} & +\partial_3 & -\partial_2 \\ -\partial_3 & -\frac{i}{c}\frac{\partial}{\partial t} & +\partial_1 \\ +\partial_2 & -\partial_1 & -\frac{i}{c}\frac{\partial}{\partial t} \end{bmatrix} \begin{bmatrix} B_1 + iE_1 \\ B_2 + iE_2 \\ B_3 + iE_3 \end{bmatrix} = 0 \tag{3.378}$$

Finally, the resulting three equations can be represented in a vectorial equation

$$-\frac{i}{c}\frac{\partial}{\partial t}(\mathbf{B} + i\mathbf{E}) - \nabla \times (\mathbf{B} + i\mathbf{E}) = 0 \tag{3.379}$$

The real part of the above equation gives

$$+\frac{1}{c}\frac{\partial \mathbf{E}}{\partial t} - \nabla \mathbf{B} = 0 \tag{3.380}$$

and the imaginary part

$$-\frac{1}{c}\frac{\partial \mathbf{B}}{\partial t} - \nabla \mathbf{E} = 0 \tag{3.381}$$

Both equations are valid for positive helicity. For negative helicity, we need to solve

$$\begin{bmatrix} +\frac{i}{c}\frac{\partial}{\partial t} & +\partial_3 & -\partial_2 \\ -\partial_3 & +\frac{i}{c}\frac{\partial}{\partial t} & +\partial_1 \\ +\partial_2 & -\partial_1 & +\frac{i}{c}\frac{\partial}{\partial t} \end{bmatrix} \begin{bmatrix} B_1 - iE_1 \\ B_2 - iE_2 \\ B_3 - iE_3 \end{bmatrix} = 0 \tag{3.382}$$

but the final "curl" equations are identical to Eqs 3.378–3.379. The transversality condition $\mathbf{k} \cdot (\mathbf{B} + i\mathbf{E})$ gives the two Maxwell "div" equations, $\nabla \cdot \mathbf{B} = 0$, and $\nabla \cdot \mathbf{E} = 0$.

3.1.26 Chirality

What distinguishes life from inanimate objects are two properties: self-replication and chiral purity of the main bio-polymers.

The diversity of life on the Earth is surprising, but not least surprising is that the basis of life is only 20 amino acids are the constituents of proteins of all living beings on this planet. Five azote constituents: adenine, guanine, cytosine, thymine, and uracil combined with sugar and phosphate are put together in long chains of nucleic acids: DNA and RNA.

Complex plasmas are a suitable medium to create helical structures, showing a richness of nature and the beauty of self-organized structures [12].

3.2 Zilch Densities Z

It was found unexpectedly that for any arbitrary electromagnetic field in vacuum, Maxwell's equations validate the differential conservation law

$$div\left[\mathbf{E}\times\frac{\partial\mathbf{E}}{\partial t}+\mathbf{H}\frac{\partial\mathbf{H}}{\partial t}\right]+\frac{\partial}{\partial t}[\mathbf{E}\cdot(\nabla\mathbf{E})+\mathbf{H}\cdot(\nabla\mathbf{H}]=0 \quad (3.383)$$

This new conservation law led to the discovery of a tensor of valence three that expresses nine additional conservation laws. The complete set of 10 new conservation equations augments the known stress-energy quantities of the electromagnetic field [13].

The contravariant components of Lipkin's Zilch densities $\mathcal{Z}$ are the following

$$\mathcal{Z}^{000}=\frac{\epsilon_0}{2}[\mathbf{E}\cdot\nabla\times\mathbf{E}]+c^2\mathbf{B}\cdot[\nabla\times\mathbf{B}]] \quad (3.384)$$

This conserved quantity is defined by the name "optical chirality." The following ones are

$$\mathcal{Z}^{0i0}=\frac{\epsilon_0 c}{2}[\mathbf{E}\times[\nabla\times\mathbf{B}]-\mathbf{B}\times[\nabla\times\mathbf{E}] \quad (3.385)$$

$$\mathcal{Z}^{ij0}=\delta_{ij}\mathcal{Z}^{000}-\frac{\epsilon_0}{2}[E_i[\nabla\times\mathbf{E}]_j+E_j[\nabla\times\mathbf{E}]_i+c^2B_i[\nabla\times\mathbf{B}]_j+c^2B_j[\nabla\times\mathbf{B}]_i] \quad (3.386)$$

$$\mathcal{Z}^{ijk}=\delta_{ij}\mathcal{Z}^{00k}+\frac{\epsilon_0 c}{2}\left[B_i\frac{\partial E_j}{\partial x_k}+B_j\frac{\partial E_i}{\partial x_k}-E_i\frac{\partial B_j}{\partial x_k}-E_j\frac{B_i}{\partial x_k}\right] \quad (3.387)$$

$$\mathcal{Z}^{\mu\nu\gamma}=\mathcal{Z}^{\nu\mu\gamma} \quad (3.388)$$

$$\mathcal{Z}^{00i}=\mathcal{Z}^{0i0} \quad (3.389)$$

$$\mathcal{Z}^{0ij}=\mathcal{Z}^{ij0} \quad (3.390)$$

They verify the balance equation

$$\partial_\gamma \mathcal{Z}^{\mu\nu\gamma} = 0 \tag{3.391}$$

and

$$Z^{\mu\nu 0} = \iiint \mathcal{Z}^{\mu\nu 0} d^3\mathbf{r} \tag{3.392}$$

represents a set of 10 conserved quantities.

3.2.1 Retrograde Flow of Zilch

We will now consider a simple case of retrograde flow of Zilch. Let us consider the propagation along the Oz axis of an electromagnetic wave. Assume

$$E_x = H_y = Ae^{-jkz} \tag{3.393}$$

$$E_y = -H_x = Be^{-jkz} \tag{3.394}$$

$$E_z = H_z = 0 \tag{3.395}$$

A, B are complex coeficients. We obtain

$$< \mathcal{Z}^{000} > = \frac{1}{T}\int_0^T \mathcal{Z}^{000} dt \tag{3.396}$$

$$= \frac{\epsilon_0}{4}[-AB^* + BA^*]jk \tag{3.397}$$

Use

$$\frac{1}{T}\int_0^T Re\,[Ae^{j\omega t}]\,Re\,[Be^{j\omega t}]dt = \frac{1}{2}Re\,[AB^*] \tag{3.398}$$

It can be shown that the only nonvanishing components are the following

$$< \mathcal{Z}^{0i0} > = < \mathcal{Z}^{00i} > = \tfrac{i}{3} < Z^{000} > \tag{3.399}$$

$$< \mathcal{Z}^{ij0} > = < \mathcal{Z}^{0ij} > = \delta_3^i \delta_3^j < Z^{000} > \tag{3.400}$$

$$< \mathcal{Z}^{ijk} > = \delta_3^i \delta_3^j \delta_3^k < \mathcal{Z}^{000} > \tag{3.401}$$

All components have a common real value $\frac{\epsilon_0}{4}jk\,(A^*B - AB^*)$. We have three cases to consider

Case i: Linearly polarized wave.
In this case, $\frac{A}{B}$ is real and $(A^*B - AB^*) = 0$ and the Zilch is not transported by linearly polarized waves.

Case ii: Right-circularly polarized wave.

$$A = +jk \Rightarrow jk\frac{\epsilon_0}{4}(A^*B - AB^*) = \frac{\epsilon_0}{4}jk2(-j)\,|\,B\,|^2 > 0 \tag{3.402}$$

Case iii: Left-circularly polarized wave, $A = -jB$.
In this case,

$$\frac{\epsilon_0}{4}jk\,(A^*B - AB^*) = -2k\,|\,B\,|^2\,\frac{\epsilon_0}{4} < 0 \tag{3.403}$$

Thus, a flow of Zilch follows any circularly polarized wave. However, the direction of the flow reverses with the screw sense of the wave. Additionally, the rate of the flow appears proportionally to the frequency of the wave since $k = \frac{\omega}{c}$.

3.3 Fractional Calculus and Its Applications to Physics

3.3.1 Historical Perspective

These notes aim to introduce an appropriate tool to solve a certain class of phenomena. From the historical perspective, the origin of fractional calculus is the same as differential calculus. In fact, M. de L'Hospital raises an interesting question about the possible meaning of $d^n f(x)/dx^n$ if $n = 1/2$ to G. Leibniz in a letter dated September 9, 1695. To which, Leibniz responds: *"[...] $d^{1/2}x$ will be equal to $x\sqrt{dx\!:\,x}$. this is an apparent paradox from which, one day, useful consequences will be drawn."*

A fractional derivative is an extension of the concept of a derivative operator from integer order n to arbitrary order α, where α can be a real or complex number, or a more general complex-valued function $\alpha = \alpha(x, t)$

$$\frac{d^n}{dx^n} \to \frac{d^\alpha}{dx^\alpha} \tag{3.404}$$

We know that

$$\frac{d^\alpha}{dx^\alpha}\frac{d^\beta}{dx^\beta}f(x) = \frac{d^{\alpha+\beta}}{dx^{\alpha+\beta}}f(x) \tag{3.405}$$

Using $f(x) = \exp(-\gamma x)$, we obtain

$$\frac{d^\alpha}{dx^\alpha}\exp(-\gamma x) = (-\gamma)^\alpha \exp(-\gamma x) \tag{3.406}$$

Also, we know that

$$\frac{1}{x} = \int_0^\infty \exp(-\gamma x)d\gamma; \ x > 0 \tag{3.407}$$

From it, we obtain through derivation

$$\frac{d^\alpha}{dx^\alpha}\left(\frac{1}{x}\right) = \int_0^\infty (-\gamma)^\alpha \exp(-\gamma x)d\gamma \tag{3.408}$$

Let us introduce now the new variable $y = -\gamma x$. Then it follows

$$\frac{d^\alpha}{dx^\alpha}\left(\frac{1}{x}\right) = \int_0^\infty \left(-\frac{y}{x}\right)^\alpha \exp(+y)\frac{1}{(-x)}dy = x^{-\alpha-1}\int_{-\infty}^0 y^\alpha \exp(y)dy \tag{3.409}$$

But by partial integration

$$\int_{-\infty}^0 y^\alpha \exp(y)dy = y^\alpha \exp(y) \,|_{-\infty}^0 \; - \alpha\int_{-\infty}^0 \exp(y)y^{\alpha-1}dy \tag{3.410}$$

Now, we may introduce the abbreviation

$$\overline{\Gamma}(\alpha) = \int_{-\infty}^0 y^\alpha \exp(y)dy \tag{3.411}$$

Then, we obtain the recursion relation

$$\overline{\Gamma}(\alpha) = -\alpha\overline{\Gamma}(\alpha - 1) \tag{3.412}$$

We now make another modification

$$\overline{\Gamma}(\alpha) = (-1)^\alpha\Gamma(\alpha) \tag{3.413}$$

This leads us toward the new recursive relation

$$\Gamma(\alpha) = \alpha\Gamma(\alpha - 1) \tag{3.414}$$

We have also

$$\frac{d^\alpha}{dx^\alpha}\left(\frac{1}{\alpha}\right) = x^{-\alpha-1}\overline{\Gamma}(\alpha) = (-1)^\alpha\Gamma(\alpha)x^{-\alpha-1} \tag{3.415}$$

We can obtain now the second fractal derivation. Let us start with

$$\frac{1}{x}\exp(\lambda x) = \int_{-\infty}^\lambda \exp(\gamma x)d\gamma; \quad x > 0 \tag{3.416}$$

Then, it follows after a second derivative

$$\frac{d^\alpha}{dx^\alpha}\left(\frac{1}{x}\exp(\lambda x)\right) = \int_{-\infty}^{\lambda} \gamma^\alpha \exp(\gamma x) d\gamma \tag{3.417}$$

When introducing the new variable $y = \gamma x$, we obtain

$$\frac{d^\alpha}{dx^\alpha}\left(\frac{1}{x}\exp(\lambda x)\right) = x^{-\gamma-1}\int_{-\infty}^{\lambda/x} y^\alpha \exp(y) dy \tag{3.418}$$

Put now $\Lambda = \lambda/x$. Then

$$\int_{-\infty}^{\lambda/x} y^\alpha \exp(y) dy = y^\alpha \exp(y) \left.\right|_{-\infty}^{\lambda/x} - \alpha \int_{-\infty}^{\lambda/x} \exp(y) y^{\alpha-1} dy \tag{3.419}$$

or

$$\int_{-\infty}^{\lambda/x} y^\alpha \exp(y) dy = y\,()^\alpha$$

3.3.2 Applications

3.3.2.1 Abel's Equation

A particle slides down on the curve $x = \phi(y)$ actuated by the gravitational field. It takes the time $T(h)$ to descend from a height h. Find a relation between $T(h)$ and $\phi(y)$.

Answer: The energy conservation states that

$$\frac{1}{2}v^2 = g(h-y) \tag{3.420}$$

The speed of the particle is given by

$$v = \frac{ds}{dt} = \frac{\sqrt{dx^2 + dy^2}}{dt} = \frac{\sqrt{1+\phi'^2}\,dy}{dt} \tag{3.421}$$

and, therefore,

$$T = \int_0^T dt = \int_0^h \frac{\sqrt{1+\phi'^2}}{\sqrt{2g(h-y)}} dy \tag{3.422}$$

For convenience, define a new function

$$f(y) = \frac{1}{2g}\sqrt{1+\phi'^2} \tag{3.423}$$

with Eq. 3.3.2.1 we rewrite

$$T(h) = \int_0^h \frac{f(y)}{\sqrt{h-y}} dy \tag{3.424}$$

that is, we have either

$$T = \sqrt{\pi}\, {}_0D_h^{-1/2} f \tag{3.425}$$

or,

$$f = \frac{1}{\sqrt{\pi}}\, {}_0D_y^{1/2} T \tag{3.426}$$

3.3.3 Fractional Integrals

Two types of fractional integrals are generally considered: the Riemann–Liouville and Liouville integrals. We start with the definition of Riemann–Liouville fractional integration on a finite interval of the real line.

Assuming that $f(x)$ is a continuous function on the real line, the definition of Riemann integrals is given by

$${}_aD_x^{-1} f(x) = \int_a^x f(x_1) dx_1 \tag{3.427}$$

We can extend to the next step

$${}_aD_x^{-2} f(x) = \int_a^x {}_aD_x^{-1} f(x_1) dx_1 = \int_a^x dx_1 \int_a^{x_1} f(x_2) \tag{3.428}$$

and even further, to an arbitrary order

$${}_aD_x^{-n} f(x) = \int_a^x dx_1 \int_a^{x_1} dx_2 \ldots \int_a^{x_{n-1}} dx_n f(x_n) \tag{3.429}$$

The Cauchy formula for repeated integration gives us the result

$${}_aD_x^{-n} f(x) = \frac{1}{(n-1)!} \int_a^x (x-z)^{n-1} f(z) dz \tag{3.430}$$

which can be demonstrated by induction. This integral can be generalized to noninteger $n = \alpha$ and we may remove the discrete nature of the factorial introducing the Gamma function $\Gamma(n) = (n-1)!$, and we can write the definition of **fractional integral**

$${}_aD_x^{-n} f(x) = \frac{1}{\Gamma(\alpha)} \int_a^x (x-z)^{\alpha-1} f(z) dz, \quad (\alpha > 0) \tag{3.431}$$

Exercise: Obtain ${}_0D_x^{-\alpha}x^{\mu}$.

Solution: We can use [14] the integrals

$$\int_0^u x^{\nu-1}(u-x)^{\mu-1}dx = u^{\mu+\nu-1}B(\mu,\nu) \tag{3.432}$$

$$B(\mu,\nu) = \frac{\Gamma(\mu)\Gamma(\nu)}{\Gamma(\mu+\nu)} \tag{3.433}$$

It results

$${}_0D_x^{-n}x^{\mu} = \frac{\Gamma(\mu+1)}{\Gamma(\mu+\nu+1)}x^{\mu+\nu} \tag{3.434}$$

3.4 Short Note on String Theory

3.4.1 Historical Perspective

Gabrielle Veneziano discovered accidentally in 1968 that Leonhard Euler's beta function seems to describe the strong nuclear force.

The beta function has this name because it is a function of the binomial distribution. It can be related to the gamma function.

In 1969, while looking for a function that fulfills the fundamental postulates for the scattering amplitudes of elementary particles in strong interaction, Veneziano found that Euler's beta function was a good choice. Although Veneziano was looking for a theory of strong interactions, it was through this first step that string theory was initiated.

So, string theory proposes to us a new vision of the universe, in which the universe is composed of vibrating filaments (strings) of energy. This physical theory also predicts the existence of other physical objects called *branes*.

String theory is at the level of high-energy theoretical physics, it is in fact a quantum field theory describing the way how particles exist and how forces are generated based on the way hidden dimensions are wrapped up into very small size (a process called compactification).

Although the theory was developed in 1968 in an attempt to explain the behavior of hadrons, it was soon abandoned due to the extra-hidden dimensions required until the mid-1980s. Later on, in the 90s, it become a more sophisticated theory, the M-theory.[22]

String theory unites all the panoply of particles but also the forces responsible for their interactions. The electromagnetic field is well fitted to the quantum theory, this type of force kind of propagates through space making it possible for objects to interact. However, gravity doesn't. And why? Mainly because gravity warps space.[23]

3.4.2 Some Properties of Veneziano and Virasoro Amplitudes

In 1968, Gabriel Veneziano postulated 4-particules scattering amplitudes $A(s,t,u)$ are given by

$$A(s,t,u) = V(s,t) + V(s,u) + V(t,u) \tag{3.435}$$

where

$$V(s,t) = \int_0^1 x^{-\alpha(s)-1}(1-x)^{-\alpha(t)-1}dx \equiv B(-\alpha(s), -\alpha(t)) \tag{3.436}$$

with B denoting the **Euler's beta function**, and $\alpha(s)$ and $\alpha(t)$ the **Regge trajectories**.

3.4.3 Regge Poles

One important particle-classification scheme states that hadrons can be grouped into families with distinct angular momentum and mass, but which are similar in others aspects. The members of each family lie on a curve, called **Regge trajectory**, a designation that came from the name of its mentor.[24] This theory of the Regge poles comes from the nonrelativistic Schrödinger equation. We consider in particular the wave-expansion for a scattering process of spinless particles

$$\Psi = e^{ikz} + \frac{e^{ikr}}{r} f(\theta, E) \tag{3.437}$$

where

$$f(\theta, E) = \sum_{l=0}^{\infty} (2l+1)F(l,E)P_l(\cos\theta) \tag{3.438}$$

the differential cross-section can be calculated through the relation

$$\frac{d\sigma}{d\Omega} = |f(\theta, E)|^2 \tag{3.439}$$

The function $F(l, E)$ is given by

$$F(l,E) = \frac{1}{2ik}(e^{2i\delta_l(E)} - 1) = \frac{1}{k}e^{i\delta_l(E)}\sin\delta_l(E) \tag{3.440}$$

Regge uses a technique developed before by Watson and Sommerfeld, transforming $f(\theta, E)$ into an integral in the complex l-plane along a contour C_1

$$f(z,E) = \frac{i}{2}\int_{C_1} \frac{(2l+1)F(l,E)P_l(-z)}{\sin \pi l} dl \tag{3.441}$$

where $z \equiv \cos\theta$ and treat l as a complex number.

Notes

1 In French it is sometimes called "direction proper."

2 We can say ultimately that this is just a change of notation, but the capacity of calculation is greatly improved.

3 See, e.g., *The Geometry of Physics: An Introduction*, Theodore Frankel, p. 50 (CUP, New York, 2004).

4 From the Greek, brachistos means the shortest, and chronos, means time.
5 Johann Bernoulli defied the readers of Acta Eruditorum in June, 1696, to solve this problem, and with its solution the calculus of variation was born
6 Obtained by Eugenio Beltrami (1835–1900), an Italian mathematician, when searching the solution for the minimal area surface of revolution about a given axis between two specified points. It gives a straightforward solution of the brachistochrone problem. See *Opere Matematiche di Eugenio Beltrami*, Vol. II (Ulrico Hoepli, Milano, 1904).
7 An algebraic form is a homogeneous rational integral function of two or more variables.
8 A bilinear form on a vector space V is a bilinear mapping $V \times V \rightarrow$ F, where F is the field of scalars.
9 We may also say a differential or a 1-form.
10 We say that a set U is open if any point x in U can be moved in any "direction" and still be in the set U.
11 Let k be a non-negative integer. The function f is said to be of class C^k if the derivatives $f', f'', \ldots, f^{(k)}$ exist and are continuous. The function f is said to be of class C^∞, or smooth if it has derivatives of all orders. f is said to be of class C^ω, or analytic, if f is smooth and if it equals its Taylor series expansion around any point in its domain. The class C^0 consists of all continuous functions. The class C^1 consists of all differentiable functions whose derivative is continuous; such functions are called continuously differentiable. Thus, a C^1 function is exactly a function whose derivative exists and is of class C^0.
12 In operator theory, a bounded operator T on a Hilbert space is nilpotent if $T^n = 0$ for some n.
13 Or, in cylindrical coordinates, having rows and columns of time t, radius r, angle θ, and height z.
14 A Hermitian matrix (or self-adjoint matrix) is a square matrix with complex entries with elements a_{ij} that are equal to the complex conjugate $\bar{a}_{ji}$.
15 Word composed by Frege from two other words, *Begriff* (concept) and *Schrift* (mode of writing).
16 The use of the word "gauge" was introduced by Hermann Weyl (1885–1955) in German (maßstab) in his work published in 1918 in "Sitzungsber. d. Preuss. Akad. d. Wissensch., 30 May 475.
17 Binary operation or a function of two variables.
18 Some groups have discrete properties.
19 But a Lie group needs them.
20 The term Lie algebras was apparently proposed in 1935 by Nathan Jacobson and reluctantly accepted by Herman Weyl.
21 This device, arguably the most popular proposed spintronic device, is based on the spin rotation of the injected carriers when transported along a low-dimensional channel with length L connecting two ferromagnetic reservoirs, the source and drain. Spin rotation is actuated by using the Rashba spin–orbit interaction actuating at the interface where a 2D electron gas (2DEG) is formed. Rashba spin–orbit can be externally tuned via proper gates electrodes. The Rashba s-o coupling constant α is proportional to the average electric field normal to the 2DEG interface and it can be shown that the rotation angle is $\theta_R = 2m\alpha L/\eta^2$.
22 The other leading alternative is known as loop quantum gravity. According to this theory, space consists of not ordinary atoms but extremely small chunks of space (like knots in a carpet).
23 Quantum theory of gravity tries to handle this.
24 After the name of the Italian physicist, Tullio Regge.

References

1. R. Becker, *Electromagnetic Fields and Interactions*. New York: Dover, 1964.
2. C. Monstein and J. P. Wesley, "Observation of scalar longitudinal electrodynamic waves," *Europhysics Letters*, vol. 59, no. 4, pp. 514–520, 2002.
3. P. M. Bellan. *Fundamentals of Plasma Physics*. Cambridge: Cambridge University Press, 2006.

4. R. M. Kiehn, *Non-Equilibrium Systems and Irreversible Processes*. volume 4. Morrisville: Lulu, 2009.
5. D. V. Dantzig, "Electromagnetism independent of metrical geometry. 2. Variational principles and further generalisation of the theory," *Proceedings of the Cambridge Philosophical Society*, vol. 30, pp. 421, 1934.
6. C. N. Yang and R. Mills, "Conservation of isotopic spin and isotopic gauge invariance," *Physical Review*, vol. 96, no. 1, pp. 191–195, 1954.
7. T. W. Barrett, "Topological foundations of electromagnetism," *Annales de la Fondation Louis de Broglie*, vol. 26, pp. 55, 2001.
8. S. Datta and B. Das, *Applied Physics Letters*, vol. 56, pp. 665, 1990.
9. K. Moriyasu, *An Elementary Primer for Gauge Theory*. Singapore: World Scientific Publishing, 1983.
10. R. Gilmore, *Lie Groups, Physics, and Geometry*. Cambridge: Cambridge University Press, 2008.
11. V. V. Varlamov, "Maxwell field on the poincaré group," *International Journal of Modern Physics A*, vol. 20, no. 17, pp. 4095–4112, 2005.
12. S. V. Vladimirov, H. M. Thomas, V. N. Tsytovich, and G. E. Morfill, *Elementary Physics of Complex Plasmas (Lecture Notes in Physics)*. Berlin Heidelberg: Springer, 2008.
13. D. M. Lipkin, "Existence of a new conservation law in electromagnetic theory," *Journal of Mathematical Physics*, vol. 5, pp. 696, 1964.
14. I. S. Gradshteyn and I. M. Ryzhik, *Table of Integrals, Series, and Products*. Amsterdam: Academic Press, 2007.

4

The World of the Tiniest Building Blocks

4.1 Understanding Nature for Improving Processes and Devices

Although nanotechnology is a very recent result of human inventive and technological capabilities, nature is full of nanoscopic architectures that support the essential functions of a variety of living and inanimate forms, from viruses to whales and galaxies. The tactful use of the principles of nanophysics, chemistry, and biology can be traced to natural structures that are very old. Now the goal is to balance molecular reactions in positionally controlled locations and orientations to obtain desired chemical reactions, and then to build systems by further assembling the products of these reactions. A framework based only on the development of nanomaterials is restricted and underestimates Feynman's vision of molecular nanotechnology, leading with miniature factories using nanomachines to build complex products on the ability to build structures to complex, atomic specifications by means of mechanosynthesis [1]. In addition, one should also consider that if nanofactories gain the ability to produce other nanofactories, production still be limited by the input raw-materials, energy, and software, which of course are not inexhaustible and must follow sustainability procedures. The products of molecular manufacturing could range from cheaper, mass-produced versions of known high-tech products to novel products with added capabilities in many areas of application. Hopefully, this means that Drexler's catastrophic scenarios involving self-replicating machines have few chances to occur. At present, the practice of nanotechnology embraces both chemical stochastic approaches and deterministic approaches wherein single molecules are manipulated on substrate surfaces by deterministic methods comprising scanning tunneling microscopy (STM) or atomic force microscopy (AFM) probes and causing simple binding or cleavage reactions to occur. Thus, the dream of a complex, deterministic molecular nanotechnology remains elusive. In the last few decades many surface scientists redefined their disciplines as nanotechnology. This has caused some confusion in the field and most of the reports are extensions of the more ordinary research done in the parent fields.

Biomolecules can be engineered for various purposes, but biomaterials are not considered nanomaterials because they are not man-made. Although biotechnology can use biomolecules (DNA, RNA, proteins) for engineering innovations or biomedical applications, nanotechnology may imitate nature by using exclusively man-made components. However, since it is already possible to synthetize biomolecules, such distinction is becoming more blurring. Nanobiotechnology deals with technology which incorporates nanomolecules into biological systems, or which miniaturizes biotechnology solutions to nanometer size to achieve greater reach and efficacy. This may result in more effective and inexpensive assays and therapies. Biomolecules are often added to the outside of

DOI: 10.1201/9781003285083-4

nanoparticles to target or make use of specific molecules for a given purpose. These hybrid nanostructures are used to make biosensors or to image certain body parts.

Nanostructures can also be engineered to incorporate them into body systems by altering their solubility in water, compatibility with biologic material, or recognition of biological systems. To give one example, DNA is typically difficult to insert into a cell nucleus because of its strand-like form. However, if it is mounted on a spherical nanoparticle, the spherical DNA may pass through the cell and nuclear membrane with ease. Antibodies and proteins may also be used to coat nanomolecules such as carbon tubes or gold nanoparticles for easy and rapid bioassays. Bionanotechnology, on the other hand, deals with new nanostructures that are created for synthetic applications, the difference being that these are based upon biomolecules. In other words, the building blocks out of which the nanostructure is made are antibodies, nucleic acids or other molecule of life. The molecules used are typically self-assembling and have a highly predictable pattern of binding. This makes them ideal for the purpose of building functional nanostructures, which can be used for various nanotechnological applications such as the manufacture of nanomachines. These molecules are being investigated because both their structure (nanocrystals, nanoshells, and nanomachines) and their properties can be tailored quite precisely. Bioconjugate chemistry thus makes good use of the differing functional properties of both biomolecules and nanomaterials, which share the same size range, for a wide range of applications such as more sensitive and specific cell markers, acquire better images, and prevent the immune system from reacting to and neutralizing targeted drug delivery systems. These benefits are due directly to the nanoscale of the structure. For instance, some nanostructures act as fluorophores or produce other optical effects in the near infrared region of the spectrum of light. In this region of the spectrum, tissues are transparent, and coating appropriate nanoparticles with specific biomolecules such as antibodies could potentially help image tissues or even test their function, using such light sources.

There are several sources of inspiration that nano-scientists could use to create the next generation of human technology and life is efficient but not necessarily the most effective. Some examples of nature-inspired phenomena are structural colors, long-range visibility, mimicry and camouflage, adhesion, porous strength, bacterial navigation, minimization of energy dissipation and wear, artificial photosynthesis, self-cleaning, self-assembling, self-repairing, carbon sequestration, among others. Biomimetics is the application of biological methods and systems found in nature to the study and design of engineering systems and modern technology. The coloration of several types of beetles and butterflies is produced by sets of carefully spaced nanoscopic pillars. Made of sugars such as chitosan, or proteins like keratin, the widths of slits between the pillars are engineered to manipulate light to achieve certain colors or effects like iridescence. One benefit of this strategy is resilience. Pigments tend to bleach with exposure to light, but structural colors are stable for remarkably long periods. Another advantage is that color can be changed by simply varying the size and shape of the slits, and by filling the pores with liquids or vapors too. In addition to simply deflecting light at an angle to achieve the appearance of color, some ultra-thin layers of slit panels completely reverse the direction of the travel of light rays. This deflection and blocking of light can work together to create stunning optical effects such as a single butterfly's wings with half-a-mile visibility, and beetles with brilliant white scales, measuring a slim five micrometers. Microstructured surfaces are for various optical applications, including nonreflective (the moth-eye effect), highly reflective, colored (in some cases, including the ability to dynamically control coloration), and transparent surfaces.

Considering the immense blue sky or the endless blue ocean, one might think that the color blue is common in nature. But it is the scarcest. Of all the shades found in rocks, plants and flowers, feathers and animal scales, blue is the hardest color to find. In the visible spectrum, red has long wavelengths, meaning it has very little energy compared to other colors. For a flower to look blue, it needs to be able to produce a molecule that can absorb very small amounts of energy, to absorb the red part of the spectrum. Generating such molecules (which are large and complex) is difficult for plants, which is why blue flowers represent less than 10% of all other species of flowering plants in the world. One possible driver of blue flower evolution is that blue is highly visible to pollinators such as bees, and blue flower production can benefit plants and ecosystems where competition for pollinators is high.

Biomimetics could, in principle, be applied in many other fields. For instance Murray's law, which in conventional form determined the optimum diameter of blood vessels, has been re-derived to provide simple equations for the pipe or tube diameter which gives a minimum mass engineering system [2]. Cuttlefish change color and pattern (including the polarization of the reflected light waves), and the shape of the skin for purposes of communication and camouflage. Its skin has chromatophores containing hundreds of thousands of pigment granules and a large membrane that is folded when retracted. Under neural control, when they expand, they reveal the hue of the pigment contained in the sac. Cuttlefish have three types of chromatophore: yellow/orange (the uppermost layer), red, and brown/black (the deepest layer). Furthermore, the chromatophores contain luminescent protein nanostructures in which pigment granules modify light through absorbance, reflection, and fluorescence (between 650 and 720 nm) as well [3]. Microelectromechanical system (MEMS) based dynamically tunable surfaces for the control of liquid/matter flow and/or coloration (for example, mimicking the coloration control in cephalopods), can be used for displays and other applications [4].

Adhesion is a general term for several types of attractive forces that act between solid surfaces, including the van der Waals force, electrostatic force, chemical bonding, and the capillary force due to the condensation of water at the surface. Adhesion is a relatively short-range force, and its effect (which is often undesirable) is significant for microsystems which have contacting surfaces. The adhesion force strongly affects friction, mechanical contact, and tribological performance of such a system's surface, leading, for example, to "stiction" (combination of adhesion and static friction [5]), which precludes microelectromechanical switches and actuators from proper functioning. It is therefore desirable to produce nonadhesive surfaces and applying surface microstructure mimicking the lotus effect has been successfully used for the design of nonadhesive surfaces, which are important for many tribological applications. The Gecko effect is the ability of specially structured hierarchical surfaces to exhibit controlled adhesion. Geckos are known for their ability to climb vertical walls due to a strong adhesion between their toes and various surfaces. They can also detach easily from a surface when needed. This is due to a complex hierarchical structure of gecko feet surface. The Gecko effect is used for applications when strong adhesion is needed (e.g., adhesive tapes) or for reversible adhesion (e.g., climbing robot) [6]. Gecko feet can bind firmly to practically any solid surface in milliseconds and detach with no apparent effort. This adhesion is purely physical with no chemical interaction between the feet and surface. The active adhesive layer of the gecko's foot is a branched nanoscopic layer of bristles called "spatulae," which measure about 200 nanometers in length. Several thousand of these spatulae are attached to micron-sized "seta." Both are made of very flexible keratin, and they operate without sticky chemical flexible keratin, or any other sticky chemicals.

The strongest form of any solid is the single crystal state which can be heavy, but nature has a solution for this in the form of nanostructured pores. Sea urchin spines and nacre (mother of pearl) are both made of meso-crystalline forms. These creatures have light-weight shells and yet can reside at great depths where the pressure is high. In theory, meso-crystalline materials can be manufactured, although using existing processes would require a lot of intricate manipulation. Tiny nanoparticles would have to be spun around until they line up with atomic precision to other parts of the growing mesocrystals, and then they would need to be gelled together around a soft spacer to eventually form a porous network.

Some bacteria have ability to sense minute magnetic fields, including the Earth's own, using small chains of nanocrystals called magnetosomes. These are grains sized between 30 and 50 nanometers, made of either magnetite (a form of iron oxide) or, less commonly, greghite (an iron sulfur combo). The strings of magnetic nanoparticles are used to orient the bacterium body along the local magnetic field lines of the Earth. If the local field lines have a large angle to the horizontal, as they do in Northern Europe, then the string of magnetic nanoparticles makes the body point downward, and what every bacterium has to do is to swim knowing that it will eventually find the bottom. Several features of magnetosomes work together to produce a foldable "compass needle," many times more sensitive than man-made counterparts. Though these "sensors" are only used for navigating short distances, their precision is incredible. Not only can they find their way, but varying grain size means that they can retain information, while growth is restricted to the most magnetically sensitive atomic arrangements. However, as oxygen and sulfur combine with iron to produce magnetite or other compounds (only a few of which are magnetic), great skill is required to selectively produce the correct form and create the magnetosome chains. If the particles are single magnetic domain particles, then they will stay magnetized forever, so forming a string of these ensures that the navigation system will naturally work. If a bacterium formed a single piece of the material the same size as the chain of particles, then a domain structure would form, and it would become magnetically dead. The nanoparticles are composed of magnetite (Fe_3O_4) rather than pure iron, but the argument is the same. The formation of single-domain magnetic particles is just one example of the great variety of size effects occurring in the nanoworld. If one continues to cut the particles into smaller pieces, other size-effects start to become apparent (Figure 4.1). In nanoparticles smaller than 10 nm (about 50,000 atoms), the energy levels of the outermost electrons in the atoms start to display their discrete energies. Therefore, the quantum nature of the particles starts to become apparent. In this size range, a lot of the novel and size-dependent behavior can be understood simply in terms of the enhanced proportion of the atoms at the surface of the particles [7].

Environmental engineers have only just started paying attention to biomimetic surfaces. Sustainable chemistry and green engineering principles should be used for the manufacturing of new components for tribological applications, coatings, and lubricants. Friction is the primary source of energy dissipation. A very relevant amount of energy consumption is spent to overcome friction. Most energy dissipated by friction is converted into heat and leads to heat pollution of the atmosphere and the environment. The control of friction and friction minimization, which leads to both energy conservation and the prevention of damage to the environment due to the heat pollution, is a primary task of tribology. Minimization of wear is the second most important task of tribology which has relevance to green tribology. In most industrial applications wear is undesirable. It limits the lifetime of components and therefore creates the problem of their recycling. Wear can lead also to catastrophic failure. In addition, wear creates debris and particles which contaminate the environment and can be hazardous for humans in certain situations.

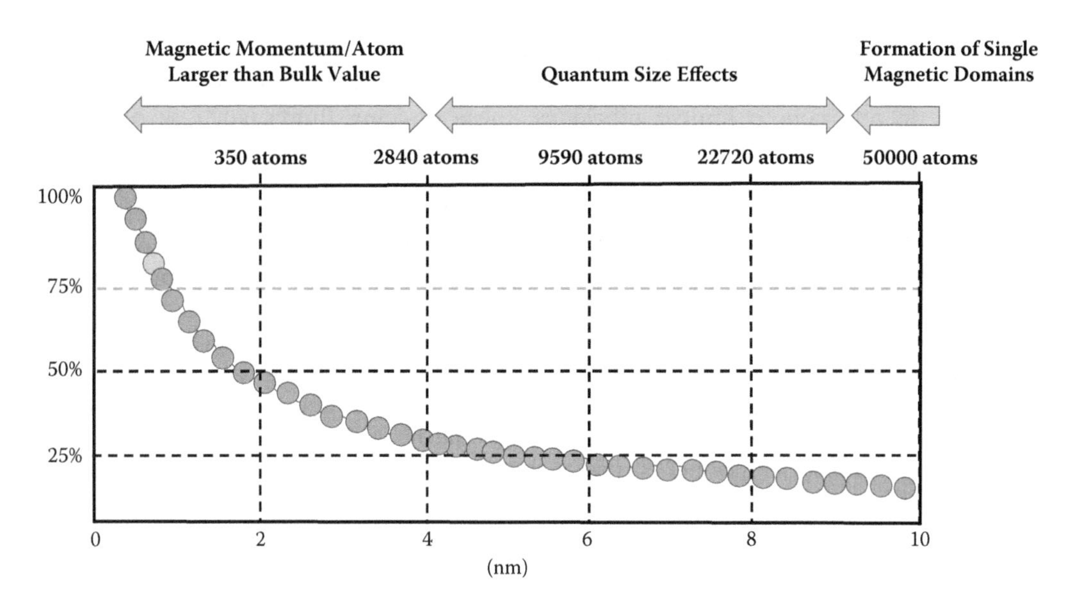

FIGURE 4.1
Size-dependent magnetic behavior in nanoparticles. For particles smaller than 10 nm, the proportion of atoms in the surface layer is becoming relevant and reaches around 50% for 2 nm particles. Quantum effects also start to dictate the magnetic performance in this range and below 3–4 nm the strength of magnetism per atom increases faster.

Many biological materials have remarkable properties which can hardly be achieved by conventional engineering methods. For example, a spider can produce huge amounts (compared with the linear size of his body) of silk fiber, which is stronger than steel without any access to the high temperatures and pressures which would be required to produce such materials as steel using conventional human technology. These properties of biomimetic materials are achieved due to their composite structure and hierarchical multiscale organization [8]. The hierarchical organization provides biological systems with the flexibility needed to adapt to the changing environment. As opposed to the traditional engineering approach, biological materials are grown without the final design specifications, but by using the recipes and recursive algorithms contained in their genetic code. Biomimetic materials are also usually environmentally friendly in a natural way, since they are a natural part of the ecosystem. Superhydrophobicity is defined as the ability to have a large water contact angle and, at the same time, low contact angle hysteresis [9]. The lotus effect based nonadhesive surfaces stands for surface roughness-induced superhydrophobicity and self-cleaning. The lotus flower is famous for its ability to emerge clean from dirty water and to repel water from its leaves. This is due to a special structure of the leaf surface (multiscale roughness) combined with hydrophobic coatings [10]. These surfaces have been fabricated in laboratory with comparable performance [11].

Oleophobic surfaces can repel organic liquids and the principle of operation is like to superhydrophobicity, but it is much more difficult to produce an oleophobic surface, because surface energies of organic liquids are low, and they tend to wet most surfaces. Underwater oleophobicity can be used also to design self-cleaning and antifouling surfaces [12].

Protecting a surface against the formation of frozen contaminants (anti-icing) is a significant issue in cold regions for many applications that need to operate below the water freezing temperature (e.g., aircrafts, runway pavements). Anti-icing is accomplished by applying a protective layer of a viscous-ice fluid, but all anti-ice fluids offer only limited protection. In addition to limited efficiency, these de-icing fluids (such as propylene glycol or ethylene glycol) can be toxic and raise environmental concerns. It is therefore suggested that a water-repellent surface can also have de-icing properties [4]. When a superhydrophobic or microtextured surface is wetted by water, an air layer or air pockets are usually kept between the solid and the water droplets. Thus, after freezing, ice will not adhere to solid due to the presence of air pockets and will be easily removed.

Naturally occurring nanoparticles are ubiquitous in land, sea, and air and come from a variety of processes. The particles in the Earth's atmosphere have an important influence on the climate and some effects on our health too. Considering the vapor pressure of a liquid drop in equilibrium with its vapor, the enthalpy of evaporation becomes dependent on the radius of the drop (because a molecule near the surface has, on average, slightly fewer nearest neighbors because of the curvature), and it can be shown that [13]

$$E_c(R) = E_f - \frac{2\gamma v}{R} \tag{4.1}$$

where R is the drop radius, $E_c(R)$ is the radius-dependent enthalpy for a curved surface, E_f is the enthalpy of evaporation (or the energy required by a molecule to escape from the flat surface), γ is the surface tension of the drop, and v is the volume of the departing molecule. This equation is derived by working out how much the surface energy of a drop changes because of losing a molecule. Since the vapor pressure follows an Arrhenius dependence with temperature, the increased vapor pressure, $p > p_0$ of a drop compared to its value p_0 on a flat surface is obtained by

$$p = p_0 e^{\frac{2\gamma v}{RkT}} \tag{4.2}$$

Therefore, if we have a vapor with a pressure $p > p_0$ (a supersaturated vapor) containing no liquid drops and if we introduce in it a drop with a radius R derived from Eq. 4.2 into this vapor, it will be stable because the rate of molecules evaporating from it will equal the rate of molecules incident on it from the vapor. If the initial drop is smaller than R, however, it will shrink because it will evaporate molecules faster than acquiring them from the vapor. On the other hand, and similarly, a larger initial drop will grow. In a highly pure vapor, getting the initial stable size drops is a bottleneck because the only way they can form is by the simultaneous collision of a sufficient number of molecules (homogenous nucleation), which is a highly improbable event. If there are, however, pre-existing particles (liquid or solid) in the supersaturated vapor, it quickly condenses onto these. In the case of clouds, these pre-existing particles are called cloud condensation nuclei.

Nanofabrication is a remarkably effective technique to create desirable nanoscale patterns and the surface nanofabrication can be tailored for altering virus adhesion to silicon substrates. Scanning electron microscopy and atomic force microscopy along with surface wettability analyses revealed that the nanofabrication had the effect of reducing not only the number of viruses attached but also the strength of virus adhesion. Smart design of surface chemical composition and nanostructure will offer a feasible solution to improve mitigations for controlling viral adhesion and transmission to and from food contact surfaces. Nanocomposite materials can be tailored in such way that they produce required surface properties, such as self-cleaning, self-lubrication, and self-healing. Natural fiber-reinforced composites are among these materials. The difference between microstructured surfaces and composite materials is that the latter have hydrophobic reinforcement in the bulk and thus can be much more wear resistant than microstructured surfaces, which are vulnerable even to moderate wear rates. Self-lubricating surfaces use various principles, including the ability for friction-induced self-organization [14]. Optical surfaces are sensitive to contamination, so the self-cleaning ability should often be combined with optical properties [4].

Self-repairing materials are able to heal minor damage (cracks, voids) [15]. In some biological systems, self-healing occurs via chemical releases at the site of fracture, which initiates a systemic response to transport repairing agents to the fracture site. This promotes autonomic healing [16]. Bio-inspired self-healing structural color hydrogels that maintain the stability of an inverse opal structure and its resultant structural colors were developed [17]. Self-healing polymers and composite materials capable of mending cracks have been produced based on biological materials [18]. The self-healing properties may also be achieved by the breaking and reforming of hydrogen bonds upon cyclical stress of the material [19].

To end up this sub-chapter, let us mention that the phenomenon of self-assembly, which occurs in the mesoscopic and macroscopic domains (from molecules, biological membranes, and nanoparticles to galaxies), can also be used to minimize friction and build advanced films nanostructures for better modern device performances. Self-assembly is a process in which components, either separate or linked, spontaneously form ordered aggregates. Although the concepts of self-assembly were developed with molecules, and self-assembling processes currently are best understood and most highly developed for molecules, components of any size (from molecules to galaxies) can self-assemble in a permissive environment [20].

Self-assembly is the autonomous organization of components into patterns or structures without human intervention and occurs when components interact with one another through a balance of attractive and repulsive interactions. Considering self-assembly is an equilibrium process, that is, the individual and assembled components exist in equilibrium, the thermodynamics of the self-assembly can be represented by a simple Gibbs free energy equation:

$$\Delta G_{SA} = \Delta H_{SA} - T\Delta S_{SA} \tag{4.3}$$

where if it is negative, self-assembly is a spontaneous process. ΔH_{SA} is the enthalpy change of the process and is largely determined by the potential energy/intermolecular forces between the assembling entities. ΔS_{SA} is the change in entropy associated with the formation of the ordered arrangement. In general, the organization is accompanied by a decrease in entropy and in order the assembly to be spontaneous, the enthalpy term must be negative and in excess of the entropy term. This equation shows that as the value of $T\Delta S_{SA}$ approaches the value of ΔH_{SA} and above a critical temperature, the self-assembly process will become progressively less likely to occur and spontaneous self-assembly will not happen. The self-assembly is governed by the normal processes of nucleation and growth. Small assemblies are formed because of their increased lifetime as the attractive interactions between the components lower the Gibbs free energy. As the assembly grows, the Gibbs free energy continues to decrease until the assembly becomes stable enough to last for a long period of time. The necessity of the self-assembly to be an equilibrium process is defined by the organization of the structure which requires nonideal arrangements to be formed before the lowest energy configuration is found.

External fields are the most common directors of self-assembly. For instance, electric and magnetic fields allow induced interactions to align nanoparticles. The fields take advantage of the polarizability of the nanoparticle and its functional groups. When these field-induced interactions overcome random Brownian motion, particles join to form chains and then assemble. At more modest field strengths, ordered crystal structures are established due to the induced dipole interactions. Electric and magnetic field direction requires a constant balance between thermal energy and interaction energies. There are two main kinds of self-assembly: static and dynamic. Static self-assembly involves systems that are at global or local equilibrium and do not dissipate energy. For example, molecular crystals are formed by static self-assembly; so are most folded, globular proteins. In static self-assembly, formation of the ordered structure may require energy (for example in the form of stirring), but once it is formed, it is stable. In dynamic self-assembly, the interactions responsible for the formation of structures or patterns between components only occur if the system is dissipating energy. The patterns formed by competition between reaction and diffusion in oscillating chemical reactions are simple examples of this. Self-assembly in microscopic systems usually starts from diffusion, followed by the nucleation of seeds, subsequent growth of the seeds, and ends at Ostwald ripening [21]. The thermodynamic driving free energy can be either enthalpic or entropic or both. In either the enthalpic or entropic case, self-assembly proceeds through the formation and breaking of bonds, possibly with non-traditional forms of mediation. The kinetics of the self-assembly process is usually related to diffusion, for which the absorption/adsorption rate often follows a Langmuir adsorption model which in the diffusion-controlled concentration (relatively diluted solution) can be estimated by Fick's laws of diffusion. The desorption rate is determined by the bond strength of the surface molecules/atoms with a thermal activation energy barrier. The growth rate is the competition between these two processes.

Regarding the *bottom-up* perspective of galaxies' self-assembling, many things need to be clarified but it seems so far that dark matter plays a crucial role. Initially, luminous and dark matter in the universe was distributed almost uniformly and the challenge for galaxy formation theories is to show how this "not quite" smooth distribution of matter developed the structures (galaxies and galaxy clusters) that we see today. It is likely that the filamentary distribution of galaxies and voids was built in near the beginning before stars and galaxies began to form. These smaller structures then merged over cosmic time to form large galaxies and clusters of galaxies. The apparent degree of similarity between the existing network of neuronal cells in the human brain, and the cosmic network of galaxies seems to suggest that the self-organization of both complex systems is likely being shaped by similar principles of network dynamics, despite the radically different scales and processes at play [22].

Fullerenes have been believed to be able to exist in space, particularly near the environment of carbon-rich stars, but evidence has been limited to inconclusive spectra and traces of the molecules in meteorites that fell to the Earth. However, the presence of polycyclic aromatic hydrocarbons (PAHs) and fullerenes in the space has been recently recorded, particularly in molecular clouds, but also in the interstellar media. These hydrocarbons have been found in dark and cold clouds, where the stars have not even begun to form [23]. The discovery and synthesis of fullerenes in laboratory led to the hypothesis that they may be present and stable in interstellar space. Fullerenes have been reported in an impact crater on the LDEF spacecraft, and investigations of fullerenes in carbonaceous meteorites have yielded only small upper limits. Fullerene compounds and their ions could be interesting carrier molecules for some of the "diffuse interstellar bands"

(DIBs), a long-standing mystery in astronomy. Astrophysicists have detected two diffuse bands that are consistent with laboratory measurements of the C_{60}^{+}, as first evidence for the largest molecule ever detected in space. Large amounts of the fullerenes C_{60} and C_{70} have been detected in a blast of gas from a dying star. The discovery suggests that fullerenes form readily in space, given the right conditions.

The ubiquity of self-assembly in the natural world is evident in every living organism, serving as a route to form organized structures. This phenomenon can be seen, for instance, when two strands of DNA (without any external guidance) join to form a double helix, or when large numbers of molecules combine to create membranes or other vital cellular structures. Everything goes to its rightful place as an imaginary invisible builder having to put all the pieces together, one at a time. Designing molecules that assemble themselves in water, with the goal of making nanostructures for biomedical applications (such as drug delivery or tissue engineering) has been pursued but such molecule-based materials tend to degrade quickly. The structures fall apart when water is removed or when some environmental force is active. This drawback has been overcome using molecules that spontaneously assemble into nanoribbons with enough strength when water is added, retaining their structure outside of water ("Kevlar-inspired aramid amphiphile" nanoribbons) [24].

Recently and surprisingly, the self-assembly of carbon nanotubes without making use of wet chemistry has been achieved in laboratory [25]. This is a milestone in nanoengineering which enable relevant future developments with implications in several technologies. It represents a new step in the dry self-assembling of carbon nanotubes, in the absence of chemicals or solvents, and in the presence of a harsh biological environment, because it makes use of the plasma-ionized medium inherent to plasma-enhanced chemical vapor deposition (PECVD) technique. This is a chemical vapor deposition process used to deposit thin films from a vapor to a solid state on a substrate. After the creation of a plasma with the input gases specific high-energy processes and reactions can occur. Basically, the plasma is generally created by discharge between two electrodes, the space between which is filled with the reacting gases. This way, plasma (the most common form of visible matter in the universe) proves that it is the cutting-edge technology providing a privileged medium for developing clean advanced nanoengineering.

4.2 From Atoms to Nano-Objects

Invisible nano-objects are widely spread in our world and used in a large list of applications. Fundamental understandings of physical and chemical properties at the nanoscale have been gained through studies of multiple procedures. Nano-objects are from natural or artificial origin and can be produced by many different processes and techniques. They are usually grouped in the two well-known classes of *bottom-up* and top-down strategies, and for both two possibilities the output result is always one ensemble of nonequal nano-objects despite their similarities. Each ensemble is characterized by a certain distribution of particle sizes, which dictates the global properties of the ensemble. In nanophysics, the size and shape of a nano-object determines their physical and chemical behaviors. This implies that each element in one ensemble do not behave necessarily in the same way, and the properties of the ensemble are dependent on the spreading of its size distribution. Single nano-object contact studies provide understanding of friction mechanisms, showing that friction is influenced by the real area of contact, roughness, and work of adhesion. For manufacturing a certain type of nano-objects, the chosen technique (and the selected values of the experimental parameters in each method) significantly influences their size distributions, their most abundant sizes, and full-widths. These two aspects together with the purity and amount of production are crucial in applications. There are different methods of generating nano-objects and tackle their properties, growth, and ageing. Erosions have many facets and constitute the core of processes in top-down manufacturing of nano-objects. Furthermore, erosions are also the main sources of aging.

Due to the nanometric dimensions of nano-objects their properties, for the same elemental or molecular composition, are significantly different from those of atoms/molecules, as well as those of bulk materials. A good deal of nanotechnology is about learning how to make various types of nano-objects (nanoparticles, nanotubes, fullerenes, clusters, proteins, and so on) and the ability to image them and probe them to study their properties. These two sides of the coin are what has only become possible in the last few decades. What has really changed in recent years is the ability to study nanostructures at the nanometer level, determine the properties of individual nanoparticles, and, as a result, manipulate them to do what we want. Thus, modern nanotechnology developed alongside the probes that enable us to image, measure, and manipulate the nanostructures. The interest in nano-objects has hugely grown because of their peculiar physical and chemical properties related to quantum size effects with respect to bulk samples and to the enhancement of surface properties due to the large surface/volume ratio. Nano-objects have at least one dimension that is between 1 and 100 nm and come in a variety of discrete geometries. Compared to their bulk material counterparts, they usually exhibit enhanced physical and chemical properties, which makes them especially attractive. For applications on the micro/nanoscale, increasing the lifetime and efficiency of individual components of systems is crucial to the commercialization of micro/nano-systems. Adhesive and friction forces can hinder device operation and reliability. Studies in friction, wear, and deformation behavior are critical to successful integration of nano-objects for the various applications.

Methods for synthesizing nanoparticles typically fall into two categories: bottom-up and top-down approaches (Figure 4.2). The first approach refers to the build-up of a material from the bottom: atom-by-atom, molecule-by-molecule, or cluster-by-cluster in a gas phase or in solution. They are spread by several techniques; in gas phase manufacturing the transformation of a gaseous precursor to the final particulate is a complex physical process, often involving nucleation, condensation, coagulation, and coalescence between particles; in solution synthesis, the main variants are hydrothermal, sol-gel, and micro-emulsion methodologies. The top-down approach involves the breaking down of the bulk material into nano-sized structures or particles. This approach may address milling or attrition, chemical methods, and volatilization of a solid followed by condensation of the volatilized components. High-energy ball milling is an example of this strategy and nanoparticles so produced usually display a large average size (close to 1 mm) and broad size distribution. For instance, in high-energy ball milling, particles of the starting material are subject to heavy deformation, cold work hardening and subsequent fragmentation; the grinding balls and the material in the grinding bowl are acted upon by centrifugal forces, which constantly change in direction and intensity. The grinding ball and the supporting disc rotate in opposite directions, so that the centrifugal forces alternatively act in the same and opposite directions. All transformations are driven by forces in a simple or more complex ways. For instance, fragmentation can be caused by simple friction while

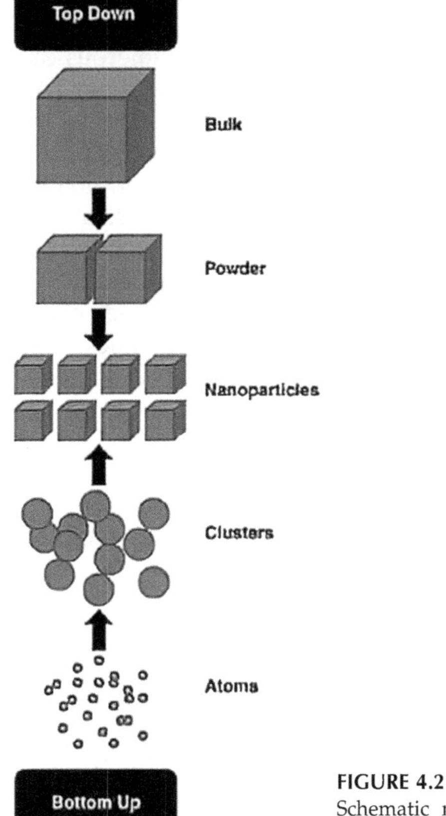

FIGURE 4.2
Schematic representation of *bottom-up* and *top-down* synthesis of nanoparticles.

oxidation is consequence of chemical reactions which are in turn an expression of quantum physics and thermodynamics in a more complex way. Physical and chemical transformations operating in a material are the causes of its ageing. Nano-objects are also subject to these kinds of transformations and ultimately their individual ageing determine the global ageing of a material made of them. Erosion is present in many top-down technologies, and several mechanisms contribute to it, with special emphasis on friction adhesion and wear. Therefore, the experimental studies of these phenomena at the nanoscale using scanning probe microscopies are inescapable.

There are nanoparticles everywhere. In fact, they are ubiquitous in land, sea, and air and come from several processes, including volcanic activity, fires, hydrothermal vents, geological processes, aerosol dust precipitation, and human industrial activity. They can also appear in certain anomalous circumstances [26] and are even found in space (coming for instance from supernova explosions [27]).

In nanophysics of clusters, there are two main approaches. One taken from molecular physics, which consists of to build the cluster from individual atoms and to minimize the energy as a function of shape; and the other coming from solid state physics, that starts with a band-type description of the bulk phase and to investigate how this breaks down as the size gets smaller. In the latter case, a primary goal is often the calculation of the density of states function of metals, especially near the Fermi level. The kinetic models attempt to simulate size distributions whereas molecular dynamics computer-based approaches attempt to describe the detailed structure of individual clusters. In addition to calculation of the electronic properties of metal clusters, there have been numerous calculations of the geometrical shape of clusters. Surprisingly, the rare gases and the metals often are predicted to have similar shapes perhaps reflecting the non-directionality of the binding forces. The more directional ionic forces, caused by the presence of both positive and negative local charges, lead to more differentiated structures. A common goal of many of these calculations is to try to predict the bulk three-dimensional crystal structure. In general, the small clusters often have quite different structures from the bulk and the transition to bulk geometries occurs only gradually and for rather large clusters (> 500 monomers). Structural calculations for rare gases have considered various pair potentials, especially the Lennard–Jones, and minimizes the energy as the coordinates are allowed to change. Such molecular dynamics calculations clearly show a lower energy for tetrahedral groupings of atoms as compared to octahedral structures. The 13-atom case is most interesting because it represents the smallest structure that can have an internal atom, that is, one that is not on the surface.

A considerable number of methods of synthesizing nano-objects (nanoparticles, nanowires, nanorings, nanotubes and their assemblies, quantum dots, 2D nanostructures) have been found, as well as novel methods for characterizing their properties. Scanning probe microscopies (SPMs) have provided powerful means for studying and modifying these nano-objects, as well as the substrates where they stand. Without the discovery and development in the last decades of the SPMs, nanophysics and nanotechnology would not certainly has become a present recognized field of science and technology with a major impact in society in many domains.

The size effects are related to the evolution of physical properties (structural, thermodynamic, electric, magnetic, optical), and consequently chemical reactivity with increasing size (e.g., gold nanoparticles with 1–2 nm diameter exhibit unexpected catalytic activity). Such effects can be classified into two categories: one dealing with specific size effects (e.g., magic numbers of atoms in metal clusters, quantum mechanical effects at small sizes) and the other involving size-scaling applicable to relatively larger nanostructures. The former includes the appearance of new features in the electronic structure.

While the largest number of current papers deal with solution-based growth, there are many other methods of making nanoparticles such as inert gas evaporation [28], atomic layer deposition [29], evaporation or sputtering [30], laser pyrolysis [31], and hydrothermal synthesis [32]. There are in all of them some commonalities, as the initial step will always be nucleation, in most cases after some incubation period. The nucleation may be homogeneous (for instance, small groups of atoms in the gas phase or in a solution with ligands around them), or it can be heterogeneous at steps on a substrate, the side walls of the container or at contaminants. Some nuclei will disperse back to single atoms, some will grow, it is a statistical process. After nucleation there will be growth, as around each nanoparticle there will be a population of monomers and perhaps dimers and trimers (diffusing on the substrate in case it exists). Whichever is the mechanism, as the size increases, transformations will become slower and eventually cease to be relevant for the timescales of the growth. While often growth is by addition from an external fluid phase, it can also proceed via reaction of solid phase such as an amorphous hydroxide particularly with hydrothermal growth methods [32]. In such cases, as the nanoparticle growths, there can be differential diffusion of species, which can lead to voids in the material which slowly collapse with time and temperature [33]. Depending upon conditions, the growth may involve local equilibration of the external surfaces via surface diffusion or not. In many chemisorbants, both small molecules and large surfactant molecules can play an important role in either accelerating or decelerating this. If surface diffusion is fast enough, the shapes will tend to be thermodynamic ones, either the global minima or the constrained minima; if it is slow then the kinetic routes will dominate [34]. The surrounding cloud of monomers may slowly drop as the growth proceeds, for instance as chemicals are depleted in solution growth. There will always be some equilibrium concentration, as monomers can leave the particle and return to the surrounding medium. At high enough temperatures the concentration of these monomers can be high enough that there will be exchange between nanoparticles, Ostwald ripening where small particles shrink and larger ones grow [35]. More rigorously, it will be those with higher chemical potential that shrink, either because of their internal structure or other contributions such as interfacial stresses. There can also be growth via coalescence; in a fluid there is random Brownian motion of nanoparticles whereas on a substrate there can be net translation due to local fluctuations in the concentrations of monomers. In solution growth, chemisorbed ligands can play an important role to prevent coalescence by acting as buffers. When two particles coalesce, there will be a substantial release of energy due to reduction of exposed surfaces, which may be large enough to lead to a full restructuring. The general process will invoke neck formation and is referred to as sintering [36].

Let us report the interesting example of nanorings, which have optical properties associated with excitons and the so-called Aharonov–Bohm effect (besides they have potential applications in light capacitors and photonic computing or communications technology). Nanorings are cyclic nanostructures with a thickness small enough to be on the nanoscale, and they are synthesized using a bottom-up approach, as top-down syntheses are limited by the entropic barriers presented by these materials. Currently, the number of different synthetics techniques used to make these particles is almost as diverse as the number of different types of nanorings themselves.

Nanorings are most synthesized aqueously by creating entropically unique conditions which force them self-assembly [37]. One common method for synthesizing nanorings involves first synthesizing nanobelts or nanowires with an uneven charge distribution focused on the edges of the material. These particles will naturally self-assemble into ring structures such that Coulomb repulsion forces are minimized within the resulting crystal [38]. Other approaches for nanoring synthesis include the assembly of a nanoring around a small seed particle which is later removed [39].

In nanophysics, the spatial-temporal window in which the relevant processes and shapes of physical reality came into play concerns a restricted range of powers of 10 (in SI units). And, if on the one hand, it is true that nanoscale stands for dimensions between the size of an atom and submicron nanostructures, it is also true that the physical processes taking place at these levels can display timescales ranging from 1 femtosecond (the response time of the electron in the atom) to 30 microseconds (average duration of lightning). Nowadays, clocks are extraordinarily stable, meaning time can be measured very accurately. In the most extreme cases, intervals have been measured (in seconds) with a precision of 15 decimal places, and this accuracy is continuously being improved. As such, time is the most precisely measured quantity we have at our disposal. In second place is the measurement of length, measured in meters – with a precision of 12 decimal places.

A scaling law allows to predict the value of a system variable as a function of some other significant variables, and it is mainly used to compare physical magnitudes at different scales or make performance comparisons between devices. Scaling laws are relevant in several domains of physics, biology, and engineering, including nanophysics, micromachines, and nanotechnology [40]. It turns out that the reasoning based on our own experience must be modified when dealing with the microworld and the nanoworld, due to the fact that various physical effects scale differently with size. Effects negligible at the macroscopic level become important at the micrometer scale, and vice versa. Scaling laws are useful for an understanding of the origin of such differences and/or to generalize results obtained at various scales. Moreover, at the nanoscale, quantum effects may become dominant which impose limits on the use the scaling laws due to the non-verification of the principle of mathematical continuity. When one goes from the macroworld to the nanoworld, one passes through two diffuse limits. When the characteristic dimensions of the elements decrease from the macroscopic to the micrometer size, the effects of gravity become negligible as compared with adhesive and friction effects. Surface tension dominates gravity. When the characteristic size decreases further to attain the nanometer range, another limit is encountered. While the macroscopic properties of matter remain generally valid at the micrometer size, surface effects become dominant at the nanometer scale. Moreover, when one reaches the interatomic distance range, quantum effects appear. When elements come close together, of the order of nanometers or below, electrons can hop from one element to the other by so-called tunneling. As it is known, the tunnel current density varies with distance L in a decaying exponential way. This effect is important when L is of the order of a few tenths of a nanometer.

At the microscopic level, adhesion forces dominate, because the details of the forces at the molecular level are much larger than the gravitational ones. In the absence of electric or magnetic effects, the forces responsible for adhesion between a solid and another solid (or a liquid) are van der Waals type forces between the atoms and molecules. The attractive force experienced by an infinite flat slab separated by a distance x (around 2 and 10 nm) from another infinite flat slab is known to be given by

$$F_{vdw}(x) = \frac{H}{6\pi x^3} \tag{4.4}$$

where H is the so-called Hamaker constant which depends on the nature of the medium between the slabs. H is of the order of 10^{-19} J in air and of 10^{-20} J in water. Since this force varies with the linear dimension L like the contact area as L^2, but the gravity force F_g varies with L^3 (because the mass scales with L^3), then the ratio

$$\frac{F_{vdw}}{F_g} \sim L^{-1} \tag{4.5}$$

which means that adhesion force dominates the gravitational force at low L. The critical value at which both forces are equal depends on x and on the nature of the medium between the two solids. However, below say $L = 1$ mm, F_g is much less than F_{vdw}. Gravitation may then be neglected at such small dimensions, both in the micro- and the nanoworlds.

When the size of elements decreases down to the nanometer scale, quantum effects become important. In most situations, they arise in the electronic properties. In other words, quantum effects must be considered at a critical size of the elements L_c, which compares with the wavelength associated with the electrons λ_{el}, that is, $L_c \approx \lambda_{el} = h/p$, where h is the Planck constant and p is the momentum of the electron. Under these conditions, nanoparticles (also called quantum dots) behave like large atoms. Since the electronic motion is restricted in three dimensions, such a system is referred to as a zero-dimensional system. The quantum mechanical calculations indicate that the electronic levels are discrete as in an atom (and contrary to a solid in which the levels are grouped in energy bands), and the spacing of the discrete electronic levels of the quantum dots, scale as L^{-2}.

In the size range of nanoparticles (dimensions of the order of some nanometers), quantum effects are particularly important. Moreover, it is possible to work with movements of single electrons. The energy for charging a small particle with one electron is given by $E_{cap} = e^2/2C$, where e is the electron charge and the capacitance $C \sim L$ (as it can be easily deduced from a parallel plate capacitor). When one applies a given voltage V_{el}, the charge on each plate is $Q = CV_{el}$, and so $Q \sim L$. Therefore, $Ecap \sim L^{-1}$. For instance, in a CdSe dot, $E_{cap} \sim 0.1$ eV for a diameter of 5.5 nm [41]. By coincidence, in CdSe, the spacing of the lower energy levels is also of the order of 0.1 eV for the same diameter. This diameter corresponds to the wavelength of the electron, via $L_c \approx \lambda el = h/p$. Indeed, $\lambda_{el} = 4$ nm when the electron energy is equal to 0.1 eV. Such devices work well at very low temperature, $T = 4.2$ K, since thermal electronic excitation rate to highest energy levels is so low that this does not perturb the behavior of the device, via the electronic transfer to an upper level. This rate, R_{el}, is proportional to exp $(-\Delta E/kT)$, where ΔE is the energy separation of the electronic levels. One can now infer what then would be the size of the dot necessary for operation at a room temperature of 300 K. For correct behavior R_{el} must be equal to its value at low temperature. In other words, $\Delta E/kT$ remains unchanged. Since $\Delta E_{dot} \sim L^{-2}$, one must have $T \sim L^{-2}$ or $L \sim T^{-1/2}$. If $L = 5.5$ nm at $T = 4.2$ K, one thus obtains $L = 0.65$ nm. To apply such small devices, it would be necessary to inject a small number of electrons. This is generally done by injecting electrons through tunnel barriers. One drawback of these small systems is that spurious electrons might transfer from one element to the other through such small barriers. In other words, the noise due to this effect might become relatively large.

Magnetic particles are very efficient magnetic resonance imaging (MRI) contrast agents which allow a high sensitivity for the early detection of different pathologies and the tracking of magnetically tagged cells in vivo through molecular and cellular imaging [42]. It is possible to find a direct relationship between the size and magnetization of the particles and their nuclear magnetic resonance relaxation properties, which condition their efficiency. This allows deriving a experimental master curve of the transverse relaxivity versus particle size and to predict the MRI contrast efficiency of any type of magnetic nanoparticles. This prediction only requires the knowledge of the size of the particles impermeable to water protons and the saturation magnetization of the corresponding volume.

Scaling and similarity techniques are also a useful tool for developing and testing reduced models of complex phenomena, including plasma phenomena [43], where dimensional analysis can be used to find confinement scaling in fusion devices. In particular, the dependences of the confinement time on the plasma parameters for a variety of plasma physics models have been found. The most complete plasma model of any normally used for plasma confinement studies is one in which the plasma distribution function is described by the Vlasov equation including collisions while the electromagnetic field is determined self consistently from the Maxwell equations, and charge neutrality [43].

Let us now pay a little more attention to the bottom-up techniques of nano-objects generation, mainly concerning the production of nanoparticles from gas phase that allows to mitigate the use of chemicals and solvents (which in general introduce undesirable contamination and size selectivity effects), in comparison with other more "wet"

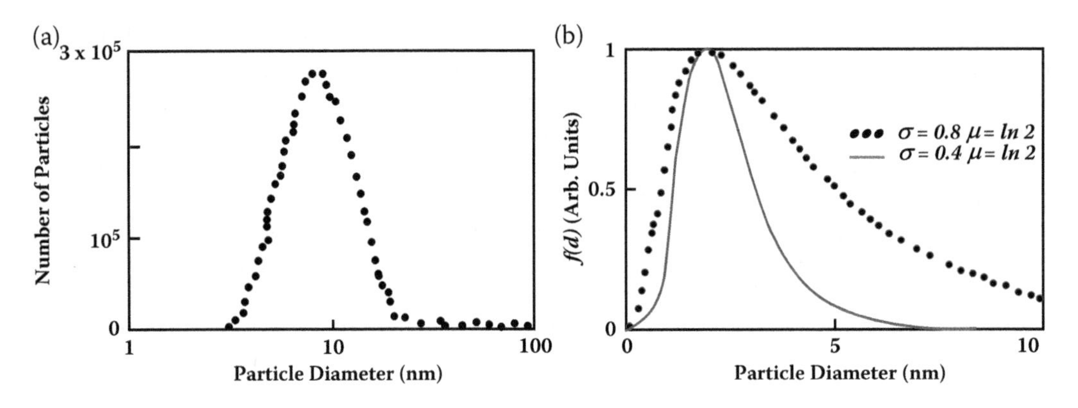

FIGURE 4.3
(a) Size distribution of nanoparticles produced by combustion in a common candle. Particles are mostly made of carbon and the distribution was measured by an aerosol particle sizer. (b) Log-normal size distribution of particle diameters formed by condensation in a supersaturated vapor; both curves have the same most probable size (2 nm) but different width parameters.

processes. Indeed, physical techniques generally make it possible to ensure these objectives in the face of chemical processing. A method process that generates nanoparticles in the atmosphere is gas-to-particle conversion. When a vapor is rapidly cooled – for example, in combustion where hot gases meet cool air – the atoms or molecules in the vapor condense to form particles. Alternatively, a vapor produced by some process may chemically react with the atmosphere to produce a less volatile compound, which then condenses into solid or liquid particles. Just about any combustion will generate a cloud of nanoparticles. For example, Figure 4.3a shows the typical carbon nanoparticles size distribution generated by a common candle. The size selection of the aerosol nanoparticles was done by a differential mobility analyzer filter where the radial velocity of the ionized particles depends on the particle size. Common detectors are condensation particle counters, where the nanoparticles act as condensation nuclei for the water vapor, and they rapidly increase in size as they travel through) and when they emerge, they are detected by the scattered light from a laser focused on the particle beam, with each one producing a flash of light at the detector that is recorded as an event.

Some nanoparticle sources produce particles by mixing an atomic vapor of the material required with an inert gas at much lower temperature to produce a supersaturated vapor. For generating the pressure and temperature at which the vapor will condense into particles, it is necessary to establish conditions in which the pressure of the atomic cloud of material is greater than its vapor pressure at the temperature maintained in the cloud. Without cloud condensation nuclei there is no way to achieve a water droplet of the critical size. Most cloud condensation nuclei are in fact nanoparticles, and the special properties of matter at the nanoscale come into play. For example, the growth of cloud droplets is profoundly affected if the cloud condensation nuclei are soluble in water. One mechanism is that soluble cloud condensation nuclei can change the surface tension of the water droplets condensing onto them and thus change the stable droplet size for a given water vapor pressure. The solubility of the cloud condensation nuclei nanoparticles becomes dependent on their size, and substances that are insoluble in the bulk can become soluble in sufficiently small particles. For a given set of pressure–temperature conditions, there is a critical size of condensed particle above which the particle will naturally grow and below which it will shrink. Under extreme conditions, this critical size can be smaller than an atom in which case the atomic vapor will naturally condense into particles with no further help. Under less extreme conditions the cloud can still self-nucleate since there is a finite probability that enough atoms will simultaneously collide to form a particle larger than the critical size, which is then stable and continues to grow. This is known as homogeneous nucleation, and the probability of it happening decreases rapidly with decreasing number density in the atomic cloud. Generally, homogeneous nucleation, if it is required, is the slowest part of the growth process and presents a bottleneck to particle formation. It is known that the critical diameter for a stable particle in the vapor increases with the volume of the particle and the surface tension of the drop and decreases with temperature and with the natural logarithm of supersaturation ratio. This is defined as the ratio between the actual pressure and the vapor pressure of the material at that temperature. For a copper vapor at 1,200°C and pressure of 150 mbar, the critical size is about 0.13 nm, which is smaller than a Cu atom. In this case, the vapor will spontaneously condense into particles as every collision between Cu atoms produces a stable condensate from the dimer upward. It is found experimentally that the distribution of particle diameters of condensed particles in a supersaturated vapor obeys to a log-normal distribution, $f(d)$ [44], as displayed in Figure 4.3b and given by

$$f(d) \propto exp\left[-\frac{((\ln d) - \mu)^2}{2\sigma^2}\right] \tag{4.6}$$

where μ and σ are, respectively, the mean and standard deviation of $ln(d)$. Thus, the most probable size is given by $ln(\mu)$ and the standard deviation of particle diameters is [45]

$$\sigma_d = exp\left(\mu + \frac{\sigma^2}{2}\right)[exp(\sigma^2) - 1]^{1/2} \tag{4.7}$$

Much of the fundamental research on pure metal nanoparticles is carried out using sources that produce the particles in a supersaturated vapor and transport this through apertures to form beams of particles in vacuum. Due to the high proportion of surface atoms, most materials when they are formed as nanoparticles are very reactive: and if they are to be studied in their pristine state, they must be produced in an ultra-clean environment. Thus, most sources of this type, rest at high or ultra-high vacuum before the supersaturated vapor is created. To operate the source, a cloud of metal vapor is created by heating (or sputtering) and mixed with an inert bath gas such as pure helium or argon in a primary, sealed chamber with a single exit aperture. Conditions for supersaturation are maintained in this chamber at a pressure typically of about 10 mbar, though the pressure in this region can, for some sources, be much higher and above atmospheric pressure. The mixture of condensed particles and bath gas escape from the aperture into a region that is pumped by a very high throughput pump, which maintains a pressure just outside the aperture of ~ 10^{-1} mbar. This effusion of gas at high pressure though a small hole into vacuum is known as a free jet expansion. The effusing beam encounters a "skimmer " with a small aperture that efficiently transfers the nanoparticles with as little of the bath gas as possible into the next chamber, which is pumped by a high-vacuum pump maintaining a pressure in the high-vacuum region of about 10^{-5} mbar. At this level of vacuum, collisions between the nanoparticles and background gas are rare and the particles are in a beam traveling freely through space. The clustering continues until the mean free path becomes too long to allow significant interactions between the condensed particles. The particle beam can then be mass-filtered and deposited onto a substrate which is made accessible to scanning probe microscopies or electron microscopies. Clusters are entities that have neither the well-defined compositions, geometries, and strong bonds of conventional molecules nor the boundary-independent properties of bulk matter. For example, an aggregate of a few atoms held together by van der Waals forces constitutes a cluster. In molecular beams the collisionless environment prevents further aggregation and clusters can be characterized free of matrix effects. The reactions of a N-size cluster are of two different types: the bimolecular reactions with the monomers (which involves energy exchange) and unimolecular reactions of spontaneous decay characterized by the lifetime of the decaying cluster. For a given cluster of order N, this first-order reaction of decay depends only on the transit time dt to pass from a temperature T to another $T+dT$. The phenomenological microscopic approach includes several types of processes involving activation energies. The total number of binary collisions experienced by a molecule during the free jet expansion is typically of the order 10^2 to 10^3. Consequently, any kinetic process that requires this number of binary collisions to approach equilibrium will be subject to kinetic lags or relaxation effects during the expansion and may "freeze." For example, the vibrational relaxation of simple diatomics may require in excess 10^4 collisions; therefore, the vibrational modes of such diatomics do not participate in the expansion.

Several variants of sources have been developed, differing mainly in the method for producing the metal vapor and the pressures existing in various parts of the source. An alternative to heating a metal to produce an atomic vapor is sputtering, where high-energy ions are accelerated into a surface and eject atoms from it. The gas that is ionized and used for sputtering is usually argon with a background pressure that is similar to that of the inert gas introduced into other types of sources to produce a supersaturated vapor of the metal. Thus, the vapor of sputtered metal atoms condenses into nanoparticles, with the important difference that the vapor is produced in relatively cool conditions. Clustering is highly efficient because the sputtered vapor consists of not only individual atoms but also a significant proportion of dimers and small clusters, which act as condensation nuclei for nanoparticles. So, the start of particle growth does not depend only on homogeneous nucleation overcoming an initial bottleneck for growth. Another characteristic is that a high proportion (up to 50%) of the emerging nanoparticles are ionized, having kept the initial charge donated from sputtering. This is an important factor if the clusters are to be size selected, since most mass analyzers filter charged particles, and apart from sources that naturally produce a high proportion of ions (such as the sputtering and laser sources) a separate ionizer is required. Ions are very effective condensation germs, and it is easy to understand why ions are so much more effective in starting a condensation than neutral atoms. The long-range part of the weak van der Waals interaction potential between two neutral particles varies as R^{-6}, where R is the distance between the two particles; the interaction between an ion and a neutral atom/molecule is much stronger and of a larger range, varying as R^{-4}.

There are other alternatives to heating or sputtering a metal to produce a metal vapor cloud, like for instance to use a pulsed Nd-YAG laser focused onto a suitable target which can vaporize even refractory materials; in this so-called laser vaporization technique, if the laser pulse coincides with a gas burst across the target produced by a pulsed valve, suitable conditions for clustering can be achieved [46]. Clustering occurs within the nozzle as the metal vapor encounters the rare gas and continues in the strong expansion as the mixture is ejected. A useful variable is the phase of the laser pulse relative to the valve opening that can be used to control the size distribution of clusters by altering the average gas pressure during the pulse. Phasing becomes particularly powerful in sources employing two targets to produce binary clusters [47] since it can be used to control the distribution of elements within the cluster. The pulsed output couples efficiently to time-of-flight mass analyzers. A pulsed arc can also be used to produce evaporated plumes of metal vapor, and in this case the source is known as a pulsed arc cluster ion source [48]. The clustering process is very similar to that found in the laser ablation source. The cluster output also contains a high proportion of ions (~10%) and is particularly suited to charged-particle mass analyzers. More recently, an arc continuous ion source (ACIS) has been developed in which a continuous arc is driven around a hollow cathode by a magnetic field [49], and this is capable of very high fluxes of clusters. A high-resolution mass spectrometer can detect magic numbers of clusters (they correspond to the number of atoms that produce an especially stable cluster, and so they are much more abundant than their neighbors containing an extra atom). In the region of a nanoparticle source where clusters are growing, each cluster has atoms impinging on it and is also re-evaporating atoms; but if there is an imbalance between the flux-on and flux-off the cluster in favor of the former, the particle will grow. When it reached the size of a magic number, which is a highly stable configuration, the first additional atom is much more likely to re-evaporate from the cluster surface than one of the atoms in the magic configuration. So, if N is a magic number, clusters containing N atoms will be more abundant than those containing $N + 1$ atoms. For Na, Cu, Ag, and Au clusters, the stability of magic clusters arises from the electronic configuration of the electrons [50]. However, electronic shell filling is not the only mechanism for the occurrence of magic numbers, and for some materials, as the transition metals, magic numbers are due to particularly stable atomic packings around a central atom [50]. The enhanced stability of Au_{55} was demonstrated by Boyen et al. [51] who exposed a series of Au_n nanocrystals to oxidation. These nanocrystals have special stability because they consist of a "magic number" of metal atoms which enables the complete closure of successive shells of atoms in a cubic close packed arrangement. The magic numbers 13, 55, 147, 309, 561, and 1,415 correspond to the closure of 1, 2, 3, 4, 5, and 7 shells, respectively [52]. Since then, several magic nuclearity nanocrystals have been prepared including stabilized Pd_{561} nanocrystals [53]. An illustration of some magic nuclearity in gold nanocrystals is shown in Figure 4.4.

Clusters have a large fraction of their atoms on the surface. In a very simplified model, clusters are considered spherical bodies of radius R that are collections of N identical spherical atoms of radius r (further, neglecting the details of the packing and the packing

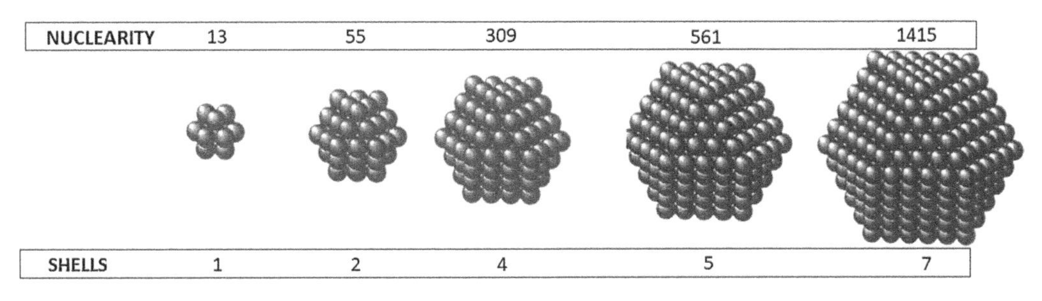

FIGURE 4.4
Gold nanocrystals in closed-shell configurations with a magic number of atoms.

fraction, which is 0.87 for the closest packed structures). With these assumptions we have, $V = 4\,\pi R^3/3 = N\,v = 4\,N\,\pi r^3/3$, where V and v are the volumes of single clusters and atoms, respectively. At this level of approximation, $N = (R/r)^3$ or $R = N\,1/3\,r$. One can then also estimate the number of surface atoms Ns by projecting the cross-sectional area of atoms onto the outer sphere. Thus, $A = 4\,\pi\,R^2 = Ns\,\pi\,r^2$ and $Ns = 4\,(R/r)^2 = 4\,N^{2/3}$. The fraction of surface atoms is then given by $Ns/N = 4\,N^{-1/3}$. Assuming an atomic radius of ≈ 0.1 nm, it comes out that even for clusters with $N = 10^3$ atoms, approximately 40% of them are surface atoms and so surface effects are bound to be large. In the realm of micro-clusters several standard thermodynamic variables, such as surface tension (surface free energy), lose their traditional meaning. Let us consider a liquid-gas interface or surface where the atoms experience an attractive force toward the interior of the liquid without a counter balancing force from the gas phase. This imbalance leads to the concept of surface tension γ, which exerts a force in opposition to attempts to increase the surface area. If the surface is curved with a radius of curvature r, the well-known Young-Laplace equation for the pressure differential across the interface gives

$$P_{in} - P_{ex} = 2\frac{\gamma}{r} \tag{4.8}$$

where P_{in} and P_{ex} are, respectively, the internal (toward the center of curvature) and external pressures. Thus, since $\gamma > 0$, the internal pressure of a liquid drop is higher than the external pressure, that is, the internal atoms are compressed, and in fact the inter-atomic distances measured for many clusters decrease with decreasing cluster size. The problem of the vapor pressure of small drops is solved by the Kelvin equation,

$$\ln\frac{P}{P_0} = \frac{2M\gamma}{RT\rho r} \tag{4.9}$$

which gives the ratio of the equilibrium vapor pressure P of a drop to that of the flat liquid P_0 in terms of the molecular weight M, the density ρ, and the drop radius r. Using the already mentioned simplifications, one can estimate the vapor pressure enhancement for water at 25°C for each size drop (as an application of the size of this effect), and we can notice that a 100-molecule water droplet might be expected to have a vapor pressure about 10 times the equilibrium vapor pressure of a flat surface. Thus, the droplet would evaporate quickly.

The usual reaction mechanism proposed to explain the kinetics of cluster and surface growth consists in the following sequence:

$$A_n + A_i + M \rightarrow A_{n+i} \tag{a}$$

$$A_n \rightarrow A_{n-i} + A_1 \tag{b}$$

Process (a) represents the growth of a cluster of size n by aggregation of a cluster of size i. It is usually assumed that $i = 1$, that is, the growth is by accretion of monomers. The involvement of a third body M is necessary to remove the energy of condensation, otherwise there is always sufficient energy to evaporate an atom from the newly formed cluster. This is especially important for very small clusters. Process (b) represents the spontaneous evaporation of monomers from the surface of a cluster. Energetics would indicate that monomers are the only particles that easily leave a homonuclear cluster.

Making the usual approximation that $i = 1$ (growth by monomers), we can express the rate of change of n-size clusters by a set of coupled differential equations (one for each n), which presents several difficulties to estimate the various rate constants, even making use of the steady-state approximation. Such rate constants for growth are often estimated by consideration of collision probabilities derived from the kinetic theory of gases with unitary sticking coefficients. The various models usually give rise to the notion of a critical-size cluster. Clusters below this critical size decay faster than they grow; we are dependent upon statistical fluctuations to get past this critical size barrier. In the case of surface adsorption, we often introduce the idea of roughening transition, below which growth is slow. A common result of most of the models is that the population of clusters decreases exponentially with number of atoms for free clusters. Most experimental studies of cluster-size distributions confirm these predictions.

Some experimental measurements on molecular (CO_2) clusters [54] indicate that growth by coagulation on larger clusters rather than by monomer aggregation dominates for clusters larger than about 100 individual units. This leads to a Gaussian distribution in the logarithm of the cluster size. The monomer population remains high, but intermediate-sized clusters disappear. This result indicates that evaporation of monomers dominates cluster cooling and that the exchange of monomers provides a means of energy (velocity) exchange among clusters. Growth by coagulation is energetically favorable since the energy released upon coagulation of an i-sized cluster on a j-sized one is significantly less than the energy released by the aggregation of i individual particles. Small clusters may well be "liquid" because of this extra energy and the fact that the melting point decreases with cluster size.

Using the supersonic expansion technique (with argon as carrier inert gas) the silver nanoparticle deposition height histograms resulting from AFM analysis is represented in Figure 4.5.

The laser vaporization source, also known as the "Smalley source," was developed in 1981 in Rice University and Bondybey at AT&T Bell Labs, although previous authors had reported the use of lasers to ablate material for mass spectrometry or atomic beams [55].

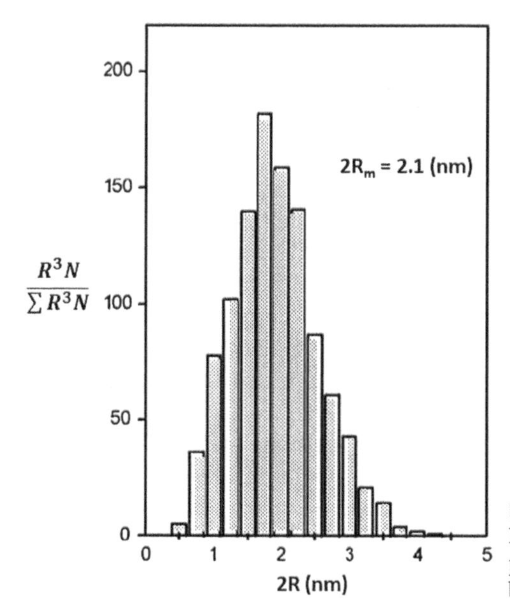

FIGURE 4.5
Histogram of the height of Ag clusters (produced by supersonic expansion with argon used as carrier), obtained by AFM.

However, this earlier work was different in that no collisional gas was employed to cool the metal or to promote condensation. In early work in the Smalley laboratory, the detection employed a time-of-flight mass spectrometer, so that the full range of clusters produced could be detected. Bondybey used laser induced fluorescence spectroscopy of metal dimers, and thus could not see all the other species present. Because the ablation process produces a hot plasma, collisional cooling is necessary to promote cluster growth as well as to quench metastable excited states and/or hot vibrations in the growing clusters. Supersonic nozzle sources provide the most rapid and efficient collisional cooling, and therefore are recommended for this application. Argon is more effective per collision than helium in removing translational and vibrational energy in so-called T-T or V-T processes. However, because the cluster source is pulsed, and because the collisions in a supersonic expansion only take place in the vicinity of the nozzle, the collisional cooling here is confined by a finite timescale.

For producing nanoparticle aerosol without the need for vacuum conditions, it is necessary to account that they have a certain fallout time, and so they are effectively suspended, and the gas/nanoparticle mixture can be piped around as if it were a fluid. If nanoparticle size selection is required in an aerosol, one can use as a filter one differential mobility analyzer and the flux of particles emerging from it is then detected by a condensation particle counter. In the spark source method, a pulsed high-voltage spark is generated between two sharp electrodes made of the material required as nanoparticles. The spark takes place in a flow of inert gas and is normally produced by charging a capacitor until it reaches a sufficient voltage to initiate a spark and discharge via the plasma generated. Each spark produces a plume of supersaturated metal vapor from the electrodes that condenses into a nanoparticle aerosol in the inert gas [56]. In the thermal plasma source method [57], an inert gas is mixed with a relatively coarse powder of the material required in nanoparticle form and blown into the coil of high-power radio frequency that heats the powder-containing gas by induction. The high-frequency field ionizes the gas, which is then induced to carry a heating current. Powers of several kilowatts are used and the powder containing gas is heated to temperatures of up to 10,000 K and converted into a plasma. Because of the very high temperatures employed, the process can produce aerosols of just about any material, including refractory metals and ceramics. The plasma is passed into a condensation chamber and is rapidly cooled, either passively by radiation into the water-cooled walls or sometimes by also injecting cold gas into the chamber. The material in the argon flow condenses into nanoparticles that are collected at the base. As already mentioned, flames can be also used to produce nanoparticle aerosols. The heat in a flame is generated by a chemical reaction that produces a vapor of reaction products, which stream away from the hot zone and condense into particles as the vapor cools [58]. An example is the industrial process used to make SiO_2 nanoparticles, where the starting material is liquid $SiCl_4$ that is mixed with oxygen and then hydrogen and passed into a combustion chamber. The flame products, which include HCl and SiO2, are passed into a cooling chamber where they condense into SiO_2 nanoparticles and hydrochloric acid. These are separated, and after further purification the SiO_2 nanopowder is collected.

Nowadays, nanoparticles of a wide range of chemical compositions and phases can be prepared by a variety of methods; however, the production of large amounts of pure, non-agglomerate nanoparticles, with desired size and narrow size distribution, still results to be an extremely difficult task [59]. The successful synthesis of nanocrystals involves three steps: nucleation, growth, and termination by the capping agent or ligand. Though the reaction temperature and reagent concentrations provide a rudimentary control of the three steps, it is often impossible to independently control them and so the obtained nanocrystals usually exhibit a distribution in size. Typically, the distribution is log-normal with a standard deviation of 10% [60]. Given the fact that properties of the nanocrystals are size-dependent, it is significant to be able to synthesize nanocrystals of precise dimensions with minimal size-distributions. This can be accomplished to a limited extent by size selective precipitation, either by centrifugation or by use of a miscible solvent–non-solvent liquid mixture to precipitate the nanocrystals. Some authors have reported the synthesis of GaAs nanoparticles by using the organometallic precursor trimethylgallium, $GaMe_3$, which on mixing in a furnace flow reactor with arsine gas, AsH3, gives crystalline GaAs particles [61]. Vapor condensation involves controlled thermal decomposition of one or more metalorganic precursors in a low-pressure flame followed by rapid condensation of the products of precursor decomposition in a cooling gas stream or on a chilled substrate. The process has been used to produce a variety of single-component ceramic nanopowders (Al_2O_3, TiO_2, Y_2O_3, ZrO_2) starting from readily available metalorganic precursors. In the optimal configuration, the process employs a flat-flame burner, which ensures uniform thermal decomposition of the precursor feed over the entire surface of the burner. Hence, the resulting nanopowder product has a narrow particle size distribution.

Gases, such as H_2, CH_4, or acetylene, are burned in oxygen to generate a steady-state combustion flame. The flat flame provides a uniform heat source with short residence time for efficient thermal decomposition and reaction of the precursor/carrier gas stream. The substantial heat release in the flame allows the burner to support a high precursor flow rate at pressures of 5–50 mbar, which ensures that the nanoparticles are minimally aggregated.

Gas-phase combustion synthesis of carbonaceous and inorganic particles (like various kinds of soot or the oxides SiO_2, TiO_2, Al_2O_3) is used routinely today in industry to make a variety of additives for commodities, and the flame reactor is today the workhorse of this technology. It is cost-effective and offers advantages over other material synthesis processes, for example, wet-phase chemistry, in particular the ability to control the particle morphology, while achieving high production yields and large production rates. The characteristic of gas-phase-made nanoparticles is self-purification with the final powder product because of normally high synthesis temperature. Radicals, intermediates, and product molecules are formed, which polymerize or nucleate to first clusters. The clusters can grow by gas kinetic collisions between each other or by the addition of monomers to the cluster surface. The coalescence of cluster–cluster ensembles is normally very fast resulting in compact nearly spherical structures, which could be called "particle." The further particle dynamics is determined by surface growth and by the interdependence of coagulation and coalescence. Nanoparticles from gas-phase flame processes are mostly synthesized in laboratory flames by adding a precursor dopant of high purity, in a gaseous or liquid state, to the unburned gas. Such precursors are often metal halides, metal-organic, or organometallic compounds. They can also be dissolved in water or in liquid hydrocarbons and sprayed into a flame, which usually leads to evaporation and subsequent decomposition of the respective precursor material. Reactants, heated by collisional transfer, decompose and particles are nucleating and growing in a pyrolysis flame; due to a quenching effect at the exit of the flame, nanoparticles are obtained.

Particles with a very narrow size distribution and well-controlled phase composition and morphology became desirable products. When typical particle diameters become larger than several nanometers, the further development of a particle is determined by surface growth and by the interdependence of collision, coagulation, and coalescence.

Brownian coagulation starts to form fractal structures, which merge again into spheres by rapid coalescence. As coalescence rates show a strong dependence on particle size and temperature the flow because, Brownian coagulation finally wins the race in the cooler parts of the characteristic times for coalescence and sintering dramatically increase. Materials properties, residence time, temperature of the flowing gas as well as the time-temperature profile and the characteristic timescales for collision, coalescence and sintering are the key properties which determine the morphology and crystallinity of the agglomerates.

Due to the lack of solvents in gas-phase processes, they have an advantage when ultraclean materials are required. Furthermore, being continuous flow processes, they provide the possibility for scaling production capacities to an industrial scale. In particle synthesis, information about particle size and morphology is required, which can best be obtained from collected particles. Molecular beam sampling is a variant of the classical mass spectrometer for gaseous flame species. Nanoparticles in a flame are partly charged either by the particle synthesis process itself or by an appropriate source. A sample of the aerosol is supersonically expanded through an electrically grounded nozzle into a first vacuum chamber. The supersonic free jet formed by the expanding flow contains both, particles, and gaseous species. The flow conditions are such that the gas temperature decreases extremely rapidly, thus, freezing any physical or chemical processes. The center of the free jet is extracted by a skimmer and moves as a particle-laden molecular beam into a high-vacuum chamber. The molecular beam is then directed through an electric mass filter where the charged particles are deflected from the beam according to their mass, velocity, and charge. Instead of a burner for flame synthesis one can use a pulsed micro-plasma cluster source coupled to an aerodynamic focusing system (Figure 4.6). The plasma produced by a magnetron cluster source contains many clusters and is called *dusty plasma*.

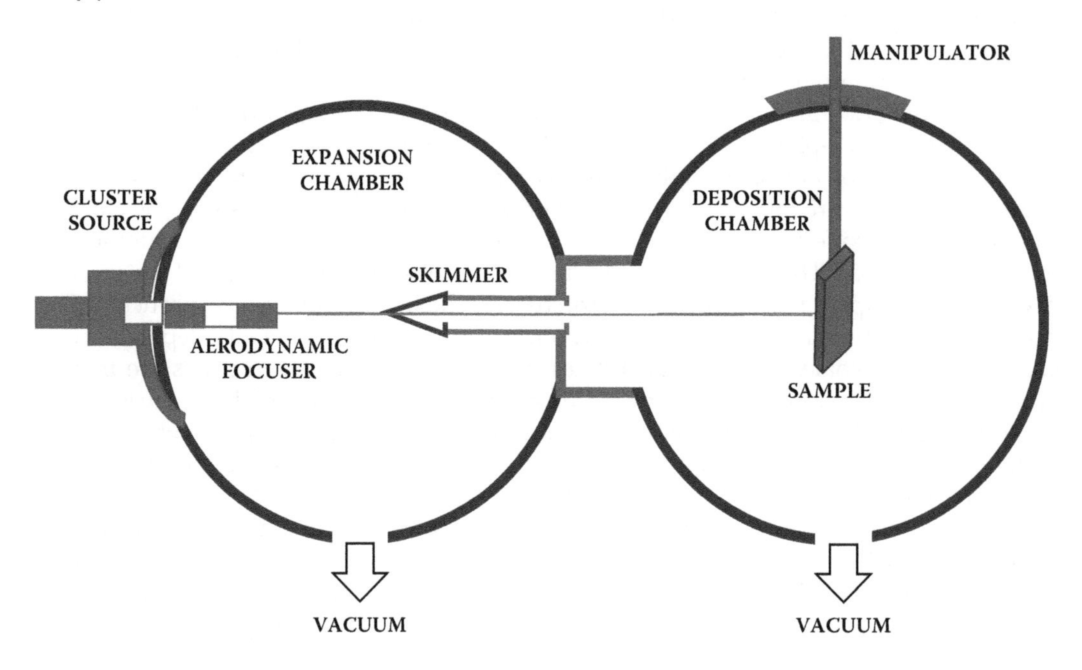

FIGURE 4.6
Sketch of a supersonic cluster beam deposition apparatus using a pulsed microplasma cluster source.

A time-of-flight (TOF) mass spectrometer can be used to measure the size distribution of the clusters, which is commonly narrow in these techniques. Therefore, the clusters are used as generated and not further mass selected, thus reducing the deposition time considerably. The fraction of charged particles is very much higher than that obtainable by photon or electron impact, and so the intensity (summed over all cluster sizes of one charge) is about 1 angstrom/s; higher intensities can be obtained, but at the cost of an inferior size distribution. The mass spectra data obtained for large metal clusters as a function of mass and cluster diameter commonly fit well the already mentioned log-normal distribution. This distribution fits most experimental data for nanoparticle production, and interestingly it can be derived in two different ways, either because of addition of one atom after the other [62] or by coalescence of already existing clusters [63].

Laser pyrolysis is usually classified as a vapor-phase synthesis technique for the production of nanoparticles, where the interaction between a laser and a gaseous flow of precursors is used to obtain homogeneous nanoparticles. It is based on the resonance between the mid-infrared emission of a CW CO_2 laser and the absorption band of a precursor or a component present in incoming reactant stream, and subsequent transfer of energy to non-laser active molecules. It is necessary to use at least one reactant having an infrared absorption band close to that of the emission wavelength of the laser. The photons are highly selective in energizing certain molecules, depending on their wavelength, and the energy is then transmitted to the rest of the reactant gas mixtures in a short time. Depending on the number of photons absorbed per molecule of the excitable species, either photolysis or pyrolysis can occur. In either case controlled chemical reactions are initiated, and depending on the precursors used, produce nanoparticles from the process. In this class of synthesis routes, nanoparticle formation starts abruptly when a sufficient degree of supersaturation of condensable products is reached in the vapor phase. Once nucleation occurs, fast particle growth takes place by coalescence/coagulation rather than further nucleation. At sufficiently high temperatures, particle coalescence (sintering) is faster than coagulation and spherical particles are formed. At lower temperatures, coalescence slows down and partially sintered, non-spherical particles and/or loose agglomerates of particles are formed. It follows that to prepare small spherical particles, it is necessary to create a high degree of supersaturation for inducing the formation of a high density of nuclei and then quickly quench the particle growth either by removing the source of supersaturation or by slowing down the kinetics. In the process of CO_2 laser pyrolysis, the condensable products result from laser induced chemical reactions at the crossing point of the laser beam with the molecular flow of gas or vapor-phase precursors.

The prerequisite for energy coupling into the system, leading to molecular decomposition, is that at least one of the precursors absorbs through a resonant vibrational mode the infrared (IR) CO_2 laser radiation tuned at about 10 micrometers. Alternatively, an inert photosensitizer is added to the vapor phase mixture. The high power of the CO_2 laser induces the sequential absorption of several IR photons in the same molecule, followed by collision assisted energy pooling leading to a rapid increase in the average temperature in the gas through V-T (vibration-translation) energy transfer processes, often accompanied by the appearance of a flame in the interaction volume. If the molecules are excited above the dissociation threshold, molecular decomposition, eventually followed by chemical reactions, occurs with the formation of condensable and/or volatile products. The laser heats the flow mixture of gases, initiating and sustaining a chemical reaction. The decomposition of gaseous reactants by the laser is followed by a quenching effect. Short millisecond scale residence times, spatial uniformity of the reaction zone, and continuous source of activation energy are the key attributes of the process. The main advantage of using this method is that it generates very small nanoparticles in a continuous way with narrow particle-size distribution and nearly absence of aggregation; in addition, due to the absence of surfactants of potential toxicity the product is ideal for the preparation of colloidal dispersions without toxicological potential. There are a broad range of available materials that absorb in the mid-infrared:

1. Precursors which directly produce the desired materials upon decomposition or chemical reactions following laser photolysis or pyrolysis (e.g., silane gas to produce silicon nanoparticles)
2. Sensitizers that only absorb and transfer energy to heat/decompose reactants without producing products and or participating in the reaction (e.g., SF_6 to decompose iron carbonyl for Fe nanoparticles synthesis)
3. Species that have the role of absorbing, decomposing, and reacting to assist decomposition of precursors (e.g., ethylene in the presence of oxygen to produce vanadium oxide from vanadium oxy-trichloride vapor)

The method has been used in the preparation of ceramic powders (Si, Si_3N_4, and SiC) by means of thermal decomposition reactions of different gas-phase organometallic precursors exposed to the laser radiation. It is possible to produce carbides, nitrides, oxides, metals, and composites nanoparticles by this process. As an example, for silicon carbide nanoparticles production at a rate around 1 kg/h by using a mixture of silane and acetylene, it has been shown that the decrease of the gas flow rate favors the increase of the mean grain size of the particles and that the increase of the laser intensity seems to provoke an increase of the mean crystal size and/or crystal number. A small reaction zone is defined by the overlap between the reactant gas stream and the laser beam. The design ensures that the reaction zone is placed safely away from the chamber walls, providing an ideal environment for the nucleation of small particles in the nanometer range, without contamination and a narrower size distribution than those prepared by conventional thermal methods [63]. Above a certain pressure and laser power, a critical concentration of nuclei is reached in the reaction zone, which leads to homogeneous nucleation that are further transported to a filter by an inert gas.

Compared with other vapor-phase synthesis methods, laser pyrolysis permits highly localized and fast heating (leading to rapid nucleation) in a volume that can be limited to a few hundred mm^3, followed by fast quenching of the particle growth (in a few milliseconds). As a result, nanoparticles with average size ranging from 5 to 30 nm and narrow size distribution are formed in the hot region. Unavoidable agglomeration, however, occurs when the nanoparticles leave the high temperature region since coalescence becomes much slower than coagulation. Some attempts have been made to understand the physical basis for the formation of nanoparticles with this technique and the effect of the process variables on particle size [64]. The size of the particles is only governed by the residence time of the reacting species in the flame, which in practice depends on the reactant's flow rates, the diameter of the capillary inlet and the cell pressure.

A coagulation model based on the aerosol theory can be used to maximize the productivity as a function of particle size, which reproduces satisfactorily the average particle sizes measured for a variety of experimental conditions [65]. To produce iron particles of larger than 20 nm size, the ruling factors are the pressure and the carrier gas flux, whereas the production of very small particles depends on the evaporation temperature of the precursor. The formation of iron nanoparticles by laser-induced decomposition of iron pentacarbonyl is likely to take place according to the following scheme [66]:

$$nC_2H_4 + IR(CO_2) \rightarrow nC_2H_4^*$$
$$Fe(CO)_5 + nC_2H_4^* \rightarrow Fe(g) + 5CO + nC_2H_4$$
$$Fe(g) \rightarrow \alpha Fe$$

The model uses only the coagulation time as adjustable parameter. Using the classical nucleation theory, the critical radius estimated for iron nuclei at the very high supersaturation levels attained in this system is always similar to – or even smaller than – the atomic iron radius. This allows us to ignore the nucleation stage and describe the condensation process of the nanoparticles as a coagulation process where pairs of particles just collide to form bigger ones. The analytical description of the coagulation kinetics of particles can be approached by analogy with the theory of aerosols [67]. Then, coagulation is supposed to occur by collisions between pairs of particles, and each impact is assumed to be effective. The model predicts a variation of the particle diameters with time proportional to $t^{2/5}$.

In order to study the photoluminescence of size-separated silicon nanocrystals with diameters between 2.5 and 8 nm, these were prepared by pulsed CO_2 laser pyrolysis of silane in a gas flow reactor and expanded through a conical nozzle into a high vacuum [31]. Using a fast-spinning molecular-beam chopper, they were size-selectively deposited on dedicated quartz substrates. Finally, the photoluminescence of the silicon nanocrystals and their yield were measured as a function of their size. It was found that the photoluminescence follows very closely the quantum-confinement model. The yield shows a pronounced maximum for sizes between 3 and 4 nm (Figure 4.7b). The shape and size of the nanoparticles is examined *ex situ* (by electron microscopy (SEM and TEM) and AFM, and their structure by X-ray diffraction), and in situ by time-of-flight mass spectrometry (TOFMS). It is worth to mention the significant agreement between the AFM and TOFMS methods of analysis.

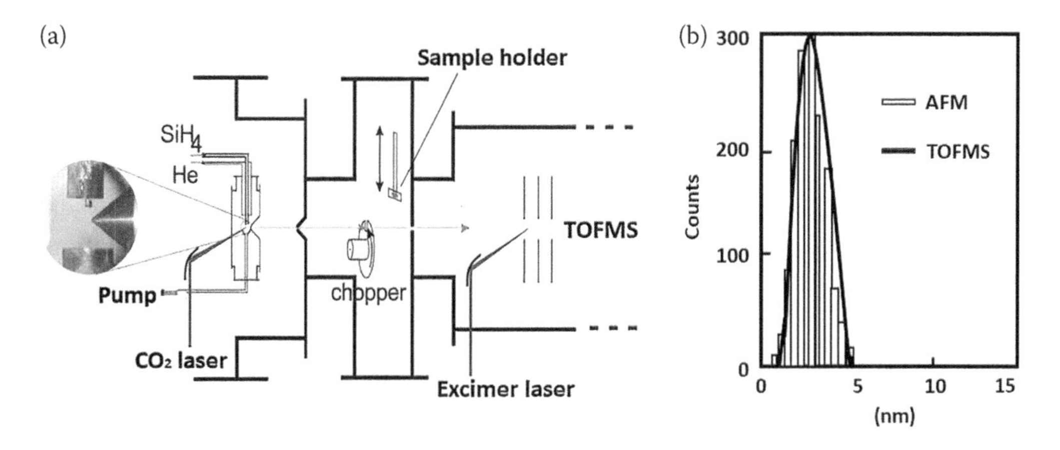

FIGURE 4.7
(a) Laser pyrolysis set-up for size-selected silicon nanoparticles production. (b) Size-selected distributions of silicon nanoparticles measured by TOFMS and AFM.

Figure 4.7a schematically describes the experimental set-up for laser pyrolysis in the absence of oxygen. Silane gas is laminarly flowing through the center of the laser pyrolysis reactor, surrounded by another laminar flow of helium. The focalized laser pulse decompose silane into silicon and hydrogen atoms which then recombine to form molecular hydrogen. Since a nozzle is placed close to the reaction zone, the silicon atoms and small silicon clusters are extracted in the majority helium supersonic expansion, giving rise by condensation to larger clusters and nanoparticles. Their formation is processing in bursts of some nanoseconds, according to the laser pulses characteristics. The nanoparticles size selection is ensured by a synchronized chopper with the laser pulses, and the size distribution is measured in situ by TOFMS. It is also possible to deposit these silicon nanocrystals on a removable target mica substrate [68].

Another interesting method of nanoparticles synthesis is the pulsed plasma where particles are obtained by electric erosion of an electrode material or a substrate which are placed in the path of the plasma flow [69]. The advantages of this method are small size of synthesized nanoparticles, high chemical purity, versatility, and a wide choice of materials for synthesis. The size of deposited carbon nanoparticles varies within the range of 20–180 nm. The experiments are usually conducted on a pulsed plasma accelerator [70] where an arc discharge plasma is ignited in the inter-electrode space (after discharging previously charged capacitor banks) and accelerated along the cylindrical chamber due to the Ampere force of its own magnetic field. Graphite plates are placed in the path of the plasma flow and the emission of dust particles occurs on the surface of the plates because of the plasma flow interactions with graphite plates, and these particles are also attracted to the plasma flow in the direction of its motion. Copper substrates are used for the dust particles that appear to settle on them. The interaction of the pulsed plasma with the surface of the carbon plates is accompanied by instantaneous heating and ejection with subsequent scattering of particles from the surface of the plates. As a result of the experiments, the nanoparticles were obtained on the surface of copper plates. By adjusting the angle of the arrangement of the graphite plates relative to the axis of movement of the plasma flow, it is possible to control the direction of the flow of the formed nanoparticles to focus and change the emission intensity of the nanoparticles.

Conventional chemical vapor deposition (CVD) is a well-known vacuum deposition method frequently used in microfabrication. Nanostructures of silicon or carbon are particularly suited to be produced by these techniques. CVD is extremely useful in the process of atomic layer deposition at depositing extremely thin layers of material. It can be implemented in a variety of formats and the involved processes generally differ in how chemical reactions are initiated. Different working pressure formats can be adopted ranging from atmospheric pressures to high vacuum. In a typical CVD, the substrate is exposed to volatile precursors, which react and/or decompose on the substrate surface to produce the desired deposit. The volatile by-products produced are removed by gas flow through the reaction chamber. CVD is suitable to be associated with the modern techniques of microwaves, lasers, or plasma processing, respectively, adopting the designations of microwave plasma-assisted CVD (MPCVD), laser chemical vapor deposition (LCVD), or plasma-enhanced CVD (PECVD). PECVD processing utilizes plasma to enhance chemical reaction rates of the precursors and allows deposition at lower temperatures, which is often critical in the manufacture of semiconductors. The lower temperatures also allow for the deposition of organic coatings, such as plasma polymers, that have been used for nanoparticle surface functionalization. Surprisingly, the self-assembly of carbon nanotubes without making use of wet chemistry has been achieved in laboratory by means of PECVD [2, 61, 71], which represents a milestone in nanoengineering. A more detailed description of CVD processing and the combination of PECVD with self-assembling will be covered in the chapter of this book devoted to carbon nanotechnology.

Bottom-up experiments of nanoparticle generation involving physical processes in liquids have also been leading to successful results. TEM analysis of self-assembled thiol-derivatized gold nanoparticle monolayers obtained by solvent evaporation have shown that such assembling occurs into nanocrystalline arrays over tens of nanometer length scale. The spacing between the particles is highly regular and roughly about 1 nm. While the larger particles of mean diameter 4.2 nm assembled into regular, close-packed domains, assemblies of smaller particles show a large fraction of voids within the domains [72]. For many applications of nanoparticles, they are required to be in a liquid suspension (hydrosol). The generic method is to use metal-containing molecules (metal salts or organometallics) that are chemically reduced to the metal in the presence of surfactants that allow metal clusters to grow but prevent already-formed clusters from agglomerating. An example is the process for producing monodispersive FePt nanoparticles [73]. The size of particles can be controlled in the size range 2–5 nm by the absolute amounts of the metal-bearing chemicals added. An advantage of having the nanoparticles dispersed in a liquid suspension is that producing ordered arrays is straightforward if the particles are monodisperse. A drop of the suspension can be dispersed onto a flat surface, and as the liquid evaporates and the particles come together, the interaction between their surfactant coatings condenses the assembly into an ordered array. Quantum dots of CdSe have been formed also by a similar method, and since they fluoresce brightly at wavelengths that depend on the particle size in the range 2–10 nm, there is a simple way to control the color of the fluorescence [74].

It has been shown that the size distribution of the metal nanoparticles is strongly dependent on the stabilizer/metal ratio and follows the expected log-normal behavior [75]. Recently an improvement in narrowing the width of metal nanoparticle size distribution produced by wet chemistry has been achieved by using dendrimers as templates to produce the metal nanoparticles in suspension [76], starting from a metal salt. With a sonochemistry technique the acoustic cavitation allows the collapse of a bubble that is formed in the liquid when it reaches its maximum critical value. Very high temperatures are obtained upon the collapse and since this occurs in less than a nanosecond [77], very high cooling rates, more than 1,011 Ks^{-1}, are obtained, which hinders the organization and crystallization of the products, and so amorphous nanoparticles are usually obtained. On the other hand, the fast kinetics do not permit the growth of the nuclei and so the product is nanostructured. If the precursor is a nonvolatile compound, the reaction occurs in a 200-nm ring surrounding the collapsing bubble [78], and in this case the sonochemical reaction occurs in the liquid phase, and the products are sometimes nano-amorphous particles and, in other cases, nanocrystalline, depending on the temperature in the ring region where the reaction takes place. The temperature in this ring is lower than inside the collapsing bubble, but higher than the temperature of the bulk [79]. Temperature affects the sonochemical reaction rate in two ways: lower temperatures cause a higher viscosity, which makes the formation of the bubble more difficult; the dominant effect is that at lower temperatures, higher rates will be achieved. Therefore, the sonic reaction involving volatile precursors is run at lower temperatures. It has been demonstrated that diluting the precursor's solution decreases the particle size of the product [80]. Metallic NPs of Ag, Pd, Au, Pt, and Rh have been produced sonochemically with a narrow distribution (e.g., about 5 nm for Pd NPs obtained from a 1.0 mM Pd (II) solution in polyethylene glycol monostearate solution) [81].

Microwave heating is another method of fabrication of nano-objects, as for instance to prepare very stable Au NPs by a microwave high-pressure procedure with alcohol as the reducing agent [82]. It has been shown that temperature affects the diameter of the NPs grown by the metal catalyst and the thermal conductivity of the ambient affects the cooling rate of the nanoclusters and thus the rate of phase separation. A chemical reaction with a low thermal activation is preferred for nanoparticle synthesis because low reaction temperatures lead to smaller agglomerates. To overcome these drawbacks connected to thermal activation, it would be of advantage if the reactants are already dissociated and, if possible, ionized in the reaction zone. This is the point where plasma processes excel. Therefore, thermal activation using high temperatures is not necessary, and with the additional advantage of giving rise to extremely small particle sizes and low degree of agglomeration. In a microwave plasma most of the energy is transferred to the electrons and it reaches a maximum at a frequency close to the collision frequency. In addition, it is possible to play with the microwave frequency to vary the ratio between positive and negative charged particles. Generally, particle formation is described by the following sequence of steps:

Nucleation → Condensation → Coagulation → Agglomeration

The clusters are formed in the condensation step and due to their thermal motion, the clusters collide and consequently coagulate and agglomerate due to van der Waals forces. If one intends to minimize coagulation and further agglomeration the microwave frequency needs to be chosen in a range where the particles carry electrical charges of the same sign. In this case, the probability of collision decreases with increasing the particle diameter, and this is exactly the behavior that is desired for the synthesis of small nanoparticles. This is different in the case of neutral particles where that probability increases with the diameter. High frequencies reduce energy transfer, allowing the utilization of higher gas pressure and lower operating temperatures. Microwave processing produces smaller particle size with a narrow particle size distribution [83] while the processing time is significantly reduced, and the average particle diameter seems to display a nonlinear increase with time [62].

Laser ablation in liquids has also been revealed a successful mean of obtaining good quality NPs. In this technique, a laser pulse is focused on a target immersed in a solvent generating a microplasma that rapidly expands, quenches, and decays in times on the order of microsecond. During such a process, the atomized material removed from the target, interacting with the species originated from the solvent, nucleates, and begins to grow up to the formation of NPs (which are hindered to aggregate due to an added surfactant). Pulsed laser ablation in liquid (PLAL) technique with respect to the most used chemical methods enables to produce NPs of whatever composition, the possibility of controlling NP properties (size, shape, and crystallinity) by tuning several experimental parameters (laser energy and wavelength, solvent composition, surfactant, etc.), and finally the production of pure colloids free of chemical agents such chemical precursors or reducing agents and, in particular cases, free of chemical stabilizers. This last point is determinant for modern application of NPs. In laser ablation several experimental parameters can be varied, including the laser energy, the lens-to-sample distance, and the concentration of surfactant. The variation of the first two parameters allows an independent adjustment of both the laser energy and the laser spot onto the target surface. The effect of laser energy and of surfactant concentration on the size distribution and the stability of Pt NPs produced in water has been investigated [84] and the stability of surfactant-free platinum NPs was explained by the adsorption of OH^- on their surface, which produces an electrostatic repulsive force impeding their coalescence. Particle size and distribution have been investigated using TEM-EDS and EDX analysis. The crystalline structure of the NPs is verified using the selected area electron diffraction (SAED) technique in TEM and high-resolution TEM. Zeta-potential data are obtained from dynamic light scattering measurements at 25°C by means of the Smoluchowski equation. A qualitative estimation of the total mass of NPs dispersed in the solution can be obtained by the analysis of absorption spectra making use of Mie theory in a quasi-static approximation. This is valid for particles much smaller than the probing wavelength (verified by successive microscopic analyses), and so the main contribution of the absorbance is the dipolar absorption, which is proportional to the total volume of the particles. The NP production rate can then be measured in this condition by weighing the target before and after the production stage and by direct measurements of NP mass via atomic absorption spectroscopy. TEM micrographs of several sets of Pd NPs colloids produced at different experimental conditions, including at distinct laser fluences, have been obtained and the corresponding histograms of particle size distribution can be recorded. The NPs produced in aqueous solution of sodium dodecyl sulfate (SDS) are less aggregated than those produced in pure water, and most of the particles are detached. The size distribution of NPs in pure water markedly depends on laser energy, where the average size value increases with fluence, according to other works dealing with NP formation by PLAL [84]. The colloids produced at small laser fluences exhibit a narrow distribution peaked at ~5 nm, but the real value could be slightly shifted toward smaller sizes because of possible underestimation of the smallest particles. Such desirable distribution, where large particles are rare, is, however, associated with a quite small production rate, which was estimated to be ~0.4 mg/h [84]. Similar narrow distributions peaked at values around 5 nm can also be obtained by using larger laser fluences with the addition of SDS to the solution, with the advantage of enhancing the NPs production rate to values larger than 1 mg/h. If NPs are formed by nucleation and successive growth via liquid-like coalescence, the logarithms of particles volumes should have a Gaussian distribution, so that particle sizes are log-normal distributed. On the other hand, if the particle growth occurs via absorption of atomic vapor by the nuclei, the log-normal function can be unsuitable for fitting the NP size distribution because of the importance of diffusion effects during particle growth [85]. A bimodal distribution associated with the different mechanisms of particle formation can be envisaged. The presence of large NPs is, however, strongly affected by the laser energy, being almost negligible at low fluence values. Relying on TEM imaging and craters observation, it is possible to infer that the two distributions are associated with different mechanisms of mass removal, where the small-size particles are produced by thermal-free ablation, and the larger ones are produced by the melt effects caused by plasma-target interaction, or plasma etching [86]. Plasma etching is expected to occur and become dominant at very large fluences, where plasma breakdown is very likely to occur directly in water; on the other hand, it is very improbable that phase explosion of the target surface would occur at such large irradiances where plasma absorption is dominant, impeding the largest part of laser pulse energy to reach the target. A two-step NP formation mechanism can be considered for the understanding of the experimental data where a rapid phase of embryonic particle formation in the hot plume is followed by a second phase where the NPs grow via liquid-like coalescence or via absorption of atomic vapor; while the first mechanism stops a few tens of nanoseconds after laser ablation, the atoms absorption, according to ends when the atoms in the liquid environment have been mostly consumed or when the NPs have been coated by surfactant molecules [84]. It has been shown that in absence of surfactant agents, the growth of NPs continues for several hours after the laser ablation time [84]. The picture agrees with the rise of the average size in the distribution as laser fluences increases, since larger pulse energies produce larger pressures in the plume and in the cavitation bubble, larger amounts of metallic atomized material and plume dimensions, and longer plume lifetimes. In the case of small laser fluence, the atomized material in the solution is limited, so that the growth via atom absorption rapidly ends because of the low collisional rates. Moreover, in that case, the decay time of temperature is smaller, so that the condensation of the plume stops earlier. In the case of high laser fluence, the decay time of plume temperature is larger, particularly in the core, so that larger NPs are formed. In addition, the atom absorption lasts for a longer time, where ions dispersed in the solution can aggregate also to NPs or nucleation seeds produced in previous laser shots. This latter mechanism, strongly affected by diffusion, cannot be adequately described by a log-normal function, and produces an excess of large-particles in the size distribution [87].

It is evident that the production of large particles, obtained by a long-time growth in the solution, is strongly affected by the presence of surfactant agents as experimentally have been found [84]. Another complication of predicting the size distribution of NPs, which can result in a deviation from a log-normal function, is inherent to the spatial inhomogeneity of the plume and to the different cooling times of its different regions, leading to the formation of smaller NPs at the plume-liquid interface and of larger ones in the core of the plume and near the target surface, as shown by Wen et al. for plasma laser ablation in a gas environment [88]. In conclusion, the model of NP formation agrees with the effect of SDS in reducing the number of large NPs and suggests that significant deviations from a single log-normal distribution can be obtained, without invoking two separate mechanisms of NP formation (which would justify a bimodal fitting).

Sometimes nanoparticles are simple, single crystals with well-defined faces and simple platonic shapes, and in other cases they show more complex shapes and internal structure. However, all the situations split into cases where thermodynamic control determines the final shape and those where kinetics control. Several shapes appear on both cases, both single crystals as well as multiply twinned particles. Every structure that is thermodynamically stable can also occur under kinetic control, but the converse is not true. Fundamental to understanding nanoparticles is the energy as a function of size, shape, stress, and the external environment. In a purely atomistic description, the total energy of an ordered, crystalline nanoparticle must be represented by the infinite series:

$$E = \sum_{i,j,k} a_{ijk} n_i n_j n_k + \sum_{i,j} b_{ij} n_i n_j + \sum_{i} c_i n_i + d + \sum_{i} e_i / n_i + \sum_{i,j} f_{ij} / n_{in_j} \tag{4.10}$$

where the n_i are positive integers indicating the number of atoms along the particular directions. To converge for an infinitely large crystal the series cannot contain any terms higher than third order in n_i, and inverse powers may be needed to achieve the proper limits for a single atom. Assuming convex shapes for all single crystal regions and replacing the above equation with a vector of normal distances for each face from a common origin for k facets with all h_i real but not necessarily positive numbers it is possible to show that each h_i is a combination of a geometrical distance from the origin to the outermost plane of atoms plus a "Gibbs distance" outside the surface, the latter being needed to properly achieve the transition from atomistic to continuum models [89].

It is useful to simplify the processes taking place during growth by considering three regions. A region containing the nucleation terraces and steps on the surface of the nanoparticle (Region N) which grow by the attachment of atoms from the adjacent region around the particle, where atoms can diffuse to the terraces and interact directly with the growing nanoparticle, both via being chemisorbed onto the surface and as part of complexes (Region D); lastly there is the region far from the particle where the concentration of the relevant materials is dictated by the environment (Region F). Therefore, there is a net flux of atoms from Region F → Region D → Region N. Diffusion limited growth occurs when the controlling term is diffusion in Region F to Region D, with atoms attaching rapidly to Region N as they arrive. This limit typically leads to complex patterns such as snowflakes with dendritic or similar structures. Interface limited growth is more important for shape-controlled nanoparticles and is when atoms moving from Region D to Region N is rate limiting. If the size is small enough, as atoms are added the cluster will be flexible and the atomic arrangement can change. Above some size this will not be energetically possible, and the particle then acts as a seed for further growth. The controlling terms are the nucleation of a small monatomic terrace on an existing flat facet, and the growth of this terrace across the facet, Region N. If nucleation is fast relative to the growth across this leads to rough surfaces. Flat surfaces occur when the nucleation is relatively slow.

Making the simple approximation of a circular single atomic high terrace of radius R and introducing an edge energy γ_e per atom, the total energy of the terrace can be written as

$$E = \pi R^2 \Delta\mu N_S + 2\pi R \gamma_e N_e \tag{4.11}$$

where $\Delta\mu$ is the chemical potential change per atom (negative for growth) for addition to the nanoparticle versus being in Region D, N_S is the number of atoms per unit area of the terrace and N_E the number of atoms per unit length of step. The height of the nucleation barrier depends upon the step energy and the chemical potential relative to the external media. From classical nucleation theory, the rate of formation of a terrace above the critical size R_c is given by

$$Rate = f_0 \quad exp\left(\frac{-\Delta E}{k_B T}\right) \tag{4.12}$$

where f_0 is the attempt frequency of an atom to add to the terrace and ΔE is the nucleation barrier which for a 2D terrace gives a critical nucleus size of

$$R = -\frac{\gamma_e N_e}{\Delta_\mu N_S} \tag{4.13}$$

and an activation barrier of

$$\Delta E = -\frac{\pi (N_E \gamma_E)^2}{N_S \Delta\mu} \sim -\frac{\pi (\gamma_E)^2}{\Delta\mu} \tag{4.14}$$

As an estimate, if one uses a broken bond model and take the bulk cohesive energy as 4 eV per atom, for a {111} surface $\gamma_e \approx 2/3$ eV. If one deals with gas phase deposition the chemical potential of an adatom will be around (in eV)

$$\Delta\mu = -3 + k_B T ln[c] \tag{4.15}$$

with $[c]$ the concentration of atoms in Region D that can add to the terrace. If the deposition rate is fast then the critical nucleus may be only 2–3 atoms, in which case there can be multiple terraces on any given facet and the growth will tend to lead to relatively rough surfaces. With slow deposition, the concentration in Region D will be small, via the concentration, and the critical nucleus could be 1 nm in radius, in which case it is unlikely that multiple terraces will be present on any surface and the growth will be layer by layer.

At reasonable temperatures and when the size is small enough, there is no single shape rather a diverse population of shapes. The transitions between the different shapes can occur on the millisecond timescale or even faster. Even at larger sizes it is possible for particles to transition between one structure and another, either because of heating, exposure to electron beam or during growth [90]. Beyond some size, transitions between shapes will become improbably under most experimental conditions, so one has a frozen seed from which a larger particle will grow without changes in structure in most cases. As a function of size and temperature there will be a phase diagram for the thermodynamically most stable shapes [91].

The experimental results have shown consistency with this general model for the evolution of the shapes of Cu nanoparticles grown and studied in situ on strontium titanate as a function of temperature [92].

In general, the result will tend to follow into one of the following three situations:

1. Survival of the fittest: thermodynamic equilibria dominate, either global or constrained local minima with the nanoparticles tending to be close to spherical with facets, surface steps, small terraces and at higher temperatures roughened regions.
2. Survival of the fastest: kinetics dominates, those which grow fastest will tend to have sharp facets with rounded corners and edges, the rounding depending upon the chemical potential during growth.
3. Survival of a population: the initial set of nuclei shapes all grow under kinetic control; the stochastics of the evolution of the combination of different nanoparticles and different local concentrations of monomers will dominate, and there will be many slightly asymmetric shapes as well as the symmetric ones.

Thermal stability of nanosystems may change in a drastic way contrarily to macroscopic systems where the melting temperature is well defined, and the structures are thermally stable up to the melting point.

In contrast to macroscopic systems, the melting temperature of nanosystems depends on the particle number and is also a function of the shape of the system. Nanosystems are disordered already at relatively low temperatures and the harmonic approximation breaks down completely, that is, even at relatively low temperatures the anharmonicities can no longer be considered as small perturbations. In other words, at the nanometer scale the standard model of solid-state physics breaks down in many cases and one needs to introduce molecular dynamics for the description of material properties.

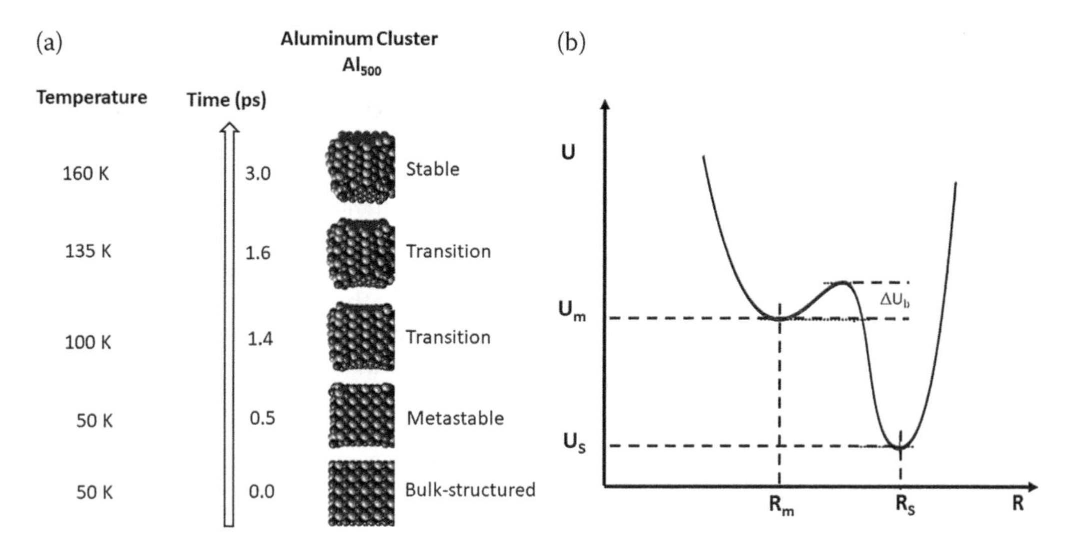

FIGURE 4.8
(a) Different phases of the Al_{500} cluster (frontview). (b) 2D projection of the potential cluster energy surface $U(\boldsymbol{R})$ in the configuration space $\boldsymbol{R} = [\boldsymbol{R}_1 \ldots, \boldsymbol{R}_{500}]$. Energy fluctuations caused by the statistical temperature motion of the atoms and enhanced by surface or body oscillations may reach a value at which the energy barrier ΔU_b has been overcome, and then the cluster transforms into the stable state at $\boldsymbol{R}_s$.

The following considerations are based on MD calculations for aluminium clusters consisting of 500 atoms [92]. The initial values for the positions have been chosen to be those of the bulk structure and the outer shape of the clusters has the form of a cube with (001) oriented surfaces. The cluster is in a meta-stable state within a first phase (T = 50 K) and in a stable state within a third phase (T = 160 K). Due to the energy conservation of the system, the potential cluster energy in the stable state must be smaller compared to that in the meta-stable. In Figure 4.8a, the various phases of an Al_{500} cluster are presented in front and cross-section view. The first row shows the initial cubic cluster before equilibration with the perfect *fcc* structure according to the bulk at 50 K. The second row shows the cluster in the meta-stable state. The third and fourth rows illustrate the cluster during the transition phase with temperatures of 100 K and 135 K. There is a structural change that starts from the faces of the cube. The final structure of the stable state is presented in the last row of Figure 4.8a. The outer shape of the cluster has changed from a cube to a polyhedron. The stable cluster has no perfect crystalline structure. It is composed of various single crystals, that is, it has a multi-crystalline structure. The stability of the Al_{500} cluster depends on its outer shape and structure, that is, dodecahedral arranged (111) faces lead to a higher stability compared to (001) oriented cubic clusters. This result is in good agreement with other authors [93], where the stability of aluminium clusters with up to 153 atoms at 0 K have been studied using the density functional theory (DFT). The results have shown that clusters larger than Al_{55} are most stable if they are *fcc* packed and if their surfaces are (111) structured. In other words, clusters with a maximum number of nearest neighbor atoms are energetically preferred. In the case of Al_{500} clusters, the average number of nearest neighbor atoms is 9.7 in the metastable state and 10.0 in the stable state (for the *fcc* structured bulk it is 12) which supports the result of [93]. The MD studies have thus shown that the occurrence of metastable states is determined by the cluster size. Furthermore, the transition into a stable configuration is a random event.

Depending on the initial conditions the meta-stable periods may vary within a broad range. But the mean duration of the meta-stable states seems to be dependent on the cluster size, since the MD results allow for a division of cubic aluminium clusters into three size categories which, in addition, are correlated to the most probable outer shapes [93].

Regardless of the lattice structure of the bulk, small nanoclusters try to form such configurations for which the potential cluster energy takes a minimum value. Since these configurations coincide with a maximum number of nearest neighbor atoms, the cluster structure often deviates significantly from that of the bulk, that is, structurally disturbed clusters are the rule. However, depending on size, shape, and material, local minima (saddle points) of different depth may appear in the potential cluster energy surface. Under certain conditions, a cluster configuration can be "locked up " in a local minimum that is surrounded by even deeper minima (see Figure 4.8b). In this case, the cluster takes a meta-stable state. Due to temperature, fluctuations of the atoms, sometimes enhanced by oscillations, the cluster configuration permanently fluctuates around the configuration point R_m. Depending on the height of the energy barrier ΔU_b and the distance between the minima, it takes some time to the meta-stable period after which a configuration point is reached by chance that enables the cluster to overcome the barrier. After that, the cluster transforms into the stable configuration R_s and the decrease of potential energy goes with an increase of temperature. For very large clusters the barrier between the local minima is too high, and in addition, the temperature fluctuations are getting smaller with increasing particle numbers. Then the cluster remains in the "metastable " state forever (at least without external influence). Therefore, it must be suspected that there is a limit above which the configurations are globally stable, independent of cluster collisions or other perturbations. Of course, this limit still must be within the nanometer scale because macroscopic single crystals are definitely stable.

Looking through the MD results, a certain analogy can be drawn between the meta-stable cluster states and the excited states of atoms (see Figure 4.9):

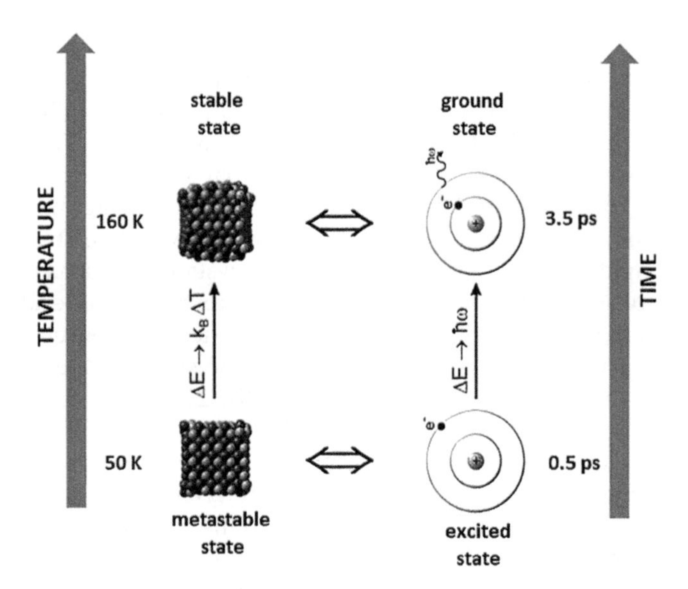

FIGURE 4.9
Illustrations of similarities between nanoclusters and atoms.

- Potential energies of the meta-stable (excited) states are larger than those of the stable (ground) state.
- After a certain period, the clusters (atoms) transit from their meta-stable (excited) states into the stable (ground) state without external influence, the atoms by emitting photons, the clusters by increasing their temperature.
- The period (lifetime) of the meta-stable (excited) states is not constant, but probability distributed. All intervals are possible, but the values with maximum probability are most likely to appear.
- The transition from metastable (excited) to the stable (ground) state can be triggered by an appropriate stimulation: the electrons by photons of certain wavelengths, the clusters by a local perturbation of the surface.

While the lifetime of the atomic excited states lasts for about 10 ns, the magnitude of the meta-stable periods of aluminium cluster configurations is 10 ps. Compared to the lifetime of excited atoms this is rather short.

However, there is at least one point for which the analogy considerations with the atomic states fails: for nanoclusters there is a huge number of stable configurations, but atoms have only one ground state. This fact leads to another analogy, the bifurcation phenomenon. When a cluster makes the transition from a meta-stable or pseudo stable state to a stable state, there are several possibilities. In other words, the local minimum (saddle point) in the potential energy surface correlated to the meta-stable or pseudo stable state is surrounded by numerous deeper minima which correspond to the stable states. Which of these stable states the cluster will finally take, cannot be in principle foreseen. Due to the unavoidable energy fluctuations of nanosystems, it is completely a matter of chance. That is, a cluster state transition is a bifurcation in the sense of chaos theory (see Figure 4.10). At the bifurcation point, nature plays dice to decide on which of the various branches (stable states) the cluster will finally rest. The already mentioned

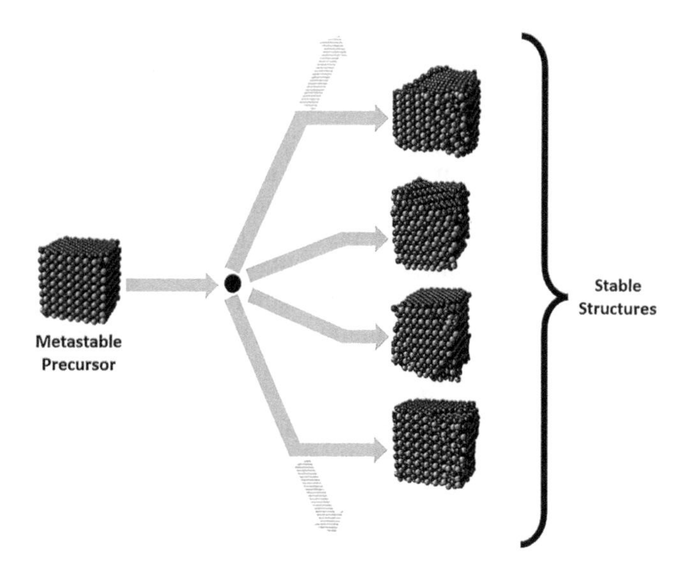

FIGURE 4.10
The stable state into which a metastable cluster will finally transit cannot be predicted. "At the bifurcation point nature plays dice" to decide on which of the various branches the cluster will finally rest.

MD studies have shown that stable cluster configurations may differ in both the inner structure as well as the outer shape. In general, the structures show multi-crystalline compositions, that is, there are grain boundaries, dislocations, and other lattice defects. And finally, the outer shape strongly depends on the arrangement of these lattice defects. With respect to the structure transformations of nanoclusters the bifurcation phenomenon has no counterpart, neither at the micrometer scale nor within the macroscopic bounds. If this would not be the case, any object which was in a meta-stable state could transform randomly into another shape. But, as we know, the only transitions of macroscopic objects result from interactions with the environment (like wear) which finally can lead to their destruction. For nanoclusters the situation is completely different: the transition at the bifurcation point is constructive (Figure 4.10), since the cluster is not to be destroyed, but transforms into a new shape that is clearly more complex than the initial meta-stable cube.

At this point, it is interesting to go further in explaining the state of the art of the so-called "superatoms." Indeed, the possibility of creating a new class of materials, composed of clusters instead of atoms as building blocks, has fueled the hope that one can synthesize particles and materials from the bottom-up with unique and tailored properties. Because of the large surface to volume ratio and quantum confinement, the structure and properties of these nanoparticles differ substantially from those of their bulk counterparts. Atomic clusters are aggregates and intermediate between atoms and bulk matter. They consist of a few to a few thousand atoms which form when a hot plume of atoms cools through collisions with rare gas atoms in near vacuum conditions. Clusters can be composed of homo- or heteroatom species. From the nanophysics point of view, the motivation to study clusters was to understand how the structure and collective properties of matter (such as electrical, magnetic, and optical) evolve as atoms come together. High-speed computers, along with efficient computer codes based on density functional theory, permitted studies of clusters, consisting of hundreds of atoms [94].

Size regimes are crucial when dealing with atomic clusters. As the size of a solid is continually reduced, the properties initially vary monotonically (scalable regime) [95]. Below a critical size (~100 nm), properties change from being monotonic to non-monotonic; however, these properties are not sensitive to the addition or removal of a single atom. Nanoparticles belong to this non-scalable regime. As the particle size becomes still smaller, the properties change abruptly and non-predictively, a stage in which even the addition of a single atom or electron may cause a drastic change. Atomic clusters, whose size falls within this range, are considered the ultimate nanoparticles. The underlying reason for this nonmonotonic behavior is quantum confinement, a regime where the electron wavelength becomes comparable to the particle size [95]. The fact that properties of matter at this length scale are fundamentally different from their bulk behavior can be effectively used to produce materials with tailored properties. For example, novel materials may be created by assembling clusters, much as conventional crystals are created by assembling atoms [96].

Stable clusters can be created to mimic the chemistry of atoms and how these "superatoms" can be assembled to form a new class of cluster-assembled materials. The pronounced peaks in the mass spectra are often interpreted as clusters that are more stable than their neighbours. These are usually referred to as "magic numbers." The relationship between the magic numbers and the underlying electronic structure of Na clusters helped bring the effect of electronic shell closure into focus, much as the magic numbers in nuclei gave rise to the nuclear shell model [96]. Geometries of atomic clusters, and the nature of their evolution with size, depend on their underlying electronic structure and rarely resemble the atomic arrangements in corresponding crystals. For example, it is well-known that dimensionality plays an important role in the underlying magnetic behavior of materials. Surface atoms of magnetic materials exhibit larger magnetic moments than those in the interior [97]. Because clusters are low-dimensional systems with a large surface to volume ratio, one can expect the magnetic moments in clusters to exceed their respective values in crystals. Similarly, the optical properties in bulk matter are governed by their energy band gap. In clusters, these gaps are usually referred to as the HOMO–LUMO (highest occupied and lowest unoccupied molecular orbitals) gaps and can be tailored [98] as they depend upon size, composition, and structure. It is also possible that two clusters, carrying like charges, attract and form a bound state [99]. Superatoms can be composed of either homo- or heteroatoms; both share the common feature that they mimic the chemistry of atoms in the periodic table.

There are two factors that govern cluster stability: atomic packing and electron shell closure [100]. Although these two factors are intertwined, they can be distinguished from one another in specific cases. In clusters composed of noble gas atoms where atomic packing is the dominant factor, simple metals where electronic structure is the driving factor, transition metals and semiconductors where it is difficult to separate between these two factors, and ionic systems where the bulk identity emerges in very small clusters. Clusters of noble gas atoms offer an extreme example where the magic numbers are governed by atomic shell closure. This is because noble gas atoms, due to their closed electronic shells (ns^2np^6), are chemically inert. Therefore, the bonding between these atoms is weak and is governed by the Van der Waal's interaction. Clusters of noble gas atoms form close-packed structures, and the most stable clusters (i.e., magic numbers) are formed when their atomic shells are closed. One such packing is given by an icosahedron (Figure 4.11). The number N of atoms, which are needed to complete successive icosahedral shells, is given by $N = (10/3)\ K3 - 5K2 + (11/3)\ K - 1$, where K is the number of shells. These correspond to 13, 55, 147, ... atoms that complete 2, 3, 4, ... shells, respectively, and this is what effectively have been observed in the experiments [101].

In gas-phase sodium clusters formation experiments, the mass spectrum peaks at cluster sizes corresponding to 8, 20, 40, and 58 Na atoms, which imply that these clusters are far more abundant, and hence are likely to be more stable than their neighbors. Similar features were seen earlier in nuclear physics where nuclei with the 8, 20, 40, ... nucleons were found to be more stable than others, hence the term magic numbers. Their origin was explained to be due to nuclear shell closure. In analogy with nuclear physics, authors also termed the most stable Na clusters as magic clusters and performed the same

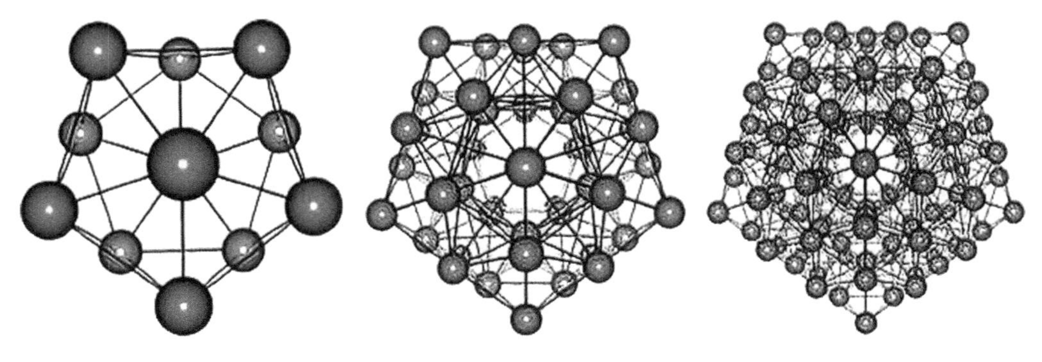

FIGURE 4.11
Icosahedric structures of noble gas atoms containing 13, 55, and 147 atoms.

analysis to see if the electronic shell closure could give rise to these magic numbers [102]. They constructed a simple model where a Na cluster is modelled by a sphere of radius *R*, with the positive charge on each of the Na atoms distributed homogeneously over this sphere. Such a model is called the spherical "jellium" model. Next, the authors calculated the corresponding energy levels of electrons, which are ordered, as in nuclear shell structure, in increasing energies as $1S^21P^61D^{10}2S^21F^{14}2P^6$. Note that the number of electrons needed to close successive electronic shells are 2, 8, 18, 20, 34, 40. The magic numbers correspond to electrons needed to close successive electronic shells. Further examination of the mass spectra counting rates showed that even numbered clusters are more abundant than odd-numbered clusters, which could not be explained within the spherical jellium model. Later one realized that [103] this phenomenon could be explained by a model, originally developed in nuclear physics. Clusters with an insufficient number of electrons, to fill the electronic shells, undergo distortion from a spherical geometry into an ellipsoidal geometry. Thus, the degenerate electronic orbitals are further split. This is referred to as the Jahn–Teller distortion. Thus, the relative stability of Na clusters can be explained purely based on their electronic structure. The electron shell-closing rule has been able to explain the magic numbers seen in the mass spectra of other simple metal clusters such as Cu, Mg, Al, etc. The jellium model has been successfully applied to study the relative stability of clusters composed of simple metals, such as the alkali metals, alkaline earth metals, and noble metals. However, this model does not apply to clusters composed of transition metal and rare earth metal atoms, or those composed of semiconductors and insulators. Unlike simple metals, where electrons are nearly free, the electrons in transition and rare earth metals are quasi-localized. On other hand, the electrons in semiconductors form covalent bonds.

Clusters formed of covalently bonded systems of the same group (as C and Si) display good stability despite they follow an entirely different chemistry. Unlike C that forms sp^1, sp^2, and sp^3 bonding, Si prefers sp^3 bonding. This is manifested in both the bulk and the cluster properties of Si. Attempts to find Si_{60} in the same fullerene structure as C_{60} have failed, and silicon clusters do not possess the chain or ring structures seen in carbon [104]. In contrast to carbon, Si_6 and Si_{10} are magic clusters that were discovered through photodissociation experiments.

It has been demonstrated both theoretically and experimentally that fragmentation of a cluster always leads to magic numbers [105]. Unique geometries set most clusters apart from any other matter; that is, they are very different from their respective bulk crystal structures, they evolve non-predictively as a function of size, and they differ from one element to another. For example, the interplay of surface tension and compressibility in metal clusters result in the preference of icosahedral or decahedral atomic structures.

It is well known that there are three general classes of materials in terms of their electronic structure. The materials composed of noble gas atoms form weakly bound systems, where electronic structures display very little overlap between atomic orbitals. In metals, the valence electrons of the atoms become free from their ions, and the energy gap vanishes at the Fermi level. In semiconductors and insulators, the valence and the conduction bands are separated by a significant energy gap at the Fermi level. In clusters, however, there are no energy bands, and the atomic energy levels form molecular energy levels as atoms come together. A schematic diagram of the electronic structure evolution is shown in Figure 4.12. As clusters form, the atomic energy levels hybridize and broaden with size. All clusters, irrespective of their elemental origin, show an energy gap between HOMO and LUMO. The HOMO–LUMO gaps of these clusters change with size and are expected to show bulk behavior at some critical size.

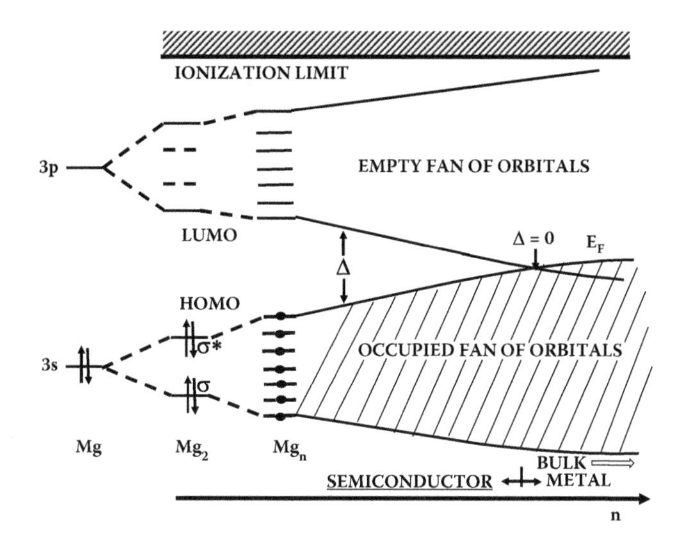

FIGURE 4.12
Schematics of the evolving electronic structure of magnesium.

Regarding the electron affinity behavior in clusters, it follows a trend that is commonly consistent with the spectral shape. For instance, in vanadium clusters containing more than 17 vanadium atoms, the electron affinity versus cluster size *N*, becomes linear as a function of $N^{-1/3}$ and approaches the bulk work function [106].

It is well known that properties of simple metals, such as the alkali metals, aluminium, and gold, are governed by *s* and *p* valence electrons. These electrons tend to be delocalized, and the atoms in the bulk phase do not carry a magnetic moment; these metals are paramagnetic. However, in clusters, the situation can be very different, as the geometry plays an important role in the underlying electronic structure. An example of such system is a Li_4 cluster. Two of the electrons occupy the $1S^2$ state, while the other two occupy the $1P^2$ state. In keeping with Hund's rule, the electrons in the $1P^2$ shell would prefer to have their spins parallel, and hence the ground state of Li_4 should be a spin triplet, with the cluster carrying a total magnetic moment of 2 μB. However, as mentioned earlier, Li_4 has a planar geometry and is nonmagnetic, which means that the magnetic moment is null. The reason for this is that energy may be lowered due to Jahn–Teller distortion. This distortion would split the degenerate energy levels, allowing the two-$1P^2$ electrons to have antiparallel spins, resulting in the ground state becoming nonmagnetic. Therefore, whether a cluster will be magnetic or nonmagnetic depends upon the competition between Hund's rule coupling and Jahn–Teller distortion.

Clusters can be charged, by either adding or removing an electron, their stability, geometry, and corresponding energy gain, or loss, will change as the size of the cluster changes, with the latter evolving toward the bulk work function. Another interesting issue is the effect on cluster's stability as the cluster is multiply charged (for example, the electrostatic repulsion between like charges would increase as the size of a cluster is decreased). When this repulsion becomes greater than the binding energy of the cluster, the cluster will fragment. This phenomenon was experimentally studied in doubly positively charged Pb, NaI, and Xe clusters [107] and is commonly referred to as "Coulomb explosion." The critical size for Coulomb explosion is dependent on the chemistry of the cluster and the nature of its binding. For example, doubly charged clusters of noble gas atoms, bound by a weak van der Waals force, will not be stable until it reaches a very large size. The situation can be different for clusters bound by strong covalent or ionic bonds, as well as for clusters bound by metallic bonds; the critical size for Coulomb explosion of Xe_n, Pb_n, and $(NaI)_n$ clusters to be 52, 30, and 20, respectively [107]. The critical size decreases as the binding strength increases. Another important question one may ask is: "What is the nature of the fragments as a cluster explodes? Are they of equal size and does each fragment carry equal charge? A considerable amount of work has been carried out on the stability and fragmentation of charged clusters [108–110].

In the case of multiply negatively charged clusters, the electron–electron repulsion will also lead to either auto ejection of the added electrons or fragmentation of the cluster. What is the critical size of a cluster that can hold two extra electrons? The answer again will depend upon the cluster composition and the underlying electronic structure. This problem could be addressed using mass spectrometry and showed that multiply charged negative ions of small molecules and clusters can exist as isolated entities in metastable states [111]. A related problem is the interaction between two equally charged atoms/clusters; while most of the doubly charged dimers were found to be metastable, those with nearly filled d-shells underwent spontaneous fission. Certain dimers, such as Cr^{2+}, W^{2+}, and Mo^{2+}, have a barrier for spontaneous fission that may be high enough for the doubly charged dimer to remain in a metastable state for an extended period. As interatomic distances increase beyond 10 au, classical Coulomb repulsion dominates, and for distances shorter than such value, quantum mechanics has the preponderant influence, due to charge polarization [112].

To use clusters as building blocks of matter, the clusters must be stable and have the ability to retain its identity. Imagine that one can design and synthesize a cluster with atomically precise size and composition such that it mimics the chemistry of an atom on the periodic table. Such a cluster may be regarded as a superatom, which can then be used to build a new three-dimensional periodic table, with superatoms constituting the third dimension. Because there are limitless ways to design these superatoms, the third dimension of this new periodic table, in principle, can be infinite. This is significant, because select elements in the periodic table, which are critical to industry, are either scarce or expensive. It will be highly desirable to identify stable superatoms composed of earth-abundant elements that can replace scarce or expensive elements. If superatoms can be assembled to build bulk matter, where the individual identities are retained, a new class of materials may emerge. Such materials, referred to as cluster-assembled materials, would have unique properties, because superatoms, instead of atoms, form the building blocks. Based on the *jellium* model calculations, Saito and Ohnishi [112] the cluster Na_8, with closed electronic shell ($1S^2 1P^6$), is chemically inert, whereas it was shown that Na_{19}, with open electronic shells ($1S^2 1P^6 1D^{10} 2S^1$), is reactive, mimicking the chemistry of alkali atoms. The authors termed Na_8 and Na_{19} as "giant atoms." However, later calculations based on the actual geometry of the Na_8 cluster, showed that while it retains its structure up to 600 K on an insulating NaCl (001) surface, it spontaneously collapses on a Na (110) surface, forming an epitaxial adlayer. In vacuum, two Na_8 clusters interact, forming a deformed Na_{16} cluster. It has been also reported that a metal atom, decorated with halogen atoms whose number exceeds the maximal valence of the metal atom by 1 (e.g., LiF_2), behave as a halogen atom; and that a cluster, with one extra electron than needed for its electronic shell closure (e.g., Li_3O), behave as an alkali atom. The former species were coined as superhalogens and the later species as superalkalis. The electron affinities of superhalogens are larger than those of the halogens, while the ionization potentials of superalkalis are smaller than those of the alkali atoms. Therefore, one can consider

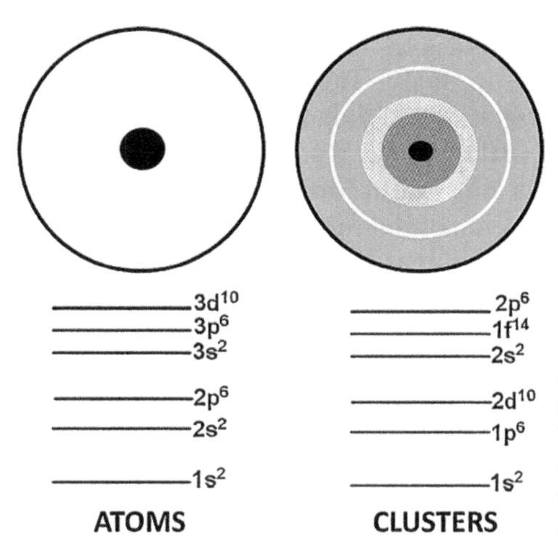

FIGURE 4.13
Schematic diagram of an atom and atomic orbitals (on the left) where the positively charged nucleus is localized at a point, and *jellium* model of a cluster where the positive charge is smeared over a sphere of finite radius with corresponding electronic orbitals (on the right).

superhalogens and superalkalis as superatoms, because they mimic the chemistry of halogen and alkali atoms, respectively. To illustrate the analogy between an atom and a superatom, one can make use of the jellium model, where a sphere carrying a uniform distribution of positive charge approximates a cluster. The electrons surround this "nucleus" of finite size, in spherical shells (see Figure 4.13), just as they surround a point nucleus in an atom. The difference between the two descriptions lies in the ordering of the shells. Therefore, the chemistry of the superatoms originates in the same manner as it does in atoms, in the order that successive shells are filled. In simple metal clusters, the super atomic orbitals, in the *jellium* model, are qualitatively like those obtained using molecular orbital theory. In clusters of transition metals and semiconductor elements, however, the geometry and electronic structure play a significant role, and the molecular orbitals are not the same as those obtained from the *jellium* model.

Another cluster, having closed electronic shells and identified as a superatom, is Au_{20}. From its photoelectron spectra it was possible to calculate that the HOMO–LUMO is 1.77 eV, about 0.2 eV greater than that of C_{60} (1.57 eV) [113]. The future possibility of synthesizing stable superatoms in solution, changing the number of atoms in the structure of a cluster, will certainly constitute a step further in the realization of the practical application of superatom clusters as substitutes for chemical elements in chemical reactions.

The superatoms with closed electronic shells have zero magnetic moment; and so, they are nonmagnetic. Clusters may also be designed to be magnetic by suitably choosing their size, shape, and composition and clusters of nonmagnetic elements can become magnetic. Two of the most fundamental quantities that control the magnetic property of a material are magnetic moment/atom and the nature of their coupling. Although one-half of the atoms in the periodic table possess a magnetic moment, there are only five elements, Fe, Co, Ni, Gd, and Dy, which show an intrinsic ferromagnetic order in the bulk. Other elements are either nonmagnetic, diamagnetic, antiferromagnetic, or ferrimagnetic. It is known that simple metals containing *sp* valence electrons are paramagnetic, while those containing localized *d* or *f* electrons show magnetic behavior. However, magnetism in atomic clusters shows very different behavior. Clusters of simple metals, as well as those of nonmagnetic and antiferromagnetic transition metals, can become ferromagnetic. Additionally, the magnetic moment per atom, in most clusters, is larger than in the respective bulk, although there are examples where the reverse is possible. Some authors have studied the influence of continuous deformation of a Li_4 cluster, from the square to the rhombus geometry, on the underlying spin multiplicity. The preferred spin multiplicity of the rhombus structure is found to be a singlet, while that of the square structure is a triplet, with singlet rhombus being the ground state. Similar results were also seen for the Na_4 cluster. Later it was found that when Li_4 is allowed to form a three-dimensional structure, the preferred spin is also a triplet, resulting in a total magnetic moment of 2 μB, as expected from the *jellium* model [114]. This suggests that there must be a spin cross over as the planar structure of Li_4 transforms into a three-dimensional structure. The conclusion that clusters of *sp* elements could be magnetic holds for Al clusters, as by measuring the deflection in a Stern–Gerlach field, even-numbered Al clusters, containing less than 10 atoms, have a spin multiplicity of 3, and thus a magnetic moment of 2 μB (one would normally expect this to be 0 μB). Furthermore, it has been shown that larger Al_n clusters, containing even number of electrons, are nonmagnetic.

Metal-organic frameworks (MOFs) are supramolecules consisting of metal ions or clusters coordinated to organic ligands to form one-, two-, or three-dimensional structures, forming a coordination network. In some cases, the pores are stable during elimination of the guest molecules (often solvents) and could be refilled with other compounds. Because of this property, MOFs are of interest for the storage of gases, gas purification, gas separation, water remediation, catalysis, and supercapacitors. In contrast to MOFs, covalent organic framework (COFs) are made entirely from light elements (H, B, C, N, and O), with extended structures, and the metal's coordination preference influences the size and shape of pores by dictating how many ligands can bind to the metal and in which orientation. MOFs have potential as heterogeneous catalysts, because their high surface area, tuneable porosity, diversity in metal and functional groups. Despite zeolites are extraordinarily useful in catalysis they are limited by the fixed tetrahedral coordination of the Si/Al connecting points and the two-coordinated oxide linkers. On the contrary, MOFs exhibit more diverse coordination geometries. It is also difficult to obtain zeolites with pore sizes larger than 1 nm, which limits the catalytic applications of zeolites to relatively small organic molecules (typically no larger than xylenes).

One of the first experiments that brought into focus clusters with hollow cage structures is that of C_{60} fullerene. The high unusual stability of C_{60} cluster originates from its geometry (a soccer ball structure with 12 pentagons and 20 hexagons). The synthesis of C_{60} by Kratschmer et al. [115] permitted direct experimental validation of this geometry. As it is well known, the smallest carbon fullerene consists of 20 carbon atoms, all arranged in pentagons. However, the fullerene structure is not the ground state of C_{20}. Furthermore, two more isomers exist; one is shaped like a bowl, while the other is shaped like a ring. Both isomers are close in energy, with the bowl shape being the lowest energy structure, followed by the cage and ring structures. As is abundantly known, these structures have been verified by photoelectron spectroscopy experiments and the relative stability of the structure has been understood theoretically. The advantage of having a cluster in the form of a hollow cage is that it can accommodate an atom inside it, and the large surface area of the cage can also be used to functionalize its exterior for applications in catalysis.

C_{60} fullerene has a cage diameter of 6.5 Å, and its ability to embed atoms in C_{60} (endohedrally or exohedrally), has opened new possibilities to explore novel science and new technologies. The large surface area of the cage can also be used to functionalize its exterior for applications in catalysis. The technological potential of cage clusters, as fullerenes and endohedral fullerenes, has led to a constant search for similar cage structures in other elements. This search has led to the discovery of cage clusters composed of B, Au, Sn, and metallocarbohedrenes (met-cars). Using a combination of photoelectron spectroscopy and density functional theory, Zhai et al. [115] identified a fullerene-like structure of B_{40}. The authors observed the added electron in B_{40}^- to have an extremely low binding energy. This is characteristic of a very stable cluster with a large HOMO–LUMO gap. Theoretical calculations revealed a cage structure with a large energy gap. While the ground state of B_{40}^- was found to have a quasi-planar structure, with two adjacent hexagonal holes, the lowest energy structure of neutral B_{40} is a fullerene-like cage with heptagonal face. In contrast to C_{60}, the authors noted that the surface of the all-boron fullerene, bonded uniformly via delocalized σ and π bonds, is not perfectly smooth and exhibits unusual hexagonal faces. Using *ab initio* calculations, Yakobson and co-workers [116] predicted B_{80} to have a hollow cage structure like that of C_{60}; however, an extra B atom occupies the center of each of the 20 hexagons. The structure was found to be lower in energy than the ring structure, which is the building block of B nanotubes. With a HOMO–LUMO gap of ~1 eV and I_h symmetry, the prediction of a B_{80} cage prompted studies, where researchers explored its potential application for CO_2 capture and separation. However, a subsequent unbiased search for the ground-state geometry of B_{80}, using simulated annealing, showed the preferred structure in fact not to be a hollow cage, but rather a B_{12} centered core-shell structure.

Unlike carbon and boron clusters, metal clusters tend to form compact structures. However, the observation of the Au_{16} cluster as a hollow cage, by Wang and co-workers [117] provided an exception to the compact structures. As pointed out earlier, Au_{20} is a compact pyramidal cluster with tetrahedral symmetry. However, when four of the vertices atoms are removed, and the structure is optimized by allowing the face-centered atoms to relax outward, Au_{16}^- forms a hollow cage structure. Wang and co-workers showed that the photoelectron spectrum of Au_{16}^- is consistent with this hollow structure. Using the electron diffraction technique, Xing et al. [118] later confirmed the validity of the structure as well as demonstrated that Au_n^-, with n = 14–17, also possesses hollow structures. Thus, Au clusters evolve, with size, from planar to hollow to compact structures. The neutral Au_{16} cluster, however, is a compact cluster with T_d symmetry. This suggests that the electron affinity (EA) and the vertical detachment energy (VDE) should be different. This is, notably, due to the former being a measure of the difference between the ground-state geometries of the neutral and the anion, whereas the latter is the energy difference between the neutral and the anion, both having the anion geometry. The experimental photoelectron spectroscopy result of Bulusu et al., [119] however, did not agree with the results above. Both the EA and the VDE were the same, within the experimental error. The discrepancy could be clarified in the aim of molecular dynamics, demonstrating that Au_{16}^-, once formed in the hollow-cage structure, is prevented from assuming the T_d symmetry of the neutral structure, due to an energy barrier. The authors also identified several isomers of Au_{16}^- that lie within a narrow energy range. However, due to the accuracy of calculations, it is difficult to predict the exact ground-state geometry of a cluster. Another metal cluster, found to have a cage structure, is Sn_{12}^{2-}, known as stannaspherene [120]. With a diameter of 6.1 Å, Sn_{12}^{2-} offers enough space to embed a metal atom.

Hollow cage clusters, composed of more than one element, have been known for a long time. A classic example of such a cluster is $B_{12}H_{12}^{2-}$, which is stable as a dianion and can be functionalized to meet the challenges of energy storage. Ti_8C_{12}, known as a "met-car," is another example of a cluster composed of two elements [100], and has a global equilibrium structure with tetrahedral symmetry [100].

This panoply of nanometric superstructures would not be complete without mentioning two other proven achievements:

1. It is possible to place fullerenes (for example C_{60}) inside MOFs and form what is called endohedral MOFs.
2. Nanocrystal superlattices formation.

The formation of nanocrystal superlattices is generally governed by the balance between enthalpic contributions and entropic interactions. Enthalpy prevails when, for instance, specific pair-wise interactions between nanocrystals are designed, such as in the aqueous-based co-assembly of DNA-coated noble-metal nanocrystals with DNA linkers [121]. But this strategy is not applicable to water-degradable perovskite nanocrystals. Entropy may prevail for steric-stabilized colloids of apolar nanocrystals when short-range repulsion is the principal component of the interparticle potential, which causes the nanocrystals to behave like hard spheres. Such colloids often undergo a transition to the densest possible periodic arrays on solvent evaporation-hexagonal close packing or face-centered cubic packing for spheres, to minimize the free energy of the system [122]. Mixtures of spherical nanocrystals have been shown to form at least 20 different binary superlattice structures and usually optimize packing density for a given ratio of their effective hard-sphere diameters. Superlattices that are isostructural with NaCl, AlB_2, $MgZn_2$, or $CaCu_5$ are typically observed [123]. Superfluorescence in lead halide perovskite nanocrystal superlattices was recently demonstrated with the simplest packing geometry-simple cubic packing of cubic nanocrystals into a three-dimensional supercrystal [124]. Unlike fluorescence, superfluorescence is a collective emission of several initially incoherently photo-excited dipoles, which are coupled by their common photon field, and is characterized by orders-of-magnitude-faster than radiative decay [124]. Those highly ordered structures are created solely by the "force of entropy." This assembly occurs because, during crystal formation, the particles tend to use the space around them most efficiently in order to maximize their freedom of motion during the late stages of solvent evaporation, that is, before they are "frozen " in their eventual crystal lattice positions. In this regard, the shape of the individual nanocrystals plays a crucial role – soft-perovskite cubes allow for a much denser packing than what is attainable in all-spherical mixtures. Thus, the force of entropy causes the nanocrystals to always arrange in the densest possible packing – as long as they are designed such that they do not attract or repel each other by other means, such as electrostatics [124].

References

1. K. E. Drexler, "Molecular engineering: An approach to the development of general capabilities for molecular manipulation," *Proceedings of the National Academy of Sciences*, vol. 78, no. 9, pp. 5275–5278, 1981. doi:10.1073/pnas.78.9.5275
2. H. R. Williams, R. S. Trask, P. M. Weaver, and I. P. Bond, "Minimum mass vascular networks in multifunctional materials," *Journal of The Royal Society Interface*, vol. 5, no. 18, pp. 55–65, 2008. doi:10.1098/rsif.2007.1022
3. L. F. Deravi, A. P. Magyar, S. P. Sheehy, et al., "The structure–function relationships of a natural nanoscale photonic device in cuttlefish chromatophores," *Journal of The Royal Society Interface*, vol. 11, no. 93, p. 20130942, 2014. doi:10.1098/rsif.2013.0942

4. M. Nosonovsky and B. Bhushan, Eds., *Green Tribology*. Berlin: Springer Berlin Heidelberg, 2012. doi:10.1007/978-3-642-23681-5
5. K. Komvopoulos, "Adhesion and friction forces in microelectromechanical systems: Mechanisms, measurement, surface modification techniques, and adhesion theory," *Journal of Adhesion Science and Technology*, vol. 17, no. 4, pp. 477–517, 2003. doi:10.1163/15685610360554384
6. M. Zhou, N. Pesika, H. Zeng, Y. Tian, and J. Israelachvili, "Recent advances in gecko adhesion and friction mechanisms and development of gecko-inspired dry adhesive surfaces," *Friction*, vol. 1, no. 2, pp. 114–129, 2013. doi:10.1007/s40544-013-0011-5
7. C. Caizer, "Nanoparticle size effect on some magnetic properties," In: *Handbook of Nanoparticles*. Cham: Springer International Publishing, 2016, pp. 475–519. doi:10.1007/978-3-319-15338-4_24
8. R. Rabiei, A. K. Dastjerdi, M. Mirkhalaf, and F. Barthelat, "Hierarchical structure, mechanical properties and fabrication of biomimetic biomaterials," In: *Biomimetic Biomaterials*. Cambridge: Elsevier, 2013, pp. 67–90. doi:10.1533/9780857098887.1.67
9. S. Parvate, P. Dixit, and S. Chattopadhyay, "Superhydrophobic surfaces: Insights from theory and experiment," *The Journal of Physical Chemistry B*, vol. 124, no. 8, pp. 1323–1360, 2020. doi:10.1021/acs.jpcb.9b08567
10. S. Latthe, C. Terashima, K. Nakata, and A. Fujishima, "Superhydrophobic surfaces developed by mimicking hierarchical surface morphology of lotus leaf," *Molecules*, vol. 19, no. 4, pp. 4256–4283, 2014. doi:10.3390/molecules19044256
11. K. Manoharan and S. Bhattacharya, "Superhydrophobic surfaces review: Functional application, fabrication techniques and limitations," *Journal of Micromanufacturing*, vol. 2, no. 1, pp. 59–78, 2019. doi:10.1177/2516598419836345
12. P. S. Brown and B. Bhushan, "Designing bioinspired superoleophobic surfaces," *APL Materials*, vol. 4, no. 1, p. 015703, 2016. doi:10.1063/1.4935126
13. A. J. Walton, *Three Phases of Matter*. Oxford: Clarendon, 1983.
14. M. Nosonovsky and V. Mortazavi, *Friction-Induced Vibrations and Self-Organization Mechanics and Non-Equilibrium Thermodynamics of Sliding Contact*. New York: CRC Press, 2014. doi:10.1201/b17497
15. S. An, S. S. Yoon, and M. W. Lee, "Self-healing structural materials," *Polymers*, vol. 13, no. 14, p. 2297, 2021. doi:10.3390/polym13142297
16. O. Speck and T. Speck, "An overview of bioinspired and biomimetic self-repairing materials," *Biomimetics*, vol. 4, no. 1, p. 26, 2019. doi:10.3390/biomimetics4010026
17. F. Fu, Z. Chen, Z. Zhao, et al., "Bio-inspired self-healing structural color hydrogel," *Proceedings of the National Academy of Sciences*, vol. 114, no. 23, pp. 5900–5905, 2017. doi:10.1073/pnas.1703616114
18. T. C. Mauldin and M. R. Kessler, "Self-healing polymers and composites," *International Materials Reviews*, vol. 55, no. 6, pp. 317–346, 2010. doi:10.1179/095066010X12646898728408
19. S. Cho, S. Y. Hwang, D. X. Oh, and J. Park, "Recent progress in self-healing polymers and hydrogels based on reversible dynamic B–O bonds: Boronic/boronate esters, borax, and benzoxaborole," *Journal of Materials Chemistry A*, vol. 9, no. 26, pp. 14630–14655, 2021. doi:10.1039/D1TA02308J
20. G. M. Whitesides and M. Boncheva, "Beyond molecules: Self-assembly of mesoscopic and macroscopic components," *Proceedings of the National Academy of Sciences*, vol. 99, no. 8, pp. 4769–4774, 2002. doi:10.1073/pnas.082065899
21. N. T. K. Thanh, N. Maclean, and S. Mahiddine, "Mechanisms of nucleation and growth of nanoparticles in solution," *Chemical Reviews*, vol. 114, no. 15, pp. 7610–7630, 2014. doi:10.1021/cr400544s
22. F. Vazza and A. Feletti, "The quantitative comparison between the neuronal network and the cosmic web," *Frontiers in Physics*, p. 8, 2020. doi:10.3389/fphy.2020.525731
23. B. A. McGuire, R. A. Loomis, A. M. Burkhardt, et al., "Detection of two interstellar polycyclic aromatic hydrocarbons via spectral matched filtering," *Science*, vol. 371, no. 6535, pp. 1265–1269, 2021. doi:10.1126/science.abb7535

24. T. Christoff-Tempesta, Y. Cho, D. Y. Kim, et al., "Self-assembly of aramid amphiphiles into ultra-stable nanoribbons and aligned nanoribbon threads," *Nature Nanotechnology*, vol. 16, no. 4, pp. 447–454, 2021. doi:10.1038/s41565-020-00840-w
25. N. T. Alvarez, P. Miller, M. R. Haase, R. Lobo, R. Malik, and V. Shanov, "Tailoring physical properties of carbon nanotube threads during assembly," *Carbon*, vol. 144, pp. 55–62, 2019. doi:10.1016/j.carbon.2018.11.036
26. R. F. M. Lobo, "Dust deposition from air with anomalous characteristics," *Rendiconti Lincei*, vol. 28, no. 4, pp. 623–633, 2017. doi:10.1007/s12210-017-0633-z
27. J. C. Sung and L. Jianping, *Diamond Nanotechnology: Synthesis and Applications*. New York: CRC Press, 2009.
28. T. Ohno, S. Yatsuya, and R. Uyeda, "Formation of ultrafine metal particles by gas-evaporation technique. III. Al in He, Ar and Xe, and Mg in mixtures of inactive gas and air," *Japanese Journal of Applied Physics*, vol. 15, no. 7, pp. 1213–1217, 1976. doi:10.1143/JJAP.15.1213
29. S. M. George, "Atomic layer deposition: An overview," *Chemical Reviews*, vol. 110, no. 1, pp. 111–131, 2010. doi:10.1021/cr900056b
30. H. Wender, L. F. de Oliveira, P. Migowski, et al., "Ionic liquid surface composition controls the size of gold nanoparticles prepared by sputtering deposition," *The Journal of Physical Chemistry C*, vol. 114, no. 27, pp. 11764–11768, 2010. doi:10.1021/jp102231x
31. G. Ledoux, R. Lobo, F. Huisken, O. Guillois, and C. Reynaud, "Photoluminescence properties of silicon nanocrystals synthetised by laser pyrolysis," In: *Trends in Nanotechnology Research*. New York: Nova Science Publishers, pp. 133–144, 2004.
32. W. Ji, W. Qi, X. Li, et al., "Investigation of disclinations in marks decahedral Pd nanoparticles by aberration-corrected HRTEM," *Materials Letters*, vol. 152, pp. 283–286, 2015. doi:10.1016/j.matlet.2015.03.137
33. Z. Yang, N. Yang, and M. P. Pileni, "Nano Kirkendall effect related to nanocrystallinity of metal nanocrystals: Influence of the outward and inward atomic diffusion on the final nanoparticle structure," *The Journal of Physical Chemistry C*, vol. 119, no. 39, pp. 22249–22260, 2015. doi:10.1021/acs.jpcc.5b06000
34. X. Xia, S. Xie, M. Liu, et al., "On the role of surface diffusion in determining the shape or morphology of noble-metal nanocrystals," *Proceedings of the National Academy of Sciences*, vol. 110, no. 17, pp. 6669–6673, 2013. doi:10.1073/pnas.1222109110
35. H. Brune, "Microscopic view of epitaxial metal growth: Nucleation and aggregation," *Surface Science Reports*, vol. 31, no. 4-6, pp. 125–229, 1998. doi:10.1016/S0167-5729(99)80001-6
36. K. Y. Niu, H. G. Liao, and H. Zheng, "Visualization of the coalescence of bismuth nanoparticles," *Microscopy and Microanalysis*, vol. 20, no. 2, pp. 416–424, 2014. doi:10.1017/S1431927614000282
37. J. K. Sprafke, D. V. Kondratuk, M. Wykes, et al., "Belt-shaped π-systems: Relating geometry to electronic structure in a six-porphyrin nanoring," *Journal of the American Chemical Society*, vol. 133, no. 43, pp. 17262–17273, 2011. doi:10.1021/ja2045919
38. X. Y. Kong, Y. Ding, R. Yang, and Z. L. Wang, "Single-crystal nanorings formed by epitaxial self-coiling of polar nanobelts," *Science*, vol. 303, no. 5662, pp. 1348–1351, 2004. doi:10.1126/science.1092356
39. H. N. Miras, C. J. Richmond, D. L. Long, and L. Cronin, "Solution-phase monitoring of the structural evolution of a molybdenum blue nanoring," *Journal of the American Chemical Society*, vol. 134, no. 8, pp. 3816–3824, 2012. doi:10.1021/ja210206z
40. B. Rogers, J. Adams, and S. Pennathur, *Nanotechnology*. New York: CRC Press, 2014. doi:10.1201/b17424
41. A. P. Alivisatos, "Semiconductor nanocrystals," *MRS Bulletin*, vol. 20, no. 8, pp. 23–32, 1995. doi:10.1557/S0883769400045073
42. P. Smirnov, F. Gazeau, J. C. Beloeil, B. T. Doan, C. Wilhelm, and B. Gillet, "Single-cell detection by gradient echo 9.4 T MRI: A parametric study," *Contrast Media & Molecular Imaging*, vol. 1, no. 4, pp. 165–174, 2006. doi:10.1002/cmmi.104

43. J. W. Connor and J. B. Taylor, "Scaling laws for plasma confinement," *Nuclear Fusion*, vol. 17, no. 5, pp. 1047–1055, 1977. doi:10.1088/0029-5515/17/5/015
44. C. Binns, K. N. Trohidou, J. Bansmann, et al., "The behaviour of nanostructured magnetic materials produced by depositing gas-phase nanoparticles," *Journal of Physics D: Applied Physics*, vol. 38, no. 22, pp. R357–R379, 2005. doi:10.1088/0022-3727/38/22/R01
45. K. O'Grady and A. Bradbury, "Particle size analysis in ferrofluids," *Journal of Magnetism and Magnetic Materials*, vol. 39, no. 1-2, pp. 91–94, 1983. doi:10.1016/0304-8853(83)90407-9
46. T. G. Dietz, M. A. Duncan, D. E. Powers, and R. E. Smalley, "Laser production of supersonic metal cluster beams," *The Journal of Chemical Physics*, vol. 74, no. 11, pp. 6511–6512, 1981. doi:10.1063/1.440991
47. W. Bouwen, P. Thoen, F. Vanhoutte, et al., "Production of bimetallic clusters by a dual-target dual-laser vaporization source," *Review of Scientific Instruments*, vol. 71, no. 1, pp. 54–58, 2000. doi:10.1063/1.1150159
48. H. R. Siekmann, C. H. Lüder, J. Faehrmann, H. O. Lutz, and K. H. Meiwes-Broer, "The pulsed arc cluster ion source (PACIS)," *Zeitschrift für Physik D Atoms, Molecules and Clusters*, vol. 20, no. 1-4, pp. 417–420, 1991. doi:10.1007/BF01544026
49. R. P. Methling, V. Senz, E. D. Klinkenberg, et al., "Magnetic studies on mass-selected iron particles," *The European Physical Journal D*, vol. 16, no. 1, pp. 173–176, 2001. doi:10.1007/s100530170085
50. I. Katakuse, T. Ichihara, Y. Fujita, T. Matsuo, T. Sakurai, and H. Matsuda, "Mass distributions of copper, silver and gold clusters and electronic shell structure," *International Journal of Mass Spectrometry and Ion Processes*, vol. 67, no. 2, pp. 229–236, 1985. doi:10.1016/0168-1176(85)80021-5
51. H. G. Boyen, G. Kästle, F. Weigl, et al., "Chemically induced metal-to-insulator transition in Au_{55} clusters: Effect of stabilizing ligands on the electronic properties of nanoparticles," *Physical Review Letters*, vol. 87, no. 27, p. 276401, 2001. doi:10.1103/PhysRevLett.87.276401
52. V. I. Kuzmin, D. L. Tytik, D. K. Belashchenko, and A. N. Sirenko "Structure of silver clusters with magic numbers of atoms by data of molecular dynamics," *Colloid Journal*, vol. 70, no. 3, pp. 284–296, 2008. doi:10.1134/S1061933X08030058
53. G. U. Kulkarni, P. J. Thomas, and C. N. R. Rao, "Mesoscale organization of metal nanocrystals," *Pure and Applied Chemistry*, vol. 74, no. 9, pp. 1581–1591, 2002. doi:10.1351/pac200274091581
54. M. Knapp, D. Kreisle, O. Echt, K. Sattler, and E. Recknagel, "Size distributions of negatively and positively charged clusters: CO_2 and N_2O," *Surface Science*, vol. 156, pp. 313–320, 1985. doi:10.1016/0039-6028(85)90589-8
55. J. M. Hollas and D. Phillips, Eds. *Jet Spectroscopy and Molecular Dynamics*. Springer Netherlands, 1995. doi:10.1007/978-94-011-1314-4
56. W. A. Saunders, P. C. Sercel, R. B. Lee, et al., "Synthesis of luminescent silicon clusters by spark ablation," *Applied Physics Letters*, vol. 63, no. 11, pp. 1549–1551, 1993. doi:10.1063/1.110745
57. S. L. Girshick, C. P. Chiu, R. Muno, et al., "Thermal plasma synthesis of ultrafine iron particles," *Journal of Aerosol Science*, vol. 24, no. 3, pp. 367–382, 1993. doi:10.1016/0021-8502(93)90009-X
58. S. E. Pratsinis, "Flame aerosol synthesis of ceramic powders," *Progress in Energy and Combustion Science*, vol. 24, no. 3, pp. 197–219, 1998. doi:10.1016/S0360-1285(97)00028-2
59. O. Masala and R. Seshadri, "Synthesis routes for large volumes of nanoparticles," *Annual Review of Materials Research*, vol. 34, no. 1, pp. 41–81, 2004. doi:10.1146/annurev.matsci.34.052803.090949
60. L. B. Kiss, J. Söderlund, G. A. Niklasson, and C. G. Granqvist, "New approach to the origin of lognormal size distributions of nanoparticles," *Nanotechnology*, vol. 10, no. 1, pp. 25–28, 1999. doi:10.1088/0957-4484/10/1/006
61. C. N. R. Rao, A. Müller, and A. K. Cheetham, Eds. *The Chemistry of Nanomaterials*. Wiley, 2004. doi:10.1002/352760247X

62. C. G. Granqvist and R. A. Buhrman, "Ultrafine metal particles," *Journal of Applied Physics*, vol. 47, no. 5, pp. 2200–2219, 1976. doi:10.1063/1.322870
63. P. Tartaj, M. del P. Morales, S. Veintemillas-Verdaguer, T. Gonz lez-Carre o, and C. J. Serna, "The preparation of magnetic nanoparticles for applications in biomedicine," *Journal of Physics D: Applied Physics*, vol. 36, no. 13, pp. R182–R197, 2003. doi:10.1088/0022-3727/36/13/202
64. P. E. di Nunzio and S. Martelli, "Coagulation and aggregation model of silicon nanoparticles from laser pyrolysis," *Aerosol Science and Technology*, vol. 40, no. 9, pp. 724–734, 2006. doi:10.1080/02786820600806522
65. O. Bomatí-Miguel, X. Q. Zhao, S. Martelli, P. E. di Nunzio, and S. Veintemillas-Verdaguer, "Modeling of the laser pyrolysis process by means of the aerosol theory: Case of iron nanoparticles," *Journal of Applied Physics*, vol. 107, no. 1, p. 014906, 2010. doi:10.1063/1.3273483
66. T. Majima, Y. Matsumoto, and M. Takami, "On the SF6-sensitized IR photodecomposition of Fe(CO)5: 5 μm transient absorption measurements and absorption energy measurements," *Journal of Photochemistry and Photobiology A: Chemistry*, vol. 71, no. 3, pp. 213–219, 1993. doi:10.1016/1010-6030(93)85002-P
67. K. Friedlander, *Smoke, Dust and Haze: Fundamentals of Aerosol Behavior*. Oxford: Wiley, 1977.
68. W. J. Stark, P. R. Stoessel, W. Wohlleben, and A. Hafner, "Industrial applications of nanoparticles," *Chemical Society Reviews*, vol. 44, no. 16, pp. 5793–5805, 2015. doi:10.1039/C4CS00362D
69. M. K. Dosbolayev, A. U. Utegenov, A. B. Tazhen, and T. S. Ramazanov, "Investigation of dust formation in fusion reactors by pulsed plasma accelerator," *Laser and Particle Beams*, vol. 35, no. 4, pp. 741–749, 2017. doi:10.1017/S0263034617000805
70. S. H. Lim, Z. Luo, Z. Shen, and J. Lin, "Plasma-assisted synthesis of carbon nanotubes," *Nanoscale Research Letters*, vol. 5, no. 9, pp. 1377–1386, 2010. doi:10.1007/s11671-010-9710-2
71. M. S. Bell, K. B. K. Teo, R. G. Lacerda, W. I. Milne, D. B. Hash, and M. Meyyappan, "Carbon nanotubes by plasma-enhanced chemical vapor deposition," *Pure and Applied Chemistry*, vol. 78, no. 6, pp. 1117–1125, 2006. doi:10.1351/pac200678061117
72. Y. Xia, Y. Xiong, B. Lim, and S. E. Skrabalak, "Shape-controlled synthesis of metal nanocrystals: Simple chemistry meets complex physics?" *Angewandte Chemie International Edition*, vol. 48, no. 1, pp. 60–103, 2009. doi:10.1002/anie.200802248
73. J. Hao, H. Liu, J. Miao, et al., "A facile route to synthesize CdSe/ZnS thick-shell quantum dots with precisely controlled green emission properties: Towards QDs based LED applications," *Scientific Reports*, vol. 9, no. 1, p. 12048, 2019. doi:10.1038/s41598-019-48469-7
74. T. Yonezawa and T. Kunitake, "Practical preparation of anionic mercapto ligand-stabilized gold nanoparticles and their immobilization," *Colloids and Surfaces A: Physicochemical and Engineering Aspects*, vol. 149, no. 1-3, pp. 193–199, 1999. doi:10.1016/S0927-7757(98)00309-4
75. F. Gröhn, B. J. Bauer, Y. A. Akpalu, C. L. Jackson, and E. J. Amis, "Dendrimer templates for the formation of gold nanoclusters," *Macromolecules*, vol. 33, no. 16, pp. 6042–6050, 2000. doi:10.1021/ma000149v
76. H. Xu, B. W. Zeiger, and K. S. Suslick, "Sonochemical synthesis of nanomaterials," *Chemical Society Reviews*, vol. 42, no. 7, pp. 2555–2567, 2013. doi:10.1039/C2CS35282F
77. A. Gedanken, "Using sonochemistry for the fabrication of nanomaterials," *Ultrasonics Sonochemistry*, vol. 11, no. 2, pp. 47–55, 2004. doi:10.1016/j.ultsonch.2004.01.037
78. N. Usen, S. A. Dahoumane, M. Diop, X. Banquy, and D. C. Boffito, "Sonochemical synthesis of porous gold nano- and microparticles in a Rosette cell," *Ultrasonics Sonochemistry*, vol. 79, p. 105744, 2021. doi:10.1016/j.ultsonch.2021.105744
79. Y. Mizukoshi, E. Takagi, H. Okuno, R. Oshima, Y. Maeda, and Y. Nagata, "Preparation of platinum nanoparticles by sonochemical reduction of the Pt(IV) ions: Role of surfactants," *Ultrasonics Sonochemistry*, vol. 8, no. 1, pp. 1–6, 2001. doi:10.1016/S1350-4177(00)00027-4
80. Y. J. Zhu and F. Chen, "Microwave-assisted preparation of inorganic nanostructures in liquid phase," *Chemical Reviews*, vol. 114, no. 12, pp. 6462–6555, 2014. doi:10.1021/cr400366s
81. D. Szabó and S. Schlabach, "Microwave plasma synthesis of materials—from physics and chemistry to nanoparticles: A materials scientist's viewpoint," *Inorganics*, vol. 2, no. 3, pp. 468–507, 2014. doi:10.3390/inorganics2030468

82. A. Bapat, C. Anderson, C. R. Perrey, C. B. Carter, S. A. Campbell, and U. Kortshagen, "Plasma synthesis of single-crystal silicon nanoparticles for novel electronic device applications," *Plasma Physics and Controlled Fusion*, vol. 46, no. 12B, pp. B97–B109, 2004. doi:10.1088/0741-3335/46/12B/009
83. F. Mafuné, J. ya Kohno, Y. Takeda, and T. Kondow, "Formation of stable platinum nanoparticles by laser ablation in water," *The Journal of Physical Chemistry B*, vol. 107, no. 18, pp. 4218–4223, 2003. doi:10.1021/jp021580k
84. A. V. Kabashin and M. Meunier, "Femtosecond laser ablation in aqueous solutions: A novel method to synthesize non-toxic metal colloids with controllable size," *Journal of Physics: Conference Series*, vol. 59, pp. 354–359, 2007. doi:10.1088/1742-6596/59/1/074
85. S. Besner, A. V. Kabashin, F. M. Winnik, and M. Meunier, "Synthesis of size-tunable polymer-protected gold nanoparticles by femtosecond laser-based ablation and seed growth," *The Journal of Physical Chemistry C*, vol. 113, no. 22, pp. 9526–9531, 2009. doi:10.1021/jp809275v
86. C. Y. Shih, M. V. Shugaev, C. Wu, and L. V. Zhigilei, "The effect of pulse duration on nanoparticle generation in pulsed laser ablation in liquids: Insights from large-scale atomistic simulations," *Physical Chemistry Chemical Physics*, vol. 22, no. 13, pp. 7077–7099, 2020. doi:10.1039/D0CP00608D
87. L. D. Marks and L. Peng, "Nanoparticle shape, thermodynamics and kinetics," *Journal of Physics: Condensed Matter*, vol. 28, no. 5, p. 053001, 2016. doi:10.1088/0953-8984/28/5/053001
88. K. Koga, T. Ikeshoji, and K ichi Sugawara, "Size- and temperature-dependent structural transitions in gold nanoparticles," *Physical Review Letters*, vol. 92, no. 11, p. 115507, 2004. doi:10.1103/PhysRevLett.92.115507
89. P. M. Ajayan and L. D. Marks, "Phase instabilities in small particles," *Phase Transitions*, vol. 24-26, no. 1, pp. 229–258, 1990. doi:10.1080/01411599008210232
90. F. Silly and M. R. Castell, "Temperature-dependent stability of supported five-fold twinned copper nanocrystals," *ACS Nano*, vol. 3, no. 4, pp. 901–906, 2009. doi:10.1021/nn900059v
91. M. Rieth and W. Schommers, "Metallic nanoclusters: Computational investigations of their applicability as building blocks in nanotechnology," *Journal of Computational and Theoretical Nanoscience*, vol. 1, no. 1, pp. 40–46, 2004. doi:10.1166/jctn.2004.005
92. R. Ahlrichs and S. D. Elliott, "Clusters of aluminium, a density functional study," *Physical Chemistry Chemical Physics*, vol. 1, no. 1, pp. 13–21, 1999. doi:10.1039/a807713d
93. C. V. Ciobanu, C. Z. Wang, and K. M. Ho, *Atomic Structure Prediction of Nanostructures, Clusters and Surfaces*. New York: Wiley, 2013.
94. P. Jena, S. N. Khanna, and B. K. Rao, Eds. *Physics and Chemistry of Finite Systems: From Clusters to Crystals*. Amsterdam: Springer Netherlands, 1992. doi:10.1007/978-94-017-2645-0
95. S. N. Khanna and P. Jena, "Atomic clusters: Building blocks for a class of solids," *Physical Review B*, vol. 51, no. 19, pp. 13705–13716, 1995. doi:10.1103/PhysRevB.51.13705
96. S. G. Frauendorf and C. Guet, "Atomic clusters as a branch of nuclear physics," *Annual Review of Nuclear and Particle Science*, vol. 51, no. 1, pp. 219–259, 2001. doi:10.1146/annurev.nucl.51.101701.132354
97. H. R. Banjade, Deepika, S. Giri, S. Sinha, H. Fang, and P. Jena, "Role of size and composition on the design of superalkalis," *The Journal of Physical Chemistry A*, vol. 125, no. 27, pp. 5886–5894, 2021. doi:10.1021/acs.jpca.1c02817
98. T. Zhao, J. Zhou, Q. Wang, and P. Jena, "Like charges attract?" *The Journal of Physical Chemistry Letters*, vol. 7, no. 14, pp. 2689–2695, 2016. doi:10.1021/acs.jpclett.6b00981
99. P. Jena and Q. Sun, "Super atomic clusters: Design rules and potential for building blocks of materials," *Chemical Reviews*, vol. 118, no. 11, pp. 5755–5870, 2018. doi:10.1021/acs.chemrev.7b00524
100. O. Echt, K. Sattler, and E. Recknagel, "Magic numbers for sphere packings: Experimental verification in free xenon clusters," *Physical Review Letters*, vol. 47, no. 16, pp. 1121–1124, 1981. doi:10.1103/PhysRevLett.47.1121

101. W. D. Knight, K. Clemenger, W. A. de Heer, W. A. Saunders, M. Y. Chou, and M. L. Cohen, "Electronic shell structure and abundances of sodium clusters," *Physical Review Letters*, vol. 53, no. 5, pp. 510–510, 1984. doi:10.1103/PhysRevLett.53.510.2
102. K. Clemenger, "Ellipsoidal shell structure in free-electron metal clusters," *Physical Review B*, vol. 32, no. 2, pp. 1359–1362, 1985. doi:10.1103/PhysRevB.32.1359
103. J. C. Grossman and L. Mitáš, "Quantum Monte Carlo determination of electronic and structural properties of Si_n clusters (n = 20)," *Physical Review Letters*, vol. 74, no. 8, pp. 1323–1326, 1995. doi:10.1103/PhysRevLett.74.1323
104. P. Jena, B. K. Rao, and S. N. Khanna, Eds. *Physics and Chemistry of Small Clusters*. New York: Springer US, 1987. doi:10.1007/978-1-4757-0357-3
105. H. Wu, S. R. Desai, and L. S. Wang, "Evolution of the electronic structure of small vanadium clusters from molecular to bulklike," *Physical Review Letters*, vol. 77, no. 12, pp. 2436–2439, 1996. doi:10.1103/PhysRevLett.77.2436
106. K. Sattler, J. Mühlbach, O. Echt, P. Pfau, and E. Recknagel, "Evidence for Coulomb explosion of doubly charged microclusters," *Physical Review Letters*, vol. 47, no. 3, pp. 160–163, 1981. doi:10.1103/PhysRevLett.47.160
107. H. Haberland, *Clusters of Atoms and Molecules II: Solvation and Chemistry of Free Clusters, and Embedded, Supported and Compressed Clusters*. Heidelberg: Springer, 1994.
108. U. Näher, S. Bjørnholm, S. Frauendorf, F. Garcias, and C. Guet, "Fission of metal clusters," *Physics Reports*, vol. 285, no. 6, pp. 245–320, 1997. doi:10.1016/S0370-1573(96)00040-3
109. O. Echt, P. Scheier, and T. D. Mark, "Multiply charged clusters," Published online, 2002.
110. M. K. Scheller, R. N. Compton, and L. S. Cederbaum, "Gas-phase multiply charged anions," *Science*, vol. 270, no. 5239, pp. 1160–1166, 1995. doi:10.1126/science.270.5239.1160
111. S. Saito and S. Ohnishi, "Stable (Na_{19})$_2$ as a giant alkali-metal – atom dimer," *Physical Review Letters*, vol. 59, no. 2, pp. 190–193, 1987. doi:10.1103/PhysRevLett.59.190
112. E. S. Kryachko and F. Remacle, "20-nanogold Au $_{20}$ (T_d) cluster and its hollow cage isomers: Structural and energetic properties," *Journal of Physics: Conference Series*, vol. 248, p. 012026, 2010. doi:10.1088/1742-6596/248/1/012026
113. P. Jena and A. W. Castleman, "Clusters: A bridge across the disciplines of physics and chemistry," *Proceedings of the National Academy of Sciences*, vol. 103, no. 28, pp. 10560–10569, 2006. doi:10.1073/pnas.0601782103
114. W. Krätschmer, "The story of making fullerenes," *Nanoscale*, vol. 3, no. 6, p. 2485, 2011. doi:10.1039/c0nr00925c
115. H. J. Zhai, Y. F. Zhao, W. L. Li, et al., "Observation of an all-boron fullerene," *Nature Chemistry*, vol. 6, no. 8, pp. 727–731, 2014. doi:10.1038/nchem.1999
116. G. Chen, Q. Wang, Q. Sun, Y. Kawazoe, and P. Jena, "Structures of neutral and anionic Au16 clusters revisited," *The Journal of Chemical Physics*, vol. 132, no. 19, p. 194306, 2010. doi:10.1063/1.3427293
117. X. Xing, B. Yoon, U. Landman, and J. H. Parks, "Structural evolution of Au nanoclusters: From planar to cage to tubular motifs," *Physical Review B*, vol. 74, no. 16, p. 165423, 2006. doi:10.1103/PhysRevB.74.165423
118. S. Bulusu, X. Li, L. S. Wang, and X. C. Zeng, "Evidence of hollow golden cages," *Proceedings of the National Academy of Sciences*, vol. 103, no. 22, pp. 8326–8330, 2006. doi:10.1073/pnas.0600637103
119. L. F. Cui, X. Huang, L. M. Wang, et al., "Sn_{12}^{2-}: Stannaspherene," *Journal of the American Chemical Society*, vol. 128, no. 26, pp. 8390–8391, 2006. doi:10.1021/ja062052f
120. C. M. Niemeyer and U. Simon, "DNA-based assembly of metal nanoparticles," *European Journal of Inorganic Chemistry*, vol. 2005, no. 18, pp. 3641–3655, 2005. doi:10.1002/ejic.200500425
121. M. A. Boles, M. Engel, and D. V. Talapin, "Self-assembly of colloidal nanocrystals: From intricate structures to functional materials," *Chemical Reviews*, vol. 116, no. 18, pp. 11220–11289, 2016. doi:10.1021/acs.chemrev.6b00196

122. E. V. Shevchenko, D. V. Talapin, C. B. Murray, and S. O'Brien, "Structural characterization of self-assembled multifunctional binary nanoparticle superlattices," *Journal of the American Chemical Society*, vol. 128, no. 11, pp. 3620–3637, 2006. doi:10.1021/ja0564261
123. I. Cherniukh, G. Rainò, T. Stöferle, et al., "Perovskite-type superlattices from lead halide perovskite nanocubes," *Nature*, vol. 593, no. 7860, pp. 535–542, 2021. doi:10.1038/s41586-021-03492-5
124. R. Bonifacio and L. A. Lugiato, "Cooperative radiation processes in two-level systems: Superfluorescence," *Physical Review A*, vol. 11, no. 5, pp. 1507–1521, 1975. doi:10.1103/PhysRevA.11.1507

5

Nanophysics and Nanotechnology: From Quanta to Weak Forces Metrology

5.1 Quantum Physics in the Nanoworld

Among the first iconic outcomes in quantum physics were the Stern–Gerlach experiments, in which the intrinsic spin angular momentum of an atom was measured making use of a beam of silver atoms. The revelation of the spin as a quantum property plays an important role in nanophysics. One can illustrate the framework starting by a point particle charge q and mass m, moving in a circular orbit of radius r with speed v. The magnetic moment μ is given by $qL/(2mc)$, and this relationship between L and μ turns out to be generally true whenever the mass and charge coincide in space. One can obtain different constants of proportionality by adjusting the charge and mass distributions independently. For example, a solid spherical ball of mass m rotating about an axis through its center with the charge q distributed uniformly only on the surface of the ball has a constant of proportionality of $5q/6mc$. When we come to intrinsic spin of a particle, we can then write

$$\vec{\mu} = \frac{gq}{2mc}\vec{S} \tag{5.1}$$

where the value of the constant g is experimentally determined for a specific particle, which in the case of the electron is $g = 2.00$. In fact, it appears that even a point particle in quantum mechanics may have intrinsic spin angular momentum. Relativistic quantum mechanics predicts that $g = 2$ for an electron. The deviations from this value can be accounted for by quantum field theory. The much larger deviations from $g = 2$ for the proton and neutron are because these particles are not fundamental but are composed of charged constituents (the quarks).

In the typical and simplest Stern–Gerlach experiments a collimated beam of silver atoms is directed through an inhomogeneous magnetic field. Since the interaction energy of a magnetic dipole with an external magnetic field is $(-\boldsymbol{\mu} \cdot \boldsymbol{B})$, when a neutral atom with a magnetic moment $\boldsymbol{\mu}$ enters the magnetic field $\boldsymbol{B}$, it experiences a force given by $\boldsymbol{F} = \nabla (\boldsymbol{\mu} \cdot \boldsymbol{B})$. If we call the direction in which the inhomogeneous magnetic field is larger than the z-direction, then

$$F_z = \vec{\mu} \bullet \frac{\partial \vec{B}}{\partial z} \simeq \mu_z \frac{\partial B_z}{\partial z} \tag{5.2}$$

DOI: 10.1201/9781003285083-5

Classically, $\mu_z = \mu \cos\theta$, where θ is the angle that the magnetic moment makes with the z-axis. Thus, μ_z should take on a continuum of values ranging from $+\mu$ to $-\mu$. Since the atoms coming from the effusive source are not polarized (with their magnetic moments pointing in a preferred direction), one should find a corresponding continuum of deflections. The visible deposit of silver on the glass plate that was used, led to the surprising well-known conclusion that μ_z takes only two values for S_z: $+\hbar/2$ and $-\hbar/2$, where h is the Planck constant. Silver atoms are composed of 47 electrons and a nucleus. According to the atomic theory the total orbital and total spin angular momentum of 46 electrons is equal to zero, and the 47th electron has zero orbital angular momentum. Moreover, the nucleus makes a very small contribution to the magnetic moment of the atom because the mass of nucleus is so much larger than the mass of the electron. Therefore, the magnetic moment of the silver atom is effectively due to the magnetic moment of a single electron. Thus, in carrying out the experiments, what is measured is the component of the intrinsic spin angular momentum of an electron along the z axis and found it to take on only the two discrete values ($+\hbar/2$ and $-\hbar/2$, commonly called *spin-up* and *spin-down*).

Rotating an electron by 2π radians negates its spin vector while rotating it by 4π, the original state is reached again. The Stern–Gerlach type experiments using different associations of magnets are a good guide to the realm of the strange world of quantum world. Quantum mechanics is more than just a collection of probabilities. We live in a world where the allowed states of a particle include superpositions of the states in which the particle possesses a definite attribute, such as the z-component of the particle's spin angular momentum, and thus by superposing such states we form states for which the particle does not have a definite value at all for such an attribute.

In quantum mechanics, the Pauli exclusion principle is responsible for the fact that ordinary bulk matter is stable and occupies volume, and it states that two or more identical fermions (particles with half-integer spin) cannot occupy the same quantum state within a quantum system simultaneously. A more rigorous statement is that, concerning the exchange of two identical particles, the total (many-particle) wave function is antisymmetric for fermions, and symmetric for bosons. This means that if the space and spin coordinates of two identical particles are interchanged, then the total wave function changes its sign for fermions and does not change for bosons. If two fermions were in the same state (e.g., the same orbital with the same spin in the same atom), interchanging them would change nothing and the total wave function would be unchanged. The only way the total wave function can both change sign as required for fermions, and remain unchanged, is that this function must be zero everywhere, which means that the state cannot exist. Thus, the identical nature of electrons has a great effect on cohesion occurring in the nanoworld. The indescribable nature of particles manifests itself when two of them are present in a system, and so the two distribution probabilities $P(x_1, x_2)$ and $P(x_2, x_1)$ are identical. No observable change occurs when the positions of the two particles are exchanged. That is, $P(x_1, x_2) = P(x_2, x_1) = |\Psi_{n,m}(x_1, x_2)|^2$. For two non-interacting electrons situated, for example, in a one-dimensional trap, with a wave function $\Psi_{n,m}(x_1, x_2)$, since the full-wave function must be antisymmetric with respect to the exchange of the two electrons, this can be achieved in two ways, after separating the spatial part $\phi(x)$ and the spin part χ of the wave function: singlet spin and triplet spin. This is the well-known antisymmetric property in fermion exchange, which gives rise to the exchange interaction between electrons and protons (both fermions), and which is at the origin of two important nanoscale aggregation processes: ferromagnetism and covalent molecular formation. Attractive interactions between electrons and nuclei stabilize the formation of a molecule but destabilize a ferromagnet. This is because the exchange integral is negative in the first situation, and positive in the second, that is, the covalent bond is established when the spins are antiparallel (singlet). This reasoning does not apply to bosons because the sign does not change. A rigorous proof was provided by Freeman Dyson, who considered the balance of attractive (electron–nuclear) and repulsive (electron–electron and nuclear–nuclear) forces and showed that ordinary matter would collapse and occupy a much smaller volume without the Pauli principle [1]. In consequence of the Pauli principle, electrons of the same spin are kept apart by a repulsive exchange interaction, which is a short-range effect, acting simultaneously with the long-range electrostatic or coulombic force. This effect is partly responsible for the everyday observation in the macroscopic world that two solid objects cannot be in the same place at the same time.

One can apply a similar formalism to another physical two-state system: the polarization of the electromagnetic field (photon polarization or photon's spin). Many polarization effects can be described by classical physics, unlike the spin of particles, which is a purely quantum phenomenon. Nonetheless, analyzing polarization effects using quantum mechanics can help to clarify the differences between classical and quantum physics and at the same time tell us something fundamental about the nature of the quantum of the electromagnetic field. Instead of a beam of spin-atoms passing through a Stem–Gerlach device, one can consider a beam of photons, traveling in the z-direction, passing through a linear polarizer. Those photons that pass through a polarizer with its transmission axis horizontal, that is, along the x-axis, are said to be in the state $| x \rangle$, and those photons that pass through a polarizer with its transmission axis vertical are said to be in the state $| y \rangle$. These two polarization states form a basis, and the basis states satisfy $\langle x | y \rangle = 0$, since a beam of photons that passes through a polarizer whose transmission axis is vertical will be completely absorbed by a polarizer whose transmission axis is horizontal. Note that in the Dirac (or bra-ket) notation of quantum mechanics, the angle brackets denote quantum states (vectors) and matrix elements, for example, $\langle \mathrm{a} | \mathrm{b} \rangle$. Thus, none of the photons will be found to be in the state $| x \rangle$ if they are put into the state $| y \rangle$ by virtue of having passed through the initial polarizer (assuming polarizers with 100% efficiency). One can also create polarized photons by sending the beam through a polarizer whose transmission axis is aligned at some angle to our original xy-axes.

Once two quantum particles interact, one can say that they are entangled, and it means that their quantum states are interdependent (there is a correlation between them). So that if one is in one particular state, then the other one has to have some other particular state depending on the kind of entanglement that they have. Since two electrons have the quantum property of spin, when they interact this may create a situation where those two spins become correlated, such that if one of the electrons has a spin up, then the other one must have a spin down (Pauli exclusion principle). And if those two electrons are separated by a certain distance after the interaction, both Einstein's hidden variables and Bohr's measurement interpretations lead to the same outcome, that is the well-known paradox of instantaneous action at a distance. After the recent successes of real experiments that materialize the former idealizations of Bell, it seems that Einstein's interpretation which fixes the properties of quantum particles before they are measured does not apply. However, that does not mean it is really a spooky action from a distance. In fact, what seems to happen is that when the particles become entangled, "they are no longer separated objects," but a single quantum entity: there is "some kind" of sharing information among them. This is called quantum nonlocality and should not be misunderstood with information transmission, otherwise it will violate relativity theory. In

Aharonov's view there is a kind of a stretching of the normal laws of cause and effect. In fact, nanorings can display the so-called Aharonov–Bohm effect. As the thicknesses of a nanoring lie at the nanoscale (although occasionally the diameters can overcome it) they are members of the sub-class of one-dimensional (1D) nano-objects. These are objects in which one of the three physical dimensions in a single unit of the material is on a length scale greater than the nanoscale (like nanowires or nanotubes).

In quantum Aharonov–Bohm effect an electrically charged particle is affected by an electromagnetic potential (φ, $\boldsymbol{A}$), despite being confined to a region in which both the magnetic field $\boldsymbol{B}$ and electric field $\boldsymbol{E}$ are zero. Remind that the vector potential $\boldsymbol{A}$ was formerly introduced in classical physical, and is simply an auxiliary field that helps to determine the physical electromagnetic fields $\boldsymbol{E}$ and $\boldsymbol{B}$. And that we can alter the function $\boldsymbol{A}$ by adding to it a gradient of a scalar function, in a way that this transformation does not affect the magnetic field, and the electric field will also be unaffected; it is the well-known so-called gauge transformation. The underlying mechanism of the quantum effect is the coupling of the electromagnetic potential with the complex phase of a charged particle's wave function, giving thus rise to interference in suitable experiments. For instance, when the wave function of a charged particle passes around a long solenoid it experiences a phase shift as a result of the enclosed magnetic field, despite the magnetic field being negligible in the region through which the particle passes and the particle's wave function being negligible inside the solenoid. This phase shift has been observed experimentally [2].

Considering a long solenoid carrying an electric current, the magnetic field inside the solenoid is uniform. So, it is possible to calculate the vector potential $\boldsymbol{A}$ inside and outside the solenoid lateral surface. Now suppose that in the famous Young's double-slit experiment we insert a small solenoid directly behind the barrier between the two slits. Recall that the intensity at an arbitrary point P on the screen arises from the interference between the amplitude Ψ_1 for the particle starting at the source point S to arrive at P after passing through one of the slits (path 1) and the amplitude Ψ_2 for it to arrive at P after passing through the other slit (path 2). Surprisingly, the phase for each path that contributes to the path integral is modified by the presence of the solenoid, even though the magnetic field may vanish at all points along the path. It was found that the relative phase between Ψ_1 and Ψ_2 is proportional to the closed line integral of the vector potential going from the source to the point P along path 1 and back to the source from point P along path 2. This is a rather startling result. Classically, we would expect that the particle must follow either path 1 or path 2. Along each of these paths, the magnetic field $\boldsymbol{B}$ vanishes everywhere. How then does the charged particle "know" about the magnetic field within the solenoid? While the classical particle responds to the magnetic field only where the particle is, that is, locally, the quantum particle takes both paths. Since the solenoid produces a vector potential that changes the phase for each of the paths, in some sense we might say that the particle compares the phase that it has picked up along the two different paths and responds directly to the phase difference. Note that this relative phase difference depends on the magnetic flux passing through the surface bounded by the paths and not on the vector potential itself. Thus, the phase difference is a gauge invariant quantity that may be measured. Even though the phase difference depends on the magnetic field $\boldsymbol{B}$ and not on vector potential $\boldsymbol{A}$ directly, the Aharonov–Bohm effect suggests that the particle learns about the magnetic field by responding to the vector potential along the path!

The Aharonov–Bohm effect can be understood from the fact that one can only measure absolute values of the wave function. While this allows for measurement of phase differences through quantum interference experiments, there is no way to specify a wave function with a constant absolute phase. The Aharonov–Bohm effect shows that the local E and B fields do not contain full information about the electromagnetic field, and the electromagnetic potential (φ, $\mathbf{A}$), must be used instead. Thus, the Aharonov–Bohm effect validates the view that forces are an incomplete way to formulate physics, and potential energies must be used instead. In fact, Feynman complained that he had been taught electromagnetism from the perspective of electromagnetic fields, and he wished later in life he had been taught to think in terms of the electromagnetic potential instead, as this would be more fundamental. In Feynman's path-integral view of dynamics, the potential field directly changes the phase of an electron wave function, and these changes in phase lead to measurable quantities.

Going back to the quantum entanglement phenomenon one can say that in the framework of nonstationary scattering theory, it is possible to study the formation of an entangled state of two identical nonrelativistic electron particles because of their elastic scattering. It is useful to recall the qualitative picture of the quantum process of the elastic scattering of two particles. In the initial channel of the process (as $t \to -\infty$), the wave packets of the particles are separated by an asymptotically large distance exceeding the action radius of their interaction potential. As the wave packets approach each other, the particles begin to interact; leaving the interaction region, they move away from each other to an asymptotically large distance in the final channel (as $t \to +\infty$). What is then the entanglement of the states of these particles in the final channel of the collision process? It is known that because of the Pauli principle, the scattering cross-section for two electrons depends on their spin states even if their interaction potential is independent of the spins. Indeed, the angular differential cross-sections behave differently in the cases of singlet and triplet fermion pairs. The triplet cross-section vanishes, and the singlet cross-section remains finite in the case where scattering occurs at the angle $\pi/2$ in the center-of-mass frame. Some authors [3], developed a theoretical approach for describing the formation of quantum entanglement in a collision of two nonrelativistic electrons based on the formalism of nonstationary scattering theory, where one-particle coordinate wave functions before ($t \to -\infty$) and after ($t \to +\infty$) collision do not overlap, and antisymmetrization hence does not lead to the appearance of the nonphysical entanglement. In addition, the particles can be regarded as distinguishable in the case of the local measurement of their spin states. One can then obtain general expressions for the Bell inequality in terms of the direct and exchange scattering amplitudes and of angular differential cross-sections for elastic scattering of the polarized particles. The authors [3] have also shown that the Coulomb interaction between the particles can lead to oscillations in the angular dependence of the measures of entanglement of the quantum state of the pair after the collision.

If a quantum system is perfectly isolated, it would maintain coherence indefinitely, but it would be impossible to manipulate or investigate it. If it is not perfectly isolated, for example, during a measurement, coherence is shared with the environment and appears to be lost with time; it is the process called quantum decoherence. Recall that in quantum mechanics, as long as there exists a definite phase relation between different states, the system is said to be coherent. Decoherence can be viewed as the loss of information from a system into the environment (often modeled as a heat bath), since every system is loosely coupled with the energetic state of its surroundings. The decoherence has irreversibly converted quantum behavior (additive probability amplitudes) to classical behavior (additive probabilities).

In classical scattering of a target body by environmental photons, the motion of the target body will not be changed by the scattered photons on the average. On the contrary, in quantum scattering the interaction between the scattered photons and the superposed target body will cause them to be entangled, thereby delocalizing the phase coherence from the target body to the whole system, rendering the interference pattern unobservable (Figure 5.1).

The wave function collapse occurs when a wave function (initially in a superposition of several eigenstates) reduces to a single eigenstate due to interaction with the external world. This interaction is called an "observation" (it is the essence of a measurement in quantum mechanics, which connects the wave function with classical observables, like position and momentum). Collapse is one of two processes by which quantum systems evolve in time; the other is the continuous evolution via the Schrödinger equation. Collapse is like a "black box" for a thermodynamically irreversible interaction with a classical environment. Collisions with residual gas molecules, the emission of heat radiation, and the absorption of blackbody radiation are among the most important decoherence mechanisms for interferometry with massive particles [4]. These authors investigate experimentally the effect of decoherence due to collisions with various gases and they found a very good quantitative agreement with decoherence theory. Furthermore, they explored the practical limits of matter wave interferometry at finite gas pressures and estimate the required experimental vacuum conditions for interferometry with even larger nano-objects. For instance, for a virus of mass 5×10^7 amu interacting with air at room temperature, they estimate that collisions would not limit quantum interference provided one can reduce the background pressure to below $\approx 10^{-5}$ Pa.

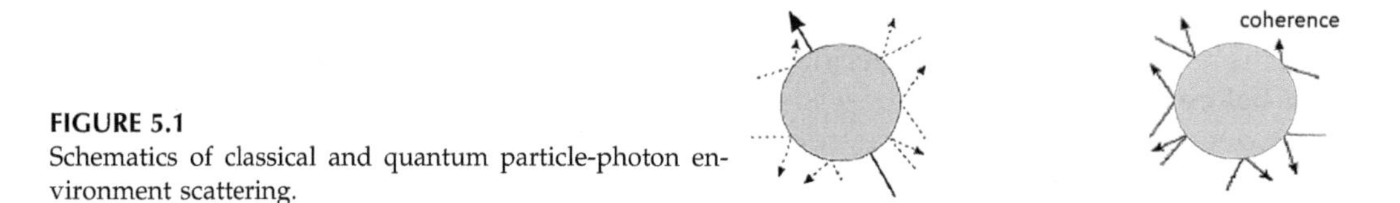

FIGURE 5.1
Schematics of classical and quantum particle-photon environment scattering.

When a measurement is made, the wave function collapses (from an observer's perspective) to just one of the basis states, and the property being measured uniquely acquires the eigenvalue of that particular state. After the collapse, the system again evolves according to the Schrödinger equation. Recall that before collapsing, the wave function may be any square-integrable function, and is therefore associated with the probability density of a quantum mechanical-system. This function is expressible as a linear combination of the eigenstates of any observable. Observables represent classical dynamical variables, and when one is measured by a classical observer, the wave function is projected onto a random eigenstate of that observable. The observer simultaneously measures the classical value of that observable to be the eigenvalue of the final state. Quantum decoherence explains why a system interacting with an environment transitions from being a pure state, exhibiting superpositions, to a mixed state, an incoherent combination of classical alternatives. When entanglement between the particles is lost, the wave function Ψ for two particles can be factorized and written as a product $\Psi = \psi_1 (X_1)\, \psi_2(X_2)$ and the probability is just $|\Psi|^2 = |\psi_1|^2 + |\psi_2|^2$.

Observable quantum effects do not usually occur in everyday macroscopic objects because their individual atoms are not in a coherent state. Therefore, any quantum effects are basically averaged out by the time we make a measurement of the macroscopic object. But there are other examples of quantum effects on a macroscopic scale, including laser beams, superconductors, and superfluids. In fact, a common misconception is that quantum physics is about size, that is, small things should behave in a quantum way, whereas large things always behave classically. This is not really the case; what gives to an object its quantum weirdness is not so much its size but its number of degrees of freedom. Small objects like elementary particles have a small number of degrees of freedom but when a massive object has many degrees of freedom, it will behave classically, as its quantumness will be averaged out over its many degrees of freedom. This is the case of Bose condensates, superfluids, superconductor wires, and there are also those cases of entangled particle states created for quantum communication, which remain entangled even across geographic distance scales of more than one hundred kilometers.

In the classical picture of two neutral particle interaction the angle of deflection is of paramount relevance, but on the atomic-molecular scale quantum theory of binary-collisions (recall that Heisenberg principle avoids exact determination of that angle) it is the phase shift of the radial wave function that assumes the critical role, as it comes out from the resolution of stationary Schrödinger equation for a particle of reduced mass μ [5]. This is the starting point where one can say that an incident plane wave representing a stationary beam of particles is scattered and at each angle a stationary gas beam can be observed. To validate this stationary procedure two conditions must be applied: the gas beam needs to be constant at least in times larger than $\hbar/\Delta E$ (where ΔE is the experimental beam energetic width), and the jet must be uniform within dimensions larger than the dispersion velocity field extension. Luckily both requirements are met in typical molecular beam experiments [6]. This allows to consider that the asymptotic form of the wave function $\Psi(r)$ is the sum of z-direction incident plane wave term with a propagating spherical wave centered at the origin of the dispersion field: $e^{ikz} + e^{ikr} f(\theta)/r$, where $f(\theta)$ is the angular dependent amplitude and k stands for the wave number. In addition, the wave function $\Psi(r)$ must be regular, that is, $\Psi \rightarrow 0$ at $r = 0$. All this framework can be applied except in eventual resonance situations or when there is a coulombic-type interaction potential. It can then be demonstrated [7] that under such conditions the differential cross-section $\sigma(\theta, E)$ is given by $|f(\theta)|^2$, and for a potential with spherical symmetry $V(r)$ it is possible to find a solution where the spherical contribution of the wave function and the amplitude are given by a suitable expansion in partial waves involving Legendre polynomials $P_l \cos(\theta)$ [7], where l stands for the angular momentum quantum number. The differential cross-section $\sigma(\theta, E)$ can be calculated for the case of a bipolar interaction potential (i.e., a Lennard–Jones type) [6], and so with the exception of the angle rainbow, the average of the quantum oscillations agrees very well with the classical result [8]. This is the reason why in most of the cases, transport properties can be classically treated. In several situations it is possible to replace numerical calculations of the cross-sections by suitable approximate analytical methods [7]. For instance, the Ford–Wheeler random-phase semiclassical approximation can be applied to calculate $\sigma(\theta, E)$ when the de Broglie wavelength is smaller than the potential variation scale; in this case, it is possible to introduce the concept of trajectory by using $l + ½ = kb$, where b is the classical impact parameter, $k = \mu v/\hbar$ being v the initial velocity, and $\hbar$ the reduced Planck constant.

For a compact comparison between classical and quantum relevant magnitudes in a binary collision the reader can rely on the analysis of Table 5.1.

TABLE 5.1

Comparison of Magnitudes in a Binary Collision

Classical	Quantum
β - Impact parameter	l - Angular momentum quantum number
$\vartheta(b, g)$ - Angle of deflection	$\eta(b)$ - Phase shift
$\mu g b$ - Angular momentum	$\hbar\ [l(l+1)]^{1/2}$ - Angular momentum
g - Initial relative speed	$\beta = \mu g/\hbar = 2\pi/l$ - de Broglie wavenumber
μg - Relative momentum	$\beta\,\hbar$ - Relative momentum

A molecular beam is generically defined as a collimated stream of molecules, moving in the absence of collisions through a vacuum. The molecular beam technique provides unique conditions for the applicability of the law of inertia and Newton's second law to atoms and molecules. The origin of such a circumstance can be understood on the basis of Ehrenfest's theorem applied to the linear momentum operator ($-i\hbar\nabla$) [9].

Mostly atomic and molecular collisions are not 100% elastic, and the greater the collision energy, the greater the number of collision channels that are opened. Thus, in its simplest way, a chemical reaction is just one of the various possibilities and cannot be seen as a phenomenon of emergence because there is no interaction with other particles. The probability for its occurrence at low collision energies can be roughly estimated by the ratio σ_R/σ, where σ_R represents the reactive cross-section. During the collision of a metal atom M with an atom X (with a certain electron affinity), the distance among them gets shorter, and a critical distance r_c is reached, where the difference between the ionization potential of M and the electron affinity of X is similar with the coulombic attraction energy between M^+ and $(RX)^-$ ionic pair species. It follows that below a certain collision energy threshold, the chemical reaction occurs with a cross-section σ_R given approximately by πrc^2, since r_c is the maximum possible impact parameter. This is the framework of the well-studied harpooning mechanism [10], which has proven to be also successfully applied to atom-molecule collisions of the type M-RX (X is an atom or group of atoms with electron affinity) [6, 11–14], and atom-cluster of the type M-C_{60} [15].

Such theoretical mechanism involving a two-state Landau–Zener approximation [6], accounts for the experimental observations with crossed molecular beams techniques, namely the reaction product's translational energy distribution function [11], as well as both ion-pair formation and chemical reaction cross-sections. For instance, in the nonelastic scattering behavior on the K + Br_2 system, studied over an energy range of 0.1–10 keV, it becomes clear that the chemical reaction (giving rise to neutral products) gives way to the ion-pair formation process above the collisional ionization threshold [6, 11, 16].

Binary collisions become possible to study by making use of atomic and molecular beam techniques which represent one of the best ways to perform single-molecule experiments (i.e., get rid of additional interactions with a third partner). The recent developments in these techniques have had a huge impact not only in physics but also in chemistry (because they allow to study the dynamics of the chemical reaction and can detect individual reactive collisions). They have reached such a level of sophistication that allows experiments of binary collisions where the molecular target is previously oriented by selecting different quantum states of the molecule in an inhomogeneous electric field [12, 17, 18]. This way, it becomes possible to focus on the influence of the target orientational effects in the binary collision, which opened the opportunity to study in more detail the reactivity of RX and the quantum electron-transfer mechanism in different collision geometries [18, 19]. The data shown in

Figure 5.2 demonstrated that there is an enormous steric effect for the case of CF_3H target. Attacking the positive end produces more F^- just above the ion production threshold, and this preference rapidly shifts to the negative end as the energy increases. On the other hand, the steric asymmetry for CCl_3H and CBr_3H is almost negligible. Any hindering effect of the hydrogen atom would be expected to be about equally effective in each of the haloforms, and hindrance cannot account for the H-end reactivity in CF_3H. Clearly, factors other than shielding by the H atom are important, and the electronic structure of the molecule plays a prominent role. The large differences between CF_3H and the two heavier haloforms are a signature of different low-lying σ^* orbitals. The preference for H-end attack in CF_3H may arise from a low-lying σ^* orbital centered on the C or the CH bond, whereas the LUMO in the heavier haloforms is believed to be σ_{CX}^*, resulting in almost no steric preference. Occupation of the σ_{CH}^* orbital in CF_3H can be responsible for the formation of significantly more CF_3^- ions and free electrons than for the heavier haloforms. Once the electron is transferred, the products formed, either ions or salt, depend on the proximity of the reactants as well as the energy. The formation of ions at very low energies is favored by "backside" attack, where the incipient ions could be formed as far apart as possible. Proximity favors salt formation which may explain the lack of ion-pair formation for attack inside the CF_3 umbrella.

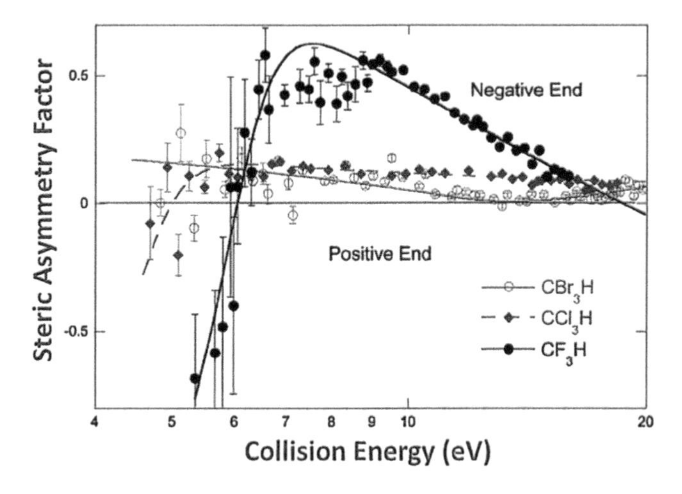

FIGURE 5.2
Steric asymmetry for formation of halide negative ions for the three haloform molecular targets. Curves are model fits to data and a log scale is used to better display the low-energy data. Adapted with permission from Evidence for Orbital-Specific Electron Transfer to Oriented Haloform Molecules. Beike Jia, Jonathan Laib, R. F. M. Lobo, and Philip R. Brooks. *Journal of the American Chemical Society*, vol. 124, no. 46, pp. 13896–13902, 2002. DOI:10.1021/ja027710k. Copyright 2002 American Chemical Society.

Inelastic collisions in beam experiments have been also carried out with clusters making use of lasers (laser molecular beam experiments) for studying quantum state-to-state inelastic collisions (including chemical reactions) [20]. Using high laser field conditions in the photoinduced van der Waals dimer process $Ba...FCH_3 + h\nu \rightarrow BaF\ (Ba^*) + CH_3\ (CH_3F)$, it has been found that two-photodissociation channel yields, i.e., the reactive BaF and nonreactive Ba^* products, exhibited opposite behavior depending upon the laser fluence. Whereas the BaF yield rises as the laser fluence is increased, the Ba^* yield decreases over the same laser fluence range. This means that the product $(Ba^*)/(BaF)$ branching ratio changes significantly from low to high laser fluence in such a way that allows to control the photoinitiated intracluster reaction by changing the intensity of the excitation laser field [20, 21].

Still related with the electron-transfer studies in binary collisions, the case of C_{60} target deserves a particular mention, because it assumes significance in the field of fullerene nanophysics. In fact, electron-transfer experiments using binary collisions with the buckminsterfullerene target have been performed, which revealed for the first time the success of the Landau–Zener harpooning mechanism applied to a nano-object [15]. Furthermore, the experiments revealed shape effects of the fullerene negative ion in the presence of an electric field, which call our attention for the rising of peculiar nano-effects when dealing with the behavior of an extended electric charged nano-object submitted to the field [22]. Apparently, when the electric surface charge of a nano-object extends within a distance not negligible compared with the uniform electric field range, the shape of the charge distribution dictates the nano-object dynamics [22, 23]. All these experiments with C_{60} are paradigmatic examples of experiments where a well-defined nano-object (the close-spherical buckyball) is studied free of any surface contact effect.

Other relevant example is the diffraction of C_{60} collimated beams through a slit. Diffraction of particles started to be studied with the well-known pioneer experiments of Davison and Germer (electron diffraction in crystals) and progressively extended to larger neutral entities, as a way of validation the limits of their quantum behavior. This was done from atoms to clusters, with different molecules [22, 24, 25], and culminating in fullerene nano-objects [26]. The researchers calculated a de Broglie wavelength of the most probable C_{60} velocity as 2.5 pm, and the essential features of the interference pattern can be understood using standard Kirchhoff diffraction theory [27] for a grating with a period of 100 nm (Figure 5.3), by taking into account both the finite width of the collimation and the experimentally determined velocity distribution. More recent experiments prove the quantum nature of molecules made of 810 atoms and with a mass of 10,123 u [28], and even molecules of 25,000 u [29].

It is intriguing that C_{60} can almost be a body obeying classical physics given its many excited internal degrees of freedom. When leaving the effusive source, it has as much as 7 eV of internal energy stored in 174 vibrational modes, and highly excited rotational states with quantum numbers greater than 100. Fullerenes can emit and absorb blackbody radiation very much like a solid and they can no longer be treated as a simple few levels system. Still going one step further than Louis de Broglie, new theories appear in which quantum mechanics eliminate the concept of a point-like classical particle and explain the observed facts by means of wave packets of matter waves alone [30]. It has been found that the quantum behavior decreases as the environmental pressure grows, following an exponential decay directly related with the collision probability; on the other hand, increasing the internal temperature of the fullerene buckyballs the quantum behavior also decreases due to the emission of radiation to the surroundings. Noise (random fluctuations) in the measurement devices is also another way of destroying interference, and when the molecules get bigger and the de Broglie wavelength shrinks, experiments become increasingly sensitive to these effects. Anyway, matter composed of a giant number of atoms can exhibit quantum behavior in certain cases at low temperatures as it demonstrated by the well-known helium superfluid experiments. Observable quantum effects do not usually occur in everyday macroscopic objects because their individual atoms are not in a coherent state. Therefore, any quantum effects are basically averaged out by the time we make a measurement of the macroscopic object. But there are other examples of quantum effects on a macroscopic scale, including laser beams and superconductors.

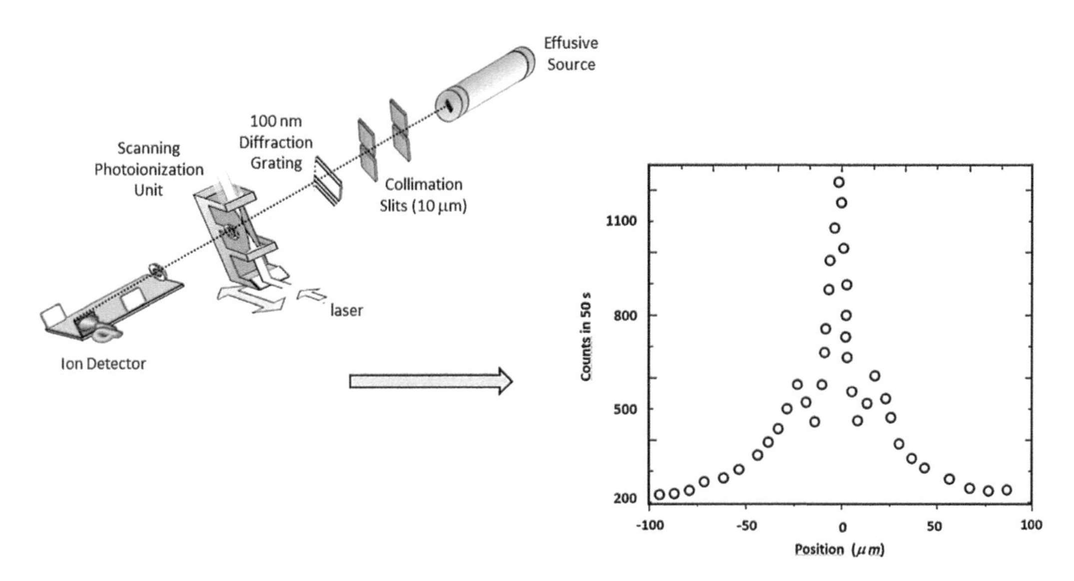

FIGURE 5.3
On the left: Diagram of the experimental set-up for diffraction of C_{60}. On the right: Interference pattern produced by C_{60} molecules, where the expected zeroth and first-order maxima are clearly detected.

The formation of stable van der Waals aggregates generally occurs in molecular beams produced by isentropic expansion. These aggregates have low binding energies, and their formation is generally attributed to a two-step kinetic mechanism, involving an intermediate orbital state, stabilized by collision with a third body, and assuming the following balances:

$$A + A \leftrightarrow A_2^*$$
$$A + A_2^* \leftrightarrow A_2 + A$$

Once the dimer is formed, it can interact again with another particle and form a trimer, and so on. The properties of larger helium clusters are of great interest for nuclear many-body theoreticians because of the many close analogies between nuclei and helium clusters. Helium atoms have the advantage over nucleons, that their interactions depend only on the interatomic distance and do not involve complicated tensorial terms. The experiments with van der Waals clusters of defined sizes are not easily possible because nozzle beam expansions used in their production yield broad size distributions. Moreover, being weakly bound they readily fragment in the commonly used electron impact-ionization mass spectrometer detectors. However, it was shown that light fragile clusters of He can be selected and identified nondestructively by diffraction from a transmission grating, and the method is universally applicable also to heavier species and well suited for spectroscopic studies [24, 25].

Conclusive results involving the observation of quantum interference with large and complex objects are important, since they open many novel possibilities, among them decoherence studies and nanolithography experiments. Structures made of light can also be used to coherently control the motion of complex molecules. Namely, it has been demonstrated the diffraction of the fullerenes C_{60} and C_{70} at a thin grating based on a standing light wave [31]. The authors could prove experimentally that the principles of this effect (known from atom optics) can be successfully extended to massive and large molecules which are internally in a thermodynamic mixed state, and which do not exhibit narrow optical resonances.

The interference experiments with large molecules, clusters, and nano-objects illustrate one of the most unusual aspects of quantum theory, namely that these entities can exist in a superposition of different states. Decoherence is not just an issue of fundamental importance but has also relevance in the development of quantum computers, which exploit the quantum-superposition principle. The absence of quantum behavior in the macroworld arises naturally because it becomes increasingly difficult to isolate objects of increasing size and complexity. It is the quantum nature of the interaction with the environment and the resulting information transfer that leads to the classical behavior of a quantum object. On the one hand, in metrology the high sensitivity of macromolecules to external perturbations complicates attempts to prepare pure quantum states, but on the other hand, it enables new ways of sensitively measuring molecular properties, by monitoring their matter-wave fringe shifts in the presence of external forces.

Experiments in quantum control of a nanoparticle optically levitated in cryogenic free space have been recently successful and represent a milestone toward the generation of macroscopic quantum states of a nanosphere, which would require extremely low levels of decoherence [32]. On the one hand, cryogenic pumping can achieve extreme-high vacuum suppressing decoherence due to gas collisions. On the other hand, silica nanospheres quickly thermalize at the temperature of the surrounding cryogenic environment once the laser is switched off. This drastically reduces the decoherence due to emission of blackbody photons and it is possible to coherently expand the quantum wave function up to a size comparable with the nanosphere itself, bringing macroscopic quantum effects within experimental reach.

Now, going on with more relevant quantum manifestations in the nanoworld, one cannot fail to mention the quantum confinement, the tunneling effect, the harmonic oscillator, van der Waals forces, and the quantum vacuum fluctuations. Indeed, it has been taken advantage of these manifestations to make new useful tools and devices.

Let us start by the quantum confinement synthesizing what is known about the one-dimensional infinite potential energy well of width a, for which the energy eigenvalue equation can be solved in position space. This potential well possesses an infinite number of bound states, and outside the well, the energy eigenfunction must vanish and the requirement that the wave function be continuous leads to the familiar discrete energies permitted:

$$E_n = \frac{\hbar^2\pi^2 n^2}{2ma^2} \quad \mathrm{n} = 1,\ 2,\ 3,\ \ldots \tag{5.3}$$

The requirement that the wave functions vanish at the boundaries of the box means that we can only fit in those waves with nodes at $x = \pm\, a/2$. Accordingly, to Eq. 5.3, the particle can never have zero energy, meaning that the particle can never "sit still." Additionally, it is more likely to be found at certain positions than at others, depending on its energy level. The particle may never be detected at certain positions, known as spatial nodes. However, if relativistic effects are taken to account via Dirac equation, then the probability density does not go to zero at the nodes [33].

To analyze what happens if one changes the potential energy well that is confining the particle and pull the walls of the well out rapidly so they are positioned at $x = \pm\, a$ instead of $x = \pm a/2$, problem 7 must be consulted. A transiently trapped particle enclosed in a potential well made of two barriers of height V, widths w, spacing L, can model several situations in nanophysics (including alpha-particle decay of radioactive nuclei), if one introduces a mean residence time τ, due to tunneling effect. If V and w are large, though finite, one may reasonably approximate the lowest members of the set of bound states in the $V = 0$ ($0 < x < L$) region between the two barriers by E_n in Eq. 5.3, and the wave functions as $\psi_n = (2/L)^{1/2} \sin(n\pi x/L)$, for $n = 1, 2, \ldots$. This approximation makes sense only for states of small n such that E_n is much smaller than the barrier height. However, for such strongly bound states, one can approximate the escape rate τ^{-1} as the product of the rate of collisions of the particle with a barrier (v/L) multiplied by the chance per encounter of transmission and escape, which is the transmission coefficient in the tunneling effect. The wave function for a quantum-mechanical particle in a box whose walls have arbitrary shape is given by the linear partial differential Helmholtz equation subject to the boundary condition that the wave function vanishes at the walls [34].

Conjugated polyene systems also can be described by using the above so-called "particle in a box" model. The conjugated system of electrons can be modeled as a one-dimensional box with length equal to the total bond distance from one terminus of the polyene to the other. In this case, each pair of electrons in each π bond corresponds to their energy level. The energy difference ΔE, between two energy levels is obtained from Eq. 5.3 and corresponds to the emission (or absorption) of a photon of wavelength $hc/\Delta E$, with c as usual, the speed of light in vacuum. Conjugated polyenes are examples of electronic delocalized molecules that are important in the development of molecular electronics.

Molecular electronics triggered by the forecast that silicon technology might reach its scalability limits in a few years, plays a major role in the forthcoming developments of nanotechnology. For molecular electronics to become a valuable alternative to silicon technology, it will not be sufficient to fabricate molecular electronic devices with special characteristics, but suitable circuit and architectural solutions will also be needed. Squeezing entire nonlinear circuit elements (diodes or transistors) into single molecules could, in principle, lead to significantly faster and smaller active electronic elements than solid-state electron devices. Integration with CMOS technology may be the first step for devices based on single molecules. Despite some progresses it is still unclear whether individual molecular devices could be integrated into a larger-scale computing circuit. One possibility to create electronic logic circuits or memory cells based on molecular devices is to synthesize complex molecules whose arms can be separately contacted to provide the same electrical input/output of conventional logic gates. However, this is extremely challenging from the chemical-synthesis point of view, and difficult to implement. Another possibility is to create a programmable interconnected network of nanoparticles and molecular entities called nanocells [35]. These are conducting metallic nanoparticles randomly deposited on a substrate (e.g., silicon) and subsequently bridged via molecular connections, to create electrical pathways between previously patterned metallic leads. The molecular linkers should exhibit nonlinear characteristics, in the form of negative differential resistance or hysteretic behavior and the network is connected to a limited number of input/output pins at the edges of the nanocell, which could then be accessed, configured, and programmed from the edges. Simulations have demonstrated the capability of the nanocell to act as logic gate [35]. Molecular electronics also includes the employment of organic materials as substitutes for the solid-state layers of traditional semiconductor devices. Their carrier mobility is relatively low but can become competitive in the market of large-area devices, because the processing technology is cheaper, and circuits can be realized on eventually any substrate.

The coherent current flowing through a molecule can be calculated from Landauer formula [36] by integrating the transmission function $T(E)$ in the energy range created by the voltage applied between the two electrodes. Materials employed as organic semiconductors are conjugated polymers, which are based on sp^2-hybridized linear carbon chains. While $1s$ orbitals of the carbons do not change when the atoms are bonded, $2s$ orbitals mix their wave functions with two of the three $2p$ orbitals. Thus, three sp^2 orbitals are formed and lie on the molecular plane, at 120° to each other, leaving one p orbital which is orthogonal to the molecular plane. Hybrid sp^2 orbitals can then give origin to different bonds; sp^2 orbitals from different carbon atoms which lie on the molecular plane form strong covalent bonds (σ bonds), whereas p orbitals which are orthogonal to the molecular plane can mix to give less strong covalent bonds (π bonds). The π bonds constitute a delocalized electron density above and below the molecular plane and are responsible for the conductivity of the molecule, as the charge carriers move through these bonds. This transport process is usually called *hopping transport* and is described by means of quantum mechanics. The energy levels of the orbitals, when multiple molecules are considered, give origin to energy bands, in analogy to what happens in inorganic semiconductors: the edge of the valence band corresponds then to the highest occupied molecular orbital (HOMO), whereas the edge of the conduction band corresponds to the lowest unoccupied molecular orbital (LUMO). The difference between the energy of the HOMO and the energy of the LUMO is the energy gap E_g of the organic material and usually $1.5 < E_g < 4$ eV.

Electronic conduction through a variety of different molecules has been studied experimentally by many research groups but the exact nature of the transport mechanisms is usually a still open question. First-principles computations are normally based on density-functional methods, which are as well known, limited to molecular systems made up of a small number of atoms. In addition, a precise estimation of energy band gaps requires sophisticated many-body corrections. Moreover, to understand the transport properties through a molecule, which is connected to metallic electrodes, one needs first to clarify how the interactions between molecule and contacts determine the energy levels of the complete system. Usually it is used as a density-functional tight-binding method, which allows a first-principles treatment of systems comprising many atoms. Electron transport through oligopolyvinylene (OPV) chains anchored to metallic contacts can serve as an example of study. OPV molecules consist of benzene rings which are interconnected by vinylene groups. The OPV terminal molecules in the chain can be anchored to gold contacts through a sulfur bonding. The electronic transport characteristics of the OPV molecules are governed by the electronic states available in the molecule. Such states can be effectively described by the local density of states (DOS), which represents the equilibrium DOS existing in the molecular device broken down into the contributions of single atoms or ensembles of atoms of the device. The density-functional tight-binding method allows to calculate the local DOS of each atom by projecting the DOS of the total structure on the molecular orbitals of that atom.

In solid state, hopping and tunneling are distinct phenomena, having the first an inherently quantum-mechanical origin (the particle can cross the potential barrier even with energy less than the height of the barrier), while the second is a classical statistical physics process where the particle has enough energy to surmount the potential energy barrier, in consequence of the approximation of using single-site wave functions as the basis. This is similar to the evaporation process, for example. In a periodic lattice every site corresponds to a harmonic potential, with the ground state wave function of a harmonic oscillator, which is nonzero everywhere and thus there is a nonzero probability of finding the particle anywhere in the space. From Bloch's theorem, in a periodic lattice the energy eigenstates are always periodic. Therefore, when choosing a nonperiodic basis of single-site wave functions, the Hamiltonian is not diagonal and the nondiagonal terms are the "hopping" terms which means that the stationary state is a particle spread throughout the lattice rather than staying at one point.

Ensuring energy transport is a fundamental condition in molecular devices, which can be of an intermolecular or intramolecular nature, and in supramolecular machine electronics the energy transport situations of greatest interest are photonic transfer and electron transfer. In the first case, let's take the example of two dye molecules D and A, where D absorbs in blue and emits in green and A absorbs in green and emits in red. If the separation d, between the two molecules is less than a certain critical value, part of the excitation energy is transferred from D (donor) to A (acceptor), and the latter will emit in red. The condition of energy transfer by resonance is that the emission spectrum of D overlaps with the absorption spectrum of A. The fluorescence probabilities of the excited molecule D and of molecule A can be estimated as a function of d, through the Förster model, which considers D and A as electromagnetic oscillators spaced by $d << \lambda$ and with $d > 2R$, where R is the radius of the molecule D, so that the dipole approximation can be applied. The electric field acting on A is the Coulomb field of the oscillator D, given by $E = E_0 \cos 2\pi\nu_0 t$, where $E_0 = 2\mu/(4\pi\varepsilon_0 d^3)$, where μ is the electric dipole moment of the oscillator. It can be shown that the average power absorbed by A is proportional to $\varepsilon_A {E_0}^2$ (where ε_A is the molar extinction coefficient of A), and that considering a Hertz antenna type emissive model for D, the energy transfer probability is given by the so-called Förster equation:

$$P = \left[1 + \left(\frac{d}{d_0}\right)^6\right]^{-1} \tag{5.4}$$

where d_0 (Förster's radius) is the distance at which the emission is reduced to half the value it would have had if the acceptor A did not exist. Thus, the probability of light emission is given by (1−P), or, in terms of the energy transfer rate from D to A:

$$k_{D \to A} = \frac{1}{\tau_D^r} \frac{{d_0}^6}{d} \tag{5.5}$$

where $({\tau_D}^r)^{-1}$ is the radiative decay rate of the donor. Typical d_0 values are in the 5 nm range for many organic molecules. If molecule D is surrounded by a layer of acceptor molecules A, then the probability P is proportional to the number of molecules in the layer. Since this number varies with d^2, if the acceptor molecules lie in a plane at a distance d, then

$$P = \left[1 + \left(\frac{d}{d_0}\right)^4\right]^{-1} \tag{5.6}$$

An interesting application of Förster energy transfer between a D donor and an acceptor A, is in single-molecule spectroscopy, using the fluorescence resonant energy transfer (FRET) technique. Thus, it becomes possible to determine D-A distances (either intermolecular or intramolecular) in a range between 1.5 and 8 nm, as well as the relative donor/acceptor orientation. This is a step toward obtaining information about conformations and interactions in macromolecules (e.g., proteins) using fluorophores as labels. Another important application is sensitized fluorescence, in which for example solid anthracene containing a 10^{-5} times lower concentration of tetracene (impurity) does not fluoresce in blue-violet but in yellow-green. The energy absorbed by the anthracene is transferred with high efficiency to the tetracene and re-emitted by the latter. In another variant, by dissolving tetracene in anthracene, the probability of its luminous excitation can be increased by several orders of magnitude. Therefore, using suitable energy transfer systems, one can preferentially direct energy toward certain molecules, in particular those where a certain photochemical reaction is desired to occur (e.g., chlorophyll). The supramolecular structures existing in photosynthetic biological organisms, for capturing solar energy, contain aggregates made up of N donor molecules D. These oscillators are tightly coupled, in such a way that when light is absorbed, the excitation is distributed over a certain domain, which can be described as an oscillation in phase of all its oscillators, with a shift of the absorption λ_{max} to longer wavelengths, which corresponds to a ΔE that is related to N (by $N = -\Delta E/k_B T$). This excitation domain composed of N oscillators is an exciton. The fluorescence time of a monomer can be estimated using the classical oscillator model and is in the order of 5 nanoseconds. In photosynthetic systems, between the reactive center and the surrounding active molecules there are also carotenes that absorb even shorter wavelengths, allowing for better efficiency in energy transfer. Carotenes (linear or branched polyenes) are molecules with a delocalized π electronic system, whose energy levels can be obtained approximately by the free electron model in a 1D box. That is, the individual π electrons are described by 1D standing waves, along a "stretched string" of length L, with de Broglie length $\lambda = 2L/n$, where $n = 1, 2, 3$ .... As previously studied, in this 1D box model, energy levels and wave functions are given, respectively, by

$$E = \frac{1}{2}m_e v_e^2 = \frac{h^2}{2m_e\lambda^2} = \frac{h^2}{2m_e L^2}n^2; \varphi_n = \sqrt{\frac{2}{L}}\sin\frac{n\pi s}{L} \quad \text{with} \quad n = 1, 2, 3 \ldots \tag{5.7}$$

This is, in fact, the same model that allows, in a simplistic way, to explain the additional stability of molecules that obey the so-called Hückel rule, according to which molecules with a number of $4j + 2$ delocalized π electrons have lower energy than would be expected by the sum of its number of single and double bonds. This is true for many conjugated linear molecules and for instance the Kékulé structures of benzene, graphene, fullerenes, and polycyclic aromatic hydrocarbons (PAHs) are in this situation. In general, as the π system increases, there is a shift in the absorption band to longer wavelengths. This effect is also observed in the absorption spectra of linear polyenes and cyanines, and the variation in the position of the absorption maximums with the increase in the number of monomers is very well described by the 1D free electron box model, in which the energy of excitation can be determined by

$$\Delta E = \frac{h^2}{8m_e L^2}(n_{\text{LUMO}}^2 - n_{\text{HOMO}}^2) \tag{5.8}$$

where LUMO and HOMO, respectively, designate the lowest-energy unoccupied orbital and the highest-energy occupied orbital. Since each level is occupied with two electrons, we have: $n_{\text{HOMO}} = N/2$ and $n_{\text{LUMO}} = N/2 + 1$, where N is the number of π electrons. Thus,

$$\Delta E = \frac{h^2}{8m_e L^2}(N + 1) \tag{5.9}$$

The so-called quantum tunneling composites (QTCs) are composites with an elastomeric polymer matrix and filled with metal particles (usually nickel), and make use of the quantum tunneling effect mostly for pressure sensors. Without pressure, the conductive elements are too far apart to conduct electricity; but under a deformation, they move closer (without actually touching) and electrons can tunnel through the insulator, which causes the electrical resistance to fall drastically. The effect is far more pronounced than would be expected from nonquantum effects alone, as classical electrical resistance is linearly proportional to distance, while quantum tunneling is exponential with decreasing distance, allowing the resistance to change by a huge factor between pressured and unpressured states. One of the major advantages of QTCs (also called piezo-resistive sensors) is that since they are insulators when in a state of rest, the electronics fitted to these devices do not consume any energy unless pressure is applied. It is possible to reckon about modeling the behavior of these materials, based on the use of a simple quantum tunneling model of a barrier to which are added effects such as the influence of the matrix viscoelasticity on the sensor response, the influence of pressure and temperature on the elasticity modulus and the dynamic viscosity of the material, as well as phenomena associated with thermal expansion [37].

In nanometric structures that form low-dimensional systems each surface/interface acts like a potential barrier, i.e., the wall of a quantum well, inducing quantum confinement and so generating new energy levels. These levels can be calculated with a model that uses the approximation of the infinite rectangular quantum wells, and can be adapted for 2D, 1D, and 0D systems, respectively [38]. In principle, all structures have three dimensions, but if their size on at least one direction is small enough (no more than one order of magnitude greater than the interatomic distance on that direction), the structure is considered approximately low-dimensional. In such structures, the ratio between the number of atoms located at the surface/interface Ns and the total number of atoms N can be roughly expressed by $2(3 - \xi)a/\xi$, where ξ, represents the dimensionality and a, the mean interatomic distance. This result is exact for 2D plane, 1D cylindrical, and 0D spherical symmetry, respectively. In a first approximation, a metal can be thought as a 3D box containing a gas of N_e free electrons from the outer most orbitals and starting for a one particle quantization due to the confinement of the wave function, this translates into discrete states and energies, which are obtained from the resolution of the time-independent Schrödinger equation. Considering the confinement of an electron of mass m_e in a box of edges L_x, L_y, L_z, the well-known results for the wave functions and energies are, respectively:

$$\psi_n = \sqrt{\frac{2}{a}} \sin\frac{n_x \pi x}{a} \sin\frac{n_y \pi y}{a} \sin\frac{n_z \pi z}{a} \tag{5.10}$$

$$E_{n_x,n_y,n_z} = \frac{h^2}{2m_e}\left(\frac{n_x^2}{L_x^2} + \frac{n_y^2}{L_y^2} + \frac{n_z^2}{L_z^2}\right) = \frac{\hbar^2}{2m_e}(k_x^2 + k_y^2 + k_z^2) \tag{5.11}$$

Considering the high number of moles of electrons, it is convenient to notice that this same result can be obtained by recognizing that each combination of the three quantum numbers corresponds to a point in a 3D grid, and then all points lying on the surface of the corresponding sphere radius have the same value of $(n_x^2 + n_y^2 + n_z^2)$, and so the same energy. In fact, one can determine the energy of the highest occupied state if one calculates the radius n_{max} of the surface which just encloses sufficient grid points to accommodate all the N electrons in a crystal. As only integer values of n_x, n_y, n_z are allowed, each grid point corresponds to a cube of unit volume, and so the number of grid points is equal to the volume under the surface. Thus, by determining the number of grid points contained beneath the surface of radius n_{max} it is possible to show that the Fermi energy (energy of the highest occupied state at $T = 0K$), is given by

$$E_F = \frac{\hbar^2}{2m_e}\left(\frac{3\pi^2 N_e}{V}\right)^{2/3} \tag{5.12}$$

where each grid point corresponds to a state which can accommodate two electrons (according to Pauli's exclusion principle) and Eq. 5.11 is used with the volume of the crystal $V = L_x, L_y, L_z$. The electron density is a property of the solid material (conductor, semiconductor, or insulator), and therefore, assuming the Fermi energy as constant, one can consider that the number of electrons is proportional to $E_F L^3$. For small volumes, the energy states cover the same range of variation as for large volumes, but with a greater ΔE spacing between them. Since average spacing between energy states can be roughly estimated by E_F/N_e one realizes that $\Delta E \propto 1/L^3$, which means that for smaller volumes the spacing for energy levels increase. In a metal, one considers that $\Delta E = 4E_F/3n_v$, being n_v the total number of valence electrons that the nanoparticle can have (remember that in a metal all valence electrons are free, which at $0K$ fill all energy levels below E_F). Smaller dimensions mean fewer atoms and therefore fewer valence electrons. Only electrons with energy $K_B T$ at the Fermi level can become conductors, and so quantum confinement can make a metallic particle nonmetallic. As soon as $\Delta E > k_B T$, the driver stops driving. An example of such metal-to-insulator transition is the Schottky diode with nanoscale control.

Now, let us remind that integrating the DOS function $g(E)$ up to the Fermi energy is performing a sum over all the occupied states, and so, the number of electrons N_e is

$$N_e = \int_0^{E_F} g(E)dE \tag{5.13}$$

or

$$g(E) = \frac{1}{V}\frac{dN_e}{dE} = \frac{1}{2\pi^2}\left(\frac{2m_e}{\hbar^2}\right)^{3/2} E^{1/2} = \frac{8\pi\sqrt{2}}{h^3} m_e^{3/2}\sqrt{E}, \text{ for } E \geq 0 \tag{5.14}$$

Combining the number of quantum states that exist at a particular energy with the probability that a state with a given energy is occupied, one can find how many electrons have a particular energy. This is called the density of occupied states and can be obtained by the product $g(E)\ P(E)$, where $P(E)$ stands for the Fermi–Dirac distribution. Therefore, in a conductor the density of conduction electrons can be calculated by

$$n_e = \frac{8\pi\sqrt{2}}{h^3} m_e^{3/2} \int_0^{E_F} \frac{\sqrt{E}}{e^{[(E-E_F)/k_B T]} + 1} dE = \frac{8\pi\sqrt{2}\, m_e^{3/2}}{h^3}\left(\frac{2}{3} E_F^{3/2}\right) \tag{5.15}$$

The DOS definition for an intrinsic semiconductor starts at the top of the valence band and the minimum energy for the electron is the one it has at the bottom of the conduction band (E_g) and so the expression Eq. 5.14 of DOS becomes modified:

$$\begin{aligned} g(E) &= \tfrac{8\pi\sqrt{2}}{h^3} m_e^{*\ 3/2} \sqrt{E - E_g}, \quad \text{for } E \geq E_g \\ &= 0, \quad \text{for } E < E_g \end{aligned} \tag{5.16}$$

with m_e^* representing the mass effective of the electron.

The band gap of semiconductors increases as the material decreases in size, which makes it possible to "tune" the optical and electronic properties of a suitable medium.

In 3D k-space each allowed state occupies a volume $(2\pi/L)^3$ and the volume of a spherical shell of radius k and thickness dk is $4\pi k^2 dk$. In this elemental layer exists a number dN of electrons given by

$$dN = 2\frac{4\pi k^2 dk}{(2\pi/L)^3} = \frac{k^2 L^3 dk}{\pi^2} \tag{5.17}$$

When the available space decreases, say within a few nanometers in the z-direction, the electrons are confined in this direction, moving only freely in the xy plane (quantum well). Therefore, for a 2D box one of the k vector components is fixed, implying that it is necessary to calculate the number of states contained in the ring between the radii k and $k + dk$; each allowed state occupies an area $(2\pi/L)^2$ and the area of the ring is $2\pi k dk$:

$$dN = 2\frac{2\pi k dk}{(2\pi/L)^2} = \frac{k L^2 dk}{\pi} \tag{5.18}$$

Thus, the DOS per unit area becomes independent of energy:

$$g_{2D}(E) = \frac{1}{L^2}\frac{dN}{dE} = \frac{m_e^*}{\pi h^2} \tag{5.19}$$

and considering the other energy levels, the DOS takes the form of a ladder function. Similarly, when the confinement takes place in two space directions, the electrons are free to move just in one direction (quantum wire), and for this case:

$$dN = 2\frac{dk}{2\pi/L} = \frac{L dk}{\pi} \tag{5.20}$$

and

$$g_{1D}(E) = \frac{1}{L}\frac{dN}{dE} = \frac{\sqrt{2m_e^*}}{\pi h}\frac{1}{\sqrt{E}} \tag{5.21}$$

In this case, using more than one energy level, the DOS transforms into a Heaviside function.

In 2D structures, the electron Hamiltonian can be exactly split as the sum of two parts: a parallel part (i.e., parallel with the layer surface), which is Bloch-type, leading to a 2D band structure; and an orthogonal part, which leads to quantum confined levels. On its turn, quantum wires act in effect as wave guides for charge carriers, permitting only a few propagating modes. Endowed with such a quality, they would provide enormous benefits for the photonics and electronics industries. They could make efficient lasers powered by far less current, offer a more economical alternative to superconducting wires as resistance-free conductors, or serve as tiny sensors designed to detect the slightest trace of chemicals. The ballistic propagation of particles in a quantum wire shows up most clearly in the quantization of the wire conductance (or inverse of resistance). Electrons that enter empty states at one end of the channel, to which we have applied a small voltage V, pass right through it without loss or gain, provided that the wire length is much shorter than the electron's mean free path. As the current flowing along the channel is independent of which subbands the electrons occupy, it must be proportional to the number of occupied subbands and the applied potential. The proportionality factor turns out to be a universal constant (the conductance quantum), $G_0 = 2e^2/h$, where e is the unit of electric charge, and 2 the number of spin states for each electron. So, the conductance increases in steps of G_0 as the wire is made wider, thereby accommodating successive higher-order wave guide modes.

Lastly, when the confinement takes place in the three space directions (quantum dot), the k values are quantized in all directions. All available states exist only at discrete energies and can be represented by Dirac delta functions. The 0D electronic systems have typically characteristic dimensions below 100 nm and are authentic artificial atoms with a discrete spectrum of energy levels. In conventional atoms the electrons are attracted to a positive central nucleus; in a 2D artificial atom electrons are typically trapped in a pit of parabolic potential. In many cases, the quantum dot (QD) resembles a "pancake" in which electrons are constrained to move in the xy plane, so the appropriate potential is that of a 2D harmonic oscillator, for which it is possible to solve the corresponding Schrödinger equation.

The technique of molecular beam epitaxy (MBE) is the most powerful technique for manufacturing with excellent precision and high quality all those three types of quantum confinement nanostructures, as well as a great variety of epitaxial heterostructures. The combination of using clean surfaces in ultra-high vacuum (UHV) with a highly controlled and slow evaporation and with the controlled temperature of the substrate makes it possible to grow ultra-thin films at rates as low as 1 μm/h, ensuring surface diffusion and relaxation, as well as controlling the thickness in the nanometer range, and the production of extremely abrupt interfaces (with fast mechanical shutters positioned in the path of the atomic effusive beams). The high substrate temperature allows a convenient mobility of the species, and the low incident flux ensures a sufficient time for surface diffusion and structural relaxation. The accommodation coefficient (ratio between the number of atoms of a certain species that are adsorbed to the number of atoms of that same species that impinge on the surface) can be obtained experimentally through time-of-flight molecular beam techniques and depends on the desorption energy and activation energy for superficial diffusion.

Quantum well lasers and QDs can be analyzed considering this confinement model as well. In quantum well lasers the quantization of the energy levels of the electrons allows the laser to emit light more efficiently than in conventional semiconductor ones. On its turn, due to their small size, QDs do not showcase the bulk properties of the specified semi-conductor but rather show quantized energy states. This quantum confinement led to numerous applications of QDs such as the quantum well laser. A nanometer scale semiconductor (QD) can be described by 3D particle-in-a-box energy quantization equations, and its energy-gap $\Delta E(r)$, is the energy gap between the valence and conduction bands. This is equal to the gap of the bulk material plus the energy equation derived particle-in-a-box, which gives the energy for electrons and holes [39].

$$\Delta E(r) = E_{\text{gap}} + \left(\frac{h^2}{8r^2}\right)\left(\frac{1}{m_e^*} + \frac{1}{m_h^*}\right) \tag{5.22}$$

where m_e^* and m_h^* are the effective masses of the electron and hole, respectively, and r is radius of the dot. It is thus clear that the energy gap of the QD is inversely proportional to the square of the "length of the box," that is, the radius of the QD. Therefore, manipulation of the band gap allows for the absorption and emission of specific wavelengths of light; the smaller the QD, the larger the band gap and thus the shorter the wavelength absorbed.

A bright red photoluminescence at room temperature has been observed for porous silicon, and apparently explained by a widening of the band gap because of quantum confinement [40]. This size effect would supposedly lead to a shifting the photoluminescence into the visible for crystallite sizes below 5 nm, and these small volumes are also responsible for radiative recombination, since the spatial confinement by potential barriers prevents the diffusion of carriers to nonradiative recombination centers. It was then shown that photoluminescence energy is shifted with respect to the band gap of bulk silicon (E_0 = 1.17 eV) and obeys a power law with an exponent equal to −1.39 for the particle diameter d measured in nanometers [41]:

$$E_{PL}(d) = E_0 + \frac{3.73}{d^{1.39}} \tag{5.23}$$

This came to reinforce the conviction that luminescence of porous silicon could be explained by quantum confinement of the electron-hole pairs in quantum crystallites with diameters lower than 5 nm. To go deep for validation of the quantum confinement model became essential to derive a correct size distribution of the Si nanocrystals. More recently it has been possible to obtain experimental results on the photoluminescence of silicon nanocrystals synthesized by laser pyrolysis as a function of their size. In fact, coupled with a molecular beam apparatus it allowed to deposit size-selected nanocrystals on a substrate and after oxidation in air (aging) they exhibit a strong red to infrared photoluminescence [42]. The oxide layer due to surface passivation plays a role not only in the position of the photoluminescence, but also in the width because of its influence on the lattice parameter. Laser pyrolysis of silane in a gas flow reactor followed by cluster beam deposition at low energy on a substrate, after a mechanical velocity and size selection, proved crucial to shed light on the phenomenon [43–45]. The evolution of the photoluminescence was measured as a function of the time; the samples were exposed to air (note that with such technique the deposition occurs in high vacuum). With gradual air exposure and oxidation, the photoluminescence increases in intensity and its peak wavelength shifts to the blue. At the same time, the photoluminescence band becomes wider.

Removing the oxide layer by hydrofluoric acid vapor exposition had no influence on the band position but leads to a smaller width. From evaluation of the size distribution obtained by atomic force microscopy it was concluded that the intrinsic photoluminescence yield of the nanocrystals can reach almost 100%. Results of correlating Si nanoparticle different size distributions with varying maximum position and width have also the potential to reproduce observations of the astrophysical phenomenon of extended red emission [38]. When the Si nanoparticles are taken out of the vacuum synthesis apparatus, no photoluminescence is observed with the naked eye upon illumination with a UV lamp (emitting at 256 nm). It takes hours until the luminescence becomes clearly visible. This effect is correlated with the progressive oxidation of the surface, leading to the passivation of the dangling bonds that are expected to play the role of nonradiative recombination centers. From this explanation, it becomes clear that the importance of having reached the ability to measure the photoluminescence in the same apparatus (Figure 4.7) where the samples are prepared, without exposing them to air (Figure 5.4) [43, 45].

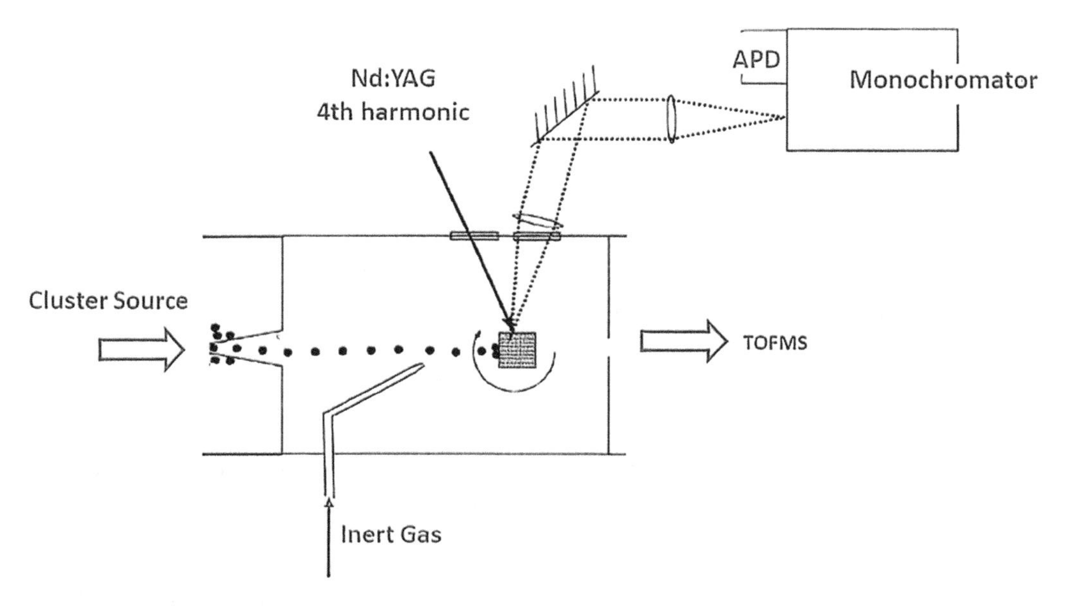

FIGURE 5.4
Detail of the spectroscopic analysis in the size-selected silicon nanoparticles deposition set-up of Figure 5.7. It is used for studying in situ photoluminescence analysis of the deposit. Nanoparticles are deposited on a substrate attached to a cold finger which can be rotated and displaced; when displaced, the particles are analyzed by TOFMS. By turning the cold finger, photoluminescence of the deposit can be analyzed.

Apart from time-of-flight mass spectrometry (TOFMS) analysis, the deposits are studied after deposition by atomic force microscopy (AFM) and high-resolution transmission electron microscopy (HRTEM) (Figure 4.7). These studies showed that the particles were very crystalline with diamond lattice structure [46].

The theoretical dependence of the photoluminescence energy on the size of the Si nanoparticles is important to predict the optical response of a given nanocrystal Si sample once the size distribution within the sample is known Figure 5.5.

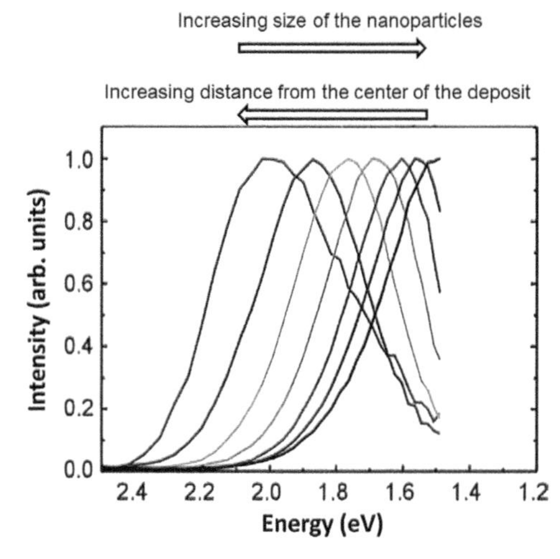

FIGURE 5.5
Photoluminescence study of size-separated Si nanoparticles. Each of the spectra is recorded at different regions of the deposit illuminated by UV lamp; peak energy increases as the analyzed region is further away from the center of the deposit (where the nanoparticles are larger). Adapted from "Photoluminescence of size-separated silicon nanocrystals: Confirmation of quantum confinement" G. Ledoux, J. Gong, F. Huisken, O. Guillois, and C. Reynaud, *Applied Physics Letters*, vol. 80, no. 25, 24 June 2002, with the permission of AIP Publishing).

Then it is possible to reproduce the measured photoluminescence spectra and thus confirm the direct correlation between size distribution and spectral response (Figure 5.6), where the samples are prepared, without exposing them to air. The photoluminescence shift with aging of the samples under normal atmosphere is also consistent with this model [42].

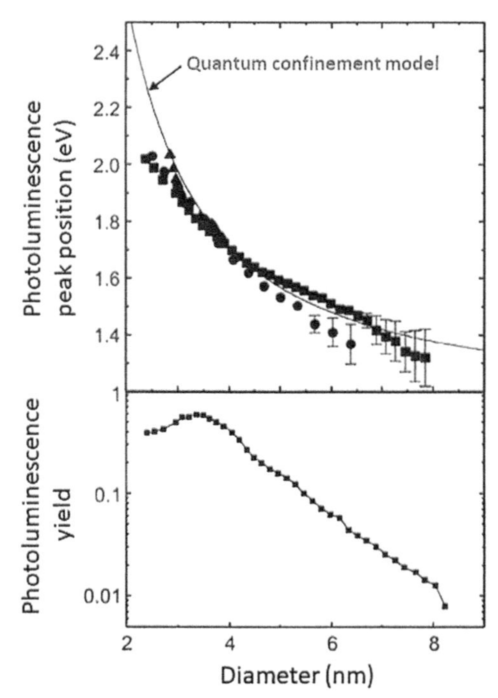

FIGURE 5.6
Evolution of the position of the photoluminescence peak maximum (dots are experimental points in the upper panel) and its intensity (bottom panel) as a function of the size of the silicon nanocrystals. (Adapted from G. Ledoux, J. Gong, F. Huisken, O. Guillois, and C. Reynaud *Applied Physics Letters*, vol. 80, no. 25, 24 June 2002, with the permission of AIP Publishing; and from G. Ledoux, R. Lobo, F. Huisken, O. Guillois, and C. Reynaud: "Photoluminescence properties of silicon nanocrystals synthesized by laser pyrolysis," in *Trends in Nanotechnology Research*, edited by E. V. Dirote (Nova Science Publishers, Hauppauge, NY, 2004), 133–144, with the permission of Nova Science Publishers, Inc.

Now regarding the tunneling phenomenon and considering particles with energy $E < Vo$ incident on a potential square energy barrier of width a and height Vo, the energy eigenfunction is

$$\psi = \begin{cases} Ae^{ikx} + Be^{-ikx} & x < 0 \\ Fe^{qx} + Ge^{-qx} & 0 < x < a \\ Ce^{ikx} & x > a \end{cases} \tag{5.24}$$

where k and q are given by

$$k = \sqrt{\frac{2mE}{\hbar^2}} \tag{5.25}$$

and

$$q = \sqrt{\frac{2m(V_0 - E)}{\hbar^2}} \tag{5.26}$$

As it is known from quantum mechanics, when satisfied the boundary conditions on the continuity of the wave function and its first derivative, the transmission coefficient T (probability that the particle will tunnel through the potential barrier) can be calculated. The result for $qa >> 1$, T is then given by

$$T \underset{qa \gg 1}{\rightarrow} \left(\frac{4qk}{k^2 + q^2}\right)^2 e^{-2qa} \tag{5.27}$$

or evaluating the natural logarithm of the transmission coefficient, we find:

$$\ln T \underset{qa \gg 1}{\rightarrow} \ln\left(\frac{4kq}{k^2 + q^2}\right)^2 - 2qa \underset{qa \gg 1}{\rightarrow} -2qa \tag{5.28}$$

Within these requirements, one can use this expression to calculate the probability of transmission through a non-square barrier, because this can be approximated by a sequence of square barriers (if the potential energy $V(x)$ does not vary too rapidly with position). Thus, if the barrier is sufficiently smooth so that we can approximate it by a series of square barriers (each of width Δx) that are not too thin for Eq. 5.28 to hold, then for the barrier, it is possible to write:

$$\ln T \approx \ln \prod_i T_i = \sum_i \ln T_i \approx -2 \sum_i q_i \Delta x \tag{5.29}$$

and so, if one assumes that we can approximate this last term as an integral:

$$T \approx \exp(-2 \sum_i q_i \Delta x) \approx \exp\left(-2 \int dx \sqrt{\frac{2m}{\hbar^2}[V(x) - E]}\right) \tag{5.30}$$

where the integration is over the region for which the square root is real. Note that these approximations break down near the turning points, where $E = V(x)$. Nonetheless, a more detailed treatment using the Wentzel–Krammers–Brillouin (WKB) approximation shows that 5.10) works well [47]. As it is well-known, quantum tunneling is present in many situations at the molecular level, in several nano-systems, in important nanotechnology applications, and it is at the base of the functioning of the revolutionary scanning tunneling microscope (STM).

In the simplest model of tunneling where one considers a rectangular unidimensional potential barrier $U(z)$, of width w and that in the zero-potential regions on the two sides of the barrier the wave function $\psi(z)$ can be written as

$$\psi = \begin{cases} e^{ikz} + re^{-ikz} & z < 0 \\ te^{ikz} & z > w \end{cases} \tag{5.31}$$

The coefficients r and t provide a measure of how much of the incident particle's wave is reflected or transmitted through the barrier, and k is given by Eq. 5.25 being m, the mass of the particle ($m = m_e$ for an electron). On its turn, inside the barrier ($E < U$) the wave function is a superposition of two decaying terms,

$$\xi e^{-kz} + \xi e^{kz} \quad \text{for} \quad 0 < z < w \tag{5.32}$$

but now κ is given by

$$k = \frac{1}{\hbar}\sqrt{2m_e(U - E)} \tag{5.33}$$

where m_e stands for the mass of the electron.

The probability current is calculated by

$$j_t = -i\frac{\hbar}{2m_e}\left\{\psi_R^*\frac{\partial}{\partial z}\psi_R - \psi_R\frac{\partial}{\partial z}\psi_R^*\right\} \tag{5.34}$$

where $\psi_R = te^{ikz}$, and so of the whole impinging particle current ($j_i = \hbar k/m_e$), only ($|t|^2 j_i$) will be transmitted, as calculated from Eq. 5.30. The transmission coefficient is obtained from the continuity condition on the three parts of the wave function and their derivatives at $z = 0$ and $z = w$, giving for $|t|^2$ the following result:

$$|t|^2 = [1 + (1 - \varepsilon)^{-1} \sin h^2 kw \varepsilon^{1/4}]^{-1} \tag{5.35}$$

where $\varepsilon = E/U$.

Since in operating STM conditions a typical barrier height is of the order of the material's surface work function W, which for most metals has a value between 4 and 6 eV, the electrons can tunnel between two metals only from occupied states on one side into the unoccupied states of the other side of the barrier. Without bias, Fermi energies are flushed and there is no tunneling. Bias shifts electron energies in one of the electrodes higher, and those electrons that have no match at the same energy on the other side will tunnel.

In experiments, bias voltages of a fraction of 1 V are used, so κ is of the order of 10 nm^{-1} and w is a few tenths of a nanometer. Since the barrier is strongly attenuating, the transmission probability reduces to

$$|t|^2 = 16\varepsilon(1 - \varepsilon)e^{-2kw} \tag{5.36}$$

and the tunneling current from a single level is

$$j_t = \left[\frac{4k\kappa}{k^2 + \kappa^2} \right]^2 \frac{\hbar k}{m_e} e^{-2\kappa w} \tag{5.37}$$

where both wave vectors, k and κ, depend on the level's energy E. Thus, the tunneling current is exponentially dependent on the separation of the sample and the tip, and typically reduces by an order of magnitude when the separation is increased by 0.1 nm. Because of this, even when tunneling occurs from a non-ideally sharp tip, the dominant contribution to the current is from its most protruding atom or orbital.

When two conductors approach each other, because of the restriction that the tunneling from an occupied energy level on one side of the barrier requires an empty level of the same energy on the other side of the barrier, tunneling occurs mainly with electrons near the Fermi level. The tunneling current can be related to the density of available or filled states in the sample. Assuming tunneling occurs from the sample to the tip, the current due to an applied voltage V depends not only on the number of electrons between the Fermi level E_F and $(E_F - eV)$ in the sample, but also on the number among them which have corresponding free states to tunnel into on the other side of the barrier at the tip. The higher the density of available states in the tunneling region the greater the tunneling current. By convention, a positive V means that electrons in the tip tunnel into empty states in the sample; for a negative bias, electrons tunnel out of occupied states in the sample into the tip. For small biases and temperatures near absolute zero, the electron density available for tunneling is the product of the density of the electronic states $\rho(E_F)$ and the energy interval between the two Fermi levels, eV. Half of these electrons will be traveling away from the barrier. The other half will represent the electric current impinging on the barrier, which is given by the product of the electron density, charge, and velocity v:

$$I_i = nev = e^2 v\rho(E_F)V/2 \tag{5.38}$$

The tunneling electric current I_t will be a small fraction of this impinging current, determined by the transmission probability T, which in the above simplest model equals $|t|^2$:

$$I_t = |t|^2 e^2 v\rho(E_F)V/2 \tag{5.39}$$

An STM model based on more realistic wave functions for the two electrodes was devised by Bardeen, who solved a time-dependent perturbative problem in which the perturbation emerges from the interaction of the two subsystems [48]. Each of the wave functions for the electrons of the sample (S) and the tip (T) decay into the vacuum after hitting the surface potential barrier, roughly of the size of the surface work function. The wave functions are the solutions of two separate Schrödinger's equations for electrons in potentials U_S and U_T. When the time dependence of the states of known energies E_μ^S and E_ν^T is factored out, the wave functions have the following general form

$$\begin{cases} \psi_\mu^S(t) = \psi_\mu^S \exp\left(-\frac{i}{\hbar}E_\mu^S t\right) \\ \psi_\nu^T(t) = \psi_\nu^T \exp\left(-\frac{i}{\hbar}E_\nu^T t\right) \end{cases} \tag{5.40}$$

If the two systems are put closer together, but are still separated by a thin vacuum region, the potential acting on an electron in the combined system is $(U_T + U_S)$. Here, each of the potentials is spatially limited to its own side of the barrier. Only because the tail of a wave function of one electrode is in the range of the potential of the other, there is a finite probability for any state to evolve over time into the states of the other electrode. Typical experiments are run at low temperature (around 4 K) at which the Fermi level cut-off of the electron population is less than a milli-electronvolt wide. The allowed energies are only those between the two step-like Fermi levels, and then:

$$I_t = \frac{4\pi e}{\hbar} \int_0^{eV} \rho_S (E_F - eV + \varepsilon) \rho_T (E_F + \epsilon) \, |M|^2 \, d\epsilon \tag{5.41}$$

where M is the tunneling matrix. When the bias is small, it is reasonable to assume that the electron wave functions and, consequently, the tunneling matrix element do not change significantly in the narrow range of energies. Then the tunneling current is simply the convolution of the densities of states of the sample surface and the tip:

$$I_t \propto \int_0^{eV} \rho_S (E_F - eV + \varepsilon) \rho_T (E_F + \epsilon) d\epsilon \tag{5.42}$$

How the tunneling current depends on distance between the two electrodes is contained in the tunneling matrix element [48]:

$$M_{\mu\nu} = \frac{\hbar^2}{2m} \int_{z=z_0} \left(\psi_\mu^S \frac{\partial}{\partial z} \psi_\nu^{T*} - \psi_\nu^{T*} \frac{\partial}{\partial z} \psi_\mu^S \right) dxdy \tag{5.43}$$

Results identical to Bardeen's can be obtained by considering adiabatic approach of the two electrodes and using the standard time-dependent perturbation theory [49].

This leads to the well-known Fermi's golden rule for the transition probability in the form given above. In fact, time-evolution can be treated in perturbation theory, as at $t \to -\infty$, the tip is far from the substrate and an electron is stationary in a state φ_μ^S of the sample. We assume that the tip is approached slowly toward the sample and thereby the tip potential is turned on adiabatically (this is reasonable since the timescale of electrons are femtoseconds while the time needed to move the tip is in seconds. Formally one describes this adiabatic switching of the perturbation via an exponential time-dependent potential, and with the presence of the combined potential the φ_μ^S state has a probability of populating the states of electrode T, φ_v^T. Such probability is directly related to the tunneling current. Note the similarities with the non-adiabatic effects in the framework of atom-molecule inelastic collisions. Single molecule vibrational spectroscopy via STM deserves mention. The technique will become even more valuable when the chemical nature of the objects can be identified. This has been achieved by measuring the current-voltage characteristics, and observation of the onset of vibrationally inelastic tunneling from a single adsorbed molecule. The mechanism of this excitation of vibration by inelastic electron scattering is intrinsically non-adiabatic.

Bardeen's model is for tunneling between two planar electrodes and does not explain scanning tunneling microscope's lateral resolution. Tersoff and Hamann used Bardeen's theory and modeled the tip as a structureless geometric point [48]. This helped them disentangle the properties of the tip – which are hard to model – from the properties of the sample surface. The main result was that the tunneling current is proportional to the local DOS of the sample at the Fermi level taken at the position of the center of curvature of a spherically symmetric tip (s-wave tip model). With such a simplification, their model proved valuable for interpreting images of surface features bigger than a nanometer, even though it predicted atomic-scale corrugations of less than a picometer. These are well below the microscope's detection limit and below the values observed in experiments.

The development of STM opened a whole new range of possibilities in nanophysics the ability to manipulate individual atoms and molecules in the vicinity of a surface, and many other microscopy techniques have been developed based upon STM. These gave rise to a numerous family of scanning probe microscopies (SPMs) which constitute a real "laboratory on a tip" (lab-on-a-tip) and the modern core of experimental nanophysics [50–52]. Among several others, they include atomic force microscopy (AFM), in which the force caused by interaction between the tip and sample is measured; scanning near-field optical microscopy (SNOM) that achieves optical resolution below light diffraction limit; photon scanning microscopy (PSTM), which uses an optical tip to tunnel photons. Remind that optical tunneling is demonstrated by frustrated total internal reflection.

For each of these techniques there are commonly different modes of operation [51, 53].

Since STM has an atomically precise positioning system it can be also used to perform very accurate atomic scale manipulation and change the topography of the sample. Furthermore, after the surface is modified by the tip, the same instrument can be used to image the resulting structures. This post-modification observation ability is, by the way, also common to the other members of the SPMs family. Aside from modifying the actual sample surface, one can also use the STM to tunnel electrons into a electron resist film, in order to perform lithography. This has the advantage of offering more control of the exposure than traditional electron beam lithography. Another practical application of STM is atomic deposition of metals (gold, silver, tungsten, etc.) with any desired (pre-programmed) pattern, which can be used as contacts to nanodevices or as nanodevices themselves.

It is well-known that IBM researchers famously developed a way to manipulate atoms adsorbed on a metal surface [54]. This technique has been used to create electron corrals with a small number of adsorbed atoms and observe quantum oscillations in the electron density on the surface of the substrate. These so-called quantum Friedel oscillations arise from localized perturbations in a metallic caused by a defect in the Fermi medium, and they are a quantum mechanical analogy to electric charge screening of charged species in a pool of ions. Whereas electrical charge screening utilizes a point entity treatment to describe the make-up of the ion pool, Friedel oscillations describing fermions in the Fermi medium require a quasi-particle or a scattering treatment. Such oscillations depict a characteristic exponential decay in the fermionic density near the perturbation followed by an ongoing sinusoidal decay resembling the *sinc* function. The electrons that move through the metal behave like free electrons of a Fermi gas with a plane wave-like wave function, $\psi_k(r) = e^{i\mathbf{k}.\mathbf{r}}$, and as they obey to Fermi–Dirac statistics every **k**-state in the gas can only be occupied by two electrons with opposite spin. The occupied states fill a sphere in the band structure **k**-space, up to the Fermi energy, E_F. If there is a foreign atom embedded in the metal (impurity), the electrons that move freely through the solid are scattered by the deviating potential of the impurity. During the scattering process the initial state wave vector $\boldsymbol{k}_i$ of the electron wave function is scattered to a final state wave vector $\boldsymbol{k}_f$. Because the electron gas is a Fermi gas only electrons with energies near the Fermi level can participate in the scattering process because there must be empty final states for the scattered states to jump to. Electrons that are too far below E_F can't jump to unoccupied states. The states around the Fermi level that can be scattered occupy a limited range of **k**-values or wavelengths. So only electrons within a limited wavelength range near E_F are scattered, resulting in a density modulation (sinusoidal function) around the impurity that depends on the Fermi wave vector k_F (radius of the sphere in **k**-space). The quantum mechanical description of a perturbation in a one-dimensional Fermi fluid is modeled by the Tomonaga–Luttinger liquid model [55]. The fermions in the fluid that take part in the screening cannot be considered as a point entity, but a wave vector is required to describe them. Charge density away from the perturbation is not a continuum but fermions arrange themselves at discrete spaces away from the perturbation. This effect is the cause of the circular ripples around the impurity. As the STM image is taken on a metal surface, the regions of low electron density leave the atomic nuclei "exposed" which result in a net positive charge. The barriers ("quantum corrals") are constructed by individually positioning adatoms using the tip of a cryogenic STM (4K). Fe adatoms strongly scatter metallic surface state electrons, and so are good building blocks for constructing atomic-scale barriers to confine these electrons. Tunneling spectroscopy performed (with STM in spectroscopic mode) inside of the corrals reveals discrete resonances, consistent with size quantization. Electrons trapped in a small two-dimensional potential well on the surface of a metal constitute a quantum corral, and the traces of "ripples" are visible by STM measurement due to the sensibility of this microscopy to DOS as well as to the height of the surface. Thus, electron ripples inside a rectangular 2D box can be predicted from the Born probability applied to the eigenfunction

$$\psi_n(x, y, z) = \left(\frac{2}{L}\right)^{3/2} \sin\left(\frac{nx\pi x}{L}\right)\sin\left(\frac{ny\pi y}{L}\right), \; n_x, n_y = 1, 2 \ldots \tag{5.44}$$

which corresponds to the solution of the 3D Schrödinger equation for an infinite trap of area L^2, whose eigenvalues are

$$E_n = \left[\frac{h^2}{8mL^2}\right](n_x^2 + n_y^2) \tag{5.45}$$

Figure 5.7 illustrates the progressive formation of a circular quantum corral ring of 48 Fe atoms on Cu (111) with a diameter of 14.3 nm, using atom-by-atom STM manipulation. The circular array of iron atoms acts as a barrier that reflects electronic waves. Note that standing waves arise when the quantum corral is already finalized. The formed electronic waves add to each other and constitute a standing wave with observable wavelength given by πk (the double of the ψ eigenfunction wavelength) as the calculated probability is given by:

$$\psi^*\psi = 4 \; |A|^2 \cos^2(kx + \delta) \tag{5.46}$$

where A and δ stand, respectively, for the amplitude and phase of the ψ wave function.

A more quantitative understanding is obtained with the above mentioned Tomonaga–Luttinger model, by accounting for the multiple scattering of the surface state electrons with the corrals' constituent adatoms. This scattering is characterized by a complex phase shift which can be extracted from the electronic density pattern inside a quantum corral. Eigler et al. [56] have also observed by STM a quantum mirage, resulting

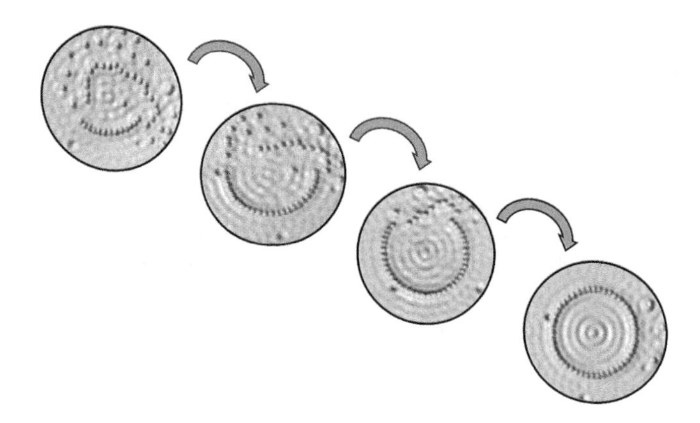

FIGURE 5.7
Progressive construction of a circular quantum corral by STM atomic manipulation. Adapted from image originally created by IBM Corporation. Used with the permission of IBM.

from elliptic quantum corral built by STM nanomanipulation. The quantum mirage is a result predicted in quantum chaos when a system of quantum dynamic billiards exhibits the so-called scarring (the quantum probability density shows traces of the paths a classical billiard ball would take). For an elliptical arena, the scarring is particularly pronounced at the foci, as this is the region where many classical trajectories converge. The scars at the foci form the so-called quantum mirage. STM manipulation also opened the possibility of controlling single-molecule synthesis, including the construction of new molecules on surfaces. Using variable temperature STMs in ultra-high-vacuum, investigators have discovered at low surface temperatures, reaction pathways that cannot be probed by other conventional surface science instruments [57]. A variety of tip-atom/molecule surface interactions, such as an electric field existing at the tip-sample junction, tunneling electrons, and tip-atom/molecule interactions forces, can be used in a controlled manner to manipulate single atoms or molecules. By combining STM manipulations with the complementary tunneling spectroscopy measurements (e.g., local I/V, dI/dV, d^2I/dV^2) and I/Z spectroscopy, the chemical, physical, electronic, and mechanical properties of single molecules can be studied at each reaction step with atomic-level resolution [58–60]. STM manipulation and spectroscopy techniques can induce basic metal-catalysed chemical-reaction, and probe how the atom or molecule moves and what kind of interactions are involved during manipulation (from the corresponding STM feedback or tunneling-current signals). Successful atom manipulation is dependent on the tip-atom distance in STM experiments. The tip-atom distance can be expressed by means of tunneling resistance, and the tip height is controlled by the STM feedback system to compensate for the exponential nature of the tunneling current over the tunneling distance. Thus, at a chosen tunneling voltage, variation of the tunneling current alters the tunneling resistance value that in turn changes the tip-atom/molecule distance. To dissociate a molecule using an STM tip, the necessary energy for the dissociation is supplied by injecting tunneling electrons into the molecule. Based on the electron energy, the STM tip–induced molecule dissociation process can be through field emission or inelastic tunneling regimes. Electron energies roughly above 3 eV are used in the field emission regime, where the tip acts as an electron emission gun. The dissociation involving the inelastic tunneling process uses low STM biases, and controlled bond breaking can be achieved. This is realized by positioning the STM tip above the location of the molecular bond at a fixed height, and then low-bias voltage pulses are applied to inject tunneling electrons into the molecule. The electrons can be injected from either the tip or the substrate, depending on the bias polarity. The corresponding tunneling current can be monitored, and current changes can be associated with the dissociation event, allowing one to determine the dissociation rate. In STM vertical manipulation occurs the transfer of single atoms or molecules between the tip and substrate and vice versa. The atom/molecule transfer process can be realized by using an electric field between the tip and sample, by exciting with inelastic tunneling electrons, or by making mechanical contact between the tip and the atom/molecule. This transfer mechanism can be explained by using a model of a double potential well (Figure 5.8). At a short imaging distance, the atom/molecule has two possible stable positions, one at the surface and one at the tip apex. Each position is represented by a potential well, and the two potential wells are more or less equal in shape, separated by a barrier. If the tip is stopped above the atom at the same distance and an electric field is applied, then the shape of the double potential well changes (dashed line in Figure 5.8). In this case, the barrier between the two wells reduces, and the potential well at the tip apex has a much lower energy level. The atom now can easily transfer to the tip. By applying a reverse polarity bias, the minimum potential well can be changed to the surface side; the dashed curve will have 180° reverses from right to the left. The atom can then transfer back to the surface. Regarding single-bond formation, the transfer of an atom/molecule between the sample and the tip in a vertical-manipulation procedure involves both bond-breaking and bond-formation processes. In such cases, the substrate atom/molecule bond is broken, and a new bond between the atom/molecule and the tip-apex atom, or vice versa, is formed [61]. An example of STM tip–induced single-molecule chemical reaction is oxidation of CO at a single molecule level via vertical manipulation on Ag(110) surface at 13 K of temperature [59]. Under the influence of the STM tip, reactions that otherwise might not occur in nature can be forced to proceed, which constitutes a sign of distinction of nanotechnology, due to synthesis of individual human-made molecules, never before made in chemical reactors or seen in nature, which establishes a clear separation from biological processes.

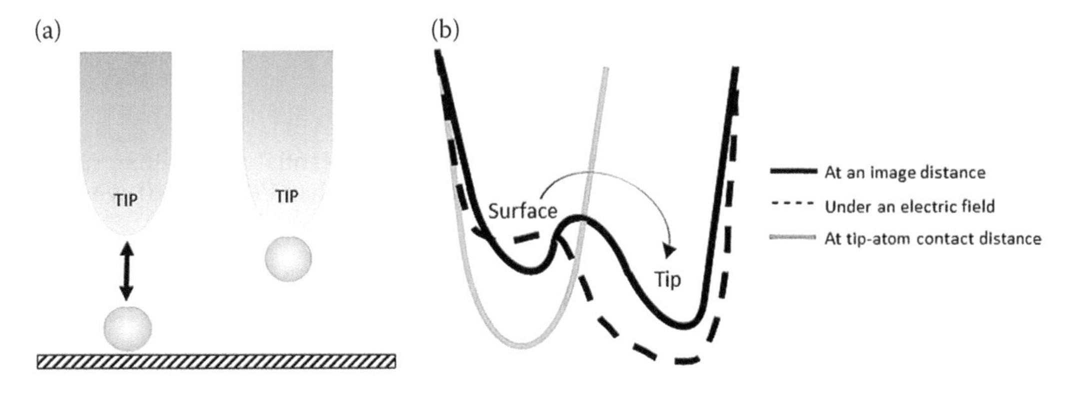

FIGURE 5.8
STM vertical manipulation (a) and respective double-potential well (b).

Let us now turn our attention for oscillators, starting with the harmonic oscillator for which there is an exact solution of the wave equation and represents a model with an extremely broad physical significance. Indeed, solving the harmonic oscillator with quantum operator methods allows for more abstract solutions in which a variable x may not be the usual position at all. In fact, the Hamiltonian of the electromagnetic field may be expressed as a collection of such abstract harmonic oscillators, and Planck's resolution of the ultraviolet catastrophe in the analysis of the blackbody spectrum (which amounted to treating these oscillators as quantum oscillators), can be considered as the starting point of quantum field theory. An arbitrary potential energy function in the vicinity of its minimum resembles a harmonic oscillator. Considering a specific example familiar from classical mechanics: a mass m attached to a massless string of length L and free to pivot under the influence of gravity about a fixed point. This is of course the simple pendulum! The energy of the system can be expressed as

$$E = \frac{1}{2}mv^2 + mgh = \frac{1}{2}mv^2 + mgL(1 - \cos\theta) \tag{5.47}$$

where θ stands for the usual offset angle from the equilibrium position. If this angle is small, we can expand cos θ in a Taylor series and retain the leading terms to obtain

$$E = \frac{1}{2}mv^2 + \frac{1}{2}mgL\theta^2 = \frac{1}{2}mv^2 + \frac{1}{2}\frac{mg}{L}x^2 \tag{5.48}$$

being the arc length $L\theta = x$. Thus, provided the oscillations are small, the system behaves like a harmonic oscillator with a spring constant $k = mg/L$, and, therefore, a spring frequency $\omega = (k/m)^{1/2} = (g/L)^{1/2}$. Note then there is no physical spring truly attached to the mass in this case. Any potential energy function $V(x)$ with a minimum at a certain position x_o, can be expanded in a Taylor series about the minimum, to obtain

$$V(x) = V(x_0) + \left(\frac{dV}{dx}\right)_{x=x_0}(x - x_0) + \frac{1}{2!}\left(\frac{d^2V}{dx^2}\right)_{x=x_0}(x - x_0)^2 + \ldots \tag{5.49}$$

As x_o is the location of the minimum of the potential energy, the first derivative vanishes there, and so

$$V(x) = V(x_0) + \frac{1}{2}k(x - x_0)^2 + \ldots \tag{5.50}$$

where k is a positive constant. Since it is only difference in potential energy that matter physically, we can choose the zero of potential energy such that $V(x_o) = 0$. If we now position the origin of our coordinates at x_o, then

$$V(x) = \frac{1}{2}kx^2 \tag{5.51}$$

Thus, provided the system is undergoing sufficiently small oscillations about the equilibrium point, we can neglect the higher-order terms in the Taylor expansion, and the effective potential energy is that of a harmonic oscillator. Note that when the spring is in a gravitational field, and its oscillation coincides in direction with the direction of the field, then the period stays the same, and the constant force simply shifts the equilibrium position of the harmonic motion. It shifts the motion by a distance mgk, and so replacing x

by $x_{eff} = x - mg/k$, and using $ma = -kx + mg$, one gets $a = a_{eff}$ and $ma_{eff} = -kx_{eff}$. Therefore, by adding a constant to x_{eff} we can get the standard harmonic motion expression. Examples of systems that behave like harmonic oscillators on a microscopic scale, are the vibrations of nuclei within diatomic molecules and the vibrations of atoms in a crystalline solid about their equilibrium positions. For a mechanical harmonic potential, the classical density probability function for the position of a particle is a Gaussian function whose positional variance is given by $k_B T/K$, where k_B is the Boltzmann constant, T the absolute temperature, and K the oscillator force constant. In quantum statistical physics, that function must be replaced by a probability distribution, considering the various discrete vibrational energy states. In turn, the energy of each state is obtained by solving the Schrödinger equation for the harmonic potential, leading to the well-known result:

$$E(n) = (n + ½)h\nu,\ n = 0, 1, 2, \ldots \tag{5.52}$$

The probability of finding the oscillator in the vibrational state n can be obtained by substituting $E(n)$ in the expression above mentioned for the probability distribution, which leads (after a few steps of mathematical handling, and further introducing the value of the oscillator average energy), to the following result for the quantum positional variance:

$$\sigma^2 = \left(\frac{h\nu}{K}\right)\left\{½ + \left[\exp\left(\frac{h\nu}{k_B T}\right) - 1\right]^{-1}\right\} \tag{5.53}$$

Therefore, it is possible to represent the ratio between the quantum variance and the classical variance, as a function of a suitable temperature dimensionless parameter (Figure 5.9).

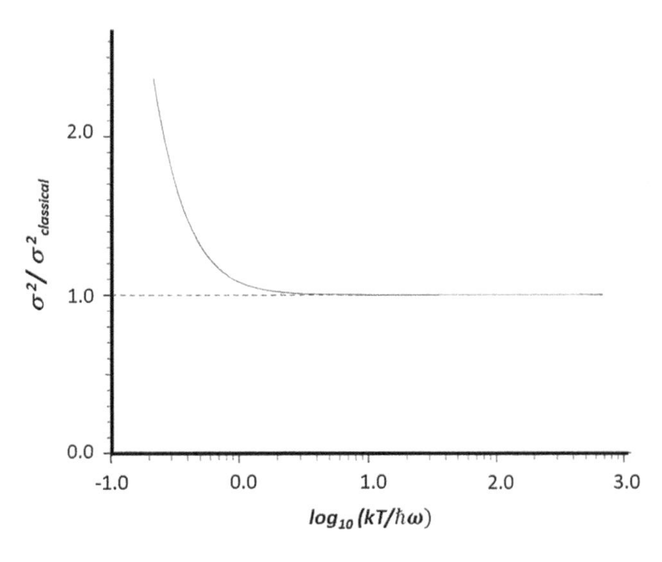

FIGURE 5.9
Dependence of the quantum/classical variance ratio.

It appears that the classical variance provides a good approximation when this parameter exceeds unity, that is, the use of quantum treatment to estimate the positional uncertainty is justified at temperatures below approximately $\nu \times 10^{-11}$, which corresponds to temperatures below 100 K for oscillation frequencies of the order of 10^{13} Hz (typical value for atoms in solids).

In some situations, anharmonic effects may occur, that is, deviations from harmonic oscillator behavior. In reality all oscillating systems are anharmonic but approximate the harmonic oscillator the smaller the amplitude of the oscillation is. Therefore, the system can be approximated to a harmonic oscillator and the anharmonicity can be calculated using perturbation theory, or if the anharmonicity is large, other numerical techniques must be used. The anharmonic oscillators display a nonlinear dependence of the restorative force on the displacement. Consequently, the anharmonic oscillator's period of oscillation may depend on its amplitude of oscillation, and the vibration frequency can change to multiples of the fundamental, depending upon the system's displacement. These changes in the vibration frequency result in energy being coupled from the fundamental vibration frequency to other frequencies through a process known as parametric coupling. In molecules, even at moderate temperatures the deviation from harmonic behavior can be caused by coupling with other active modes, namely the rotational ones. Spectroscopic experiments have shown that the potential-energy function for the lowest electronic states of diatomic molecules can be represented quite accurately by the following Morse analytic function that contains three adjustable parameters U_0, R_0 and a [62]:

$$U(R) = U_0 \left(e^{-2(R-R_0)/a} - 2e^{-(R-R_0)/a}\right) \tag{5.54}$$

where a considers the force constant at the minimum of the well. This Morse anharmonic potential approaches zero for large internuclear distances R, has the minimum value $-U_0$ at $R = R_0$, and its most notorious deviation from the harmonic parabolic shape occurs at internuclear distances further away from equilibrium R_0. U becomes negative at first as R decreases, because of van der Waals attraction, and for smaller R this is replaced by the much stronger Heitler–London resonance attraction. The Morse potential can also be used to model many situations of interaction between an atom and a surface. Even in cases where van der Waals interaction is stronger, they can be described through its application because there is a mathematical relation between the Morse and the Lennard–Jones potentials in the intermediate region of the potential curve [63].

It is worth to mentioned that in neutral clusters, including dimers, the strong attraction does not occur, and so Morse potential does not apply but instead the Lennard–Jones pair potential is mostly used as in general neutral intermolecular situations including several hydrogen bonds. In a hydrogen bond, the charged H atom between two negative entities is a bit like the sharing of a negative electron in a covalent bond between two protons. Indeed, for nonbonding interactions such as van der Waals, the Lennard–Jones potential is commonly used [64, 65]. Due to the large mass difference between the electron and the proton, the hydrogen bond interaction is much weaker than the covalent bond. Together with the van der Waals forces and the Casimir force, they constitute the nano physical basis for the formation of molecular superstructures, including biomolecules and self-organizing assemblies. All of them establish much weaker interactions than ionic and covalent bonds.

Usually in physics of many-body problems, the concept of quasiparticle allows to simplify the corresponding quantum mechanical treatment. A quasiparticle consists of the original real, individual particle, plus a cloud of disturbed neighbors. It behaves very much like an individual particle, except that it has an effective mass and a lifetime. But there also exist other kinds of fictitious particles in many-body systems, that is, collective excitations. These do not center around individual particles, but instead involve collective, wave-like motion of all the particles in the system simultaneously. Electrons and electron holes (fermions) are typically called quasiparticles, while phonons and plasmons (bosons) are commonly called collective excitations. For example, as an electron travels through a semiconductor, its motion is disturbed in a complex way by its interactions with other electrons and with atomic nuclei. The electron behaves although it has a different effective mass traveling unperturbed in vacuum (electron quasiparticle). From another situation, the aggregate motion of electrons in the valence band of a semiconductor or a hole band in a metal behave as though the material instead contained positively charged quasiparticles (electron holes). Light impinging on a semiconductor with energy greater than or equal to its gap creates a so-called exciton, in which an electron from the valence band is promoted to the conduction band, leaving an electron behind. This exciton (electron/hole pair) has a certain lifetime, and a size that can be estimated from the modified Bohr radius a_{MB}, given by

$$a_{MB} = \frac{h^2\varepsilon'}{\pi m_e e^2} \tag{5.55}$$

assuming that the effective mass is equal to the electron mass, and where ε' stands for the relative permittivity of the semiconductor. There are commonly two types of excitons in semiconductors, depending on their sizes; if they are close to the dimension of the atoms in the bulk or if they enclose several of these atoms. In single wall carbon nanotubes, they both can be formed.

When the characteristic dimension of a semiconductor nanoparticle D, approaches the exciton radius, the intrinsic gap energy ΔE^0 changes, and can then be estimated by

$$\Delta E = \Delta E^0 + \frac{h^2\pi^2}{2D^2}\left(\frac{1}{m_e^*} + \frac{1}{m_h^*}\right) - \frac{1.8e^2}{\varepsilon D} \tag{5.56}$$

The second term is due to the exciton confinement in a box whose width equals 2D, and the third term is a correction resulting from Coulomb repulsion. Therefore, the energy of the photons that fluoresce increases by a factor proportional to $1/D^2$ in relation with the bulk as far as the dimension of the QD decreases.

Collective excitations include the phonon, which derived from the vibrations of atoms in a solid, and the plasmon derived from plasma oscillation. This is a rapid oscillation of the electron density in conducting media (plasmas or metals in the ultraviolet region) and can be described as an instability in the dielectric function of a free electron gas. The frequency only depends weakly on the wavelength of the oscillation. The quasiparticle resulting from the quantization of these oscillations is the plasmon. In fact, considering an electrically neutral plasma in equilibrium, if one displaces by a tiny amount an electron or a group of electrons with respect to the ions, the Coulomb force pulls the electrons back, acting as a restoring force. If the thermal motion of the electrons is ignored, it is possible to show (from the Maxwell equations in the aim of the free electron model) [66], that the charge density oscillates at the plasma frequency given by

$$\omega_P = \sqrt{\frac{n_e e^2}{m^*\varepsilon_0}} \tag{5.57}$$

where n_e stands for the number density of electrons and m^* the effective mass of the electron. This formula is derived under the approximation that the ion mass is infinite, which is a good approximation, as the electrons are much lighter than ions. Plasmons can be described in the classical picture as an oscillation of electron density with respect to the fixed positive ions in a metal. To visualize a plasma oscillation, imagine a cube of metal placed in an external electric field pointing to the right. Electrons will move to the left side (uncovering positive ions on the right side) until they cancel the field inside the metal. If the electric field is removed, the electrons move to the right, repelled by each other, and attracted to the positive ions left bare on the right side. They oscillate back and forth at the plasma frequency until the energy is lost in resistance or damping. Plasmons are a quantization of this kind of oscillation. Thus, plasmon energy corresponds to the quantum ($\hbar\omega_P$).

Turning our focus to nanophotonics (i.e., nanoscale interaction of photons and matter), one can now draw our attention to the interaction of light with matter. Plasmons play a relevant role in the optical properties of metals and semiconductors. Frequencies of light below the plasma frequency are reflected by a material because the electrons in the material screen, the electric field of the light. Light of frequencies above the plasma frequency is transmitted by a material because the electrons in the material cannot respond fast enough to screen it. In most metals, the plasma frequency is in the ultraviolet, making them reflective in the visible range. Some metals, such as copper and gold, have electronic interband transitions in the visible range, whereby specific light energies are absorbed, yielding their distinct color. In semiconductors, the valence electron plasmon frequency is usually in the deep ultraviolet, while their electronic interband transitions are in the visible range, whereby specific light energies are absorbed, yielding their distinct color, which is why they are reflective. It has been shown that the plasmon frequency may occur in the mid-infrared and near-infrared region when semiconductors are in the form of nanoparticles with heavy doping [67].

Surface plasmons are those plasmons that are confined to surfaces and that interact strongly with light, resulting in a polariton [68]. They occur at the interface of a material exhibiting positive real part of their relative permittivity, that is, dielectric constant (e.g., vacuum, air, glass, and other dielectrics) and a material whose real part of permittivity is negative at the given frequency of light, typically a metal or heavily doped semiconductors.

The position and intensity of plasmon absorption and emission peaks are affected by molecular adsorption, which can be used in molecular sensors. Graphene has also been shown to accommodate surface plasmons, observed via nearfield infrared optical microscopy techniques and infrared spectroscopy. Potential applications of graphene plasmonics mainly addressed the terahertz to mid-infrared frequencies, such as optical modulators, photodetectors, and biosensors.

The mean field of electrons in metallic aggregates can be caused to oscillate by a suitable external disturbance, like visible light, as, for example, a green laser. In fact, collective oscillations of the dipole type have discovered in metal clusters. Theoretical predictions based on the mean field approximation describe well most experimental observations. Electrons in metals can form density waves called plasmons. When you have a plasmon in a metal, the electron density periodically increases and decreases at every point in space. Cluster plasmons appear located close to the surface, as due to light excitation, surface electrons move back and forth against the positively charged fixed background. A giant dipolar resonance can be used to check the ellipsoidal shape of aggregates with semi-filled shells. By sending light to an aggregate with a number of atoms between two magic numbers, the photo absorption peak is found to be double. This reveals the existence of two typical wavelengths. The observation of the absorption spectrum can reveal the shape of the cluster (Figure 5.10).

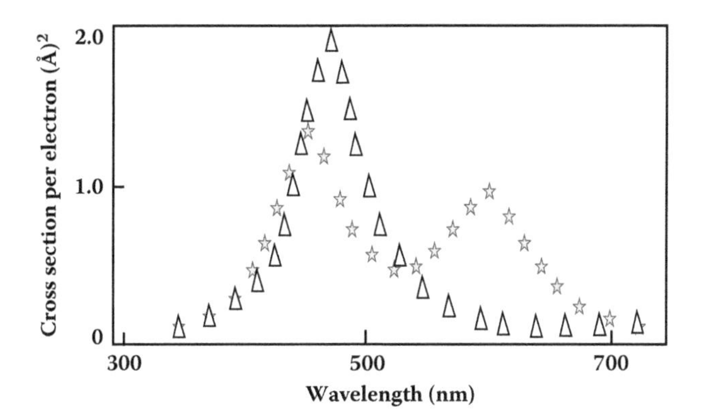

FIGURE 5.10
Theoretical description of a sodium cluster dipolar excitation: (a) with 8 (spherical) and (b) 10 atoms (deformed).

Plasmons are electron-density waves that propagate at the surface of a metal (and of some metamaterials), resulting from the oscillation of electrons stimulated by electromagnetic radiation. These oscillations correspond to those in the field of the incident electromagnetic wave.

In addition to various applications in optics, the development of plasmonics also improves the sensitivity of chemical and biological detectors. The use of metallic nanoparticles for the preparation of matrices with analytes also allows to substantially improve the detection limit through laser spectrometry [69]. However, without yet being aware of the facts, man has used, since antiquity, and in some artistic glass artifacts, the optical effects that certain tiny metallic particles give them. As it is well known, in Raman spectroscopy it is possible to characterize the composition of a material by analyzing the luminous scattering of laser light through the sample and knowing that this scattered radiation conveys important information about molecular vibrations. Plasmonics also assumes a relevant role here, since incorporated metallic nanostructures (nano-objects or nanoparticles) can substantially improve the intensity of the detected signal, since surface plasmons increase the transmission of electromagnetic energy. The ability of plasmons to transmit electromagnetic radiation (and so information) makes them a promising alternative to replace electronic circuits in microprocessors. At first, this might seem unrealistic as metals have strong optical losses resulting from collisions between free electrons and the surrounding atomic lattice. However, it happens that for a plasmon located at the metal-dielectric interface, the electromagnetic energy propagates mainly in the dielectric, without suffering a relevant absorption there (since it does not contain free electrons). However, this makes it possible to confine the propagating plasmons in the thin layer of the interface that works as a waveguide. It is possible to build plasmonic waveguides with dimensions identical to tiny electronic circuits, and capable of carrying much more information than those. It should be noted, however, that such transport does not normally exceed tens of micrometers in distance. For example, when a linear chain of gold nanoparticles is excited by laser light at 570 nm, it triggers resonant oscillations (in these particles), that remain confined, but can propagate only over very short distances. It will be possible to mitigate this difficulty by reducing said losses by making the plasmonic waveguides return on themselves, that is, placing the dielectric in the center and surrounding it with metal. In this case, it is possible to change the plasmon wavelength by adjusting the thickness of the dielectric core. Recently, it has been possible to design the metal-dielectric interface properly so that surface plasmons have the same frequency as the stimulating electromagnetic waves but with shorter wavelengths. This allows the plasmons to be routed along nanometric wires so that they carry information within the microprocessor. In plasmonic waves, the alternating electric charge density can be compared to alternating electric current, but the frequency of the optical signal is vastly higher by a factor of the order of 10^5 GHz. Scientists at Rice University designed silicon nanospheres with a diameter of 100 nm and nanometer-thick gold plating. Its exposure to electromagnetic waves causes oscillation of electrons in the gold shell, and due to the interactions of the fields present on the internal and external surfaces of that layer, and the relationship between the diameter of the nanosphere and the thickness of the shell, determine the wavelength at which the particle absorbs energy by plasmonic resonance.

Therefore, it is possible to build a nanosphere that absorbs at a certain desired wavelength, playing with the relationship between those dimensions. These nanoshells can be functionalized to become biocompatible and directed to specific cellular targets, such as cancer cells. Subsequently, they can be irradiated with infrared laser radiation through the skin, which favors the resonant absorption and consequent heating of cancerous tissues, which destroys them without causing damage to the surrounding tissues. Plasmonics is not restricted to 3D metallic nanostructures. For example, graphene also has surface plasmons when irradiated with an infrared laser. Even certain crystalline materials formed by at least two elements can strongly interact with the electromagnetic field when its frequency approaches the proper oscillation frequency of the atoms in the crystal lattice.

Scattering occurs when a photon changes direction after it strikes a bit of matter. This is usually a type of scattering called "elastic scattering," where no energy is exchanged between the photon and the matter. Fewer than one in a million of the scattered photons will instead undergo "inelastic scattering." As opposed to absorption, where the photon is annihilated and gives up the whole of its energy, inelastic scattering involves either the transfer of a fraction of the photon's energy to the matter – often in the form of heat – or the transfer of some of the matter's energy to the photon. If the photon loses energy, it carries on with a lower frequency and in a new direction; if it gains energy, it carries on with a higher frequency and so a shorter wavelength. A photon comprises oscillating electric and magnetic fields as waves. Since these waves encounter a material, polarities within the material oscillate in response. The more polarizable the material, the stronger the field that can be established inside it. So, in this case, permittivity has both real and imaginary parts, both of which are functions of the frequency f, of the field:

$$\varepsilon(f) = \varepsilon'(f) - i\epsilon''(f) \tag{5.58}$$

The imaginary part ε'' expresses the energy lost in the material as the waves move through, and the real part ε' expresses the relationship between the forward speed of the wave and the material's capacitance. If the real part is positive, the material is a dielectric. A good dielectric material will have a high permittivity (dielectric constant). If instead the real part of the permittivity is negative, the material is considered a metal. The free electrons in a metal determine its permittivity, which is given in Drude model by

$$\varepsilon = 1 - \frac{f_p^2}{f^2 + i\gamma f} \quad (5.59)$$

This equation can also be expressed by real and imaginary parts:

$$\varepsilon' = 1 - \frac{f_p^2}{f^2 + \gamma^2} \text{[real part]} \quad (5.60)$$

$$\varepsilon' = \frac{f_p^2 \gamma}{f(f^2 + \gamma^2)} \text{[imaginary part]} \quad (5.61)$$

where f stands for the frequency of the electric field (photon) and γ is the damping (determined by the inverse of the time between collisions for the free electrons). This swarm of collective electrons is a kind of plasma made exclusively of electrons and oscillates with the plasmon frequency f_p of Eq. 5.57, where $\omega_p = 2\pi f_p$. It is worth to mention that while Eq. 5.57 provides quite accurate results for many metals (e.g., Al, alkali metals, Mg, Pb, …), it does not take into account the bound electrons' effect on the plasma. In some cases, like gold and silver, electrons bound within the crystal lattice can inhibit free electrons' motion. In gold, this effect lowers the plasmon frequency from ultraviolet to visible wavelengths, and in silver it lowers f_p from far ultraviolet to near-ultraviolet wavelengths.

Photons with a f_p lower than the plasma frequency are reflected because the rapidly oscillating, high-energy electrons screen the more slowly oscillating, low-energy electric field; on the contrary, photons of frequency greater than f_p are transmitted because the electrons do not respond quickly enough to take the energy from the electric field. From the above equations, one can see that when $f \gg f_p$, the real part of the permittivity is positive, the material behaves like a dielectric; meanwhile, the imaginary part is very small, indicating that minimal energy lost as the EM waves move through the material. On its turn, when $f \ll fp$, the real part is negative, and the material behaves as a metal; the imaginary part is large so the EM field cannot penetrate very far into the material; photons are either reflected back out by free electrons near the surface or give up their energy to the electron plasma deeper in. With most metals, f_p is in the ultraviolet range, and they display their characteristic metallic sheen in the visible. However, certain metals, like copper and gold, have f_p in the visible range, leading to reflectance of red light more than blue light and so these metals show a more golden glow.

In general, the collective oscillation of the electron plasma in a metal is unaffected by boundaries, when objects are large enough, so the damping coefficient γ is determined by the inverse of the time between collisions for the free electrons. However, when an object becomes smaller than the mean free path of the electrons (Λ) it contains, these ones will have to collide more often with its surface, and so the collision frequency needs to be adjusted to account for the additional collisions. Consequently, the modified damping frequency for the free electrons become:

$$\gamma_p = \gamma + \frac{v_F}{\Lambda_p} \quad (5.62)$$

where Λ_p represents the adjusted mean free path and v_F the electron velocity at the Fermi energy. For a spherical metal particle with a diameter smaller than its Λ in the bulk, the adjusted mean free path generally varies with its radius r_p accordingly with

$$\Lambda_p = \frac{4}{3}R \tag{5.63}$$

Therefore, as the radius decreases, the damping frequency of the particle increases, and this has important consequences in the photonic properties of a particulate medium, namely in its extinction coefficient C_e. This accounts for the fraction of incident photons that are lost per unit of distance in a certain direction due to absorption and scattering (elastic and inelastic). For small spherical particles, these phenomena are described by Mie theory [70] and involves solving the Maxwell electromagnetic wave equations in the presence of a spherical boundary. In the Mie theory the total cross-section is a sum of two terms representing absorption and scattering. It is thus shown that for particles whose radius R is much smaller than the wavelength of the incident light λ, the extinction efficiency Q_e (defined by C_e divided by the geometrical cross-section of the particle πR^2) is given by

$$Q_e = \frac{24\pi R \varepsilon_m^{3/2}}{\lambda} \frac{\varepsilon''}{(\varepsilon' + 2\varepsilon_m)^2 + \varepsilon''^2} \tag{5.64}$$

where ε_m stands for the dielectric constant of the surrounding medium. Worth to remark that despite the maximum value of Q_e is expected to be one corresponding to an opaque disk with the same radius as the particle, this never happens because the particle diffracts light away from the original direction of propagation as well as block it, and so a detector along the line of incidence can register an extinction efficiency larger than one. The dielectric function for the particle material in the required wavelength region can be obtained from published tables [71]. Eq. 5.64 predicts a maximum in the extinction at a wavelength for which the condition $\varepsilon' = -2\varepsilon_m$ is satisfied. Since this depends only on the material and not the size of the particle, so the wavelength of the strong surface plasmon resonance peak (SPR) does not shift with particle size; however, for particle sizes such that $R > 0.1\lambda$ the equation starts to become invalid, and it is experimentally demonstrable that the position of such resonance peak is size-dependent for larger nanoparticles. As the particle size decreases, two effects appear. First, the location of maximum extinction shifts toward smaller wavelengths, and second, the peak height shrinks. It also happens that the peaks broaden. This is the effect of the increased damping due to the increased interactions of the electron plasma with the particle surface already mentioned. On its turn, non-spherical metallic particles, shaped more like rods, have both a longitudinal and a transverse plasma frequency (along and perpendicular to the long axis of the rod), giving the extinction coefficient curve two peaks instead of one. Typically, the transverse plasma frequency is like that of a particle, while the longitudinal is lower. The SPR can be tuned over a certain range by changing the aspect ratio (length/radius) of the nanorods. In addition, a particle consisting of a dielectric core like silica surrounded by a metal film (the so-called shell particle), also have photonic characteristics similar to purely metallic particles. However, the plasma resonance wavelength of these shells (as well as the peak in the extinction coefficient curve) is often higher than those of solid metal particles. As the thickness of the metal shell increases, leaving the size of the dielectric core constant, the resonance wavelength decreases. This nano-engineering of colors based on the nanophysics of shape, combined with the photoemission properties of semiconductor QDs (recall that dot's size determines its band gap) allows for various applications in chemistry, bio-diagnosis, and energy efficiency improvements namely in photovoltaic solar cells. In fact, typical solar cell acts as a low pass filter, absorbing photons with greater energy than the semiconductor band gap, while photons with lesser energy are not absorbed.

Usually, elastic scattering of electromagnetic radiation by particles much smaller than the wavelength of the radiation is well described by the so-called Rayleigh scattering, where for light frequencies well below the resonance frequency of the scattering particle (normal dispersion regime), the amount of scattering is inversely proportional to the fourth power of the wavelength:

$$I_R = I_0\left(\frac{1+\cos^2\theta}{2R^2}\right)\left(\frac{2\pi}{\lambda}\right)^4\left(\frac{n^2-1}{n^2+2}\right)^2\left(\frac{d}{2}\right)^6 \tag{5.65}$$

I_0 is the intensity of the incident radiation; R is the distance between the particle and the observer; θ is the scattering angle; n is the refraction index of the particle; and d is the particle diameter.

Scattering by particles with a size comparable to or larger than the wavelength of the light is typically treated by the already mentioned Mie theory, which is slightly dependent on the incident wavelength.

Usually, illuminating a diluted solution of macromolecules, with light polarized in the direction perpendicular to the scattering plane, and with a wavelength λ greater than the molecular dimensions, the intensity of scattered light in that same plane, is given by

$$I_d = 4\pi^2 I_0\left[\frac{n(n-n_0)}{m/V}\right]^2\frac{mM}{(d^2\lambda^4 N_A)} \tag{5.66}$$

where m is the dissolved mass in the volume V; M is the molar mass; N_A is the Avogadro number; d is the distance between the sample and the observer; n is the refraction index of the solution; and n_0 is the refraction index of the solvent. Thus, from experimental measurements of I_d it is possible to obtain the value of molar mass. In addition, if the size of the molecule (or nanoparticle), although smaller, approaches this, in a way that the scattered light intensity in the forward direction If, still is larger than the backscattered intensity I_b, it is possible to show that

$$\frac{I_b}{I_f} = 1-\left(\frac{2\pi a}{\lambda}\right)^2 \tag{5.67}$$

where a stands for the distance between two dispersive centers in the molecule (considered as a Hertz oscillating dipole). Therefore, by measuring the ratio (I_b/I_f) it is possible to get an estimative of the molecular dimension.

In addition, making use of a different optical technique called dynamic light scattering (DLS) or photon correlated spectroscopy (PCS), one can get a more precise information about the size of nanoparticles. In this case, a dispersion of spherical nanoparticles in Brownian motion is irradiated with monochromatic light and the resulting scattered light at 90° is detected in very short periods of time. The Brownian motion causes fluctuations in light under detection, and as larger particles move slowly than the smaller ones, a correlation function is established. When light strikes particles smaller than λ, it undergoes Rayleigh scattering and its intensity fluctuates due to the Brownian motion of the particles in the solution which causes the distance between the scattering centers to vary in time. This scattered light suffers constructive or destructive interference by neighboring particles, and this intensity fluctuation contains information that can be mathematically decoded. From a suitable fitting of the experimental results the diffusion coefficient D come out, and from this, the radius of the particles can be inferred through the well-known Stokes–Einstein equation. In fact, in Brownian motion conditions, the probability density function is

$$P(r, t) = (4\pi Dt)^{-3/2}\exp\left(\frac{-r^2}{4Dt}\right) \tag{5.68}$$

Since the particles move independently of each other, the scattered light spectrum of frequency is a known as Lorentzian function (due to Doppler effect) whose full width at half maximum (FWHM) depends on D, dispersion angle and refraction index of the solution. Once the process is random, the spectrum can be obtained through the correlation function. In this perspective, the space correlations are transformed in phase correlations and the FWHM of the new Lorentzian curve corresponds to the correlation time, which also depends on D, dispersion angle and refraction index of the solution.

One can consider that the intensity of electromagnetic scattering is given by

$$I(q) = I_0 N(\rho - \rho_0)^2 F^2(q) \tag{5.69}$$

being N the number of noninteracting dispersive particles, ρ_0, ρ, respectively, the electronic densities of the particle and the medium, and $F(q)$ a shape factor (which is the Fourier transform of the object's shape), that in the case of spheres of radius R can be obtained by

$$F(q) = 4\pi R^3\left\{\frac{\sin(qR) - qR\cos(qR)}{(qR)^3}\right\}\text{with } q = 4\pi\sin(\theta/2)/\lambda \tag{5.70}$$

For large values of q, the intensity can be simplified:

$$I(q) = 8N\pi^2(\rho - \rho_0)^2 R^2/q^4 \tag{5.71}$$

These expressions can be used for visible light if $R > 80$ nm.

When the light frequency is far from the plasmon resonance of a targeted nanoparticle, the energy flow is only slightly disturbed. On the other hand, at the plasmonic resonance frequency, the strong induced polarization introduces energy into the particle. This effect is observed as a strong increase in the scattering cross-section in optical extinction measurements. Plasmon resonance is more pronounced for particles much smaller than the wavelength of the excitation light, as in this case all the conduction electrons are excited in phase. The resonant frequency is dictated by the material type, particle shape and refractive index of the surroundings. Let's look at a simple argument that allows us to explain the spectral redshift. Consider the Drude dielectric function in the limit $\omega_p\tau \to\infty$. Suppose we have a suspension with a volume fraction p of metal, and $1 - p$ of insulator with a dielectric constant of unity. If the composite is isotropic and does not form an aggregate of infinite dimensions, it can be shown that the centroid of the surface plasmon band will occur at the frequency $\omega = \omega_p[(1 - p)/3]^{1/2}$. For $p \rightarrow 0$, this result becomes what is expected for spheres, that is, $\omega = (3)^{-1/2}\ \omega_P$. Therefore, the surface plasmon frequency shifts the more toward the red, the greater the value of p. This can be achieved by agglomeration in a suspension as long as the metallic nanoparticles remain separate. On the other hand, at the end of last centuries, it was found that by creating a metal-dielectric interface of appropriate geometry, surface plasmons could be generated with the same frequency as the incident waves but with much shorter wavelengths. This could allow plasmons to pass through nanowires, which could prove decisive in producing faster and smaller transistors. Because the frequency of an optical signal is much higher than that of an electrical signal, a future plasmonic circuit could carry much more information. In addition, the circuit will not be subject to resistive and capacitive effects inherent to the transport of electrical charge, which limit the transmission capacity. The development of a plasmonic device equivalent to the transistor ("plasmonster") with three terminals could revolutionize all ultrafast electronics. Although an optical signal may experience greater loss in a metal than in a dielectric, a plasmon can pass through a thin metallic film before extinguishing. The plasmonic losses are smaller, because the field extends it for the dielectric where there are no free electrons to oscillate, and therefore there are no dissipative collisions of the electrons with the metallic lattice. Because the electromagnetic fields on the upper and lower surfaces of the metallic film interact with each other, the frequencies of the plasmons can be adjusted by just varying the thickness of the film. Additionally, it was found that when illuminating a thin gold film perforated with microscopic holes, the transmitted light was higher than expected based on the number and size of holes, and that this could be explained through the plasmonic effect.

It is also good at this stage to remember that due to the optical interference effect between two nearby waves, the diameter of an optical fiber must be at least half the wavelength of the light that falls on it so that it can pass through. Since it is common to use infrared signals in integrated optics on chips, the minimum diameter of optical fibers is much greater than the size of current integrated circuits. A chain of equally spaced metallic nanoparticles allows waves of surface plasmons to propagate through the chain.

Turning now to surface plasmonic sensors that have been widely used in many fields, one can say that they exhibit extraordinary sensitivity based on SPR or localized surface plasmon resonance (LSPR) effects. Recent progresses in these sensors have been made, mainly in the configurations of planar metastructures, optical-fiber waveguides, and quantum plasmonic sensing beyond the classical shot noise limit [72]. Nanorings are useful in plasmonic sensors since their plasmon resonances are highly sensitive to changes in surrounding analyte concentration and the local field inside the ring cavity is relatively homogeneous [73]. Moreover, a wide range of resonance wavelengths are available by varying the ratio between inner and outer radii. Traditionally, solid nanorings have been fabricated from noble metals and support resonances across the visible or near-infrared range [73, 74]. Recently, however, planar nanorings based on graphene have emerged [75], extending the range of available resonances far into the deep infrared. Plasmon hybridization and electrical doping in nanorings of suitably chosen nanoscale dimensions are key elements for bringing the optical response of graphene closer to the near infrared.

SPR is the resonant oscillation of conduction electrons at the interface between a negative and positive permittivity material stimulated by incident electromagnetic radiation (or electron bombardment). The incoming beam has to match its momentum to that of the plasmon. The resonance condition is obtained when the frequency of incident photons matches the natural frequency of surface electrons oscillating against the restoring force of positive nuclei. A non-radiative electromagnetic surface wave that propagates in a direction parallel to the negative permittivity/dielectric material interface is called a surface plasmon polariton. Since the wave is on the boundary of the conductor and the external medium, such oscillations are very sensitive to any change of this boundary (e.g., adsorption of molecules onto the conducting surface). For describing the existence of an electronic surface plasmon, the real part of the dielectric constant of the conductor must be negative and its magnitude must be greater than that of the dielectric. This condition is met in the infrared-visible region for air/metal and water/metal interfaces. On its turn, a localized SPR (LSPR) is a coherent collective electron charge oscillation in metallic nanoparticles that are excited by incident electromagnetic radiation. It exhibits enhanced near-field amplitude at the resonance wavelength, and this field is highly localized at the nanoparticle and decays rapidly away from the nanoparticle/dielectric interface into the dielectric background, though far-field scattering by the particle is also enhanced by the resonance. LSPR displays a very high wavelength resolution, limited only by the size of the nanoparticles. There are two configurations typically used for exciting surface plasmonic waves with light. In one called Otto setup, the light illuminates the wall of a glass prism and is totally internally reflected. A thin metal film (e.g., gold) is positioned close enough to the prism wall so that an evanescent wave can interact with the plasma waves on the surface and hence excite the plasmons. In another one more used (Kretschmann configuration), the metal film is evaporated onto the glass, and an evanescent wave penetrates through the metal film. The plasmons are excited at the outer side of the film. Typical metals that support surface plasmons are silver and gold, although others have also been used (Cu, Ti, Cr). When the surface plasmon wave interacts with a local particle or a rough surface (irregularity), part of the energy can be re-emitted as light, which can be detected behind the metal film from various directions.

Nanoplasmonic sensing (NPS) is an optical technique that exploits metal nanoparticles as local sensing elements and has emerged as a suitable technology for probing functional nanomaterials, biomaterials, and catalysts in situ and in real time. An increasing number of studies focus on using LSPR as an experimental tool to study a process of interest in a nanomaterial. Two different generic experimental strategies have been developed. In a direct nanoplasmonic sensing experiment the plasmonic nanoparticles are active and simultaneously constitute the sensor and the studied nano-entity on/in which a process of interest is occurring. In an indirect nanoplasmonic sensing experiment the plasmonic nanoparticles are inert (typically gold) and adjacent to the material of interest to probe a process occurring in/on the adjacent material. Advantages with both nanoplasmonic sensing are their remote readout, non-invasive nature, high temporal resolution, high sensitivity, single particle experiment capability, and versatility in terms of compatibility with all material types (particles and thin/thick layers, conductive or insulating). Nanofabricated plasmonic gold discs embedded in a custom-made dielectric material offering optimal protection and tailored surface chemistry of the sensor. The gold nanodiscs act as optical antennas, which respond to processes at the sensor/sample interface. When white light passes through the plasmonic sensor, a peak in the extinction spectrum emerges, due to absorption and scattering of light by the particles. The resonance peak position is determined by the size, shape, and material of the nanoparticle, and it also depends on the refractive index of the medium in close proximity to the nanoparticle. Therefore, by monitoring changes in the resonance peak, one can detect and monitor processes influencing the dielectric environment of the nanoparticles on the sensor surface.

In most metals, the plasmonic absorption band is in the ultraviolet, hence its metallic luster characteristic of reflected visible light. However, in some metals the plasmon frequency is in the visible (Ag-410 nm; Au-530 nm; Cu-565 nm), leading to greater reflectance of red light than blue light (more golden glow). The damping frequency γ is related to the mean free path of the conduction electrons and the electron velocity at the Fermi level:

$$\gamma = \frac{v_F}{\lambda_{\text{bulk}}} \tag{5.72}$$

If the particle radius R is smaller than this mean free path, the conduction electrons are also scattered across the surface and the effective mean free path, λ_{ef}, becomes size dependent:

$$\frac{1}{\gamma_{ef}} = \frac{1}{R} + \frac{1}{\lambda_{\text{bulk}}} \tag{5.73}$$

This expression was verified experimentally by Kreibig for silver and gold particles up to a minimum size of 2 nm. Particles measuring only a few tens of nanometers tend to absorb and scatter most photons at a frequency that is inversely proportional to the particle size. As we have already seen in semiconductors, the band gap (inversely proportional to the particle size) determines which photon energies the material will absorb and emit. Therefore, in both cases (metals and semiconductors), one can "tune" the photonic properties (colors it absorbs or emits) of a nanomaterial, by changing its size.

Worth to note that the widely used Drude model applies only to bulk material and has been modified for nanostructures to include effects due to the finite size of the system. One of the corrections considered by Kreibig and Fragstein [76] is the inclusion of an additional damping due to the scattering on the physical particle boundaries. This is important in particles of sizes equal or smaller than the mean free path of electrons in bulk metal. In such a case, the electrons will experience (in the classical picture) additional scattering from the boundary of the system. In addition, pure damping models, such as the Kreibig damping and Lorentz friction, show a dependence on the particle size, where the material influences relevant size regimes. This results in a correction of the spectral position of resonant phenomena and introduces additional, implicit damping contributions. The phenomenological Kreibig damping does yield a plasmon broadening that agrees with experiments; however, it also introduces a redshift of the resonance with respect to the classical Mie result contrary to measurements on nanoparticles [77]. The inclusion of electro-optical effects at the nanoscale into (metal) nanoparticle systems is of importance in nanostructures employed for photovoltaics and catalysis as well as in spectroscopy and sensing applications.

5.2 Measuring Increasingly Weak Forces

When replacing the massless string (or spring) by an oscillating beam of length L fixed at one end, we came across the situation of the bending of a cantilever, where one should consider its deformation and the radius of bending. At equilibrium, the external moment M will equal the internal moments generated by the stress that are distributed throughout any given cross-section. Using both definitions of the area moment of inertia I and the moment, it is possible to arrive at the result $1/R = M/(YI)$, where Y is the modulus of elasticity (or Young modulus), and R stands for the radius of bending [78]. This result can be combined with the expression for the curvature of a plane curve

$$\frac{1}{R} = \frac{d^2z/dy^2}{[1 + (dz/dy)^2]^{3/2}} \tag{5.74}$$

where z represents the direction normal to plane xy of the cantilever in equilibrium. Therefore,

$$\frac{\partial^2 z}{\partial y^2} = \frac{M}{YI} \tag{5.75}$$

if the angle of bending is small enough (neglecting the derivative in the denominator). The moment M at any given point $y \le c$, produced by a force F acting at point c is equal to $F(y - c)$. Thus, inserting in Eq. 5.34, integrating twice, and setting, $y = c = L$, the deflection of the cantilever beam is given by

$$z = -\frac{L^3}{3YI}F \tag{5.76}$$

Now, defining a spring constant k (compliance) as the ratio $|F/z|$, and assuming that the bending is small enough so that F is not a function of z, one finally gets

$$k = 3\frac{YI}{L^3} \tag{5.77}$$

The fundamental resonance frequency of the lever can be computed approximately by equating its strain energy (when the deformation is at a maximum) to the kinetic energy when the deformation is zero. Doing this, it is possible to show [78] that the angular frequency can be estimated by

$$\omega = k^2\sqrt{\frac{YI}{A\rho}} \tag{5.78}$$

with $k = 1.875/L$, for a lever of cross-section, A, and mass density ρ. We have shown how to calculate the resonance frequency and deflection of levers of a known geometry in terms of their material parameters Y and ρ. Conversely, it is possible to derive the material parameters Y and ρ if we know the geometry of a lever, its resonance frequency, and its spring constant [78]. When the lever is in a gravitational field, and its oscillation is perpendicular to the direction of the field, then the period stays the same, and the constant force simply shifts the equilibrium position of the harmonic motion. Therefore, one gets again the standard harmonic motion, which within the limits of the classical mechanics scaling laws (and considering L the linear dimension), has frequency and oscillation period that vary respectively with $L^{-3/2}$ and $L^{3/2}$.

For a lumped system consisting of both a concentrated and a distributed mass, m_c and m_d, respectively, it is possible to show that if $m_c = 0$ the fundamental mode is given by [60]:

$$\omega = \sqrt{\frac{\kappa^4 EI}{\rho A}} \tag{5.79}$$

and then using Eq. 5.40, one gets

$$\omega = \sqrt{\frac{k}{0.24 m_d}} \tag{5.80}$$

Given so, we can define an effective mass by

$$m_{\text{eff}} = m_c + 0.24 m_d \tag{5.81}$$

to obtain the general expression for the frequency of the fundamental mode of a lever having both distributed and concentrated masses:

$$\omega = \sqrt{\frac{k}{m_{\text{eff}}}} \tag{5.82}$$

Continuing in this path we are gradually permeating the world of the nano-oscillators. Suppose a linear chain of N identical particles, with mass m each one, spaced by springs of equal force constant, K, and length a. Thus, the total length of the chain is $L = Na$. If u_n is the longitudinal displacement of the nth mass from its equilibrium position, the equation that describes its motion is

$$m\left(\frac{d^2 u_n}{dt^2}\right) + K\left(u_{n+1} - 2u_n - u_{n-1}\right) = 0 \tag{5.83}$$

and the wave solution for this equation can be written as

$$u_n = u_0 \cos\left(\omega t + kna\right) \tag{5.84}$$

being k the wave vector, and $kna = kx = 2\pi x/\lambda$. By replacing Eq. 5.47 in Eq. 5.46 one finds the following dispersion relation:

$$m\omega^2 = 4K \sin^2(ka/2) \tag{5.85}$$

The allowed frequencies depend on the wave number k, according with

$$\omega = 2\left(\frac{K}{m}\right)^{1/2}\left|\sin\left(\frac{ka}{2}\right)\right| \tag{5.86}$$

and as for small values of x, sin $x \approx x$, they can be simplified:

$$\omega = ak\sqrt{\frac{K}{m}} \tag{5.87}$$

which represents a velocity for the wave equal to $v = \omega/k = a\,(K/m)^{1/2}$.

One may compare it with the sound velocity in the material of density ρ, which is $(Y/\rho)^{1/2}$. This means that for a nanowire it is fair to establish a correspondence between Y/ρ and Ka^2/m, and so it becomes possible to express the Young modulus dependence on microscopic magnitudes as the following:

$$Y \leftrightarrow \rho Ka^2/m \tag{5.88}$$

This makes sense when assuming that the atoms of mass m are spaced apart by a, and the interatomic interactions are described by a spring constant K. If one considers the ideal massless spring equivalent to a beam of cross-section A, and $Y = (F/A)(L/\Delta L)$, subject to a compressive force, then the deformation is $\Delta L/L = -F/(AY)$. So, $K \equiv AY/L$, as it was demonstrated from the elasticity theory, in Eq. 5.40. Table 5.2 displays some typical Young's modulus values for distinct very stiff materials.

TABLE 5.2

Young Modulus of Some Relevant Cantilever Materials

Material	Diamond	C_{60}	MWNT	SWNT	SiC	Al_2O_3	Si_3N_4	W
Y(GPa)	1050	310	900	1000	700	532	385	350

Note that the resonance frequency of an AFM cantilever of length L, thickness b, and width a, varies with L^{-2}, since $\omega = 2\pi\,(0.56/L^2)\,(YI/\rho A)^{1/2}$, where $I = ab^3/12$ is the moment of inertia of area in the direction of the beam deflection [53].

Let us now consider for simplicity a hard-sphere collision between an atom A and another B which is attached to a fix point by a harmonic spring of force constant k. This is a special case of the A + BC collision when the mass of C is infinite. Consider that the kinetic energy of A is $E_A = ½\, m_A v_A^2$ and B vibrates about its equilibrium position with a displacement $x(t)$ that varies in time according to a sin $(\omega t + \phi)$, where a, ω, and ϕ are, respectively, the amplitude, frequency, and phase of the harmonic oscillator. Therefore, the oscillator energy will be ½ $m_B\, a^2 w^2$. Using the conservation of energy and conservation of momentum and since the change in energy of B on collision (ΔE_B) is entirely in its kinetic energy, if one takes the collision to occur at time zero, we will arrive to the expression of $\Delta E_B/E_A$ as a function of ϕ. Thus, by differentiating such expression one can find the maxima and minima points. After easy mathematical handling the solutions are $\phi = 0$, $\phi = \pi$, and

$$\varnothing = \cos^{-1}\left[\frac{m_B - m_A}{2(m_A m_B)^{1/2}}\left(\frac{E_A}{E_B}\right)^{1/2}\right] \tag{5.89}$$

The first two solutions correspond to the collision occurring as the oscillator goes through its equilibrium position.

Let us now turn slightly about van der Waals interactions, which are universal and ultimately responsible for cohesion in objects. Van der Waals' attractive forces are felt between atoms and molecules at close range and include in general three components: permanent dipole-permanent dipole, permanent dipole-induced dipole, and induced dipole-induced dipole. Even when neither partner has a permanent electrical dipole, there is an attractive force of quantum origin (London dispersion forces). At any given instant, any atom or molecule has an asymmetric distribution of charge that corresponds to an instantaneous electrical dipole moment $\boldsymbol{p}$. The energy of the attractive interaction between a dipole and a nonpolar molecule is proportional to r^{-6}. It happens that the so-called van der Waals effect (London dispersion forces) is a dynamic interaction of completely quantum origin, resulting from the fact that the "electron in the atom does not have a regular orbit, giving rise to a floating electric dipole." As we will see next, the van der Waals interactions between two neutral nonpolar particles, are essentially quantum mechanical in origin and can be understood by considering first the simple interaction energy of two single valent atoms in a vacuum. Fixing the centers of each atom at $\boldsymbol{r} = 0$ and $\boldsymbol{r} = R$, respectively, the total Hamiltonian is the sum of the Hamiltonians of the two isolated atoms with the interaction potential, $H = H_1 + H_2 + V$, where V accounts for the interaction of the two electrons with the two nuclei. The interaction of the electron of atom 1(2) with the nucleus of atom 1(2) is already included. Labeling the two electron coordinates by $\boldsymbol{r}_1$ and $\boldsymbol{r}_2'$, one can then write

$$V = e^2\left[\frac{1}{|\boldsymbol{R}|} - \left(\frac{1}{|\boldsymbol{R} - \boldsymbol{r}_1|} + \frac{1}{|\boldsymbol{r}_2'|}\right) + \frac{1}{|\boldsymbol{r}_1 - \boldsymbol{r}_2'|}\right] \tag{5.90}$$

The interaction energy ΔU, to the second order in perturbation theory is

$$\Delta U = \langle 0|V|0\rangle + \sum_n \frac{|\langle 0|V|n\rangle|^2}{U_0 - U_n} \tag{5.91}$$

where the two-atom, unperturbed ground-state ($|0\rangle$) energy is U_0, and the higher lying states ($|n\rangle$) have an energy U_n. When the distance between the atoms is not too large ($R < c/\omega$, where ω is a typical atomic frequency), relativistic corrections do not have to be considered. However, one assumes that R is large enough so that the overlap of the wave functions is small and one expands V in the small quantities $(|\mathbf{r}_1|) << R$ and $(|\mathbf{r}_2' - \mathbf{R}|) << R$. Defining $\mathbf{r}_2 = \mathbf{r}_2' - \mathbf{R}$, the result of the expansion to the lowest order is then

$$V \approx -\frac{e^2}{R^3}[3(\boldsymbol{r}_1.\boldsymbol{R})(\boldsymbol{r}_2.\boldsymbol{R}) - \boldsymbol{r}_1.\boldsymbol{r}_2] \tag{5.92}$$

For spherically symmetric charge densities (i.e., nonpolar atoms), the first-order term in the perturbation theory for ΔU vanishes, and the second-order term yields an interactive attraction that decays with $1/R^6$. Since the ground state has the lowest energy, $\Delta U < 0$, and for distances between the atoms larger than 5 nm, relativistic corrections cause a decay faster than $1/R^6$ and not always an attractive interaction. Similarly, in a dielectric medium the interaction can be either attractive or repulsive. An upper bound to ΔU can be found by noting that each term in the Eq. 5.54 is negative and approximating the sum over excited states by the state with the energy closest to the ground state (i.e., $n = 1$). Therefore, the energy of the excited state can be written as, $U_1 = U_0 + \hbar\omega_1 + \hbar\omega_2$, where the two last terms are the energy differences for the two atoms, respectively. In this approximation and for isotropic systems one finds

$$\Delta U = \frac{-6e^4}{R^6} \frac{|\langle g_1|z|e_1\rangle\langle g_2|z|e_2\rangle|^2}{\hbar\omega_1 + \hbar\omega_2} \tag{5.93}$$

where $< g_i|$ and $< e_i|$ refer to the wave functions for the *i*th atom, and $i = 1, 2$. Therefore, atom fluctuations are correlated with the fluctuations in neighboring atoms, resulting in an attractive interaction proportional to r^{-6}, and the additional interaction energy correction arises from the finite transit time for the propagation of light between the two interacting systems.

The matrix elements and energy denominators in the van der Waals energy ΔU of Eq. 5.56 are related to the polarizability α, which relates the electric dipolar moment to an applied electric field. For isotropic systems, α is a scalar quantity and the perturbation energy is $V = -(1/2)\,\boldsymbol{p}.\boldsymbol{E} = -1/2\,\alpha\,E^2$. Thus, in the approximation of a single excited state, one finds

$$\alpha_i = \frac{e^2\,|\langle g_i|z|e_i\rangle|^2}{\hbar\omega_i} \tag{5.94}$$

and so, one can write the van der Waals energy in terms of the polarizabilities of each atom:

$$\Delta E \approx -\frac{6\beta_{12}\alpha_1\alpha_2}{R^6} \tag{5.95}$$

where $\beta_{12} = \hbar\,\omega_1\omega_2/(\omega_1 + \omega_2)$ is $\hbar$ a factor coming from the difference in the energy denominators. The quantum theoretical calculations have shown that the helium dimer (He_2) represents the longest and weaker molecular bond (on the order of 10^{-7} eV). However, the traditional techniques of spectroscopy, as well as diffraction, or electronic scattering have not proved able to detect it, given its extreme fragility. Toennies et al., succeeded in obtaining such experimental evidence, using the molecular beam technique in which cold helium atoms (4.5 K) were isentropically expanded through a nanometric diffraction grid [79]. The peak intensities obtained revealed a bonding length 5.2 nm for He_2 and a bond energy of 9.5×10^{-8} eV, in clear agreement with theoretical predictions.

When the distance between atoms is greater than the distance that light can travel during the characteristic lifetime of the fluctuations, London dispersion forces are weakened. This is the so-called retardation effect, which following Casimir and Polder can be neglected, if such distance is less than 5 nm. If, otherwise, the decay of the attractive force ceases to do with r^{-7}, following then a behavior with r^{-8} (Casimir–Polder equation) [80].

For the case just treated, namely a vacuum, the polarizability is related to dielectric constant by $\alpha_i = (\varepsilon_i - 1)/\rho_i$, where $(1/\rho_i)$ is the molecular volume of species *i*. In a medium of dielectric constant ε_m, it is possible to generalize making use of the excess polarizability, $\alpha_i = (\varepsilon_i - \varepsilon_m)/\rho_i$. Therefore, one can rewrite Eq. 5.96 for the dispersion interaction, $U = \Delta E$ between two molecules separated by a distance $\boldsymbol{r}$:

$$U \approx -\frac{1}{r^6}\frac{6\beta_{12}}{\rho_1\rho_2}\frac{(\epsilon_1 - \epsilon_m)(\epsilon_2 - \epsilon_m)}{\epsilon_m^2} \tag{5.96}$$

One notices from this expression that the sign of the dispersion interaction depends on the differences between the molecular and medium dielectric constants. Note that the only case where the interaction becomes repulsive is when ε_m lies between ε_1 and ε_2. Usually, one defines the Hamaker constant A_{12} (or H), for the interaction of two molecules in a given medium by

$$U = -\frac{A_{12}}{\pi^2 \rho_1 \rho_2} \frac{1}{r^6} \tag{5.97}$$

where ρ_i is the density of molecule i. Typically the Hamaker constant is of the order $25k_BT$, in units of energy.

For a many-particle system the interaction energy can be obtained if one assumes that the van der Waals interactions are pairwise additive, and so, it is possible to obtain the interaction U, between two parts of the system with densities $\rho_1(\mathbf{r})$ and $\rho_2(\mathbf{r})$. If an atom or a molecule is interacting with a larger body such as a plane, the dependence of the interaction with the distance to the plane, can be obtained through integration of the interaction energy to all atoms in the plane, assuming additivity of potential pairs of the type $V(r) = -C/r^6$ (which is justified at distances where the three-body interaction can be neglected, i.e., greater than the equilibrium distances). The total interaction energy can thus be estimated by

$$U \approx \int_{r_{eq}}^{r} V(r)\rho(r) 4\pi r^2 dr \tag{5.98}$$

where $\rho(r)$ is the number density of the interacting bodies. The long-range nature of these interactions can be seen by calculating U for a single molecule above a surface (with a known density of particles) and at a distance r from it. The output result is proportional to $1/r^3$, and so the attractive force that the molecule feels decays as $1/r^4$. Several geometries have been analyzed based on r^{-6} interaction, and some representative results are displayed in Table 5.3 [81, 82]. In these expressions n_v represents the number of atoms per unit of volume and H (or A) is the Hamaker constant, that depends on the material properties (it can be positive or negative in sign depending on the intervening medium) being approximately given by $\pi^2 C\rho^2$. The Hamaker constant is not completely invariable (with a variation between 0.2 and 2.0 eV), since constant C is proportional to the square of atomic (or molecular) polarizability, and thus to the square of the volume.

TABLE 5.3

Selected Van Der Waals Interaction Geometries

Interaction atom-plane separated by a distance r	$U_{VdW} = -\frac{\pi n C_V}{6r^3}$
Interaction sphere-plane separated by a distance $r << R$ (R is the sphere radius)	$U_{VdW} = -\frac{HR}{6r}$
Interaction between two spheres of radius R_1, R_2 $Z = R_1 + R_2 + r$	$U_{VdW} = -\frac{H}{6}\left[\frac{2R_1R_2}{z^2-(R_1+R_2)^2} + \frac{2R_1R_2}{z^2-(R_1-R_2)^2} + \left(\frac{z^2-(R_1+R_2)^2}{z^2-(R_1-R_2)^2}\right)\right]$
Interaction between two spheres of radius R_1, R_2 separated by a distance $r << R_1, R_2$	$U_{VdW} = -\frac{H}{6r}\frac{R_1R_2}{R_1+R_2}$
Interaction between two parallel planes of area S separated by a distance r	$U_{VdW} = -\frac{HS}{16\pi r^2}$

London dispersion forces have been used in a more detailed way than before to explain many phenomena, as for instance the general stability of colloidal suspensions. However, it fails to interpret the experimental observations when the distances between the particles in suspension become larger and close to 1 μm. Casimir pointed a reason for that, by noticing how such distances become comparable to the wavelengths of the involved electromagnetic field; thus, the field dynamical properties assume a larger relevance than the interacting particles. Therefore, it seems that van der Waals and Casimir forces are only one and the same interaction, just playing a role in different scales. Given that, Casimir reasoned that such forces' mechanisms must be understood from the electromagnetic field own properties, namely its fluctuations. He could experimentally prove that two macroscopic parallel plane mirrors in vacuum attract each other due to those field fluctuations, as expected from the application of wave theory. In fact, Casimir experiments have shown that the mirror's attraction force is independent of its material properties but varies instead with its separation and surface morphology [83]. Furthermore, the expression for the force includes the Planck constant and the speed of light in vacuum. These are the iconic constants involved in quantum physics and relativity theory, respectively. As we already described, oscillators always display a residual quantum of energy even at zero absolute temperature. Therefore, the position and velocity of a particle exhibit fluctuations which obey Heisenberg uncertainty principle, and for a null temperature they are the so-called zero-point fluctuations. They also exist for the electromagnetic fields since each mode is equivalent to a mechanical oscillator. Therefore, vacuum is full of field fluctuations, and for each field mode they travel at velocity c and transport an energy $h\nu$. This causes a radiation pressure which is a function of the photon momentum. When two mirrors are face to face in vacuum, Casimir verified that their mutual attraction varies with their separation D, according to $1/D^4$ (note the similarity with the van der Waals force that we already explained). To estimate the Casimir force between the two perfect flat mirrors, one requires the inclusion of a very large number of pressure contributions on the mirrors for all the present modes in the electromagnetic spectrum. The pressures from each of them are different at the external and internal mirrors surfaces. The optical cavity confines the vacuum inside and modifies its properties in contrast with the external vacuum. Some internal modes are resonant, and others are anti-resonant. The resonance occurs when the distance between the mirrors is an integer multiple of half-wavelength, and so the pressure is stronger inside than outside; for the anti-resonance modes the opposite occurs. The net global effect is a slight preponderance of the latest ones, and consequently the Casimir force is attractive. It is expected to increase strongly for distances shorter than 1 μm, until the theoretical formula ceases to apply at the atomic and molecular scales. The zero-point energy density must be lower in the region between the mirrors than outside, and this difference in energy will depend on the distance between the mirrors. As two surfaces are brought together, the power law describing how the force varies with distance changes, and this is the crossover between the two regimes. This crossover can be thought of as the minimum distance between two components before they stick together. This distance is about 10 nm. The force can be quite large at very small distances: for a 10-nm separation of the mirrors, the pressure on them is the same as atmospheric pressure (10^5 Pa). However, it drops very rapidly (as the inverse 4th power of separation for perfect mirrors), and so it has been very difficult to accurately measure such force between two sufficiently flat and sufficiently smooth parallel surfaces approaching from micron to submicron distances. A special emphasis should be addressed to the distinction between the Casimir and van der Waals forces since they both have origin in the same source: the zero-point electromagnetic field. Their theoretical descriptions are different, because in general the van der Waals force is expressed in terms of surface charges on the two bodies whereas the Casimir force is described in terms of the zero-point electromagnetic field. The surface charges, however, are a result of fluctuations in the zero-point field; hence fundamentally the two forces arise from the same source.

The recent successful measurement of the Casimir force [84] reinforces the experimental evidence of vacuum fluctuations predicted by quantum field theory. Indeed, it has been successfully measured at the micrometer scale, making use of revolutionary scanning probe microscopes, and the results have confirmed the quantum prediction of the Casimir effect, which reflects the variation of the vacuum energy as a function of the cavity linear dimension [84].

Two parallel mirrors spaced at a distance L allow the creation of stationary electromagnetic waves (propagating in the z-direction perpendicular to them) when $L = n\lambda/2$. The state of lowest energy of electromagnetic modes correspond to $h\nu/2$, as we already mentioned. Thus, even an empty cavity between mirrors has an energy equal to $Nh\nu/2$, where N is the number of allowed modes in the cavity of width L. The allowed frequencies are given by $nc/(2L)$, where $n = 1, 2, 3, \ldots$, and so frequencies lower than $nc/2L$ are not allowed, that is, wavelengths longer than $2L$. Therefore, as L decreases, more wavelengths (and more zero-point energies) are excluded. This gives rise to an attractive force obtained from $(-dU/dz)$. Thus, Casimir force, F_c, has been calculated for some simple geometries [85] as for instance between parallel mirrors, giving the result:

$$F_c = -\frac{\pi hc}{480}\frac{1}{z^4} \tag{5.99}$$

or in the the case of a sphere of radius R, at a distance d, from a flat surface. In this situation, the Casimir force exerted on the ball is

$$F_c = -\frac{\pi^2 hc}{720}R\frac{1}{d^3} \tag{5.100}$$

This result was fully confirmed using an experimental microdevice [86], basically outlined in the Figure 5.11a. A polycrystalline silicon microelectromechanical (MEM) device is composed by a swing that can be rotated by an angle θ, and a sphere of radius 100 μm (via a piezoelectric stage), approaches at 75 nm. Both the sphere and the surface of the plates are coated with gold. The force between the ball and the right portion of the blade is determined by angle θ, which is measured. The torsion constant is calibrated through the torsion angle observed at a distance such that the Casimir force is negligible. The Coulomb force between the sphere of radius R and the plane is a known function of (R, z), according to which the measures were carefully adjusted, and so calibrating this way the force versus angle θ. The magnitude of the force, for a given value of distance d is less in the sphere-plate configuration than with parallel flat mirrors, but the experimental problem of maintaining perfect parallelism at sub-micron separations is largely removed. The films of metallic gold deposited on surfaces are not perfect mirrors, thereby allowing high-frequency electromagnetic waves passing through them. This effect can be corrected using tabulated optical constants for gold. A second disturbance effect come from the fact that gold surfaces are not perfectly flat, having a surface roughness amplitude of 30 nm, which is previously measured by atomic force microscopy.

Van der Waals forces and the Casimir effect both rely on virtual photons. It is strong evidence for the existence of virtual particles associated with quantum field theory and zero-point vacuum energy (which may be the origin of the so-called dark energy driving the expansion of the universe). Since light is an electromagnetic wave, any field theory of electromagnetism needs to be consistent with the theory of special relativity that describes the behavior of light. Quantum electrodynamics is a relativistic field theory and very successful in explaining how electromagnetic force is transmitted between electrons. It considers the exchange of virtual photons which act as gauge bosons and so they did not be detectable in a measurable sense. They can pop into existence for brief periods due to the time-energy uncertainty principle, that allows the use of an amount of energy ΔE from the vacuum of space for a very short period $\Delta t < \hbar/2$; this creates a virtual particle-antiparticle pair that acts as gauge boson between interacting particles of real matter. This does not defy conservation of energy because after that very short duration, the pair annihilates giving back the quantum of energy. Thus, when the already mentioned plates are separated by a fraction of the wavelength of light, a pressure from the energy density difference pushes plates together from outside, and inside the gap there are fewer virtual photons. A difference in the energy density on each side of the plate arises from the difference in the number of modes, and this results in an attractive force.

Casimir force generates a fundamental and ever-present stickiness of components in micromachines and nanomachines. Indeed, as the size of these devices has decreased, they have become full of boundaries with submicron gaps where the Casimir force becomes dominant. It is a significant problem because while one can take measures to prevent things like capillary and electrostatic forces, there is nothing that can be done to prevent the Casimir force as it arises from the fundamental properties of the vacuum.

The utilization of a modified AFM to perform measurements of the Casimir force was pioneered by Mohideen et al. [87], using a 200-μm-diameter gold-coated polystyrene sphere (Figure 5.11b).

Casimir effect has also the potential to transmit force at short distances without physical contact, and Capasso et al. have been successful in modifying the motion in a micromechanical system [88, 89]. They built a standard micromechanical device consisting of a flat silicon plate with dimensions of a few hundred microns, which is suspended by a torsion wire above a surface. By applying an ac voltage to the pads underneath the plate, it can be made to oscillate in seesaw fashion at a frequency of a few kHz. Then, using an AFM-type manipulator, they lowered a 100-μm gold-coated sphere toward one side of the oscillating plate approaching it to within a few hundred nanometers. The Casimir force between the sphere and the plate causes a shift in the frequency of the oscillator. They found that the amplitude and frequency shift of the oscillator were measurable with only a few nanometers change in the height of the sphere [85].

Presently, it is possible to fabricate nano-resonators from atomically thin graphene or carbon nanotubes. Such nano-resonators consist of suspended graphene sheets (made by etching of the silicon oxide sacrificial layer or by the transfer of graphene to a pre-patterned substrate) that are actuated either electrostatically or optically, and their resonant frequencies and quality factors are measured [71, 90]. To measure the resonant frequency and the quality factor of graphene resonators, optical and electrical methods can be used [91]. For instance, a dc voltage gate Vg is applied to the device in combination with a radiofrequency gate voltage of frequency f, that drives the motion [$V = Vg + a\cos(2\pi ft)$]. A second rf voltage, at a frequency $f + \Delta f$, is applied to the source ($\Delta f/f \ll 1$). During the motion of the resonator, its conductance changes with the distance from the substrate and the motion is detected as a mixed-down current at the difference frequency Δf. The resonant frequency of

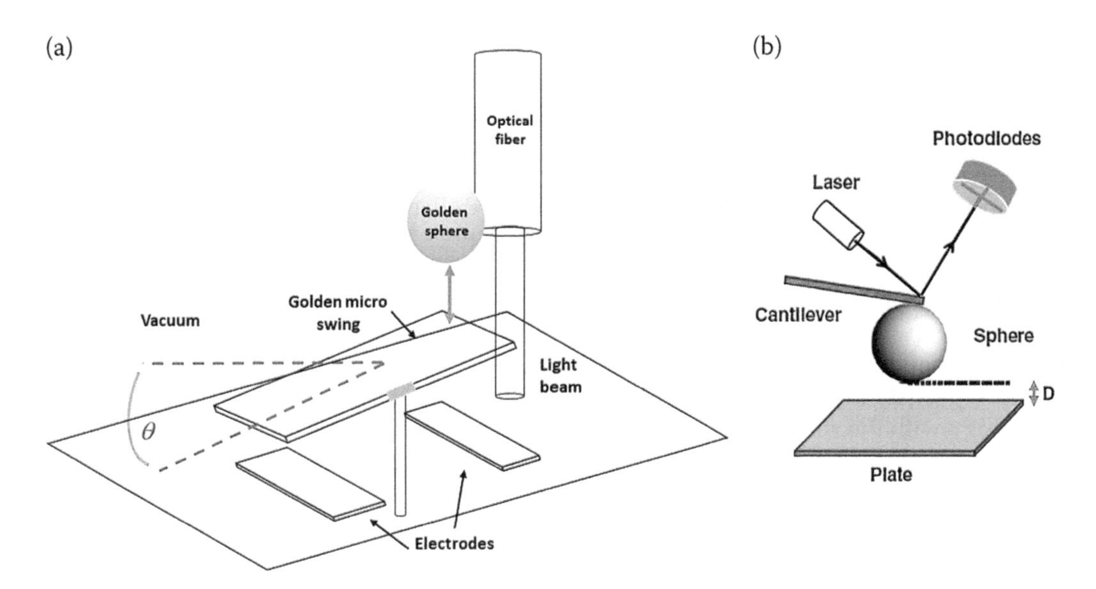

FIGURE 5.11
(a) A microelectromechanical device composed of a micrometer swing is under the action of two electrodes, and a sphere attached to an optical fiber approach (at a known distance measured by the laser beam). The Casimir force is exerted between the sphere and the swing, causing the change in the oscillation frequency. The measurement of this frequency leads to the value of the Casimir force; (b) AFM configuration for measurements of the Casimir force by optical deflection of a cantilever (with a gold-coated sphere glued onto its end).

a graphene monolayer depends on the physical state of the beam, in particular strain and absorbed mass, and can be modeled by [90]:

$$f_0 = \frac{k}{2}\sqrt{\frac{Y}{\rho_0 L^2}}\sqrt{\frac{s}{\alpha}} \tag{5.101}$$

where L is the graphene beam length, Y and ρ_0 are the Young modulus and the mass density of graphene monolayer, respectively, $k = 1, 2, \ldots$, s is the in-plane strain and α is the absorbed mass factor, that is, $\alpha = \rho/\rho_0$, where ρ represents the density (including absorbed mass). This $1/L$ dependence on the natural frequency of graphene beam monolayer has been confirmed experimentally. It is possible to calculate the resonance frequency shift δf for a small variation of mass δm:

$$\delta f = f_0 - f = f_0\left(1 - \frac{1}{\sqrt{1 + \frac{\delta_m}{m}}}\right) \approx f_0 \frac{\delta_m}{2m} = R\delta_m (\delta_m \ll m) \tag{5.102}$$

where the responsivity R is defined as $f_0/2m$. Consequently, to increase the beam responsivity, a small mass and a high resonant frequency are desired. Such requirement can be achieved with nanotubes as their mass is extremely low ($\approx 10^{-21}$ kg) and their mechanical properties allow for very high resonant frequencies. For instance, carbon nanotube mass sensors achieving atomic resolution have been fabricated (i.e., a responsivity of 0.104 MHz zg^{-1}, which corresponds to a few tens of gold atoms) [92]. These devices displayed a sensitivity of 0.4 gold atoms/$Hz^{1/2}$ which is extremely low considering that the measurements are performed at room temperature. Beams clamped at one end have a larger travel range when bent than the travel range of clamped-clamped beams, allowing increased dynamic range. Another advantage of using cantilevers is that the loss owing to clamping is less than with doubly clamped ones. Thanks to their excellent mechanical properties as well as interesting electrical characteristics, carbon nanotubes are among the most widely used materials for the study of electromechanical properties.

Nowadays, physicists and engineers are trying to measure quantum effects in macroscale objects (which could enable supersensitive sensors) and they have already observed true quantum behavior in macroscopic objects. One way is to find a suitable method to measure persistent currents in normal metal rings, that is to search for a residual current in analogy with the zero-point energy of an oscillator. If one wants to measure the electric current in a metal ring, this could be accomplished by attaching an ammeter; however, in this case, the addition of an ammeter would add too much resistance. Instead, Harris et al. [93] used one extreme sensitive force detector; when a magnetic field was applied to the predicted persistent current, a magnetic force proportional to the strength of the field and the velocity of the charge was felt. Since the persistent current is small, they needed to apply a very strong magnetic field. To ensure necessary measurements could be taken, they also used an extreme sensitive force detector: a microcantilever, capable of measuring forces as small as 10^{-18} N. They placed the ring on the microcantilever and, when the magnetic field was turned on, they could detect wobbles in the microcantilever, indicating a force was present. Another method developed by Harris group made use of a stable optical resonator [94] that has been built using a metal-coated microcantilever as one mirror, and the second mirror was a 12.7-mm-diameter concave dielectric mirror. By positioning the two mirrors 75 mm apart in a near-hemispherical configuration, a Fabry–Pérot cavity was implemented. This way, the optical stability of the cavity implies that its finesse can be further improved with better mirror coatings. Both the increased finesse and cavity length will be important for studying quantum optical effects in optomechanical cavities. The motion of micro- and nanomechanical resonators can be coupled to electromagnetic fields, and such optomechanical setups allow one to explore the interaction of light and matter in a new regime at the boundary between quantum and classical physics. The authors proposed an approach to investigate nonequilibrium photon dynamics driven by mechanical motion in a developed setup with a membrane between two mirrors, where photons can be shuttled between the two halves of the cavity [95]. For large drive, they showed that the system displays Landau–Zener dynamics originally known from atomic two-state systems, as we already explained in this chapter. It is possible to envisage ultraprecision quantum sensors for measuring very weak signals, based on hybrid optomechanical systems, in which a quantum nano-mechanical oscillator can be coupled to the electromagnetic radiation pressure, through ensembles of ultracold atoms. In fact, in any precise quantum measurement, the competition between the shot noise and the back action noise of measurement executes a limitation on the measurement precision, which is the so-called standard quantum limit. In the case where the intensity of the signal is even lower than such limit, one needs to perform an ultraprecision quantum sensing to beat it.

In optical detection setups to measure the deflection of micro-cantilevers, part of the sensing light is absorbed, heating the mechanical probe. Some authors have studied strong heating of cantilevers in vacuum using light absorption of a laser focused on its free end. They presented experimental evidence of a frequency shift of the resonant modes, when the light power of the optical measurement set-up is increased, and this frequency shift is a signature of the temperature rise and presents a dependence on the mode number [96]. Each mode can be pictured as a harmonic oscillator of mass m_{eff}, whose stiffness k_n is proportional to the Young's modulus Y of the cantilever material. The increase of temperature, due to light absorption, induces a softening of the cantilever, because the temperature coefficient $\alpha_Y = (1/Y)(dY/dT)$, is negative. The resonance frequencies of the normal modes should, therefore, decrease as light intensity increases, because since the frequency for each mode is proportional to $Y^{1/2}$, then, for a certain light intensity I one finds that

$$\frac{\delta f_n}{f_n^0} \approx \frac{1}{2}\frac{\delta Y}{Y} \approx \frac{1}{2}\alpha_Y \frac{dT}{dI} I \tag{5.103}$$

A proposed model that describes the shift of the mechanical resonance frequencies of a cantilever submitted to a temperature profile quantitatively describes the experimental data for a raw silicon cantilever heated locally in vacuum from room temperature up to the melting point [97]. In addition, when light reflects off a material or absorbs within it exerts radiation pressure through the transfer of momentum. Micro/nano-mechanical transducers have become sensitive enough that radiation pressure can influence these systems. The photothermal effects often accompany and overwhelm the radiation pressure, complicating its measurement, but some experimentalists investigated the radiation force on microcantilevers and could identify and separate the radiation pressure and photothermal forces through an analysis of the cantilever's frequency response. Then, by working in a regime where radiation pressure is dominant, they were able to accurately measure the radiation pressure [98]. For an ideal mirror, the radiation pressure is independent of the wavelength of light and depends only on the incident power. However, in real situations for a constant input optical power, wavelength-dependent radiation pressure can be observed due to coherent thin film Fabry–Perot interference effects. And so, it has been demonstrated that tunable wavelength excitation measurement can be used to separate photothermal effects and radiation pressure [99]. This ability to control the involved interactions opens the way to the development of opto-mechanical devices and studies of the coupling between optical and thermal effects in such systems.

Photothermal sensing of chemicals using microcantilevers is also a matter of interest due to its potential applications. Microfabricated cantilevers have several advantages as sensors, including miniature size, low power consumption, array-based detection, and potential low cost. They have been used for detecting a variety of chemicals with high sensitivity, although their selectivity still needs to be improved [100]. This setback is due to the weak selectivity of chemical interfaces immobilized on cantilever surfaces. In principle, there are two possibilities for enhancing the selectivity. One is based on photothermal deflection spectroscopy where selectivity is achieved through a mechanical response due to optical absorption by the adsorbed molecules on a cantilever; and in the other one, the position of the resonance frequency peak of the cantilever is monitored for shifting due to mass loading, which allows the precise measurement of the mass of the surface-adsorbed molecules. A micro-cantilever sensor undergoes a weak bending when molecules are preferentially adsorbed on one of its surfaces. Such bending is due to changes in surface energy due to molecular adsorption on one of its surfaces. Simultaneously, due to mass loading, the resonance frequency of the cantilever suffers a variation. Both variations, bending and frequency, can be detected simultaneously for cantilevers that have about 100–400 μm in length with a spring force constant of around 0.1–1 Nm^{-1}. One advantage of the microcantilever technique is that both the bending and resonance frequencies can be measured simultaneously.

The selectivity of detection is on its turn, directly related to the chemical selectivity of the coatings since this is basically accomplished by using interfaces immobilized on their surfaces. Unfortunately, chemically selective coatings capable of reversible interaction with the analyte molecules do not have very high selectivity, and the response from an array of cantilevers coated with different chemical coatings and analyzed with pattern recognition can provide limited selectivity. However, when mixtures of chemical vapors are present, the array response fails to identify the individual mixture components in the targeted vapor due to the lack of orthogonal response. Therefore, despite its many advantages, the cantilever platform remains unattractive for small molecule detection due to lack of selectivity. The desired increase of selectivity can be achieved for small molecules by incorporating photothermal deflection spectroscopy (PDS) based on the principles of vibrational spectroscopy and using the extremely high thermal sensitivity of a cantilever made of two materials (dimaterial) [101]. Such cantilever with adsorbed molecules shows bending when exposed to infrared radiation. The adsorbed molecules absorb the radiation energy and heat the dimaterial microcantilever, which causes the cantilever to bend due to differential thermal expansion. In general, most of the heating caused by non-radiative decay is lost to the massive base of the cantilever, and only a small fraction of the heat is lost to the ambient air due to the large difference in thermal conductivity between the silicon and the air.

PDS has demonstrated highly sensitive detection of sub-nanogram quantities of adsorbed molecules from volatile chemicals, pointing its potential for detection of sub-monolayer amounts of molecules with high selectivity and sensitivity. The expected deflection of the microcantilever beam due to the above mentioned dimaterial effect can be calculated from knowing the geometric dimensions of its two-layers tip, the thermal expansion coefficients for the two layers, the thermal conductivities of the two layers, and the total power absorbed by the cantilever. With the PDS technique, the microcantilever with adsorbed chemical species is sequentially exposed to different infrared radiation wavelengths from a monochromator, and generally, this radiation is chopped at a frequency chosen to improve the sensitivity of detection. Therefore, the microcantilever deflection as a function of the radiation wavelength resembles the infrared absorption spectrum of the adsorbate molecules. Additionally, the resonance frequency of the cantilever is monitored, before and after vapor exposition, because from the resonance frequency shift it is possible to calculate the mass of adsorbates on the cantilever:

$$\delta m = m\left(\frac{\nu_1^2 - \nu_2^2}{\nu_2^2}\right) \tag{5.104}$$

where ν_1 and ν_2 are the initial and final resonant frequencies of the microcantilever during mass adsorption and m is the cantilever mass.

Figure 5.12 schematically displays an example of PDS experimental set-up. For the sensor, a silicon-based microcantilever, made dimaterial by depositing a submicrometric layer of gold with a few tens of nanometers of chromium, as an adhesion layer (using an e-beam evaporator), can be used. The typical optical beam deflection method applied in scanning force microscopes can thus be used for monitoring the bending of the cantilever. To increase the sensitivity, a lock-in amplifier can be used to obtain the cantilever signal generated by the pulsed infrared light, and the resonance frequency of the cantilever is determined by using a spectrum analyzer. The PDS spectra must be obtained by dividing the infrared profile of each analyte by the baseline of the bare silicon cantilever taken prior to analyte vapor adsorption. The resulting spectra were then normalized for qualitative comparison (normalized bending versus wavelength). The observed peaks are expected to reasonably match the vibrational modes in the conventional infrared absorption spectra of the pure analyte, and the limit of detection of the technique can eventually reach values above 500 pg by optimization of the dimaterial cantilever. It is expected that thermal sensitivity of the cantilever can be improved by controlling the thicknesses of the cantilever and the metal layer as well as the thermal expansion coefficient and Young's moduli of constituent elements of the dimaterial beam.

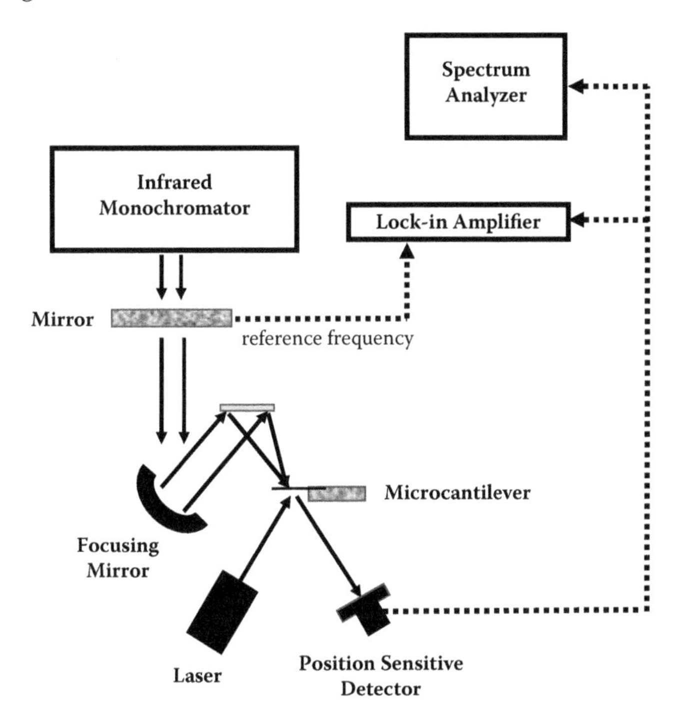

FIGURE 5.12
Brief schematics of an experimental set-up used for photothermal deflection spectroscopy.

The use of quartz tuning fork resonators (QTF) for accurate measurement of mass significantly improves the conventional gravimetric determinations, namely those making use of quartz-crystal microbalance (QCM). The unperturbed resonance frequency for QTF can be written as

$$f_0 = \frac{1}{2\pi}\sqrt{\frac{K}{m_{ef}}} = \frac{1}{2\pi}\sqrt{\frac{3YI}{m_{ef}L^3}} \tag{5.105}$$

Here I represent the moment of inertia for rectangular cross-section, Y the Young's modulus of quartz, and the effective mass m_{ef} for the cantilever beam is 0.2429 m, where $m = \rho\omega tL$ (L, ω, t are the length, width, and thickness, ρ stands for quartz density). The applied external perturbations will affect the resonant characteristics, and one can assume that the external mass loading is quite smaller than the tuning fork mass and the absorbed mass is distributed uniformly over all surfaces. Therefore, differentiating and keeping L constant, the frequency shift can be expressed as

$$\frac{\Delta f}{f_0} = \frac{1}{2}\frac{\Delta(YI)}{YI} - \frac{1}{2}\frac{\Delta m_{ef}}{m_{ef}} \tag{5.106}$$

The changes in (YI) on frequency shift can be ignored for most systems of interest, and so it is enough to consider the frequency shift due to mass loading effect. Providing that a small effective mass is added on the tuning fork arm, the resonance frequency will change into

$$f_0' = \frac{1}{2\pi}\sqrt{\frac{K}{m_{ef} + \Delta m_{ef}}} = f_0\left(1 - \frac{1}{2}\frac{\Delta m_{ef}}{m_{ef}}\right) = f_0\left(1 - \frac{1}{2}\frac{\Delta m}{m}\right) \tag{5.107}$$

Therefore, the frequency shift due to the mass loading can be expressed as

$$\Delta f = f_0' - f_0 = -\frac{f_0}{2m_{ef}}\Delta m_{ef} = -\frac{f_0}{2m}\Delta m \tag{5.108}$$

The mass loading sensitivity of the tuning fork is defined as

$$S = \frac{\Delta f}{\Delta m} = -\frac{f_0}{2m} = -\frac{1}{\pi}\frac{1}{\omega L^3}\sqrt{\frac{Y}{\rho^3}} \tag{5.109}$$

This means that the sensitivity of the tuning fork is independent of the thickness of the arm and inversely related to the beam length L and width ω. For a rigid thin solid film adsorption, the film can oscillate with the same phase of tuning fork, and no power was dissipated in this case. So, the frequency shift due to the mass loading can be

$$\Delta f_0 \approx -\frac{f_0}{2m}\Delta m \tag{5.110}$$

where m and Δm are the actual mass of tuning fork beam and the actual added mass, since one ignores the influence due to some other parameters (temperature, stress, etc.). For a certain damp coefficient b,

$$f = f_0\sqrt{1 - \frac{1}{2Q^2}} \tag{5.111}$$

where Q stands for the quality factor.

From the known equation of motion for a damped system forced to oscillate through a driving force $F_0 \cos 2\pi f_{drive}t$, the largest amplitude is achieved at resonance, that is, when the drive frequency equals the damped natural frequency f_d for the damp coefficient b.

$$f_{\text{drive}} = f_d = \frac{1}{2\pi}\sqrt{\frac{[K - (b^2/4m_{ef})]}{m_{ef}}} \tag{5.112}$$

with $m_{ef} = 0.24$ m. Still, the peak height and location of this resonance curve can vary, depending on the amount of damping. Thus, the quality factor Q is

$$Q = \frac{\sqrt{Km_{ef}}}{b} \approx \frac{f_d}{\Delta f} \approx \frac{f_d}{FWHM} \tag{5.113}$$

where Δf represents the bandwidth or full width at half maximum. The quality factor mirrors the sharpness of the resonance peak and relates oscillation frequency to energy dissipation for a given system, that is, how damped the system is. Nano- and micrometer-scale beams have quality factors of about 100 when oscillating in air, and in vacuum up to 1,000 times higher.

The oscillations can be detected by reflecting a laser off the oscillator's surface into a photosensitive detector that counts each time the beam moves up and down, or integrated electrical components can also be used to track oscillation frequency. For example, in a typical AFM contact mode feedback loop, this keeps the cantilever's deflection (and hence the tip-sample force) constant by adjusting the vertical position of the cantilever, and in tapping mode the feedback loop keeps the cantilever's oscillation amplitude constant. Phase data is the result of the mismatch between the AC input and the cantilever oscillation output. Phase contrast images, resulting from differences between properties of different materials in the sample, can provide very useful information. Stationary solution for the forced damped motion equation $z(t) = A\cos(\omega t + \phi)$, accompanies external excitation with amplitude A and phase difference ϕ, which can be thus monitored:

$$A = \frac{A_d Q\omega_0^2}{\sqrt{\omega^2\omega_0^2 + Q^2(\omega_0^2 - \omega^2)^2}} \tag{5.114}$$

$$\phi = \arctan\left[\frac{\omega\omega_0}{Q(\omega_0^2 - \omega^2)}\right] \tag{5.115}$$

The nanomechanical mass sensors are prone to detect low concentrations of molecules. In fact, when they adsorb species mass is added to the system and so the natural frequency is lowered. This decrease will be even lower if the adsorbates move out to the end of a cantilever. Since the mass of a cantilever is proportional to the cube of its characteristic dimension, if it is made small enough one can achieve a very high sensitivity. Because the natural frequency is inversely proportional to the mass of the beam, the added mass of adsorbed material brings about a detectable shift of the resonance peak to lower frequencies. From this one can infer the adsorbed mass. It is even possible to further increase the sensitivity of detection by bridging a wire fixed at both ends like a string.

For nanomechanical sensor applications, the goal is to make the sensor surface functionalized with some coating, such as an adsorptive polymer or a chemical that will bind specifically to whatever molecule the sensor is designed to collect and sense. Despite, it is important to minimize the size of the sensor in order to make it more mass sensitive, there are other issues to consider. First, the smaller the sensor, the less likely the target material is to come in contact with it. Also, smaller devices at the nanoscale tend to have lower quality factors as surface effects begin to dominate damping. This dulls the resonance peak, making it difficult to accurately measure natural frequency shifts. For these reasons we can say that minimum detectable mass scales approximately with the square of the characteristic dimension of the beam, or D^2.

Also, making use of the fact that a depositing thin film usually adds stress to the surface of the substrate, the presence of a chemical can add a distributed force (stress) to the cantilever's top surface. If the cantilever is small enough, the added stress will force it to bend. Surface stress is a force distributed within the outermost layer of a solid. Note that the units of surface stress are the same as those for the spring constant. Therefore, for simplification purposes, it may help to visualize the stress as a distribution of springs among the atoms. The stress can stretch or compress the surface. Since micro-scale beams tend to have very high surface-to-volume ratios, surface effects are quite pronounced and can lead to noticeable deflections. A simple relationship between a beam's deflection, z, and the differential surface stress (the difference in stress between the top and bottom surfaces of the beam) is the so-called Stoney relationship:

$$\Delta z = \frac{CL^2(1-\nu)\Delta\sigma}{\varepsilon t^2} \tag{5.116}$$

Here, C is a geometrical constant based on the beam's shape ($C = 3$ for a rectangular beam), L is the length of the beam, t is its thickness, ν is Poisson's ratio (a measure of a material's tendency to get thinner as it elongates), and ε is the elastic modulus of the beam material. The radius of curvature, R, is also related to the change in surface stress:

$$R = \frac{\varepsilon t^2}{6(1-\nu)\Delta\sigma} \tag{5.117}$$

It is then possible to use the so-called Shuttleworth equation to relate the total stress σ in the surface with its surface free energy γ:

$$\sigma = \gamma + \frac{d\gamma}{d\xi} \tag{5.118}$$

where $d\xi = dA/A$, is the surface strain (ratio of change in the surface area). For small deflections, one assumes that essentially the same number of atoms stay on the solid's surface and only the distances separating the atoms change as atomic bonds stretch and compress like springs. The amount of bending is proportional to the number of molecules collected, giving us a means of quantifying the concentration of a particular molecule in the cantilever's environment. Since bending can also be caused by environmental factors, including temperature changes, such effects can often be accounted for using a non-functionalized reference beam and subtracting its response from that of the functionalized beam. There are numerous sources of adsorption-induced stresses, as for instance electrostatic repulsion among adsorbed molecules leading to cantilever bending. If one assumes that the deflection of the cantilever due to the adsorbing of molecules is negligible (although measurable), we can neglect the second term of the Shuttleworth equation. This leaves a direct relationship between the surface stress of the cantilever and the surface free energy. Then, if one knows how much mass of a certain type of molecule has adsorbed on the microcantilever and the molar mass of this molecule, one can estimate the energy associated with the adsorbed molecule. For example, hydrogen can be detected by its adsorption on a Pt-coated sensor because a change in surface stress causes a static bending of the sensor.

Some authors have built chemical sensors for identification of gases and vapors based on sequential position readout via a beam-deflection technique from a microfabricated array cantilever-type sensors [102]. Each of the cantilevers can be coated on one side with a different sensor material to detect specific chemical interactions.

Disturbances from vibrations and turbulent gas flow can be effectively removed in array sensors by taking difference signals with reference cantilevers. It is also possible to fabricate large arrays of suspended single-layer graphene membrane resonators using CVD growth followed by patterning and transfer, and then measuring the resonators using both optical and electrical actuation and detection techniques. The resonators can be modeled as flat membranes under tension, and that clamping the membranes on all sides improve agreement with the model and reduces the variation in frequency between identical resonators [90].

Most of the nanophysics quartz tuning fork applications are in scanning probe microscopies, where it is currently used for measuring shear forces in scanning near field optical microscopy or to measure normal forces in magnetic force microscopy. The conversion of the frequency shift signal into a force value is also important in the application of QTFs in Casimir force measurements. The resolution of a tuning fork is higher compared with a silicon cantilever. Moreover, the tuning fork is self-sensing, and the measuring only needs electronic parts and is supposed to be more robust than the optical laser reflection used in the light silicon cantilever setup. As it was already mentioned, the force measurement with QTFs is based on the frequency shift that occurs if a force is applied to the tuning fork when it is operated close to its resonance frequency. The piezoelectric effect of the quartz yields an electric signal proportional to the deformation. The tuning fork can be driven by applying an ac voltage directly to the electrodes of the tuning fork. The response to a driving frequency is well known, and the amplitude and phase response is recorded. The resonance frequency is shifted when an external force is applied. It can also be suitable to track the phase difference between the driving force and the system because at the resonance frequency the gradient of the phase is much larger than the gradient of the amplitude. The shift of the resonance frequency f_0, is related to the derivative of the force on the tuning fork:

$$\Delta f \approx \frac{f_0}{2k}\frac{\partial F}{\partial x} \tag{5.119}$$

A quasi-static regime is required to make Casimir measurements, as the position resolution should be of a few nanometers. If the driving voltage is sufficiently small the displacement is less than a few nanometers and can be as low as picometers [103]. The frequency is fixed as well as the detectable frequency shift. It is desirable for Casimir force measurements to use tuning forks with a low spring constant (3,000–1,000 N/m, in agreement with the above formula), and with a golden surface approached to the side of the tuning fork. The quality of the surfaces is controlled by atomic force microscopy ensuring that surface roughness needs to be less than 30 nm.

Still to add some more words about manifestations of the tunneling phenomenon in nanophysics, we are going to explain in what the so-called Klein tunneling effect, and afterward the working principle of the single-electron transistor (SET), for the importance this latter assumes in nanoelectronics. In graphene, the behavior of the mobile electrons can be described by equations like those of quantum electrodynamics in 2D, considering that they display a null mass. This means that we are in the domains of a quantum and relativistic theory, which could predict that an electron wave traveling at the speed of light c (corresponding to a massless electron) and hitting a potential barrier, will cross this with a probability of 100%. It is likely that the huge conductance of graphene is a confirmation of this Klein's tunneling. On its turn, regarding one SET device, it may also be assembled from graphene components. The SET is a fast device with promising features in nanotechnology. It comprises of source, gate, drain electrodes and an island which is located between source and drain but not connected to them, as shown in Figure 5.13. A graphene nanoribbon (GNR) island with the length of L can be assumed as a quantum well and the potential profile along the device easily modeled. Its operation speed depends on the island material, so a carbon-based material such as GNR can be a suitable candidate for using in SET island. The GNR band gap, which depends on its width, has a direct impact on the Coulomb blockade and SET current. The current–voltage characteristic for the SET utilizing GNR in its island can be modeled.

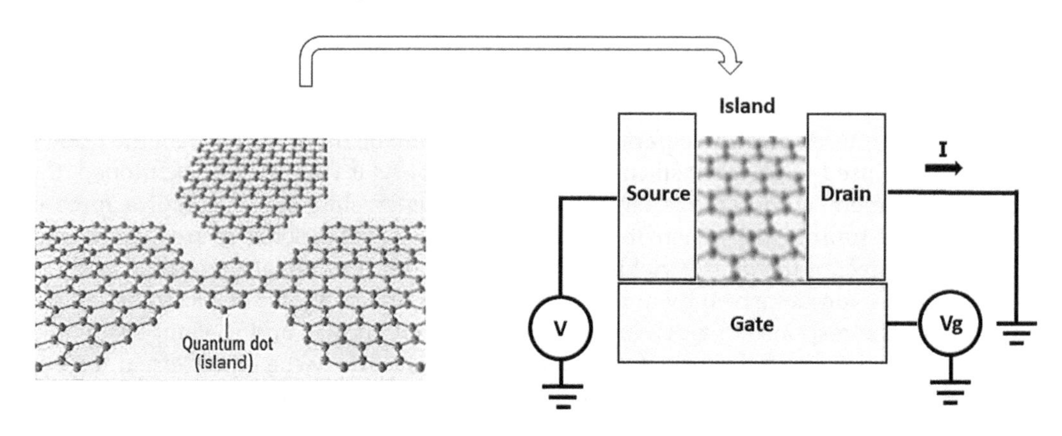

FIGURE 5.13
Schematics of a graphene single electron transistor (SET).

Schrödinger equation can be employed to derive the electron wave function and the corresponding current for this device. The wave functions for the different three different regions can be easily found, and so, the transmission coefficient of electrons in GNR SET can be calculated.

In a typical SET, when a positive voltage is applied to the gate, the island energy levels drop. The electron tunnels toward the island occupying a previously empty level. From there it can tunnel to the drain where it is inelastically dispersed and goes to the Fermi level of the drain. It will pass only one electron at a time for that voltage value. With the QD empty, the process repeats itself. The energy levels of the island are spaced by ΔE, which gives to it a self-capacitance C defined by $C = e^2/\Delta E$, where ΔE is the charging energy that varies inversely with QD linear dimension. To achieve the Coulomb blockade, the following three requirements must be fulfilled: (1) the polarization $V_{\text{bias}} < e/C$; (2) the sum of the thermal energies of the source and the island $k_B T < e^2/C$, otherwise the electron would pass to the QD via thermal excitation; (3) the tunneling resistance, $R_t > h/e^2$, so that according to the Heisenberg uncertainty principle, the uncertainty in ΔE is smaller than the ΔE itself. The unique transport property of the single-electron transistor is the basis of operation for several devices, such as electrometers and electron turnstiles, which require extreme charge sensitivity. Even more promising in terms of new physics and new applications are artificial molecules, consisting of two or more closely spaced QDs, and artificial solids, made of large arrays of linked QDs. The exciting feature here is that the researchers can control the strengths of interactions between neighboring artificial atoms by varying at will the voltages on electrodes, and thereby learn to engineer desired properties into artificial matter.

In the domain of nanoelectronics, graphene and carbon nanotubes are also very suitable to be incorporated in several nanodevices (sensors, transistors, diodes). By controlling the CNT production/separation processes they can display either the electrical conductivity of metals or act as semiconductors. CNT current carrying densities can reach 10^9 A/cm^2 (copper is limited to 10^6 A/cm^2). Besides acting to interconnect very small devices, the CNTs can be used to fabricate field-effect transistor (FET), single-electron transistor (SET), among others. The CNT-FET uses one nanotube that is laid across two gold contacts and plays the role of source and drain. The nanotube serves as a current carrying channel for the FET, and the characteristic I_{SD}-V_{gate} experimental curves of the device (for each source-drain gate voltage) is carried out just as with any FET. The conductance-voltage curve (G-V_{gate}) displays a typical quarter of quasi-circumference shape. Usually, the dedicated I-V characterization devices require low level current measurements in the nA-fA range, and in many cases the measurements are made at low temperatures.

Let us now make a few considerations about thermodynamics and transport properties at the nanoscale. As well known, the requirement of consistency between thermodynamics and electromagnetism led to the conclusion that light is quantized in photon entities. Currently quantum thermodynamics addresses the emergence of thermodynamic laws from quantum mechanics. It differs from quantum statistical mechanics in the emphasis on dynamical processes out of equilibrium. The new quantum technologies would require knowledge of quantum thermodynamics to describe the operation of quantum machines, and new instrumentation to verify, validate, and control the operation of the machines. Quantum thermodynamics acknowledges the insufficiency of conventional equilibrium-physics for quantum technologies. Statistical fluctuations in single quantum systems satisfy new laws known as quantum fluctuation theorems, and unlike classical thermodynamics, quantum properties cannot be measured (and work determined) without any perturbation to the dynamics. There is an intimate connection of quantum thermodynamics with the theory of open quantum systems. An open quantum system is a quantum-mechanical system that interacts with an external quantum system, which is known as the environment or a bath. In general, these interactions significantly change the dynamics of the system and result in quantum dissipation, such that the information contained in the system is lost to its environment. Because no quantum system is completely isolated from its surroundings, it is important to develop a theoretical framework for treating these interactions, in order to obtain an accurate understanding of quantum systems. A complete description of a quantum system requires the inclusion of the environment. Completely describing the resulting combined system then requires the inclusion of its environment, which results in a new system that can only be completely described if its environment is included, and so on. The eventual outcome of this process of embedding, is the state of the whole universe described by a wave function. The fact that every quantum system has some degree of openness also means that no quantum state can ever be in a pure state. A pure state is unitary equivalent to a zero temperature ground state forbidden by the third law of thermodynamics. Even if the combined system is a pure state and can be described by a wave function, a subsystem in general cannot be described by a wave function. This observation motivated the formalism of density matrices, or density operators, introduced last century by von Neumann, Landau, and Bloch. There is no way to know if the combined system is pure from the knowledge of the observables of the subsystem. In particular if the combined system has quantum entanglement, the system state is not a pure state. For a combined system bath scenario, the global Hamiltonian can be decomposed into the system Hamiltonian, the bath Hamiltonian, and the system-bath interaction. The state of the system is obtained from a partial trace over the combined system and bath. Another common assumption that is used to make systems easier to solve is the assumption that the state of the system at the next moment depends only on the current state of the system. In other words, the system doesn't have a memory of its previous states. Systems that have this property are known as Markovian systems. This approximation is justified when the system has enough time for relax to equilibrium before being perturbed again by interactions with its environment. For systems that have very fast or very frequent perturbations from their coupling to their environment, this approximation becomes much less accurate. The second law of thermodynamics typically applies to systems composed of many particles interacting; quantum thermodynamics resource theory is a formulation of thermodynamics in the regime where it can be applied to a small number of particles interacting with a heat bath. For processes which are cyclic or very close to cyclic, the second law for microscopic systems takes on a very different form than it does at the macroscopic scale, imposing not just one constraint on what state transformations are possible, but an entire family of constraints. Nanoscale allows for the preparation of quantum systems in physical states without classical analogue. There, complex out-of-equilibrium scenarios may be produced by the initial preparation of either the working substance or the reservoirs of quantum particles, the latter dubbed as "engineered reservoirs." There are different forms of engineered reservoirs. Some of them involve subtle quantum coherence or correlation effects, while others rely solely on non-thermal classical probability distribution functions. The latter are dubbed nonequilibrium incoherent reservoirs. Interesting phenomena may emerge from the use of engineered reservoirs, such as maximum efficiency of nanomachines, violations of Clausius inequalities, or simultaneous extraction of heat and work from the reservoirs.

Now it would be also of interest to turn into the fluid's domain and search an explanation for the experimental fact that there is a lower bound for viscosity's transport property [104]. Viscosity η denotes a liquid's resistance to a shear force and governs important properties such as diffusion and dissipation. In a dilute gas-like fluid, η is set by molecules moving at distances up to the mean free path ξ, and transferring momentum during collisions; specifically, $\eta = \rho v \xi/3$, where ρ and v are the density and average velocity of molecules, respectively. Therefore, the viscosity of a gas must increase with temperature because molecular velocity increases with temperature. That prediction is counterintuitive because fluids usually thin when they are heated. Unlike gases, dense liquids have a viscosity set by their molecules vibrating around quasi-equilibrium positions before jumping to neighboring sites. The frequency of those jumps increases with temperature, and viscosity consequently decreases with temperature, following an activated Arrhenius trend. The increase of viscosity at high temperature and its increase at low temperature imply that it has a minimum. That minimum arises from the crossover between two different viscosity regimes: a gas-like regime, where the kinetic energy of higher-temperature particles provides a larger momentum transfer, and hence larger η, and a liquid-like regime, where lower temperature decreases the frequency of site-jumping particles and slows down the liquid flow, also resulting in larger η.

Thus, it is possible to realize that it might exist a quantum quantity setting, the minimal kinematic viscosity of a fluid. Since we are not dealing with superfluidity, viscosities stop decreasing, when they hit their minima. How close to zero can be this minimum is the so-called Purcell question. It could be possible to answer this question if viscosity can be calculated at its minimum. Since molecular interactions are strong and system specific, it would be a complex task, but fortunately, the minimum of viscosity at the crossover is a special point where viscosity can, in fact, be evaluated, if only approximately [105]. It has been found that the minimal kinematic viscosity of fluids v_m is related to just two basic properties: the interatomic separation and the system's Debye frequency. These two parameters can, in turn, be related to the radius of the hydrogen atom and a characteristic bonding strength set by the Rydberg energy. Then v_m becomes

$$v_m = \frac{1}{4\pi}\frac{\hbar}{\sqrt{m_e m}} \tag{5.120}$$

where m_e is the mass of an electron and m_m is the mass of the molecule. The minimal viscosity turns out to be quantum! This equation draws attention to the nature of interactions in condensed matter which is quantum mechanical, with $\hbar$ affecting both the Bohr radius and Rydberg energy. Therefore, the answer to the Purcell question is that viscosities stop decreasing because they present minima, and those minima are fixed by fundamental constants. If m is replaced by the proton mass, the result is 10^{-7} m^2/s. It is possible to show that the same happens to another property, the thermal diffusivity, which governs how well liquids transfer heat. It also exhibits minima because thermal diffusivity depends on the same parameters. Planck's constant sets the minimum below which viscosity cannot fall, regardless of temperature. Were Planck's constant to take on a different value, for instance, water viscosity would change too, which is relevant to water flow, and sets internal friction and diffusion. If the viscosity minimum were to increase because of a higher value of $\hbar$, for instance, water would become more viscous, and life might not exist in its current form or even at all. Water and life are indeed well attuned to the degree of quantumness of the physical world.

Mesoscopic phenomena in transport processes occur when the wavelength or the coherence length of the carriers becomes comparable to, or larger than, the sample dimensions. The quantization of electrical conduction, observed in a quasi-one-dimensional constriction formed between reservoirs of two-dimensional electron gas [106] is determined by the number of participating quantum states or channels within the constriction. It can be demonstrated that in the ideal case, each spin-degenerate channel contributes a quantized unit of $2e^2/h$ to the electrical conductance. In fact, in a 1D electric conducting wire connecting two reservoirs at different potentials, adiabatically, the DOS $(dn/d\varepsilon)$ is equal to $2/(hv)$, where the factor 2 comes from electron spin degeneracy and v represents the electron velocity. Then, the voltage is $V = -(\mu_1 - \mu_2)/e$, where e stands for the electron charge and the μs are the respective chemical potentials. Therefore, the current density across the wire is given by

$$j = -ev(\mu_1 - \mu_2)\frac{dn}{d\varepsilon} \tag{5.121}$$

Given so, the resulting electric conductance is $G_0 = j/V = 2e^2/h$. This quantum conductance does not mean that the conductance of any system must be an integer multiple of G_0. Instead, it describes the conductance of two quantum channels (one channel for spin up and one channel for spin down) if the probability for transmitting an electron that enters the channel is unity, that is, if transport through the channel is ballistic. If the transmission probability is less than unity, then the conductance of the channel is less than G_0. The total conductance of a system is equal to the sum of the conductances of all the parallel quantum channels that make up the system. The STM has been proven a suitable tool to measure quantized metallic conduction [107]. As the tip in a STM is moved toward the surface, the attractive force between tip and sample increases rapidly, all the way up to the point where the attractive force gradient surpasses the spring constant of the combined system of tip/sample. Then a sudden jump to contact takes place, and this can lead to a metallic contact, which can be stretched to a connective neck by a subsequent retraction of the tip. The relation between the adhesive tip-sample force and the conductance can then be investigated, and under clean and well-controlled conditions, the conductance in a constriction drawn between an STM tip and a metal surface is indeed quantized in units of $2e^2/h$.

The electrical conductivity of a molecular wire is closely related to the nature of its surroundings. In metals, the Fermi electron wavelength (corresponding to the Fermi velocity) has dimensions like the molecular ones. In molecules, molecular orbitals constitute available conduction paths, that is, each of them is a charge transport channel.

Three limiting cases can be distinguished:

1. Coherent resonant tunneling of the carriers of charge. It is a particular case of ballistic conduction (which occurs when the length of the molecule is less than the mean electron free path). It can occur when the Fermi energies of the conduction bands are the same at the two electrodes and isoenergetic with a molecular orbital of the molecule, in particular with the LUMO. The electrical resistance must then be independent of the wire length, as it occurs only at the contacts. The conductance of a channel can be calculated by the Landauer formula, from the dispersion of the electron in the channel

$$G = G_0 \sum_{i=1}^{N} T_i \tag{5.122}$$

   where G_0 is the quantum conductance (=12 900 Ω^{-1}), N the number of conduction channels and T_i the transmission coefficient for the ith channel in the wire ($0 < T < 1$), which constitutes a measure of the efficiency with which the molecule embraces the electronic wave functions of the metallic electrodes ($T = 0$ corresponds to a situation of complete retro-dispersal). The total conductance is expressed as the sum of all individual conductivities (Eq. 5.123), where the number of active channels is determined by the number of molecular orbitals that at a given applied electrical voltage are in resonance with the Fermi level of the metal. The change in applied electric potential difference causes displacements of the LUMO level of the molecule relative to the Fermi level of the metallic electrodes, and therefore, changes in the transmission probability.

2. If the Fermi levels of the electrodes are both at a different energy than the LUMO and HOMO orbitals and this LUMO-HOMO gap covers the difference between

the Fermi levels of the electrodes, electronic flux will not occur; then only non-resonant coherent movement of the electrons can occur, by electronic exchange through the orbitals in the molecule. Changing the electric potential difference applied to electrodes, the probability of transmission can be varied by activating different molecular orbitals as conductive channels. In this case, the conductance must decrease exponentially with the length of the molecular "bridge."

3. The transport of charges through the molecular "bridge" can also become incoherent and dissipative, as they undergo inelastic dispersion, caused by molecular vibrations coupled to the electronic states of the wire. This case presents similarities to the conduction that occurs in a common electrical wire, in which the conductivity obeys Ohm's law and is inversely proportional to the length of the molecular wire. This dissipative mechanism will be more favored the longer the molecule and the higher the temperature.

It is important to mention that in any of the three cases, what is measured is a combination of the properties of the molecule and the electrodes, using the concept of "electronic transparency." The incoherent leap from atom to atom within the molecule (dissipative conductivity by multi-step hopping), and coherent single-step tunneling through molecular orbitals are two limiting cases of the conduction mechanism in a molecular wire. For the second case, the electron velocity in the electronic flux through the molecule from a donor D to an acceptor A depends on distance, whereas in the first case of hoping transfer along the molecule is less dependent on distance.

To obtain electrical conductivity measurements of individual molecules, it is necessary to establish two electrical contacts at the ends of the molecule. Using 1,4-dithiol benzene molecules, these tend to form a chemical bond with the surface of a manipulated gold nanowire electrode. For this, a self-assembled monolayer of these molecules is produced on gold, which is covalently bonded through the sulfur atom, releasing the hydrogen atom. Thus, gold functions as one of the electrodes, the other being secured by the tip of an STM after breaking the gold wire immersed in the 1,4-dithiol benzene solution. Molecular resistance can be obtained by measurements I-V, and this plot is not linear as in a true ohmic contact, because now the quantization of the conductivity plays an important role. However, in electrical conductivity measurements of C_{60}, between a flat gold electrode and the tungsten tip of an STM, the I-V ratio obtained is usually linear, which allows obtaining a resistance of the order of 5.5 MΩ [108]. The STM probe also makes it possible to measure the I-V curves as a function of the force exerted by the tip, if it is mounted on an AFM microbeam with a metallic coating. Also, macromolecules, such as a single strand of DNA with 600 nm in length and carotenes, have been probed in this type of experiments, using a method in which STM measurements are carried out with the molecule embedded in a monomolecular layer of an alkanethiol in gold. The resistance value obtained for DNA is 10^6 times lower than the alkanethiol chains with the same length.

The first single-molecule electromechanical amplifier used the C_{60} molecule sandwiched between the probe tip of an STM and a flat conductive surface. It is based on the modification of the "tails" of the LUMO and HOMO wave functions close to the Fermi level by molecular mechanical deformation, caused by the pressure of the STM probe [109]. On the other hand, an array of crossed nanowires can serve as a memory device. If molecules are sandwiched between them, and if each molecule has a resistance that varies with the applied electrical voltage, then a specific electrical voltage is used to decrease the resistance of the molecule, so that it represents bit 1 and another electrical voltage specific is used to increase electrical resistance so that the molecule represents bit 0. Carbon nanotubes exhibit characteristics of molecular wires with fascinating electronic properties, and with ballistic conduction, constituting good candidates for molecular electronic devices. In fact, armchair configuration SWNTs are long metallic molecular wires, which, because they have a rigid structure, do not show Peierls transitions even at low temperatures [110]. These one-dimensional conductors present I-V curves with discrete steps, as a result of the electronic states that contribute to the conduction being quantified and that number increases with the increase of the applied voltage. This resonant tunneling through discrete electronic states allows the study of these states.

A similar quantum behavior is expected to be observable for thermal transport in mesoscopic phonon systems, and indeed experiments attempted in this regime have so far report the observation of a quantized limiting value for the thermal conductance, G_{th}, in suspended insulating nanostructures at very low temperatures. The behavior observed is consistent with Landauer predictions in a ballistic 1D channel phonon transport at low temperatures, for the quantum of thermal conductance [111]:

$$G_{th} = \frac{\pi^2 k_B^2 T}{3h} \tag{5.123}$$

an expression devoid of any material parameters, as it happened with the quantum electrical conductance.

The spin of the electrons can also be exploit in nanoelectronics as a further degree of freedom with implications in the efficiency of data storage and transfer. Spin can flow along with electric charge. In a solid, the spins of many electrons can act together to affect the magnetic and electronic properties of a material, for example endowing it with a permanent magnetic moment as in a ferromagnet. In many cases, electron spins are equally present in both the up and the down state, and no transport properties are dependent on spin. A spintronic device requires generation or manipulation of a spin-polarized population of electrons, resulting in an excess of spin up or spin down electrons. Since the polarization of any spin dependent property X is given by $(X\uparrow - X\downarrow)$ / $(X\uparrow - X\downarrow)$, a net spin polarization can be achieved either through creating an equilibrium energy split between spin up and spin down. Recall now that magnetism is a cooperative behavior among atoms which derives from a purely quantum-mechanical process effective only over distances of a few atomic spacings. That is why magnetic materials structures are much harder to design and control than semiconductors; they must be manipulated at the length scale of a nanometer or less to have any impact on their behavior, whereas semiconductors can exhibit novel properties already at carrier lengths of tens of nanometers. Due to the characteristics of the density of electronic states (DOS) in a ferromagnetic metal, this may be used as a source of spin-polarized carriers injected into a normal metal, a semiconductor, or a superconductor, or made to tunnel through a non-magnetic insulating barrier. Despite the description of magnetic materials is complex in the absence of any applied field, the situation becomes simpler at the nanometric scale, because the magnetization prefers to align in a way that tends to occupy the whole volume as a single domain. Therefore, it can remain highly stable against the influences of external fields, and the existence of two stable magnetization states makes such structures suitable for data storage applications.

Ending up these explanations about the cross-fertilizing binomial quantum physics/ nanophysics, it is worth of note to mention that the new findings about radiation transfer at the nanoscale. Recently it has been checked the validity of Planck's radiation law at the smallest length scales, and some experimental results show that it does not apply for objects smaller than a certain length scale, and the result is about 100 times higher than what such law would predicts [112]. Planck's equation reflects the relationship between the temperature of an object and the energy emitted from that object in the form of electromagnetic radiation. Planck based his theory on the hypothesis that a photon's energy depends on its frequency, meaning the energy of electromagnetic waves is also quantized. Despite the success of the theory in explaining many different radiation phenomena, Planck himself already knew that it is likely to be completed or replaced by a more general theory when dealing with very small objects. In fact, it has been found that Planck's law does not hold up when the objects were in the near field [113], neither in the far field (farther apart than a wavelength of radiation) with objects that were smaller than a wavelength in thickness [112]. The wavelengths of infrared light that are relevant for testing the law were only about 10 microns, and so it was required to create objects even smaller, like membranes only a few hundred nanometers thick. For instance, placing apart two identical membranes, heating one of them, and measuring the heat increase in the second, it is possible to check the heat increase in this one. In fact, the researchers found that the rate of radiative heat transfer was 100 times higher than what Planck's law would have predicted. It seems that if the material and geometry of an object are such that electromagnetic waves can couple more effectively to it, then it will emit and absorb radiation more effectively.

If a radiating object is smaller than the thermal wavelength, it behaves according to different rules and cannot radiate heat efficiently. This has been confirmed by studying the heat radiation of ultra-thin optical fibers [114]. When an object is smaller than the typical length over which the radiation is absorbed, the body cannot fully absorb the incoming radiation, and so part of it can pass through. As a result, the thermal radiation of the body is altered, and Planck's law is not verified. The authors sent light through ultra-thin optical fibers with a diameter of only 500 nanometers, and they measured the amount of optical energy which was converted into heat and subsequently radiated away into the environment. Their findings agree with a more general theory of fluctuational electrodynamics, which allows one to take the geometry and the size of the body into account [114]. The results are in excellent agreement with a theoretical model that considers heat radiation as a volumetric effect and takes the emitter shape and size relative to the emission wavelength into account.

References

1. A. Lenard and F. J. Dyson, "Stability of matter II," *Journal of Mathematical Physics*, vol. 9, no. 5, pp. 698–711, May 1968. doi:10.1063/1.1664631
2. H. Batelaan and A. Tonomura, "The Aharonov–Bohm effects: Variations on a subtle theme," *Physics Today*, vol. 62, no. 9, pp. 38–43, Sep. 2009. doi:10.1063/1.3226854
3. K. A. Kouzakov, "Quantum entanglement in the nonrelativistic collision between two identical fermions with spin 1/2," *Theoretical and Mathematical Physics*, vol. 201, no. 2, pp. 1664–1679, Nov. 2019. doi:10.1134/S0040577919110102

4. K. Hornberger, S. Uttenthaler, B. Brezger, L. Hackermüller, M. Arndt, and A. Zeilinger, "Collisional decoherence observed in matter wave interferometry," *Physical Review Letters*, vol. 90, no. 16, p. 160401, Apr. 2003. doi:10.1103/PhysRevLett.90.160401
5. C. Joachain, *Quantum Collision Theory*. New York: North-Holland, 1975.
6. R. Levine, *Molecular Reaction Dynamics and Chemical Reactivity*. Oxford: Oxford University Press, 1987.
7. M. S. Child, *Molecular Collision Theory*. New York: Academic Press, 1974.
8. G. Scoles, "Molecular scattering: General principles and methods," in *Atomic and Molecular Beam Methods*, vol. 1, New York: Oxford University Press, 1988.
9. H. Smith, *Introduction to Quantum Mechanics*. World Scientific, 1991.
10. D. R. Herschbach, "Molecular dynamics of elementary chemical reactions (Nobel Lecture)," *Angewandte Chemie International Edition in English*, vol. 26, no. 12, pp. 1221–1243, Dec. 1987. doi:10.1002/anie.198712211
11. R. B. Bernstein, Ed., *Atom - Molecule Collision Theory*. Boston, MA: Springer US, 1979. doi: 10.1007/978-1-4613-2913-8
12. P. R. Brooks, "Reactions of oriented molecules," *Science (1979)*, vol. 193, no. 4247, pp. 11–16, Jul. 1976. doi:10.1126/science.193.4247.11
13. R. F. M. Lobo, A. M. C. Moutinho, K. Lacmann, and J. Los, "Excitation of the nitro group in nitromethane by electron transfer," *The Journal of Chemical Physics*, vol. 95, no. 1, pp. 166–175, Jul. 1991. doi:10.1063/1.461472
14. R. F. M. Lobo, P. L. Vieira, S. S. M. C. Godinho, and M. J. Calhorda, "Mono-halobenzenes anion fragmentation induced by atom–molecule electron-transfer collisions," *The Journal of Chemical Physics*, vol. 116, no. 22, pp. 9712–9720, Jun. 2002. doi:10.1063/1.1476697
15. R. F. M. Lobo and N. T. Silva, "Neutral C60 effusive source for atomic collisions with fullerene," *Review of Scientific Instruments*, vol. 72, no. 9, pp. 3505–3506, Sep. 2001. doi:10.1063/1.1396662
16. E. A. Gislason, A. W. Kleyn, and J. Los, "Time-dependent theory of electronic-to-vibrational energy transfer. Application to metastable argon atoms with nitrogen molecules," *Chemical Physics Letters*, vol. 67, no. 2–3, pp. 252–257, Nov. 1979. doi:10.1016/0009-2614(79)85157-X
17. P. R. Brooks, "Orientation effects in electron transfer collisions," *International Reviews in Physical Chemistry*, vol. 14, no. 2, pp. 327–354, Sep. 1995. doi:10.1080/01442359509353313
18. B. Jia, J. Laib, R. F. M. Lobo, and P. R. Brooks, "Evidence for orbital-specific electron transfer to oriented haloform molecules," *Journal of the American Chemical Society*, vol. 124, no. 46, pp. 13896–13902, Nov. 2002. doi:10.1021/ja027710k
19. P. R. Brooks, P. W. Harland, and C. E. Redden, "Electron transfer from sodium to oriented nitromethane, CH_3 NO_2: Probing the spatial extent of unoccupied orbitals," *Journal of the American Chemical Society*, vol. 128, no. 14, pp. 4773–4778, Apr. 2006. doi:10.1021/ja058206t
20. E. R. Bernstein, Ed., *Chemical Reactions in Clusters*. Oxford: Oxford University Press, 1996. doi:10.1093/oso/9780195090048.001.0001
21. A. González Ureña, K. Gasmi, J. Jiménez, and R. F. Lobo, "Excitation laser fluence effects on the Ba ... FCH3+hv→BaF+CH3 intracluster reaction," *Chemical Physics Letters*, vol. 352, no. 5–6, pp. 369–374, Feb. 2002. doi:10.1016/S0009-2614(01)01481-6
22. R. F. M. Lobo, B. N. Vicente, F. V. Berardo, I. V. Gouveia, J. H. Ribeiro, and P. Pereira, "Detection of negative fullerene conformers," *Journal of Experimental Nanoscience*, vol. 1, no. 3, pp. 317–332, Sep. 2006. doi:10.1080/17458080600967247
23. R. E. Grisenti, W. Schöllkopf, J. P. Toennies, J. R. Manson, T. A. Savas, and H. I. Smith, "He-atom diffraction from nanostructure transmission gratings: The role of imperfections," *Physical Review A*, vol. 61, no. 3, p. 033608, Feb. 2000. doi:10.1103/PhysRevA.61.033608
24. A. D. Cronin, J. Schmiedmayer, and D. E. Pritchard, "Optics and interferometry with atoms and molecules," *Reviews of Modern Physics*, vol. 81, no. 3, pp. 1051–1129, Jul. 2009. doi:10.1103/RevModPhys.81.1051

25. B. S. Zhao, W. Zhang, and W. Schöllkopf, "Universal diffraction of atoms and molecules from a quantum reflection grating," *Science Advances*, vol. 2, no. 3, Mar. 2016. doi:10.1126/sciadv.1500901
26. M. Arndt, O. Nairz, J. Vos-Andreae, C. Keller, G. van der Zouw, and A. Zeilinger, "Wave–particle duality of C60 molecules," *Nature*, vol. 401, no. 6754, pp. 680–682, Oct. 1999. doi:10.1038/44348
27. M. Born et al., *Principles of Optics*. Cambridge: Cambridge University Press, 1999. doi:10.1017/CBO9781139644181
28. S. Eibenberger, S. Gerlich, M. Arndt, M. Mayor, and J. Tüxen, "Matter–wave interference of particles selected from a molecular library with masses exceeding 10 000 amu," *Physical Chemistry Chemical Physics*, vol. 15, no. 35, p. 14696, 2013. doi:10.1039/c3cp51500a
29. Y. Y. Fein et al., "Quantum superposition of molecules beyond 25 kDa," *Nature Physics*, vol. 15, no. 12, pp. 1242–1245, Dec. 2019. doi:10.1038/s41567-019-0663-9
30. R. Horodecki, "Dé broglie wave and its dual wave," *Physics Letters A*, vol. 87, no. 3, pp. 95–97, Dec. 1981. doi:10.1016/0375-9601(81)90571-5
31. O. Nairz, B. Brezger, M. Arndt, and A. Zeilinger, "Diffraction of complex molecules by structures made of light," *Physical Review Letters*, vol. 87, no. 16, p. 160401, Sep. 2001. doi:10.1103/PhysRevLett.87.160401
32. F. Tebbenjohanns, M. L. Mattana, M. Rossi, M. Frimmer, and L. Novotny, "Quantum control of a nanoparticle optically levitated in cryogenic free space," *Nature*, vol. 595, no. 7867, pp. 378–382, Jul. 2021. doi:10.1038/s41586-021-03617-w
33. P. Alberto, C. Fiolhais, and V. M. S. Gil, "Relativistic particle in a box," *European Journal of Physics*, vol. 17, no. 1, pp. 19–24, Jan. 1996. doi:10.1088/0143-0807/17/1/004
34. D. Bohm, "A suggested interpretation of the quantum theory in terms of 'hidden' variables. I," *Physical Review*, vol. 85, no. 2, pp. 166–179, Jan. 1952. doi:10.1103/PhysRev.85.166
35. C. P. Husband, S. M. Husband, J. S. Daniels, and J. M. Tour, "Logic and memory with nanocell circuits," *IEEE Transactions on Electron Devices*, vol. 50, no. 9, pp. 1865–1875, Sep. 2003. doi:10.1109/TED.2003.815860
36. S. Datta, *Electronic Transport in Mesoscopic Systems*. Cambridge, UK: Cambridge University Press, 1995.
37. A. D. Lantada, P. Lafont, J. L. M. Sanz, J. M. Munoz-Guijosa, and J. E. Otero, "Quantum tunnelling composites: Characterisation and modelling to promote their applications as sensors," *Sensors and Actuators A: Physical*, vol. 164, no. 1–2, pp. 46–57, Nov. 2010. doi:10.1016/j.sna.2010.09.002
38. M. L. Ciurea and V. Iancu, "Quantum confinement in nanometric structures," in *New Trends in Nanotechnology and Fractional Calculus Applications*, Dordrecht: Springer Netherlands, 2010, pp. 57–67. doi:10.1007/978-90-481-3293-5_5
39. M. L. Landry, T. E. Morrell, T. K. Karagounis, C.-H. Hsia, and C.-Y. Wang, "Simple syntheses of CdSe quantum dots," *Journal of Chemical Education*, vol. 91, no. 2, pp. 274–279, Feb. 2014. doi:10.1021/ed300568e
40. M. P. Vinod and K. Vijayamohanan, "Silicon based light emitting gels," *Applied Physics Letters*, vol. 68, no. 1, pp. 81–83, Jan. 1996. doi:10.1063/1.116765
41. C. Delerue, G. Allan, and M. Lannoo, "Theoretical aspects of the luminescence of porous silicon," *Physical Review B*, vol. 48, no. 15, pp. 11024–11036, Oct. 1993. doi:10.1103/PhysRevB.48.11024
42. G. Ledoux, J. Gong, F. Huisken, O. Guillois, and C. Reynaud, "Photoluminescence of size-separated silicon nanocrystals: Confirmation of quantum confinement," *Applied Physics Letters*, vol. 80, no. 25, pp. 4834–4836, Jun. 2002. doi:10.1063/1.1485302
43. G. Ledoux, J. Gong, and F. Huisken, "Effect of passivation and aging on the photoluminescence of silicon nanocrystals," *Applied Physics Letters*, vol. 79, no. 24, pp. 4028–4030, Dec. 2001. doi:10.1063/1.1426273

44. G. Ledoux et al., "Photoluminescence properties of silicon nanocrystals as a function of their size," *Physical Review B*, vol. 62, no. 23, pp. 15942–15951, Dec. 2000. doi:10.1103/PhysRevB.62.15942
45. G. Ledoux, R. Lobo, F. Huisken, O. Guillois, and C. Reynaud, "Photoluminescence properties of silicon nanocrystals synthesised by laser pyrolysis," in *Trends in Nanotechnology Research*, E. V. Dirote, Ed. New York: Nova Science Publishers, 2004.
46. O. Guillois, N. Herlin-Boime, C. Reynaud, G. Ledoux, and F. Huisken, "Photoluminescence decay dynamics of noninteracting silicon nanocrystals," *Journal of Applied Physics*, vol. 95, no. 7, pp. 3677–3682, Apr. 2004. doi:10.1063/1.1652245
47. J. Ankerhold, *Quantum Tunneling in Complex Systems. The Semiclassical Approach*. Berlin: Springer, 2007.
48. J. Chen, *Introduction to Scanning Tunneling Microscopy*, 2nd ed. New York: Oxford University Press, 2007.
49. C. J. Chen, "Origin of atomic resolution on metal surfaces in scanning tunneling microscopy," *Physical Review Letters*, vol. 65, no. 4, pp. 448–451, Jul. 1990. doi:10.1103/PhysRevLett.65.448
50. E. Meyer, R. Bennewitz, and H. J. Hug, *Scanning Probe Microscopy*. Cham: Springer International Publishing, 2021. doi:10.1007/978-3-030-37089-3
51. B. Voigtländer, *Scanning Probe Microscopy*. Berlin, Heidelberg: Springer Berlin Heidelberg, 2015. doi:10.1007/978-3-662-45240-0
52. B. Bhushan, Ed., *Scanning Probe Microscopy in Nanoscience and Nanotechnology*. Berlin, Heidelberg: Springer Berlin Heidelberg, 2010. doi:10.1007/978-3-642-03535-7
53. B. Bhushan, Ed., *Springer Handbook of Nanotechnology*. Berlin, Heidelberg: Springer Berlin Heidelberg, 2010. doi:10.1007/978-3-642-02525-9
54. M. F. Crommie, C. P. Lutz, and D. M. Eigler, "Imaging standing waves in a two-dimensional electron gas," *Nature*, vol. 363, no. 6429, pp. 524–527, Jun. 1993. doi:10.1038/363524a0
55. D. Vieira, H. J. P. Freire, V. L. Campo, and K. Capelle, "Friedel oscillations in one-dimensional metals: From Luttinger's theorem to the Luttinger liquid," *Journal of Magnetism and Magnetic Materials*, vol. 320, no. 14, pp. e418–e420, Jul. 2008. doi:10.1016/j.jmmm.2008.02.077
56. M. F. Crommie, C. P. Lutz, and D. M. Eigler, "Confinement of electrons to quantum corrals on a metal surface," *Science (1979)*, vol. 262, no. 5131, pp. 218–220, Oct. 1993. doi:10.1126/science.262.5131.218
57. S.-W. Hla and K.-H. Rieder, "STM control of chemical reactions: Single-molecule synthesis," *Annual Review of Physical Chemistry*, vol. 54, no. 1, pp. 307–330, Oct. 2003. doi:10.1146/annurev.physchem.54.011002.103852
58. S.-W. Hla, L. Bartels, G. Meyer, and K.-H. Rieder, "Inducing all steps of a chemical reaction with the scanning tunneling microscope tip: Towards single molecule engineering," *Physical Review Letters*, vol. 85, no. 13, pp. 2777–2780, Sep. 2000. doi:10.1103/PhysRevLett.85.2777
59. J. R. Hahn and W. Ho, "Oxidation of a single carbon monoxide molecule manipulated and induced with a scanning tunneling microscope," *Physical Review Letters*, vol. 87, no. 16, p. 166102, Sep. 2001. doi:10.1103/PhysRevLett.87.166102
60. Y. Sainoo, Y. Kim, T. Okawa, T. Komeda, H. Shigekawa, and M. Kawai, "Excitation of molecular vibrational modes with inelastic scanning tunneling microscopy processes: Examination through action spectra of *cis*-2-butene on Pd(110)," *Physical Review Letters*, vol. 95, no. 24, p. 246102, Dec. 2005. doi:10.1103/PhysRevLett.95.246102
61. L. Bartels et al., "Dynamics of electron-induced manipulation of individual CO molecules on Cu(111)," *Physical Review Letters*, vol. 80, no. 9, pp. 2004–2007, Mar. 1998. doi:10.1103/PhysRevLett.80.2004

62. K.-P. Huber, *Molecular Spectra and Molecular Structure: IV. Constants of Diatomic Molecules*. New York: Springer Science & Business Media, 2013.
63. R. J. le Roy, N. S. Dattani, J. A. Coxon, A. J. Ross, P. Crozet, and C. Linton, "Accurate analytic potentials for Li2(X Σ1g+) and Li2(A Σ1u+) from 2 to 90 Å, and the radiative lifetime of Li(2p)," *The Journal of Chemical Physics*, vol. 131, no. 20, p. 204309, Nov. 2009. doi:10.1063/1.3264688
64. G. C. Maitland and E. B. Smith, "The intermolecular pair potential of argon," *Molecular Physics*, vol. 22, no. 5, pp. 861–868, Jan. 1971. doi:10.1080/00268977100103181
65. J. A. Northby, "Structure and binding of Lennard-Jones clusters: 13≤ N ≤147," *The Journal of Chemical Physics*, vol. 87, no. 10, pp. 6166–6177, Nov. 1987. doi:10.1063/1.453492
66. J. D. Jackson, *Classical Electrodynamics*, 2nd ed. New York: JohnWiley & Sons, 1975.
67. X. Pi and C. Delerue, "Tight-binding calculations of the optical response of optimally P-doped Si nanocrystals: A model for localized surface plasmon resonance," *Physical Review Letters*, vol. 111, no. 17, p. 177402, Oct. 2013. doi:10.1103/PhysRevLett.111.177402
68. S. Zeng et al., "Size dependence of Au NP-enhanced surface plasmon resonance based on differential phase measurement," *Sensors and Actuators B: Chemical*, vol. 176, pp. 1128–1133, Jan. 2013. doi:10.1016/j.snb.2012.09.073
69. C. Montero, J. M. Orea, M. Soledad Muñoz, R. F. M. Lobo, and A. González Ureña, "Non-volatile analysis in fruits by laser resonant ionization spectrometry: Application to re-sveratrol (3,5,4′-trihydroxystilbene) in grapes," *Applied Physics B: Lasers and Optics*, vol. 71, no. 4, pp. 601–605, Oct. 2000. doi:10.1007/s003400000380
70. C. F. Bohren and D. R. Huffman, *Absorption and Scattering of Light by Small Particles*. New York: Wiley, 1998. doi:10.1002/9783527618156
71. E. D. Palik, *Handbook of Optical Constants*. Orlando, FL: Academic Press, 1985.
72. Q. Duan, Y. Liu, S. Chang, H. Chen, and J. Chen, "Surface plasmonic sensors: Sensing mechanism and recent applications," *Sensors*, vol. 21, no. 16, p. 5262, Aug. 2021. doi:10.3390/s21165262
73. J. Aizpurua, P. Hanarp, D. S. Sutherland, M. Käll, G. W. Bryant, and F. J. García de Abajo, "Optical properties of gold nanorings," *Physical Review Letters*, vol. 90, no. 5, p. 057401, Feb. 2003. doi:10.1103/PhysRevLett.90.057401
74. E. M. Larsson, J. Alegret, M. Käll, and D. S. Sutherland, "Sensing characteristics of NIR localized surface plasmon resonances in gold nanorings for application as ultrasensitive biosensors," *Nano Letters*, vol. 7, no. 5, pp. 1256–1263, May 2007. doi:10.1021/nl0701612
75. Z. Fang et al., "Gated tunability and hybridization of localized plasmons in nanostructured graphene," *ACS Nano*, vol. 7, no. 3, pp. 2388–2395, Mar. 2013. doi:10.1021/nn3055835
76. U. Kreibig and C. V. Fragstein, "The limitation of electron mean free path in small silver particles," *Zeitschrift für Physik*, vol. 224, no. 4, pp. 307–323, Aug. 1969. doi:10.1007/BF01393059
77. J. A. Scholl, A. L. Koh, and J. A. Dionne, "Quantum plasmon resonances of individual metallic nanoparticles," *Nature*, vol. 483, no. 7390, pp. 421–427, Mar. 2012. doi:10.1038/nature10904
78. D. Sarid, *Scanning Force Microscopy: With Applications to Electric, Magnetic, and Atomic Forces*, vol. 5. Oxford: Oxford University Press, 1994.
79. R. E. Grisenti, W. Schöllkopf, J. P. Toennies, G. C. Hegerfeldt, T. Köhler, and M. Stoll, "Determination of the bond length and binding energy of the helium dimer by diffraction from a transmission grating," *Physical Review Letters*, vol. 85, no. 11, pp. 2284–2287, Sep. 2000. doi:10.1103/PhysRevLett.85.2284
80. R. L. Jaffe, "Casimir effect and the quantum vacuum," *Physical Review D*, vol. 72, no. 2, p. 021301, Jul. 2005. doi:10.1103/PhysRevD.72.021301
81. J. Israelachvili, *Intermolecular and Surface Forces*, 3rd ed. Amsterdam, NY: Academic Press, 2011.
82. H. C. Hamaker, "The London—van der Waals attraction between spherical particles," *Physica*, vol. 4, no. 10, pp. 1058–1072, Oct. 1937. doi:10.1016/S0031-8914(37)80203-7

83. H. B. G. Casimir, "On the attraction between two perfectly conducting plates," in *Proceedings of the Koninklijke Nederlandse Akademie van Wetenschappen*, vol. 51, pp. 793–795, 1948.
84. S. K. Lamoreaux, "Demonstration of the Casimir force in the 0.6 to 6 μm range," *Physical Review Letters*, vol. 78, no. 1, pp. 5–8, Jan. 1997. doi:10.1103/PhysRevLett.78.5
85. H. B. Chan, V. A. Aksyuk, R. N. Kleiman, D. J. Bishop, and F. Capasso, "Quantum mechanical actuation of microelectromechanical systems by the Casimir force," *Science (1979)*, vol. 291, no. 5510, pp. 1941–1944, Mar. 2001. doi:10.1126/science.1057984
86. R. S. Decca, D. López, E. Fischbach, G. L. Klimchitskaya, D. E. Krause, and V. M. Mostepanenko, "Precise comparison of theory and new experiment for the Casimir force leads to stronger constraints on thermal quantum effects and long-range interactions," *Annals of Physics*, vol. 318, no. 1, pp. 37–80, Jul. 2005. doi:10.1016/j.aop.2005.03.007
87. U. Mohideen and A. Roy, "Precision measurement of the Casimir force from 0.1 to 0.9 μm," *Physical Review Letters*, vol. 81, no. 21, pp. 4549–4552, Nov. 1998. doi:10.1103/PhysRevLett.81.4549
88. H. B. Chan, V. A. Aksyuk, R. N. Kleiman, D. J. Bishop, and F. Capasso, "Nonlinear micromechanical Casimir oscillator," *Physical Review Letters*, vol. 87, no. 21, p. 211801, Oct. 2001. doi:10.1103/PhysRevLett.87.211801
89. Y. Xu et al., "Radio frequency electrical transduction of graphene mechanical resonators," *Applied Physics Letters*, vol. 97, no. 24, p. 243111, Dec. 2010. doi:10.1063/1.3528341
90. A. M. van der Zande et al., "Large-scale arrays of single-layer graphene resonators," *Nano Letters*, vol. 10, no. 12, pp. 4869–4873, Dec. 2010. doi:10.1021/nl102713c
91. B. Witkamp, M. Poot, and H. S. J. van der Zant, "Bending-mode vibration of a suspended nanotube resonator," *Nano Letters*, vol. 6, no. 12, pp. 2904–2908, Dec. 2006. doi:10.1021/nl062206p
92. K. Jensen, K. Kim, and A. Zettl, "An atomic-resolution nanomechanical mass sensor," *Nature Nanotechnology*, vol. 3, no. 9, pp. 533–537, Sep. 2008. doi:10.1038/nnano.2008.200
93. A. C. Bleszynski-Jayich, W. E. Shanks, B. R. Ilic, and J. G. E. Harris, "High sensitivity cantilevers for measuring persistent currents in normal metal rings," *Journal of Vacuum Science & Technology B: Microelectronics and Nanometer Structures*, vol. 26, no. 4, p. 1412, 2008. doi:10.1116/1.2958247
94. J. G. E. Harris, B. M. Zwickl, and A. M. Jayich, "Stable, mode-matched, medium-finesse optical cavity incorporating a microcantilever mirror: Optical characterization and laser cooling," *Review of Scientific Instruments*, vol. 78, no. 1, p. 013107, Jan. 2007. doi:10.1063/1.2405373
95. G. Heinrich, J. G. E. Harris, and F. Marquardt, "Photon shuttle: Landau-Zener-Stückelberg dynamics in an optomechanical system," *Physical Review A*, vol. 81, no. 1, p. 011801, Jan. 2010. doi:10.1103/PhysRevA.81.011801
96. F. Aguilar Sandoval, M. Geitner, É. Bertin, and L. Bellon, "Resonance frequency shift of strongly heated micro-cantilevers," *Journal of Applied Physics*, vol. 117, no. 23, p. 234503, Jun. 2015. doi:10.1063/1.4922785
97. B. Pottier, F. Aguilar Sandoval, M. Geitner, F. Esteban Melo, and L. Bellon, "Silicon cantilevers locally heated from 300 K up to the melting point: Temperature profile measurement from their resonances frequency shift," *Journal of Applied Physics*, vol. 129, no. 18, p. 184503, May 2021. doi:10.1063/5.0040733
98. D. Ma, J. L. Garrett, and J. N. Munday, "Quantitative measurement of radiation pressure on a microcantilever in ambient environment," *Applied Physics Letters*, vol. 106, no. 9, p. 091107, Mar. 2015. doi:10.1063/1.4914003
99. D. Ma and J. N. Munday, "Measurement of wavelength-dependent radiation pressure from photon reflection and absorption due to thin film interference," *Scientific Reports*, vol. 8, no. 1, p. 15930, Dec. 2018. doi:10.1038/s41598-018-34381-z

100. N. V. Lavrik, M. J. Sepaniak, and P. G. Datskos, "Cantilever transducers as a platform for chemical and biological sensors," *Review of Scientific Instruments*, vol. 75, no. 7, pp. 2229–2253, Jul. 2004. doi:10.1063/1.1763252
101. J. R. Barnes, R. J. Stephenson, M. E. Welland, Ch. Gerber, and J. K. Gimzewski, "Photothermal spectroscopy with femtojoule sensitivity using a micromechanical device," *Nature*, vol. 372, no. 6501, pp. 79–81, Nov. 1994. doi:10.1038/372079a0
102. H. P. Lang et al., "A chemical sensor based on a micromechanical cantilever array for the identification of gases and vapors," *Applied Physics A: Materials Science & Processing*, vol. 66, no. 7, pp. S61–S64, Mar. 1998. doi:10.1007/s003390051100
103. Y. Seo, P. Cadden-Zimansky, and V. Chandrasekhar, "Low-temperature high-resolution magnetic force microscopy using a quartz tuning fork," *Applied Physics Letters*, vol. 87, no. 10, p. 103103, Sep. 2005. doi:10.1063/1.2037852
104. E. M. Purcell, "Life at low Reynolds number," *American Journal of Physics*, vol. 45, no. 1, pp. 3–11, Jan. 1977. doi:10.1119/1.10903
105. K. Trachenko and V. V. Brazhkin, "Minimal quantum viscosity from fundamental physical constants," *Science Advances*, vol. 6, no. 17, Apr. 2020. doi:10.1126/sciadv.aba3747
106. B. J. van Wees et al., "Quantized conductance of point contacts in a two-dimensional electron gas," *Physical Review Letters*, vol. 60, no. 9, pp. 848–850, Feb. 1988. doi:10.1103/PhysRevLett.60.848
107. M. Brandbyge et al., "Quantized conductance in atom-sized wires between two metals," *Physical Review B*, vol. 52, no. 11, pp. 8499–8514, Sep. 1995. doi:10.1103/PhysRevB.52.8499
108. J. K. Gimzewski, T. A. Jung, M. T. Cuberes, and R. R. Schlittler, "Scanning tunneling microscopy of individual molecules: Beyond imaging," *Surface Science*, vol. 386, no. 1–3, pp. 101–114, Oct. 1997. doi:10.1016/S0039-6028(97)00301-4
109. C. Joachim and J. K. Gimzewski, "An electromechanical amplifier using a single molecule," *Chemical Physics Letters*, vol. 265, no. 3–5, pp. 353–357, Feb. 1997. doi:10.1016/S0009-2614(97)00014-6
110. S. J. Tans, A. R. M. Verschueren, and C. Dekker, "Room-temperature transistor based on a single carbon nanotube," *Nature*, vol. 393, no. 6680, pp. 49–52, May 1998. doi:10.1038/29954
111. K. Schwab, E. A. Henriksen, J. M. Worlock, and M. L. Roukes, "Measurement of the quantum of thermal conductance," *Nature*, vol. 404, no. 6781, pp. 974–977, Apr. 2000. doi:10.1038/35010065
112. D. Thompson et al., "Hundred-fold enhancement in far-field radiative heat transfer over the blackbody limit," *Nature*, vol. 561, no. 7722, pp. 216–221, Sep. 2018. doi:10.1038/s41586-018-0480-9
113. K. Kim et al., "Radiative heat transfer in the extreme near field," *Nature*, vol. 528, no. 7582, pp. 387–391, Dec. 2015. doi:10.1038/nature16070
114. C. Wuttke and A. Rauschenbeutel, "Thermalization via heat radiation of an individual object thinner than the thermal wavelength," *Physical Review Letters*, vol. 111, no. 2, p. 024301, Jul. 2013. doi:10.1103/PhysRevLett.111.024301

6

Lab-on-a-Tip and Plasmas for Sustainability

6.1 Lab-on-a-Tip in Applied Nanophysics

Since the beginning of microscopic techniques a few centuries ago, they have developed themselves in virtue of what they allowed to observe for later modification. In fact, the so-called cycle *observe–transform–observe* has been the main driver of their developments and milestones. However, until the end of the 20th century, this was being done through the successive invention of different instruments, for instance, optical microscopes, electronic microscopes, and different types of surface analysis microscopies (electron or neutron diffraction, field emission, or ionization). When Feynman in the middle of the last century realized that there was no theoretical impediment based on the laws of physics to manipulate atoms and molecules at the nanoscale, he was exactly reckoning about the lack of such a kind of instrument at that time, when he considered that the right tools required to perform were missing. So, imagine the reader that there may be a single instrument capable of observing and modifying "in situ" a given sample at the nanoscale. In fact, this was exactly what happened when Heinrich Rohrer developed the first instrument belonging to the family of future scanning probe microscope (SPMs) variants, the scanning tunneling microscope (STM), which led him to be awarded with the Nobel Prize in Physics in 1986. Since then, more and more variants have been developed and successively incorporated into a single instrumental kit. Nowadays we can work with instrumental kits that include probes based on multiple types of physical interactions and that allow us to put in practice many different areas of experimental physics at the nanoscale. As a mere slight comparison, we can say that a sample observed by each of these variants leads to information as more valuable as the different spectroscopy variants can collect information from the same sample. They are all different ways of looking at reality not being the true reality. But a great advantage relies on that one can access experimental information at the nanoscale, in a compact way, and using a kit of different nearby probes all based on a single tip: hence the name of laboratory on a tip (*lab-on-a-tip*). In this regard, it is also worth remembering the potential that the already mentioned tunneling spectroscopy performed with an STM brings to a physics Nanolab. Other SPM variants also can also work in spectroscopic modes extending their unique potentialities of studying different physical phenomena at the nanoscale.

Atomic resolution with STM was first obtained in 1983 [1] and with atomic force microscopy (AFM) in contact mode in 1987 [2]. The addition of the force modulation technique into AFM in the year 1991 led to the AFM microscopes in FM dynamic mode, to the variants of magnetic force (MFM) and electrostatic force (EFM), and even to the "true" atomic resolution obtained later, with noncontact AFM (NC-AFM) also known as dynamic force microscopy (DFM) [3]. As with the AFM in contact mode, the NC-AFM is

DOI: 10.1201/9781003285083-6

sensitive to a combination of forces, the main ones being the electrostatic contributions (long range), vdW (long/intermediate range), and covalent (short range). The mathematical expressions of these contributions for the frequency shift Δf, are well established and mainly depend on the resonance frequency and the force constant of the cantilever, the oscillation amplitude, tip radius, the polarization applied between the tip and the sample, the Hamaker constant of the system (taken as the geometric mean of the tip–sample Hamaker constants), the potential parameters, and the distance between the sample and the tip.

The SPMs that are based on the measurement of an interaction force have the generic designation of SFMs. The EFM uses a conductive cantilever to which a polarization is applied, in a noncontact AC regime, and allows to measure the electrostatic interaction between the tip and the sample, through variations in amplitude and phase. Since electrostatic forces are made sensing at greater distances than van der Waals forces, electrostatic mapping can be decoupled from topographical information simply by correctly fitting the sample-type distance. Its ability to visualize the surface electric charge distribution makes this mode particularly useful in detecting defects in integrated circuits, as well as measuring static electric charges in dielectric films, or even local variations of the dielectric constant. On the other hand, the MFM has an operation mode identical to the EFM, but the probe uses a ferromagnetic tip magnetized along its main axis, allowing to identify the surface magnetic zones of the sample. Depending on the type of interactions between the probe tip and the sample surface, that are used for characterization, various types of SPMs have been developed, and today, more than twenty SPM variants are known, including the scanning thermal microscope (SThM), the scanning capacitance microscope (SCM – which maps the capacitance variations between the tip and the sample), the Kelvin probe force microscope (KPFM), the magnetic resonance force microscope (MRFM – where the local composition can be analyzed by the NMR-signal) [4], the scanning polarization force microscope (SPFM), the scanning near-field acoustic microscope (SNAM), and the scanning chemical potential microscope (SCPM – which maps surface chemical potential distribution), just to name a few. The combination of STM and SFM became available, since the use of conductive SFM tips, such as metallic tips, doped silicon, or doped diamond, gives the opportunity to measure both forces and electrical currents. Local conductivity measurements are done in the repulsive contact. The electrochemical STM and AFM options that operate with a fluid cell allow to follow electrochemical reactions making the tip a microelectrode. They include a potentiostat and transparent probe and electrode holders. KPFM allows us to measure the contact potential difference between different materials at the nanoscale and to obtain the surface potential with lateral resolution below 50 nm and potential resolution below 10 mV. As local surface potential is related to work function, this one can also be determined.

Curiously, in a variant of STM called photon scanning tunneling microscopy (PSTM), under certain circumstances and for some samples, electronic inelastic tunneling may occur, with part of the electronic energy transferred to different electromagnetic radiation modes, causing electroluminescence. With spatial confinement, it is even possible in certain cases, as in polycrystalline Ag films and in porous silicon (natural confined system), the occurrence of tip-induced plasmonic modes. The emitted light can be locally collected at 45° by a microlens/fiber optic system and analyzed spectrally. The recordings of light intensity versus photonic energy, emitted at various tip polarization values (optical spectra at current constant), allow obtaining information about the inelastic processes of interaction surface electron–plasmon.

The SThM uses a thermocouple-type probe to map the temperature at the surface of the sample. In its original form, the AFM probe is replaced by a Wollaston wire, whose apex is eroded to expose the platinum filament. By applying an electrical voltage to the tip, the filament can be heated to a certain temperature allowing for local thermal measurements of the sample. Currently, the resolution of this variant is already below 200 nm (Nano-TA), which allows it to identify unresolved structures, for example, by Raman microscopy (which is limited to 500 nm). After the sample has been visualized normally by AFM, a zone can be selected for thermal analysis. The probe is moved to that location, and an applied force is preselected. Then the probe heating starts, and a deflection of the cantilever is observed, due to thermal expansion, as well as an indentation if the sample melts locally. Through prior calibration with known melting point samples, it is possible to map the melting points of different components in a composite material through phase contrast.

Another interesting SPM variant is the molecular recognition force microscope (MRFM) or "chemical force microscope" (CFM), which made it possible to map specific chemical interactions, from individual recognition of simple molecules to ligand-receptor or even antibody–antigen-type systems, through functionalization. In the force–distance cycle, the tip is brought close to the surface with the adsorbed molecules that are to be "recognized," thus forming the specific chemical interaction. During the subsequent retraction of the tip, an increasing force is exerted on the interacting system, leading to the failure of the system to a certain critical force value. When the interacting molecules include a flexible chain linker, the attractive force–distance profile is, in contrast to the repulsive force–distance profile of the contact region, nonlinear. Its shape is determined by the elastic properties of the flexible linker and shows parabolic characteristics that reflect the increase in the spring constant of the linker during extension. The attractive force profile therefore contains the characteristics of an individual molecular recognition.

In 1984, D. Pohl combined the technique of near scanning with the concept of scanning imaging in the near electromagnetic field (already known since 1972 for microwaves) and developed a revolutionary optical microscope with nanometric resolution called *Scanning Near-Field Optical Microscope* (SNOM, or NSOM in the US) [5]. SNOM probes the rapidly decaying optical near-field objects at surfaces. In consequence, the resolution limit of classical optical microscopy at one-half of the wavelength is overcome. Furthermore, SNOM allows a straightforward local spectroscopy for chemical identification and other optical techniques such as polarization analysis or second harmonic generation. The near field and far field are regions of the electromagnetic field around an object (such as a transmitting antenna or the result of radiation scattering off an object). Far-field E (electric) and B (magnetic) field strength decreases as the distance from the source increases, resulting in an inverse-square law for the radiated power intensity of electromagnetic radiation. By contrast, near-field E and B strength decrease more rapidly with distance: the radiative field decreases by the inverse-distance squared, the reactive field by an inverse-cube law, resulting in a diminished power in the parts of the electric field by an inverse fourth power and sixth power, respectively. The rapid drop in power contained in the near-field ensures that effects due to the near-field essentially vanish a few wavelengths away from the radiating part of the antenna. Nonradiative near-field behaviors dominate close to the object, while electromagnetic radiation far-field behaviors dominate at greater distances. The optical near field describes the nonradiating part of the electromagnetic field close to a light emitter or scatterer. A well-known example is an evanescent field outside a glass prism, where a light beam is totally reflected at the inner surface. When a second body is brought into the range of the evanescent wave, the electromagnetic field can couple to this body and the total reflection is weakened. The near field and radiative far field are always coupled. Whenever an object is interacting with an optical near field, the far-field radiation is changed. This fact is exploited in SNOM: the near field evanescing from an optical probe is used to excite or scatter surface objects, or the optical probe collects, or scatters near-field radiation emitted or scattered from objects at the surface.

The propagation of electromagnetic waves and the retardation phenomena become irrelevant in nano-optics, and therefore the wave equation referring to the electric field of classical optics becomes the Laplace equation: $\nabla^2 V = 0$. In a discrete approach to matter, and considering it as composed of oscillating elementary electric dipoles, classical electromagnetic theory provides a general expression for the field in terms of the inverse of the power of the distance to the object; thus, both the near and far fields are simply a limiting case of the general formula that one obtains for fields produced by a single oscillating electric dipole, apart from the time dependence, and from which it follows that space can be divided into three regions distinct, defined by the value of the product kr (where $k = 2\pi/\lambda$ and r the variable distance to the dipole): $kr << 1$ (near-field zone), $kr \sim 1$ (intermediate zone), and $kr >> 1$ (field zone distant). A small emitting aperture can also be seen as a dipole source. According to the Bethe model, this aperture corresponds to a system of two oscillating dipoles (one electric and the other magnetic), the optical source being characterized by the electric dipole, $\boldsymbol{p} = -\,4\pi\varepsilon_0 a^3/3\,\boldsymbol{E}_\perp$, where a represents the radius of the aperture, and $E_\perp$ the normal component of the electric field at the center of the aperture. The near (or evanescent) fields are nonpropagating, unlike the far fields which are radiative. On the other hand, the near-field region is predominantly electrical in nature. So, when detecting the far field, one loses information about the near fields, which define the charge configuration that determines the dipole. The fields that are detected by conventional optics are the radiant ones, even when close to an illuminated object, because although propagating and evanescent fields coexist in this area, the latter are incapable of transmitting information at a distance. The evanescent field is characterized by high spatial frequencies, reflecting the surface structure, but being unable to propagate.

There are two fundamental difficulties that limit the optical resolution of the submicroscopic world:

1. An optical system is always needed between the object and the image, which collects the information and projects it into a specific detector (e.g., darkroom with an orifice); this optical system, however, perfect it may be, is spatially limited (aperture) and, therefore, works as a "low-pass" spatial frequency filter of the propagating fields, thus delimiting the spectrum of information coming from the object. Abbe (1873) determined in terms of diffraction theory the limit of resolution of an optical instrument and Rayleigh (1896) stipulated that two-point objects will be seen separately if the distance between them is greater than or equal to the relationship between the wavelength of the beam light, the index of refraction of the medium and the semi-aperture angle of the objective lens used to collect and focus the light beam on the detector; generally, a distance of the order of half the wavelength of light is taken. However, if we can detect non-propagating near fields on the surface of the object, the diffraction limit problem, which is a strictly undulating phenomenon of spatial fields, can be circumvented. A plane wave when incident on an object is scattered in all the Fourier components that compose the object through the scattering density function that only depends on the distance; there is thus a spectral decomposition of this function in the space of the scattering vector (an analogous situation occurs in the scattering

of X-rays by a solid). Therefore, this experience of diffraction by a finite object is equivalent to the diffraction of light through a slit. In fact, according to Babinet's principle (1850), if we exchange an object for an opening of the same shape in a screen, the same diffraction pattern will be obtained. That is, the diffraction phenomenon in a hair is the same as in a slit of equal thickness. The image of a point object is not a point but an Airy diffraction pattern. According to Fourier theory, the domain of space Δk_x in which the wave amplitude is different from zero is related to the finite extent of the object, by $\Delta k_x \, \Delta x \geq 2$, where $k_x = 2\pi/\lambda$ is the component of the wave vector in the x direction. Thus, the smaller the object, the greater the number of spatial high-frequency Fourier components that are needed to resolve the object (i.e., the wider its spatial spectrum), and therefore the more incident radiation fields. will be disturbed (i.e., more scattered or diffracted). According to Rayleigh, two nearby points will be resolved if the distance between the two diffraction maxima is equal to or greater than twice the distance between the maximum and the first peak minimum. Looking at an object through a classical microscope, its objective of refractive index n, will only accept values of k within a restricted opening angle equal to 2θ, that is, only components inside the interval $\Delta k_x = 2k\, n \sin\theta$, will reach the observer. So, one has $\Delta x \geq \lambda/(2n \sin\theta) \approx \lambda/2$, which corresponds to the Abbe criterion.

2. Thus, in the observation phase, there is a filtering of high spatial frequencies which leads to a decrease in resolution or loss of definition of the object edges. Every optical method has a given optical aperture which produces diffraction when one detects only propagating or far fields. If somehow one manages to break the technological limitation of the optical aperture, one could detect higher spatial frequencies and thereby increase the optical resolution; but what is the largest value that the k_x component can take? As the electromagnetic fields are scattered by an object and observed at a point at a distance greater than its largest dimension, it can be described by propagating spherical waves having an exponential term, then the components of the wave vector are required to be real so that the incident waves are diffracted by the object. If these components assume complex values, then one can extend the wave vector values to the limit of high spatial frequencies, which is possible if the spatial spectrum of the object contains frequencies greater than $2\pi/\lambda$; in this domain, evanescent waves will appear in the corresponding direction. It is a situation equivalent to that of a plane wave impinging on a slit of dimensions smaller than the wavelength; at high spatial frequencies, the light beam is strongly perturbed by the slit so that there will be total reflection on one side and evanescent waves on the other side of the slit. The spatial spectrum of a circular slit, for example, with a radius greater than or equal to $\lambda/2$, is characterized by values of k situated in the domain of propagating fields (between $-2\pi/\lambda$ and $2\pi/\lambda$), and most of the spectrum can be detected by a classical optical microscope. In this case, the image of the object can be recovered in good condition. However, if the slit radius is less than $\lambda/2$, most of the spectrum becomes evanescent, with part of the image information lost when detected in the far-field regime. Even under these conditions, it will become impossible to distinguish the image from an even smaller object. On the other hand, the fields detected under these conditions will appear identical in the far-field region, as if they originated from a single source of dimension $\lambda/2$. To obtain a better resolution it is necessary to detect a wider spectrum in values of k, that is the

evanescent fields or close to the object. We should get as close to the object as possible, in order to obtain an optical resolution below the Rayleigh limit, but even so, how far can the values of Δk_x be extended?

There is a mathematical analogy between the Heisenberg uncertainty principle and the Rayleigh limit of resolution, which is due to the fact that both have a common origin in the mathematical method established by Fourier at the end of the 19th century, which implies that certain physical quantities, such as position and momentum (wave vector) cannot be determined simultaneously up to one uncertainty. According to Rayleigh, optical resolution is dictated by the fact that instruments in use are restricted in aperture and also by the phenomenon of diffraction, while the Heisenberg uncertainty concerns the intrinsic measurement process and expresses the fact that there is always a nondeterminable interaction between the observer and what is observable. Thus, the uncertainty principle does not limit the values assumed by k, simply stipulating an acuity limit in Δx for a given Δk_x and, moreover, says that it is impossible to specify precisely and simultaneously the values of the momentum and position of a particle (including the photon). Rayleigh's criterion and the uncertainty principle are equivalent in the range of k corresponding to propagating waves, but is the latter applicable in the region of evanescent waves where the components of k can assume complex values? Since k_x can take both positive and negative values, Δk_x represents twice the maximum value of this component in the interval; being θ the value of the projection angle of vector k on the x-axis, then $\Delta k_x = 2k_x \sin\theta$. In a medium of refractive index n, one has that $k = 2\pi n/\lambda$. As the Heisenberg uncertainty relations can be written in the form $\Delta k_x \Delta x \geq 2\pi$; $\Delta k_y \, \Delta y \geq 2\pi$; $\Delta k_z \, \Delta z \geq 2\pi$, then, $\Delta x \geq \lambda/(2n \sin\theta)$, that is, the uncertainty relations lead us to the Rayleigh criterion for the diffraction of propagating waves if the angle θ and the vector $\mathbf{k}$ assume real values. One could decrease the uncertainty in x if we make $\sin\theta > 1$, allowing the components of $\mathbf{k}$ to assume complex values (or imaginary θ). This situation corresponds to that of the evanescent waves created by the interaction between light and an object of dimensions smaller than the wavelength. The fields (as well as the wave vectors) are not well defined in this transition region and their dependencies in space and time are not straightforward. They contain propagation components and surface-bound components whose properties are closely related to the electronic states of the object. The existence of a component depends on the existence of the other (and vice versa) and there is no clear separability of the components. On the other hand, this transition region corresponds to the domain of nonpropagating static fields. If there is no wavy spatial distribution, there is no diffraction, and then the Fourier method is not applicable in the near-field region. If we cannot define k_x in this domain (remember that the vector $k = p/h$ is defined in the context of a propagating plane wave), we also have no reason to consider an uncertainty Δk_x. This suggests that the Heisenberg uncertainty principle is not applicable in the near-field domain and is therefore not a limiting factor for optical resolution. This nonapplicability to near fields means that these fields can vary in this interval smaller than the wavelength of light, making no restrictions on the type of electromagnetic field. In fact, it is these evanescent fields that relate to the structure of the object and that contains the information of details of dimensions smaller than the wavelength. On the other hand, the frustrated total reflection with two prisms mutually inverted in the vicinity of each other presents an optical analogy with the quantum tunnel effect, where the amplitude of the electric wave assumes the role of a wave function. As is known, in a situation of total internal reflection on the flat surface of a prism, an evanescent wave appears that propagates only along the surface. Imagining that the flat surface closes to produce a spherical surface with a diameter smaller than the wavelength of the incident radiation, a near-field around the sphere (evanescent wave) will be generated, of thickness equal to the sphere's radius. From this point of view, the plane configures a special case of the sphere with an infinite radius. Close to the surface of the sphere, the evanescent field and the radiative field coexist, and a disturbance in one of them implies a disturbance in the other. In fact, it can be shown that light incident on a limited object is always converted into a propagating field and an evanescent field, with the incident field being either propagating or evanescent [6].

In SNOM, by scanning the probe over the surface, lateral variations in the optical properties can be recorded. The optical near field exhibits variations on the same length scale as the size of the emitting structure. In contrast, the propagating far field cannot vary on a scale smaller than half of the wavelength. Indeed, SNOM can overcome the resolution limit of classical optical microscopy only by bringing the probe within a distance from the surface which is of the same order as the distance of the objects to be resolved. Small apertures at the apex of tapered optical fibers are the most widely used optical near-field probes. The fiber must be covered by a metal coating to confine the light and to prevent light from leaking out of the tapered end. Then a sub-wavelength aperture must be made in the metal coating. The aperture can be produced by shadow effects during the metal evaporation, by cutting the fiber apex after evaporation with a focused ion beam, or by an electrochemical etching process. The distance control in SNOM is more difficult than in other SPM methods because the light intensity versus distance curve is often nonmonotonic. Therefore, it cannot be used for the distance control. Instead, the damping of a lateral oscillation of the fiber due to shear forces between the apex and sample is exploited. The fiber is excited into oscillation by means of a small quartz tuning fork attached to it, and the change in resonance parameters of the oscillation (usually the phase between excitation and oscillation) is used for the distance feedback controller. The SNOM operation is always carried out in the absence of background radiation, using a hood with optical isolation, in addition to the usual isolation from acoustic and mechanical vibrations of any SPM. One way to control the probe-sample distance is by detecting the shear forces existing in the presence of van der Waals forces acting on the tip of the optical fiber, which is set to vibrate at the frequency of mechanical resonance in a direction parallel to the surface of the sample (shear force mode), as shown in the Figure 6.1A, for a micro-tuning fork probe. The variation in the oscillation resonance parameters (generally the phase between excitation and oscillation) is used for distance feedback control (Figure 6.1B). There are several ways of directly detecting the vibration of an oscillating optical probe (Figure 6.1C). The simplest method is to project the light emitted or scattered from an optical probe onto a suitably positioned aperture and detect the transmitted light intensity. The modulation amplitude of the optical signal at the dither frequency of the tip will reflect the amplitude and phase of the tip oscillation. In a near-field optical microscope, this method interferes with the detection path of the optical signal and thus can be influenced by the optical properties of the sample. Therefore, alternative optical detection schemes were developed which employ a beam path perpendicular to the optical detection path of the microscope. An auxiliary laser can be pointed to the probe and the resulting diffraction pattern is detected by a split photodiode. This scheme works well but it can suffer from mode hopping of the laser diode or drifts in the mechanical setup leading to changes in the (interference) pattern on the photodiode. Another less restrictive and more compact way to control the distance between the probe and the sample is through a method based on the piezoelectric detection of the interaction between the probe and the sample. The oscillation of the arms of the piezoelectric tuning fork has a maximum amplitude at resonance and generates an electrical signal proportional to the amplitude of oscillation (on the order of a few microvolts maximum resonance), measured on electrodes placed appropriately, due to the creation of an alternating piezoelectric potential proportional to this oscillation. The resonant frequency is accurately measured using a phase-sensitive (lock-in) amplifier. Once the tuning fork is vibrated at the resonant frequency, this signal is used to position the tip of the optical fiber a few tens of nanometers above the sample, using an electronic feedback loop. The oscillating signal is monitored as the probe approaches the surface perpendicularly, using the "lock-in" detection technique synchronous with the oscillation frequency of the bimorph. In general, it is verified that below 50 nm the value in Volts of the signal begins to slowly decrease in relation to its free resonance oscillation value, reaching around 80% of this value at 20 nm; below 20 nm the decrease is usually much more pronounced, reaching around 10% at a couple of nanometers. These piezoelectric signals can be used in conjunction with the electronic feedback loop to obtain topographical images of the surface of the sample, as the piezoelectric resonance signal is used as a signal in the feedback loop to keep constant the distance z between the tip of the probe and the surface.

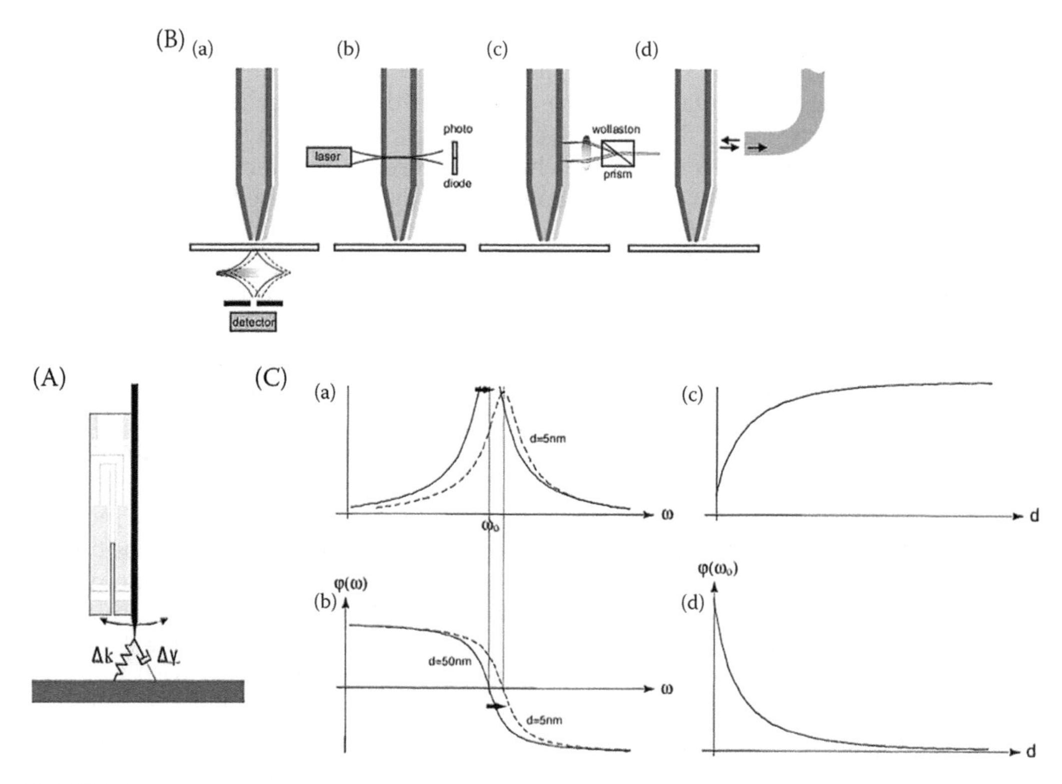

FIGURE 6.1
(A) Micro-tuning fork oscillation in shear force mode; (B) Different methods for detection of the oscillation of an optical probe: (a) light from the aperture is detected and modulated at the mechanical resonance frequency of the probe, (b) laser deflection, (c) differential interferometry, (d) fiber optic interferometry; (C) Resonance of the probe: amplitude (a) and phase (b) of a beam driven at an angular frequency w; as the tip starts to interact with a surface, the resonance shifts and the amplitude drops; amplitude (c) and phase (d) at $w = w_0$ as a function of the tip-sample distance.

SNOM combines the chemical sensitivity of optical spectroscopy techniques with the high spatial resolution of scanning probe microscopy and with the noninvasive character of light as well. As an example, the fluorescence emission from fluorophore-doped fullerene reduced films was investigated by fluorescence spectroscopy and SNOM [7]. SNOM instruments can provide highly correlated topography and optical images and can be used to determine fluorescence spectra of individual nanoparticles or even molecules. The resolution in illumination mode is primarily limited by the size of the aperture and the aperture-sample distance. The signal processing between the SNOM and the tunning-fork probe is achieved with a lock-in phase-sensitive signal detector/amplifier. The phase angle is used as a control signal for the proportional integral derivative.

In another configuration, the integration of the optical aperture into the tip of a typical SFM cantilever allows the use of SFM techniques for distance control. In a completely different approach called apertureless SNOM, the variation in light intensity scattered in the near field between a standard SPM tip and surface objects is recorded. This technique displays a poor signal-to-noise ratio and still presents difficulties in the discrimination of scattering in the near-field region and scattering in the rest of the setup. In fact, in the apertureless SNOM mode, it is possible to map the optical signal coming from the surface of the sample, by scattering this signal in an opaque type, which allows reaching lateral resolutions of less than 1 nm. However, due to topographical coupling, the interpretation of the images is difficult. The probe without optical aperture is a metallic tip (silver or gold) that serves as a scattering center. The optical field is strongly increased by the tip and serves to make it possible to perform surface-augmented Raman spectroscopy (SERS) or to excite nonlinear local effects, such as 2nd harmonic generation. The resulting signals are detected in the far field. In probes without aperture, a confined electromagnetic field is created by light scattering at the apex of the metallic tip. Among the numerous applications, SNOM enables the study of the propagation of plasmons excited in structured metal films [8], and the combination of an SFM tip as scattering probe and Raman spectroscopy exploits the surface enhancement of the Raman effect to study surface chemistry with high lateral resolution [9]. In another variant called scattering SNOM (s-SNOM), an atomic force microscope (in tapping mode or intermittent contact mode) drives a metallic or metal-coated tip to scan over the sample of interest; an external light source is coupled to the tip apex and the scattered light from the tip with sample underneath is collected, then converted into an electric signal by an optical detector. The signal is routed to a lock-in amplifier, where the nonfundamental harmonic demodulations can be extracted with the tip oscillation frequency as the reference frequency. The optical frequency of the external radiation is chosen to match the energy of a particular physical process (electronic, vibrational, or polaritonic resonances). Several light sources can be used in s-SNOM, with frequencies ranging from visible light to terahertz.

Worth to notice that evanescent fields have also been applied for trapping and guiding atoms with possible applications to quantum information. There, the evanescent field of a linear waveguide can be used to strongly enhance the field present in a ring resonator for light standing waves control of trapped atoms. The fact that a SNOM uses optical fiber as a sensor, allows to expand its optical potential, namely in the fields of spectroscopy and atomic manipulation. A photonic crystal fiber (PCF) is a type of nanostructured optical fiber, where light is guided by structural modifications, and not only by refractive index differences. Recall that a photonic crystal is an optical nanostructure in which the refractive index changes periodically, and this allows the propagation of light in the same way that the structure of natural crystals gives rise to X-ray diffraction and that semiconductor crystal structure affects the electronic properties (instead of electronic bandgap, an optical bandgap is present). Optical components (e.g., lenses) cannot focus light to scales smaller than $\lambda/2$ due to the diffraction limit. This can be achieved, as we have already seen, with sub-micrometer optical apertures (e.g., SNOM), but also with photonic crystals, or with photonic tunneling and photo-assisted STM, or even using surface plasmons located in metallic nano-objects. Also, concepts have been formulated for manipulating the thermal motion of a neutral atom in a vacuum by means of an optical near field. When an atom absorbs light whose frequency corresponds to the energy difference ΔE, between two atomic energy levels, an electric dipole is induced in the atom. Thus, a dipole force, $\boldsymbol{F_d}$, is generated when the gradient of the electric field of light acts on the electric dipole. If the distribution of the light intensity is spatially inhomogeneous, the atom is driven by the dipole force. The direction of the dipole force, in a two-level energy approximation (for simplicity), depends on the difference between the angular frequency of the incident light and the atomic resonant angular frequency ($=\Delta E\ /\hbar^2$). Thus, an atom in vacuum can be reflected by the repulsive dipole force induced by an optical near field (evanescent light) on a planar surface of an inverted dielectric prism. This way, the dielectric surface works as a kind of mirror to reflect the atom. The evanescent optical field can easily be studied by PSTM or SNOM probes located above.

It is known that the Raman spectrum of molecules adsorbed on metallic colloids exhibits a considerable increase in the measured signal intensity. This effect is called surface-enhanced Raman scattering (SERS). To understand the phenomenon, it is good to remember that the Raman effect is a scattering effect of electromagnetic radiation. The incident sinusoidal electric field induces an electric dipole in the molecule, which due to its natural vibrational motion has a polarizability term that is also sinusoidal in time:

$$\alpha = \alpha_0 + \alpha_v \cos(2\pi\nu_v t) \tag{6.1}$$

Thus, the induced electric dipole will result from the multiplication of the two time-varying terms, leading to a cross term that represents the scattered light with two different energies (one higher and one lower than the Rayleigh elastic scattering). These increments are given by the vibrational frequencies ν_v of the molecule. When Raman-active molecules are adsorbed to the surface of metallic nanoparticles, there will be an increase in the intensity of their Raman signal, because, since this intensity is proportional to the induced dipole moment, this increase can come either from α or from the incident electric field. However, the contribution of the latter is dominant. Due to the resonance effect between colloidal metals and incident radiation, the field that effectively induces Raman scattering in the adsorbed molecules is the sum of the incident electric field and the resonant field induced at the surface of the particles. The effect of increasing the Raman signal is independent of the wavelength of the incident radiation and follows the usual dependence on the wavelength of the resonant scattering. As an example, when a silver colloid is prepared, the color observed is yellow, but after adding a salt, the metallic particles tend to aggregate, causing an increase in the size of the particles as well as the dielectric function $\varepsilon(\lambda)$, causing a shift in the induced plasmon resonance wavelength. Studies of sodium citrate addition in silver colloids show Raman signal increases that can range from 10^2 to 10^6, depending on the incident wavelength. Choosing, for example, pyridine as the molecule to be adsorbed on Ag nanoparticles, the Raman spectrum of pure pyridine must be compared with that of the pyridine solution and with that of pyridine adsorbed on silver colloids, making the necessary volumetric correction for the adsorbed pyridine. It was then verified that the relative intensities of the peaks vary, but their positions on the wavelength scale do not. The different vibrational modes of the molecule can be affected differently by a charge exchange effect between the adsorbed molecule and the metal particle, which will change the polarizability characteristics. However, this is a small effect compared to the strong increase in signal strength. In this way, plasmonic materials can also greatly increase the brightness of LEDs, using adsorbed dye molecules. Furthermore, covering a GaN LED with metallic nanoparticles results in a strong increase in light intensity. On another hand, by coupling metallic nanoparticles with arrays of quantum dots, the light emitted by these quantum dots can be increased several times. In addition, it is possible to tune the frequency of light emissions by adjusting the diameters of the nanoparticles. It should be also mentioned that the light emitted by photoluminescence of nanowires (because of confinement) is strongly polarized along its longitudinal direction, due to the strong dielectric contrast between the nanowire and its surroundings.

Optical antennas have been explored in the visible region to exploit surface plasmon resonance (SPR). They are like radio frequency (RF) antennas in that they are resonant structures that respond to specific wavelengths through the geometrical and material characteristics of the antenna as well as the surrounding environment. However, while RF antennas focus on the optimization of far-field characteristics, optical antennas emphasize the near-field behavior to for example create a spot size that is smaller than the wavelength of incoming light. Worth to mention that a detailed mapping of the optical force is critical for plasmonic sensing and optomechanical switching. For example, a noncontact mode atomic force microscope can be used with a lock-in method to produce the optical force mapping near the hot-spot of a metal–dielectric–metal bow-tie infrared nanoantenna, or the optical force intensity created by a plasmonic nanoantenna [10]. The minimum force sensitivity of an AFM system limited by Brownian motion can be estimated by

$$F_{min} = \sqrt{\frac{4k_B TKB}{\omega_0 Q}} \tag{6.2}$$

where K stands for the spring constant, ω_0 is the resonant frequency of the tip, Q the quality factor of the cantilever, and B the bandwidth of measurement (set by lock-in amplifier) used in the force measurement [11]. The optical force measurement setup uses the modulation of the laser to measure the effect on the amplitude change of the AFM tip at the frequency of modulation of the laser when the AFM tip is brought near the region of optical confinement. An apertureless near-field scanning optical microscope can simultaneously measure the topography and near-field intensity of the antenna.

When SPMs are operating in the tunneling regime, frequency shifts can be detected by Dynamic Force Microscopy (DFM) [12], and variations of the tunneling current can be observed. In DFM, the cantilever is oscillating while it approaches the sample surface. The force between probe and sample is reflected by cantilever amplitude change and maintained to be constant while scanning and observing the sample surface. Also in close analogy, combinations of SNOM and force detection are possible, where shear force detection is used to control the probe sample distance. Other combinations are SFM with thermal probes [13] or SNOM with capacitance measurements [14]. Examples of advances in the field are the implementation of microfabrication procedures, where combinations of SPMs are realized in one structure; for instance, thermal sensors, which are integrated in the tip of a force sensor or microfabricated photodiodes at the apex of the probing tip have been achieved [15]. Furthermore, the combination of probe microscopy with mass spectroscopy (Scanning Tunneling Atom Probe – STAP) [16] or photoemission microscopy with a scanning aperture [17] appears promising. The use of optical spectroscopy of

SNOM or the inelastic tunneling spectroscopy [18] are other methods that are candidates to perform local chemical analysis.

Let us focus a little bit on STAP where the combination of STM with mass spectroscopy is challenging since the high lateral resolution of STM topography can be combined with chemical identification [18]. In STAP a sharp STM tip is scanned over a surface, and a short, positive voltage pulse is applied to the tip above the region of interest, causing some atom transfer from the sample to the tip. Then the STM tip is retracted from the surface and positioned in front of a time-of-flight mass spectrometer. Like it happens in atom probe technology, a large, negative DC voltage and short voltage pulse of a few nanoseconds are applied to the tip, leading to field desorption of atoms from the tip. The flight time of the ions from the tip to the detector is measured, and the mass-to-charge ratio (m/n) can be determined through

$$\frac{m}{n} = \alpha (V_{DC} + \beta V_P) \frac{(t - t_0)^2}{d^2} \tag{6.3}$$

where V_{DC} is the DC voltage on the tip, Vp is the pulse height, t the flight time, t_0 the delay time, and d the drift distance. It has been possible to acquire mass spectra from STAP probing tips [16], where the elements W, Si, and Ge were detected, which were previously picked up on a sample surface. To mitigate the diffusion of atoms on the probing tip, this may have to be cooled.

In principle, an SPM presents the possibility of operating in a controlled environment, liquid environment, high vacuum, or ultra-high vacuum, provided that the appropriate adaptations are carried out, which are in many cases already commercialized. The SPM technique, in addition to providing 3D images in real space, also allows measuring and modifying the surface of the sample (nanomanipulation and nanolithography). The images obtained in each of the SPM variants depend on the type of interaction between the probe and the sample. Scanning in x, y, and z is done by making use of piezoelectrics and a feedback loop. A 3D surface image $[z(x,y)]$ is obtained through multiple scans $[z(x)]$ shifted laterally from each other in the y direction. The absence of mechanical, thermal, and acoustic vibrations is an essential condition for obtaining the optimal functioning of an SPM. The maximum scanning and sampling rates are mainly determined by the resonance frequencies of the support structure, the tip, and the piezoelectric system. In the best instruments, these frequencies go up to about 1 MHz. High scanning speeds help to reduce the drift effect (thermal agitation of the sample surface and of the interacting system) but such a procedure can only be performed for small scanning areas and low roughness samples. More complex functionality seems to be desirable, like the combination of high lateral resolution with analysis of the local chemical composition. In contrast to the material-specific imaging, which is observed in friction force microscopy, phase contrast in tapping mode or local adhesion measurements, chemical analysis prerequisites that a fingerprint is created, which determines the elemental composition in a unique way.

Mechanical, thermal, electrostatic, and chemical interactions, or some combinations among them, are exploited to modify molecules, nanostructures, or surfaces with SPMs. The SPM modification and manipulation approaches include the sophisticated control of attractive van der Waals forces to move atoms. It is even possible that some of the AFM manipulation methods based on the control of a chemical reaction are able to modify and/or pattern surfaces with nanoscale accuracy. The small size of the AFM tip's apex is often used to confine a variety of chemical reactions and/or physical processes. In some cases, the AFM probe acts as a carrier of molecules that, upon mechanical contact with the surface, will be transferred to the sample surface to form new chemical bonds. In other cases, an electric field is a driving force that promotes the field-induced evaporation of atoms from the tip. Other processes are mediated by the presence of a liquid meniscus which provides both the chemical species and the spatial confinement for a chemical reaction to occur. In some cases, a hot tip facilitates the breaking of chemical bonds on certain polymer surfaces. On another perspective, using field-induced chemistry, unusual chemical reactions have been observed in the presence of very high electric fields (10–50 V/nm) with field ion microscopes. Since these fields are of the same order of those inside atoms and molecules, they are strong enough to induce the rearrangement of molecular orbitals leading to new chemical reactions. A force microscope interface offers a precise control and manipulation of high electric fields by changing the conductive tip-surface separation and/or the voltage. In contrast to field-ion microscopy that operates in UHV, the AFM experiments can be performed in ambient or liquid environments, which increases the number of available chemical species. Worth to note that high electric fields can be achieved in AFM by applying moderate voltages. For a hyperboloidal conductive tip of radius R, at a distance D from a flat surface, the electrostatic field is easily estimated and varied by either modifying the tip-surface distance or the voltage. Since the tip-surface separation is in the nanometer or sub-nanometer range, very high fields (1–30 V/nm) can be generated in the presence of tens of volts. In the presence of a gas, liquid, or solid material, those fields could be used to promote the evaporation of atoms from the tip, the breaking of chemical bonds in the molecules of the surrounding environment or the formation of new products. Thus, the success of the relationship between field-induced chemistry and nanofabrication relies on the control and manipulation of electric fields at the nanoscale. Electric field-induced dissociation of carbon dioxide can be used for the generation of a carbonaceous pattern with an AFM conductive tip. The transformation of CO_2 in the presence of an electric field is considered to occur in three major steps: the first one is the capture of the molecules into the gap region between the two electrodes, the second step is the activation of the molecules, and the third corresponds to the fabrication of a carbonaceous compound. The first step represents a field-induced diffusion process, which polarizes the carbon dioxide, and the resulting dipole moment interacts with the field. Quantum chemical calculations have shown that the field modifies the potential energy surface of dissociation into CO + O and shifts the energy of the molecular orbitals which leads to carbon dioxide splitting. The field shifts the energy of the molecular orbitals and modifies the dipole moment. As the field increases, the molecular dipole moment builds up and different local charges form on the atoms. The carbon atom becomes more positive, while the charge difference between the two oxygen atoms increases. Also, the two bond lengths become asymmetric. The field also greatly affects the LUMO, which becomes nearly degenerate with the HOMO at 40 V/nm. At high fields, the HOMO becomes located outside the molecule. Closing the HOMO-LUMO gap promotes the detachment of the oxygen atom and the formation of two fragments. Calculations, at fields higher than 40 V/nm, have shown that the molecule spontaneously breaks down into CO + O. SPMs have also been used to investigate other types of carbon dioxide processes such as the photocatalytic dissociation of CO_2 on titanium dioxide surfaces.

Probe microscopes are indeed going to overcome the limits of microscopy and will become tools for nanoscience and nanotechnology, for the construction of complex structures (transistors, logical elements based on molecular electronics, or complete memory devices). As proposed by Feynman, microrobots may be constructed by these tools, which will start to fabricate other structures. Thus, we are currently at the stone age of nanotechnology, where instruments, such as SPMs, were developed to perform simple manufacturing exercises. Previous local manipulations by probe microscopy from Eigler and co-workers have shown that single atom manipulation by STM is possible. The probing tip could also be designed to perform more complex tasks, such as nano-tweezers, where clusters are picked up by the nano-tweezer and electrical characterization of the nanostructures is performed [19]. Apart from the movement of individual atoms or molecules, it can be also of interest to induce self-directed growth by the probing tip. For example, it is possible to remove single hydrogen atoms from hydrogen passivated surfaces [20]. These artificially created defects may then act as nucleation sites for the growth of molecular wires [21]. On another perspective, the use of parallel operation of microscopes may lead from pure laboratory experiments to practical devices. One of the long-term goals is the molecular supercomputer with clock frequencies in the THz-regime.

In the aim of molecular electronics, molecular single electron transistors operating in the THz-regime have been proposed [22]. A possible three-terminal device incorporates source, drain and gate electrodes on a molecular level. An alternative approach is the atom relay transistor, where the mechanical motion of an atom causes conductance changes in an atom wire [23]. In single-molecule transistors, molecules are chemically attached to electrodes of nanometer-sized break junctions [24]. Other examples are the transistor-like behavior of single carbon nanotubes, which represents a molecular wire [25]. However, one of the major problems is to address individual molecules. Now precisely the STM and SFM can be used as imaging tools and as local electrodes to study the electrical properties of the molecules. Furthermore, these SPMs may also be used to manipulate the molecules, for example, move the molecules between designed electrodes. It is then possible to idealize an atomic switch to control the conductance through an atomic wire [23]. To get the topographical as well as the chemical information about a molecule with an SPM, the technique called Tip-Enhanced Raman Scattering (TERS) is a convenient achievement. Here, the molecular vibrations are suitable to get a fingerprint to identify subgroups in a molecule.

The confinement of light in small spaces is also important in ultra-sensitive spectroscopy.

The Raman effect is the inelastic scattering of photons by matter, which means that there is an exchange of energy and a change in the direction of light. However, most incident photons are elastically dispersed (Rayleigh scattering), so they retain energy even though they change direction. Because of energy conservation, the material gains or loses energy in the process. Raman spectroscopy is based on the inelastic scattering of photons (Raman scattering) in which energy change provides information about the internal modes of the system. The elastically dispersed radiation is filtered out while the rest of the scattered light goes to a detector. This Raman scattering is generally very weak. SERS enhances Raman dispersion by molecules adsorbed onto conductive surfaces. The increase factor is very high, and in some cases, it can even allow the detection of a single molecule. This increase is due to plasmonic excitation when the plasmonic frequency of the nanoparticles enters resonance with that of the incident radiation. Plasmonic oscillations must be perpendicular to the surface; if they are in the plane with the surface, there will be no dispersion; therefore, surfaces must be rough (e.g., nanoparticle arrays). In the TERS (tip-enhanced Raman scattering) technique, the increase in Raman scattering occurs only in a tip normally coated with gold. The maximum resolution achievable using an optical microscope, including confocal Raman microscopes, is limited by the Abbe criterion (approximately half the wavelength of incident light). Therefore, the use of

SNOM is of great relevance. While with SERS spectroscopy the signal obtained is the sum of a relatively large number of molecules, TERS overcomes these limitations as the signal comes mainly from molecules within a few tens of nanometers of the tip. TERS uses locally enhanced fields where the tip apex in close vicinity to the metallic surface gives a strong enhancement of the electromagnetic field. By matching the resonance of the nanocavity plasmon to the molecular vibrations one can increase the sensitivity. This matching is achieved by changing the distance with the STM feedback. Under these conditions, it becomes possible to image, for instance, a single porphyrin molecule on silver with sub-molecular resolution [26]. The Raman spectra display characteristic differences for the different positions on the porphyrin molecule on Ag (111).

Most nano-objects used in many applications alter their physical and chemical properties over time. This process is responsible for the apparent aging of these materials and is a primary determinant of their longevity [27]. The friction and wear behavior of nano-objects can be influenced by the nano-object size, shape, and mechanical properties. In single nano-object contact studies, there is a friction force dependence on the real area of contact. In multiple nano-object contacts, friction and wear reduction occur due to the lowering of the real area of contact between sliding surfaces. The ability to control and manipulate friction during sliding is also extremely important for a large variety of applications, and development of novel efficient methods to control friction requires understanding the microscopic mechanisms of frictional phenomena. The frictional aging at the nanoscale may result from the nucleation of capillary bridges and strengthening of chemical bonding, and it imposes serious constraints and limitations on the performance and lifetime of micro- and nanomachines. Performing friction experiments on single nano-objects is a powerful way of tracking their aging, and a variant AFM called friction force microscopy (FFM) or lateral force microscopy (LFM) has proved to be a powerful tool to measure the friction force on the nanoscale since its invention [28]. This technique incorporates an optical detection system of the lateral deflection of the microbeam, which is of great interest in nanotribology. During contact mode scanning, a variation of local friction causes a lateral force on the tip, which is registered by the photodetector and used to construct a lateral force image, analogous to the normal force deflection image. Since both friction and surface topology contribute to lateral deflection, the AFM and FFM images must be recorded simultaneously, to deconvolute the LFM from the topographic effects, and to distinguish the two effects. The FFM operating in UHV has allowed considerable advances in the understanding of atomic-scale friction mechanisms, namely the transition from a *stick-slip* regime to continuous slip. Despite the use of the FFM is practically limited by the tip material and its geometric shape, it is possible to apply it to directly measure the force between a surface and a colloid particle (by attaching the particle to the force sensor in the microscope), and this way, the uncertainties in force measurement caused by, for example, the irregular shape of the AFM tip could be avoided because the properties of the attached particle such as the size, shape, and material were controllable [29]. The colloidal probe technique has become a well-established and powerful tool for the study of surface forces and nanotribology. Further, the AFM can be used in both of its dynamic and contact modes for the manipulation of nanoparticles. The maximum sliding friction force between polymer latex spheres and highly oriented pyrolytic graphite (HOPG) surface, was obtained by using the dynamic force microscopy (DFM) [30]. The main disadvantage of the DFM is that the interaction between the tip and the particle can result in vibration instability of the tip, which consequently causes the difficulty in controlling the movement direction of particles. Besides, the method can only obtain the threshold force separating the particle and the surface but cannot continuously characterize the sliding behavior of the particle.

The great potential of SPMs to modify surfaces has proved to be critical in nanolithography, being able to assume multiple variants and even include combinations such as STM/AFM, using electrically conductive and piezoresistive cantilevers that are simultaneously sensors of tunneling current and of interaction force with the sample. Another possibility already discussed is the combined ability of the tip to be heated and record its temperature. This allows the recording and reading of digital information on a thermally sensitive surface. The tip of an AFM can also be used to catalyze a surface reaction, with the aim of chemically synthesizing nanostructures; for instance, a monolayer containing terminal azide groups is contacted with a platinum-coated AFM tip, and in the presence of hydrogen, the azide groups are reduced to amine groups. Thus, nanostructures are formed by attaching aldehyde-modified latex nanospheres to the amino groups. On the other hand, applying polarizations greater than 5 eV to an STM, it starts to operate in field emission mode (since the electric field in the metallic tip is in the order of MeV/cm); that is, applying the well-known Fowler-Nordheim equation, it implies that for sufficiently high fields the emitted electronic current is proportional to the square of the electric field. Thus, in the constant current mode of operation, the sample-type distance d, is proportional to the polarization V, and the electronic current is proportional to $(V/d)^2$ in the uniform field zone. The only lens that focuses the electrons is the electric field at the tip, which diverges from it; therefore, the smallest possible distance d should be used. Using vacuum and introducing TMA (trimethylamine), given that electric fields of 80 MeV/cm correspond to about 7 eV of kinetic energy, this value is not enough to ionize the TMA molecules (whose ionization energy is 9 eV). Therefore, the lithography observed with this method is not due to the ionization of the TMA by the field, but due to the field emission electronic current from the tip. These emitted electrons are accelerated by the field and collide with the TMA, ionizing it, which is also accelerated by colliding with the substrate. Within another concept, dip-pen lithography (DPN) technique uses the tip of an AFM as a pen, in which the "molecular ink" is a solution adhering to it and which flows through the meniscus onto the substrate, which plays the role of "paper." If the tip has a metallic coating it is still possible to apply a voltage between it and the substrate, performing what is called electric field assisted DPN. With it, it is possible to deposit metallic nanoparticles by controlling the applied polarization. The solution can be made from single molecules or macromolecules. Using DPN lithographic software, the width of the deposited dots or stripes is controlled by the sweep speed, the diffusion coefficient D, and the dwell time at each dot on the substrate, tp. Thus, the area of a "painted" point is given by Dtp, and the linear scratch width will be equal to D/v. To visualize the image of the drawn patterns, the LFM operating mode is used, as the contrast will be sharper, since the height of the scratches is generally not sufficient to generate a good contrast in normal contact mode and even much less in tapping mode. The DPN system includes a controlled-environment chamber and two substrates, one of which is made of paper and the other that is made of ink (reservoir with solution). The reservoir is supplied by microfluidic channels. There are alignment markers and precise control through piezos, which allow you to reposition the tip exactly after refilling the solution. Each refill consists of a liquid drop of volume on the order of 10^{-18} liters.

Let us now delve a little deeper into the mechanisms of field evaporation and surface diffusion induced by field gradients (electromigration). The field evaporation (or desorption) process is the basic operating process of Field Electron Microscopy (FEM). If the sample-tip distance d is large, the atom-tip, V_{at}, and sample-atom V_{as} potential curves do not overlap significantly. But when d decreases, there begins to overlap and the curve of the total potential energy of the atom interacting with the tip and with the sample, $V_a = V_{at} + V_{as}$, will present an activation barrier of height E^* on the tip side, and $E^{**} = E^* + (E_s - E_t)$ on the sample side, where E_t is the binding energy of the atom to the tip and E_s the binding energy of the atom to the sample. At room temperature, the atom transfer rate from tip to sample, $k = \nu \exp(-E^*/k_BT)$, is 1 s^{-1} for $E^* = 0.772$ eV and $\nu = 10^{13}$ s^{-1}; on the other hand, the atomic transfer rate in the opposite direction (substrate → tip), $k' = \nu \exp(-E^{**}/k_BT)$, will therefore be smaller. This explains the possibility of either depositing an atom from the tip to the surface (if $E_t > E_s$), or removing an atom from the surface to the tip (if $E_t < E_s$). As there is no field applied, all these considerations also apply to handling by AFM with a dielectric tip. When there is an applied electric field (STM or AFM with polarization in conductor tip) and the sample-type distance decreases, there will be, according to the charge exchange model [31], a variation of the ionic and atomic potential energy curves. Note that in the absence of a field, these curves are $(V_{is} + V_{it})$ and $(V_{as} + V_{at})$. When a positive electric field is applied to the tip, an applied external potential equal to $-neE_z$ is added to the ionic potential, and therefore the potential curves change to $V_I = V_I(0) - neEz$. Here, n is the state of charge of the ions and z is the distance to the tip. Thus, the field evaporation effect translates into the energy barrier that an atom of the tip has to overcome in order to deposit itself in the sample is strongly reduced in the case of positive ions. Field evaporation of negative ions is more complicated, as electronic field emission starts at 3 V/nm. When the field increases to 6 V/nm, the field emission current density will be sufficient to melt the tips of most metals by resistive heating.

On the other hand, relatively to the surface diffusion process, it occurs in the presence of a field. In fact, in the absence of a voltage sample type, the adsorbate does not diffuse, due to the periodic potential of the substrate network (electron density corrugation). In the presence of the field, there will be a polarization energy, which is position dependent and equal to $(-\mu E^{-1/2} \alpha E^2)$, where α is the polarizability. When this energy is added to the periodic surface potential, the potential energy becomes skewed in the direction that the field is stronger. Then, diffusion begins in which the adsorbate moves to the region directly below the tip. Thus, the activation energy of the thermal diffusion process is reduced by the field gradient. Furthermore, by applying voltage, the tunneling current increases and adds to the field emission current, heating the sample and promoting diffusion. This effect can also serve to sharpen the tip of the tip due to heat melting. By applying voltage pulses to the tip, adsorbate can then be manipulated. The interaction that moves the adsorbate involves the divergent electric field that the type of the tip causes, which induces an electric dipole moment p in the electron cloud of the adsorbed atom. Since the electric field is strongest near the tip the induced dipole is attracted to it by a force equal $-dU/dz = pz\, dEz/dz$, where $U = -p \bullet E$ is the electrostatic interaction energy. The strength of this force depends on the field, which in the first approximation is $E = V/z$, where V is the polarization imposed on the tip and z is the tip-adsorbate distance. Both parameters can be adjusted via STM control. However, if the atom jumps to the tip's surface, control is lost. Let us recall that in Rutherford's atomic model, the electron cloud can be considered as a charge $-Ze$ of uniform volume and radius r_a, and the nucleus with charge Ze is located in the center. It is shown that the force on the nucleus is zero when it is displaced a distance δ from the center of a negative charge, where $\delta = 4\pi\varepsilon_0 r_a^3 E/(Ze)$, where E represents the contribution to the field, at the position of the nucleus, due to charges external to the atom. Once the dipole moment is $p = \alpha E = q\, d$, then if $d = \delta$ and $q = Ze$, and the electrical polarizability $\alpha = 4\pi\varepsilon_0 r_a^3$. In the simple case of the hydrogen atom, it is $\alpha = 18\pi\varepsilon_0 a_0^3$. In other atoms, polarizability increases with the number of electrons in a complicated way. An estimate of the effect [32] gives $\alpha \approx 2ne^2 \langle z^2 \rangle / I$, where n is the number of electrons, $\langle z^2 \rangle$ is the average of the square of the electron's position along the applied electric field, and I the first ionization potential. The relevant fact that polarizability becomes very high with the increase in the number of electrons in the atom initially led to the choice of Xenon (Z = 54) as the atom that moves along the Cu (111) surface in the first studies of atomic manipulation with STM.

The size dependence of the friction has been investigated, and it was shown that the particle can steadily be manipulated according to a preset path if the normal force is above a threshold load. The friction between the particle and the substrate can be obtained quantitatively. The experimental data indicates the friction between particles and substrate is related by: $F_f \propto R^{2/3}$, where R is the radius of the nanoparticle [33]. This gives a side reason that the Hertzian theory is the suitable description of the studied system, in the case of particle sizes down to the nanoscale. It was noted that the ratios between F_f-static and F_f-kinetic are in the range of 0.3–0.6, depending on the size of the particles. The ratio is not changed, whether the particles were in different areas of the surface, the tip normal force was varied, or even whether the surface was modified.

High-resolution AFM is the most suitable instrument for friction measurements of nano-objects [33, 34]. Carbon nanotube tips having a small diameter and high aspect ratio are good for high-resolution imaging of surfaces and deep trenches, in the tapping mode or noncontact mode. For that purpose, it can be used a carbon nanotube tip, ProbeMax, commercially produced using mechanical assembly by Piezomax Technologies. Once the nanotube is attached to the tip, it is usually too long to image with. It is shortened using an AFM applying voltage between the tip and the sample. In *method A* (using the height mode with parallel scans) in addition to topographic imaging, it is also possible to measure friction force when the scanning direction of the sample is parallel to the y-direction (parallel scan). If there were no friction force between the tip and the moving sample, the topographic feature would be the only factor to cause the cantilever to deflect vertically. However, the friction force between the sample and the tip will also cause a cantilever deflection. Assuming that the normal force between the sample and the tip is W_0 when the sample is stationary (W_0 is typically in the range 10–200 nN) and the friction force between the sample and the tip is W_f as the sample scans against the tip. The direction of friction force (W_f) is reversed as the scanning direction of the sample is reversed from the positive (y) to the negative ($-y$) direction:

$$\text{direction} \quad \overrightarrow{W}_{f\,(y)} = -\overrightarrow{W}_{f\,(-y)} \tag{6.4}$$

When the vertical cantilever deflection is set at a constant level, it is the total force (normal force and friction force) applied to the cantilever that keeps its deflection at this level. Since the friction force is in opposite directions as the traveling direction of the sample is reversed, the normal force will have to be adjusted accordingly when the sample reverses its traveling direction so that the total deflection of the cantilever will remain the same. One can calculate the difference of the normal force between the two traveling directions for a given friction force W_f. First, by means of a constant deflection, the total moment applied to the cantilever is constant. If we take the reference point P to be the point where the cantilever joins the cantilever holder (substate), we have that

$$(W_0 - \Delta W_1)L + W_f l = (W_0 + \Delta W_2)L - W_f l \tag{6.5}$$

Thus

$$W_f = \frac{(\Delta W_1 + \Delta W_2)L}{2l} \tag{6.6}$$

where ΔW_1 and ΔW_2 are the absolute value of the changes of normal force when the sample is traveling in (–*y*) and (*y*) directions; *L* is the length of the cantilever and *l* is the vertical distance between the end of the tip and point *P*. The coefficient of friction (μ) between the tip and the sample is then given as

$$\mu = \frac{W_f}{W_0} = \frac{(\Delta W_1 + \Delta W_2)}{W_0}\left(\frac{L}{2l}\right) \tag{6.7}$$

If adhesive and attractive van der Waals forces (and the indentation effect as well, which is small for rigid samples) can be neglected, the normal force W_0 is then equal to the initial cantilever deflection H_0 multiplied by the spring constant of the cantilever. The sum ($\Delta W_1 + \Delta W_2$) can be measured by multiplying the same spring constant by the height difference of the piezo tube between the two traveling directions ($-y$ and y) of the sample. Denoting this height difference by $\Delta H_1 + \Delta H_2$ one can thus write:

$$\mu = \frac{W_f}{W_0} = \frac{(\Delta H_1 + \Delta H_2)}{H_0}\left(\frac{L}{2l}\right) \tag{6.8}$$

Since the vertical position of the piezo tube is affected by the surface topographic profile of the sample in addition to the friction force being applied at the tip, this difference must be taken point-by-point at the same location on the sample surface. In addition, precise measurements of L and *l* (which should include the cantilever angle) are also required. If the adhesive forces between the tip and the sample are large enough that it cannot be neglected, one should include it in the calculation. However, there could be significant uncertainty in using this method. It should also be pointed out that the equations of this method were derived under the assumption that the friction force W_f is the same for the two scanning directions of the sample. This is a first approximation since the normal force is slightly different for the two scans and there may also be a directionality effect in friction. However, this difference is much smaller than W_0 itself, so one can ignore the second-order correction.

Using *method B* (mode with perpendicular scan) to measure nanoscale friction, the sample is scanned perpendicular to the long axis of the cantilever beam (scanning along the *x* or –*x* direction) and the output of the two horizontal quadrants of the photodiode detector is measured. In this arrangement, as the sample moves under the tip, the friction force will cause the cantilever to twist. Therefore, the light intensity between the left (L) and right (R) detectors will be different. The differential signal between these two detectors is denoted as FFM signal *[(L–R/(L+R)]* and it can be related to the degree of twisting, and so to the magnitude of friction force. Despite *method A* (using height mode with parallel scan) is very simple to use, the piezo scanner often displays a hysteresis when the traveling direction of the piezo is reversed, which would make it difficult to measure the local height difference of the piezo tube for the two scans. Therefore, *method B* is more commonly used. Regarding calibrations, to calculate the absolute value of normal and friction forces using the AFM and FFM signals, it is necessary to first have an accurate value of the spring constant of the cantilever, which can be estimated using the geometry and the physical properties of the cantilever material, and experimentally measured through resonant frequencies of the beams [33, 34].

Nanoparticle properties are not only determined by their shape and dimensions, but they also depend on the detailed topography of single particles, in addition to their charge and chemical composition. An example is the exposed surface area, which is highly dependent on the particle surface topography, and will influence chemical reactivity. A quantitative approach to describing particle surface topography is by roughness analysis, and the same concepts generally used in 2D surfaces can be applied to describe surface structures of 3D objects, if the overall particle shape can be separated from the short-scale roughness. Such an analysis requires detailed data for the surface topography, and AFM gives highly resolved spatial information in the form of topographical maps. AFM supplies direct measurements without elaborate sample preparation as in the case of electron microscopy. AFM requires the immobilization of particles to a solid support which, for most particles, can be achieved via electrostatic interaction or an immobilization procedure.

The friction coefficient of nano-objects is proportional to the contact area as opposed to the macroscopic dry sliding friction. Results of several studies showed that the friction of nanoparticles increases with particle size [35, 36]. Further, the friction of nano-objects increased logarithmically with the sliding velocity as opposed to the macroscopic friction which is independent of sliding velocity [37]. Amontons' law states that friction is independent of load, while Coulomb's law states that dynamic friction is lower than static friction but independent of velocity. The static friction tends to increase with time, and the coefficient of friction is a continuous function of state and time. Referring to the 1D schematic in Figure 6.2 with a puller at position x_p moving at velocity v, a spring with constant k and block mass m subject to normal load N, one can write the block's equation of motion as

$$m\ddot{x} = k(x_p - x) - F_0 \tag{6.9}$$

where the friction $F_0 = \mu N + c(dx/dt)$, and c is a viscous damping coefficient. The positioning of the tip above the particle can be performed in AFM in noncontact mode (or contact mode at low normal force F_N) as to not prematurely push the nanoparticle away. The feedback switch is then set to operate in contact mode, and a comparatively high normal force set point is selected. If the static friction between tip and particle exceeds the static friction threshold between sample and substrate, $F_{S(t\text{-}p)} > F_{S(p\text{-}s)}$, then the tip can initiate sliding of the particle. The contact between tip and particle can be approximated as a contact between an elastic sphere and a flat surface. According to Hertzian contact mechanics, the contact area tip-particle, $A_{t\text{-}p}$ increases with the normal force F_N as [36]: $A_{t\text{-}p} \propto F_N^{2/3}$. Since friction increases with the contact area, it follows that $F_{S(t\text{-}p)}$ increases with normal force F_N, while $F_{S(p\text{-}s)}$ is constant in F_N. Hence, it becomes possible to tune the fulfillment of the former equation by adjusting the normal force, thus selectively rendering the particle mobile or immobile relative to the substrate.

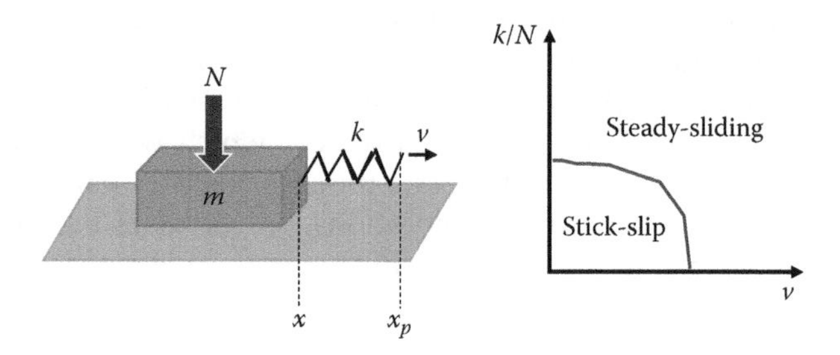

FIGURE 6.2
Left: A puller moving at velocity v causes a block of mass m, subject to normal load N to slide via a spring with constant k. Right: A schematic kinetic friction phase diagram indicating the block response as a function of v, k, and N in the stick-slip or steady-sliding regimes.

Due to the impressive z resolution (<1 nm) and lateral resolution (≈ 10 nm) of the SPM technique, it has been used also to investigate microporous and mesoporous carbons, demonstrating the effect of pore size on the ion insertion kinetics, in electrochemical energy storage devices [27, 36]. During charge/discharge of porous carbon electrodes, ions move in and out of the pores resulting in minuscule volume changes. It is then possible to monitor the volume changes *in situ* under electrochemical control. Nanostructured carbon materials have been widely investigated and utilized for energy storage and conversion devices, especially as electrode materials for supercapacitors because of their high electrical conductivity, high specific area, huge corrosion resistance, high-temperature stability, tunable porous structures, low weight, and long cycling stability and lower costs than transition metal oxides electrodes.

The parallel operation of cantilevers gives the opportunity to construct cantilever array sensors that operate as artificial electronic noses [38]. A variety of coatings is deposited on the cantilevers. Each cantilever reacts in a different way to the exposure to chemicals. These different responses are then used to discriminate or distinguish between different gases or liquids. All operation modes of SPMs can be employed with an array of cantilevers, where each cantilever is functionalized in a specific way. The goal is to achieve "orthogonal" information, which allows to distinguish between different analytes. A designed experiment is orthogonal if the effects of any factor balance out (sum to zero) across the effects of the other factors. Orthogonality guarantees that the effect of one factor or interaction can be estimated separately from the effect of any other factor or interaction in the model. Another major advantage of arrays is the differential measurement, where a coated cantilever is compared with an uncoated cantilever. The differential measurement is much less affected by disturbances from the environment.

6.2 Plasmas and *Nano* for Sustainability

The field of applications of plasmas in nanophysics and nanotechnology is extensive, with many possibilities covering both top-down and bottom-up fabrication approaches, ranging from the growth and functionalization of aligned nanotubes, and sheaths of graphene, through to the growth of gas-phase nanocrystals and dust particles, surface deposition of different structures, functionalization of the surface and conditions for developing means to prepare superhydrophobic surfaces, and plasma catalysis.

Under equilibrium conditions, very high temperatures are necessary to produce a significant concentration of electrons in a gas plasma. Thermal plasmas are those in which the heavy species temperature is approximately equal to the electron temperature (typically in the range 10,000–25,000 K). The plasmas are at or close to atmospheric pressure, and the degree of ionization is high, with electron densities reaching around 10^{23} m^{-3}. Thermal plasmas can be formed by DC or AC electric fields (electric arcs), inductively coupled rf energy, microwave energy, or laser energy. They can be achieved in two ways, either at high temperature, typically ranging from 4,000 K to 20,000 K, depending on the ease of ionization, or at high gas pressure. It is the square of the ratio of the electric field (E) to the pressure (P), i.e., $(E/P)^2$, which is proportional to the temperature difference in gas discharges [39].

A nonequilibrium plasma (also called nonthermal or cold) is a plasma which is not in thermodynamic equilibrium because the electron temperature is much hotter than the temperature of heavy species (ions and neutrals). As only electrons are thermalized, their Maxwell–Boltzmann velocity distribution is very different from the ion velocity distribution. The more usual way of producing highly ionized gases is an electric discharge, where the passage of an electric current results in a nonequilibrium situation in which the electrons are usually at a very high temperature with respect to the heavier species. Langevin equation describes the motion of the electrons and can be simplified in the form:

$$\vec{J} = \sigma_0 (\vec{E} + \vec{u} \times \vec{B}_0) \tag{6.10}$$

This represents a generalized Ohm's law, where $\boldsymbol{u}$ is the average electron velocity and $\boldsymbol{E}$ and $\boldsymbol{B_0}$ are, respectively, the electric and magnetic fields acting on the electron; the current density is equal to $\boldsymbol{J} = -n_{ion}e\boldsymbol{u}$, and the conductivity of the plasma is given by $\sigma_0 = n_{ion} e^2/m_e v$, where v is the electron collision frequency (number of collisions per second that the average electron undergoes with the heavier particles in the plasma).

The approximation that the electronic partition function of gases is equal to the unity is only acceptable at ordinary temperatures if the electronic ground state is nondegenerate. However, at temperatures that are attained in plasmas, it is necessary to take into account the population of the excited electronic states, and their density becomes very great as the ionization energy is approached. Thus, the electronic partition function, in this case, can be expressed in the form

$$Z_{electrons} = \sum_{i=0}^{\infty} g_i e^{-\varepsilon_i/kT} = g_0 + g_1 e^{-\varepsilon_1/kT} + g_2 e^{-\varepsilon_2/kT} + \dots \tag{6.11}$$

as the energy of the ground electronic state is taken equal to zero. This series is not convergent but in practice, since the energies of the electronic states all approach the first ionization potential value I_1, the system can be considered to consist of two energy levels: the ground state with degeneracy g_o, and the ionized state of degeneracy

$$g_{ion} = \sum_{i=1}^{m} g_i \tag{6.12}$$

where m represents a somewhat arbitrary limit, and in any case, $g_{ion} >> g_0$.

In a plasma described only by two energy levels, the ratio between the number of ionized and unionized species is thus given by

$$\frac{N_{ion}}{N_{union}} = \frac{g_{ion}}{g_0} = e^{-(\varepsilon_{ion}-\varepsilon_0)/kT} \tag{6.13}$$

where g_0 and g_{ion} are the statistical weights and $\varepsilon_0 = 0$, $\varepsilon_{ion} \approx I_1$ are the corresponding energies. Therefore, the percentage of ionized species is 50% at a temperature equal to $I_1/[k\ln(g_{ion}/g_0)]$, which is well below that which corresponds to the ionization potential. The electrons, being much lighter than either the positive ions or neutral species, move much more rapidly and the transfer of their kinetic energy to the heavier species is very inefficient. Thus, it is the translational partition function for the electrons that determines the ratio g_{ion}/g_0:

$$\frac{g_{ion}}{g_0} = \frac{1}{n_{ion}}\left(\frac{2\pi m_e kT}{h^2}\right)^{3/2} \tag{6.14}$$

in which n_{ion} stands for the ionic density. Consequently, one can write that:

$$\frac{N_{ion}}{N_{union}} = \frac{1}{n_{ion}}\left(\frac{2\pi m_e kT}{h^2}\right)^{3/2} e^{-I_1/kT} \tag{6.15}$$

often called the Saha equation. If the "electron gas" is considered ideal (n_{ion} = P/kT) the equilibrium constant for the ionization process, K_P, depends on the fraction of ionized species, $\alpha = [N_{ion}/(N_{ion}+N_{union})]$ in the following way:

$$K_P = \frac{\alpha^2 P}{1-\alpha} = kT\left(\frac{2\pi m_e kT}{h^2}\right)^{3/2} e^{-I_1/kT} \tag{6.16}$$

This means that extremely high temperatures are required to produce a significant ionization of a gas under equilibrium conditions. However, the ionization is enhanced at low pressures, and assuming electrons do not have a significant effect on the pressure, it is possible to estimate the degree of ionization by the following expression

$$\frac{\alpha^2 P}{1-\alpha} = 2.5 \times 10^{-4} \quad T^{5/2} e^{-1.16\times 10^4 I_1/T} \tag{6.17}$$

when the pressure is given in Torr and the first ionization potential in electron volt. It is interesting to notice that the transition from the unionized to the total ionized state occurs within a very limited temperature range (Figure 6.3). This is suggestive of a phase transition and so led Crooks in 1879 to consider a plasma as the fourth state of matter. Due to the very long-ranged Coulombic interactions between the charged particles, an electron in the plasma interacts with many of its neighbors, resulting in a cohesive property analogous to that of a jelly.

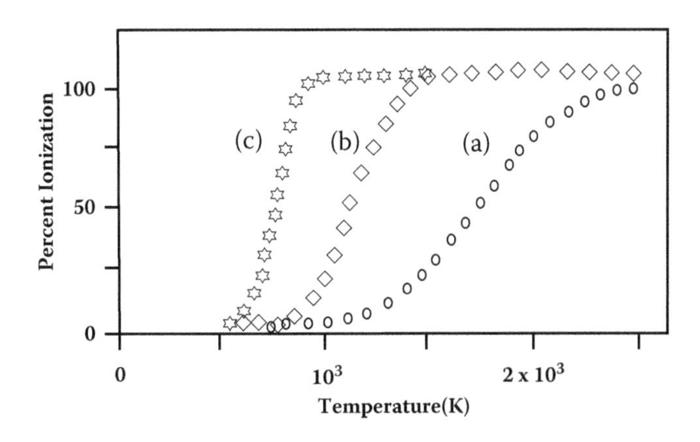

FIGURE 6.3
Fraction of ionized species as a function of temperature in hydrogen plasmas, as calculated from the Saha equation: (a) P = 760 Torr; (b) P = 1 Torr; (c) $P = 10^{-3}$ Torr.

In plasmas, electrolytes, and colloids, the measure of a charge carrier's net electrostatic effect in a solution and how far its electrostatic effect persists is called Debye length (or Debye radius) [40]. The analogous quantity at very low temperatures is known as the Thomas–Fermi length and is of interest in describing the behavior of electrons in metals at room temperature. In the Thomas–Fermi screening theory the electric potential decays in a metal as $\phi(r) \sim e^{-ik0r/r}$, where k_0 can be related to the electronic compressibility through

$$k_0^2 = 4\pi^2 \frac{\partial n}{\partial \mu} \tag{6.18}$$

where n stands for the density and μ for the chemical potential. So, the more compressible the electron gas is, the more efficient screening will be. In an analogous way, every Debye-length, the electric potential will decrease in magnitude by 1/e. The Debye length arises from the thermodynamic description of large systems of mobile charges of different species where they are distributed in a continuous medium that is characterized only by its relative static permittivity. This distribution of charges within this medium gives rise to an electric potential $\Phi(\mathbf{r})$ that satisfies Poisson's equation. If we further assume the system to be in thermodynamic equilibrium with a heat bath at absolute temperature T, then the concentrations of discrete charges, $n_j(\mathbf{r})$, maybe considered to be thermodynamic (ensemble) averages and the associated electric potential to be a thermodynamic mean field. Thus, the concentration of the jth charge species is described by the Boltzmann distribution, considering, n_j^0 the mean concentration of charges of species j:

$$n_j(r) = n_j^0 exp\left(-\frac{q_j \Phi(\mathrm{r})}{k_B T}\right) \tag{6.19}$$

Identifying the instantaneous concentrations and potential in the Poisson equation with their mean-field counterparts in Boltzmann's distribution yields a nonlinear equation (Poisson–Boltzmann equation), whose solutions are known for some simple systems and leads to the definition of Debye length, that for the case of electrons in a gas plasma is given by:

$$\lambda_D = \sqrt{\frac{\varepsilon_0 k_B T}{2 n_0 e^2}} \tag{6.20}$$

For an electrically neutral system, the Poisson equation becomes

$$\nabla^2 \Phi(\boldsymbol{r}) = \frac{\Phi(\boldsymbol{r})}{\lambda_D^2} - \frac{\rho_{ext}(r)}{\varepsilon} \tag{6.21}$$

To illustrate Debye screening, the potential produced by an external point charge $\rho_{ext} = Q\delta(\rho)$ is

$$\Phi(r) = \frac{Q}{4\pi\varepsilon r} e^{-r/\lambda_D} \tag{6.22}$$

The bare Coulomb potential is exponentially screened by the medium, over a distance of the Debye length. A plasma should exceed the mean free path between collisions and the Debye length.

When charged particle densities become sufficiently high, Coulomb interaction couples electrons and ions more strongly and there is sufficient energy transfer between the two groups so that some of the energy gained by electrons is transferred to ions. Eventually, their temperatures will become identical. At the same time, ions can transfer their momentum to the gas molecules so the energy eventually ends up in the heating of the gas and the walls of the chamber. For the same given mean energy, thermal (equilibrium) plasmas are much more efficient in producing radicals (if the mean energy is sufficient). However, overall, when total energy input is taken into account, nonequilibrium plasmas have a greater efficiency in using the energy as one may choose conditions so that electrons have a very high probability for a given process and at the same time no energy is wasted in heating of the gas and vessel. Thus, two types of plasmas may be defined. Those with lower densities of charged particles are known as the so-called nonequilibrium plasmas or low-temperature plasmas (sometimes even cold plasmas). When charged particle densities are higher usually one has equilibrium or thermal plasmas. For an isothermal plasma (where the temperatures of the electrons and the heavy species are similar) the electron density makes the magnitude of the Debye length to vary substantially, and so this may reach macroscopic values for densities below 10^{13} m^{-3}; for instance in the ionosphere ($T = 10^3$ K, $n_0 = 10^{12}$ m^{-3}, $\lambda_D = 10^{-3}$ m) and in solar wind ($T = 10^5$ K, $n_0 = 10^6$ m^{-3}, $\lambda_D = 10$ m). For a monovalent electrolyte or a colloidal suspension, with dielectric constant ε_r, and ionic strength I (in mol/m^3), the Debye length takes the new form

$$\lambda_D = \sqrt{\frac{\varepsilon_0 \varepsilon_r k_B T}{2N_A e^2 I}} \tag{6.23}$$

that in the case of a symmetric monovalent electrolyte converts into

$$\lambda_D = \sqrt{\frac{\varepsilon_0 \varepsilon_r RT}{2 \times 10^3 F^2 C_0}} = \frac{1}{\sqrt{8\pi\lambda_B N_A \times 10^{-24} I}} \tag{6.24}$$

where R and F are, respectively, the gas and Faraday constants and λ_B is called the Bjerrum length of the medium that represents the separation at which the electrostatic interaction between two elementary charges is comparable in magnitude to the thermal energy k_BT. This means that the Bjerrum length of the medium is computed using the following equation:

$$\lambda_B = \frac{e^2}{4\pi\varepsilon_0 \varepsilon_r k_B T} \tag{6.25}$$

As it was already mentioned, rapid oscillations of the electron density in conducting media (such as plasmas or metals) in the ultraviolet region give rise to oscillations which can be described as an instability in the dielectric function of a free electron gas. The frequency only depends weakly on the wavelength of the oscillation. The quasiparticle resulting from the quantization of these oscillations is the plasmon. If one displaces by a tiny amount an electron or a group of electrons with respect to the ions, the Coulomb force pulls the electrons back, acting as a restoring force. Remind that if the thermal motion of the electrons is ignored, the charge density oscillates at the plasma frequency. Plasma oscillations may give rise to the so-called effect of "negative mass". The mechanical model explaining it considers a core with mass m_2 connected internally through the spring with constant to a shell with mass m_1; the system is subjected to the external sinusoidal force $F(t) = F \sin(\omega t)$, and if one solves the equations of motion for the two masses and replace the entire system with a single effective mass m_{eff}, one obtains

$$m_{eff} = m_1 + \frac{m_2 \omega_0^2}{\omega_0^2 - \omega^2} \tag{6.26}$$

where $\omega_0 = (k_2/m_2)^{1/2}$. When the frequency ω approaches ω_0 from above the effective mass m_{eff} will be negative. The negative effective mass (density) becomes also possible based on the electro-mechanical coupling exploiting plasma oscillations of a free electron gas (Figure 6.4). The negative mass appears as a result of the vibration of a metallic particle with a frequency ω which is close to the frequency of the plasma oscillations of the electron gas m_2 relatively to the ionic lattice m_1. The plasma oscillations are represented with the elastic spring $k_2 = \omega_P{}^2 m_2$, where ω_P is the plasma frequency. Thus, the metallic particle vibrating with the external frequency ω_P is described by the effective mass

$$m_{eff} = m_1 + \frac{m_2 \omega_P^2}{\omega_P^2 - \omega^2} \tag{6.27}$$

which is negative when the frequency ω approaches ω_P from above. This negative mass effect in the vicinity of the plasma frequency has been exploited in some metamaterials [41, 42].

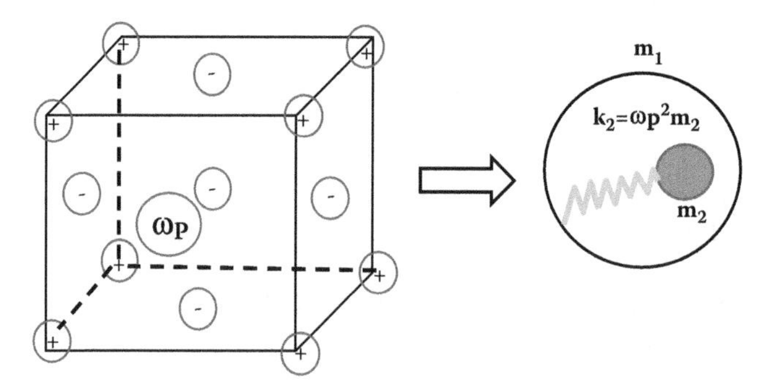

FIGURE 6.4
Free electrons gas m_2 is embedded into the ionic lattice m_1; ω_p is the plasma frequency (on the left), and the equivalent mechanical scheme of the system (on the right).

An interesting functionalization of a microplasma array is to construct a photonic crystal or a metamaterial for controlling the propagation of electromagnetic waves. In general, the permittivity of a plasma ε_p, for electromagnetic waves of angular frequency ω, is given by

$$\varepsilon_P = 1 - \frac{\omega_P^2}{\omega^2\left(1 + \frac{i\nu_m}{\omega}\right)} = 1 - \frac{e^2 n_e}{\varepsilon_0 m_e \omega^2\left(1 + \frac{i\nu_m}{\omega}\right)} \tag{6.28}$$

where ν_m is the electron collision frequency, and m_e is the electron mass. From this relation, it is seen that ε_p can be modified from unity to negative values according to n_e. Then, when one arranges microplasmas in space with a pitch considerably less than the wavelength of propagating electromagnetic waves, it is possible to create a medium whose effective permittivity ε is periodically modulated.

There are also relevant natural and artificial situations in which plasmas contain micro- or nanoparticles suspended in them. Dust particles are charged, and the plasma and particles behave as a plasma. These are called dusty (or complex) plasmas. The presence of particles significantly alters the charged particle equilibrium and can lead to new phenomena. Electrostatic coupling between the grains can vary over a wide range so that the states of the dusty plasma can change from weakly coupled (gaseous) to crystalline. The system of interacting particles enables to study physics of self-organization, pattern formation, phase transitions, and scaling. Surprisingly, complex plasmas may naturally self-organize themselves into stable interacting helical structures that exhibit features normally attributed to organic living matter [43].

The temperature of the dust in the plasma usually is much lower than its environment, for example, 10 K for dust, 10^3 K for ions, and 10^4 K for electrons. The electric potential of dust particles is typically 1–10 V (positive or negative). The potential is usually negative because the electrons are more mobile than the ions. In laboratory setups, the dust particles can be grown inside the plasma or can be inserted. Usually, a low-temperature plasma with a low degree of ionization is used. The particles then become the dominant component regarding the energy and momentum transport, and they can essentially be regarded as a single-species system. Above a certain size, it is possible to follow their movement because this is slow enough to be able to be observed with ordinary cameras, and so the kinetics of the system can be studied. However, above a critical micro-size, gravity becomes the dominant force that disturbs the system, and thus, experiments need to be performed under microgravity conditions. The particle charge is one of the most important parameters of complex plasmas. It determines the particle interactions with plasma electrons and ions, with electromagnetic fields, and between the particles themselves. The particle charging is achieved in gas-discharge plasmas, where it is due to the collection of electrons and ions from the plasma, so that the charge is determined by the competition between the electron and ion fluxes on the particle surface. Since electrons have a mobility larger than positive ions, the total charge acquired by particles is negative.

During charging of dust particles leads to them being monodisperse. During their growth, conditions are reached when the particles become too heavy and cannot remain airborne. They then fall to the bottom of the vessel leaving the plasma suddenly void of negatively charged dust particles and at this point the plasma properties change abruptly. The cloud of charged dust particles may become self-organized through Coulomb crystallization; voids may appear as well as complex structures. It is hoped that the fact that size may be controlled very accurately leads to efficient production, self-organization, and then manipulation by electric fields of the particles which may then be deposited on the surface with regular geometric patterns.

The potential $\phi\ (r)$ around an isolated spherical particle in an isotropic plasma satisfies the Poisson equation with the boundary conditions $\phi\ (\infty) = 0$ and $\phi\ (a) = \phi_s$. The relation between the surface potential and the particle charge is given by $d\phi\ /dr\,|_{r=a} = -Q/a^2$. For

Boltzmann electrons and ions their distributions can be linearized provided the condition $e\,|\,\phi_s\,|\,/T \leq 1$ is satisfied. The surface potential is $\phi_s = (Q/a)\,(1+a/\lambda_D)^{-1}$, and for small particles, $a << \lambda_D$, we have

$$\varphi(r) = \frac{Q}{r}exp\left(-\frac{r}{\lambda_D}\right) \tag{6.29}$$

where $\lambda_D^{-2} = \lambda_{De}^{-2} + \lambda_{Di}^{-2}$ is the linearized Debye length. This equation for the potential is the Debye–Hückel (Yukawa) potential which is often used in dusty plasmas. If the surface potential is not small compared to the temperatures of electrons and/or ions, then it is still possible to use such expression at sufficiently large distances from the particle; in this case, Q should be replaced by some effective value (smaller than the actual particle charge).

High-purity semiconductor nanoparticles can be synthesized in low-pressure (2.6×10^3 Pa, or below) nonthermal plasma flow tube reactors. The nonthermal plasma synthesis approach can be scaled to produce nanoparticles in quantities sufficient for device manufacture, and has several advantages including a high yield, solvent-free synthesis approach, narrow nanoparticle size distribution, high growth rate, and efficient doping is possible during the process. It is also beneficial to use low-temperature plasma reactors to transform crystalline metal nanoparticles into amorphous metal nanoparticles via a high nonequilibrium quenching process.

Examining and optimizing nanoparticle synthesis in a flow tube system would ideally be facilitated by online/in-flight measurement of the nanoparticle size distribution function at the reactor outlet. The produced nanoparticles in such reactors are typically below 20 nm in diameter (often below 10 nm), hence they are too small for single particle optical detection. Aerodynamic particle size measurements are difficult to apply in detecting them, and as already mentioned in the previous chapter, a common approach for gas phase nanoparticle size distribution analysis in the sub 20 nm range is ion mobility spectrometry (IMS). This is an analytical technique for the separation and identification of ionized particles in the gas phase based on their mobility, widely utilized in aerosol science to examine particle size distribution functions and particle growth/evaporation dynamics in atmospheric pressure systems. It can also be used in examining liquid phase synthesized particles through appropriately designed aerosolization techniques. Thus, IMS has the potential to be an efficient technique to study particle size and growth in plasmas.

With the emergence of nanophysics and technology, nanoparticle formation in chemically reactive plasmas and the ability of dusty plasmas to produce nanoparticles with controlled physical and chemical properties has been developed. Particularly for the synthesis of materials that require high synthesis temperatures, such as covalently bonded semiconductors and ceramic nanoparticles, nano dusty plasmas provide an essential synthesis route. The mechanisms that lead to the formation of nanoparticles are not completely understood, particularly whether particle nucleation is driven by neutral–neutral reactions involving radicals or ion–neutral reactions is still an open question. Particle nucleation in plasmas is favored by the high concentrations of reactive radicals. In some situations, it can also be enhanced by the faster rates of ion–neutral clustering compared with neutral–neutral reactions [44]. Better knowledge of growth mechanisms will be essential in being able to control particle properties, composition, and purity of nanoparticle materials. Expanding the methods of chemical kinetics to plasma environments will require determining the thermophysical properties of radicals as well as ionic clusters and determining clustering reaction rates [45].

Nanoparticles have a mutual interaction with charge carriers in the plasma. By collecting carriers, particles get charged, and at the same time, their presence modifies the charge carrier densities in the plasma. In fact, nanoparticles immersed in a plasma carry a unipolar negative charge once their size grows to several nanometers during synthesis [46]. This prevents nanoparticle agglomeration and enables the growth of nanoparticles with highly monodisperse size distributions. It also reduces or eliminates diffusional particle losses to the reactor walls.

The interaction of plasma species with the nanoparticle surfaces is another largely unexplored area. Plasma–nanoparticle surface reactions are the source of energy for nanoparticle heating that plays a crucial role in the particles' microstructure, for example, whether particles are crystalline or noncrystalline. Details of the particles' interaction with impacting ions may also explain why the defect densities of nanocrystals prepared with seemingly similar plasmas can differ by two or more orders of magnitude. Particularly in low-pressure plasmas, the combination of energetic surface reactions and slow nanoparticle cooling can cause a strong nonequilibrium [47], in which the particle temperature can exceed the gas temperature by several hundreds of Kelvins. This feature is important for the growth of crystalline nanomaterials of high melting point substances.

On the experimental side, models need to be validated with measurements of growth species concentrations and particle nucleation and growth rates. This will require measurements of ionic and radical species with methods such as optical emission and absorption spectroscopy, laser fluorescence methods, mass spectroscopy, as well as measurements of nanoparticles in plasmas, for instance, by laser scattering. For very small particles, light scattering may be inefficient and may require the development of novel particle probes. While measurements of the average nanoparticle charge have been reported, there are currently no experimental studies of the particle charge distribution, which becomes increasingly important for smaller nanoparticles that may become bipolarly charged.

As is well known, in the present century the environmental component of sustainability three-pillar description has become a dominant issue. This is due to the increasing consequences of climate changes, pollution, loss of biodiversity, and health risks. Given so, among the broad domain of low-temperature plasmas technological applications, those aspects that directly impact nanophysics and nanotechnology for sustainability assume a particular relevance. Plasmas are considered as one of the best possible methods to bring together nanoscience and nanotechnologies with industrial production. Applications of plasma in general in nanotechnologies are extensive, and the nonequilibrium plasmas in the same context is almost as large, with many variants covering both *top-down* and *bottom-up* fabrication approaches. In many ways, plasmas are ideal for creating and allowing self-assembly of the building blocks in the *bottom-up* approach.

From a general point of view, the control of plasma processing methodologies must be preceded by a clear understanding of the physical and chemical properties of plasmas [48]. Keeping this in mind, and without claiming to have an exhaustive approach, it is worth dedicating a few more paragraphs to subjects such as plasma etching for ultra-large-scale integration, carbon nanotechnology plasma processing, plasma functionalization, plasma catalysis, and plasmas for environmental applications.

Thermal plasmas are more productive if large quantities are required, but at the same time they are much more costly in terms of energy invested as most of the energy goes to heat the chamber and the background gas. Nonequilibrium plasmas, on the other hand, may have a very efficient use of energy as one can have most of the energy used up by electrons. In addition, by tailoring the process one may even have a reasonably productive process or a very specific process with fine tuning that cannot be achieved in thermal plasmas. At the same time nonequilibrium plasmas, if used properly, have advantage for treatment of materials (and living tissues) that are sensitive to high temperatures.

Nonequilibrium plasmas are easily produced at low pressures (around 1 Torr) and under those conditions one may achieve large volumes, great uniformity, and easily controllable properties. At high pressures, however, the ionization rate becomes quite high thereby producing large quantities of charged particles in a small distance crossed by charged particles. Thus, thermal equilibrium plasmas are generally created. In order to achieve nonequilibrium operation of plasma at elevated pressures one needs to interrupt the charged particle production in some way and limit it. It can be achieved in very inhomogeneous fields where breakdown condition is satisfied only in a very small volume (corona discharges), by spatial interruption of the field with an inserted dielectric barrier where charges from plasma are deposited and may shield the field (dielectric barrier discharges), by high-frequency fields or localized microwave plasma production in a flow of rare gases mixing with the atmosphere or sometimes even in the air. Finally, as the optimum gap for breakdown at 1 Torr is 1 cm, nonequilibrium operation at atmospheric pressure of around 760 Torr may be achieved with very small discharges with gaps of the order of 100 μm or less. Achieving flexibility and the ability to tailor the process and treat thermally sensitive targets, while still being able to operate at atmospheric pressure, is of paramount importance, and all aspects of atmospheric pressure nonequilibrium plasmas are of great interest for research and applications.

Future integrated circuits will likely be made by using the same principles that are implemented today and will include the two steps that are critical for further miniaturization: photolithography and plasma etching. Plasma etching technology has led to the efforts to shrink the pattern size of ultra-large-scale integrated devices involving the fabrication of sub-22 nm patterns on Si wafers. Plasmas used for etching operate at radio frequencies between 13.56 and 200 MHz and in recent years two modes of operation have become dominant: capacitively coupled plasmas (CCP) and inductively coupled plasmas (ICP). Plasma sources may also be enhanced by the application of external magnetic fields. As the size of the opening decreases and as aspect ratios increase in order to accommodate more interconnect layers, it may prove that the pulsing strategy is not sufficient. A general solution to the problem of charging would be to use fast neutrals instead of ions. Those fast neutrals may be formed in the gas phase by charge transfer collisions or may be produced by surface neutralization (of both positive and negative ions). Contribution of hyperthermal neutrals with energies of few electron volts to the etching has been recognized [49].

High-density plasma sources, such as inductively coupled plasma and electron–cyclotron resonance plasma, are key technologies for developing precise etching processes. However, inherent problems in the plasma processes, such as charge build-up and UV photon radiation, limit the etching performance for nanoscale devices. To overcome these problems and fabricate sub-10-nm devices in practice, tens-of-microsecond pulse-time modulated plasma etching, and neutral-beam etching has been developed [50]. In short, sub-10 nm devices require defect-free and charge-free atomic layer etching processes and these techniques can perform atomically damage-free etching and surface modification of inorganic and organic materials. Economic considerations in plasma processing for the manufacturing of semiconductor chips, flat panel displays, and solar panels have led to a continuous increase in substrate dimensions in those industries. Capacitively coupled plasmas are currently used in the processing industry for etching and deposition applications, but several critical applications use very high-frequency sources (above 60 MHz) where electromagnetic nonuniformities become prominent on substrates larger than 300 mm. Another important technology for plasma processing applications is inductively coupled plasmas that rely on the electromagnetic power coupling between the current in the RF coils and the plasma, being the plasma generated nonuniformly close to the coils. Any RF voltage or current variations along the coils due to electromagnetic effects lead to nonuniform plasma production along the coil. These nonuniformities have generally been addressed through careful antenna design for distributed plasma production and by appropriate selection of RF frequency to minimize the voltage and current variation along the lines. With increasing plasma dimensions, antenna design also needs to ensure that voltage on the coils does not become excessive leading to reliability issues and undesirable capacitive coupling.

Ultimately, the plasma species interact with the walls. Adsorption of species will occur depending on the chemical affinity and surface temperature. Adsorbed species may react with the surface to form a product or desorb without or after a surface reaction. If the product is volatile, it will desorb into the plasma phase and cause etching (removal) of the material from the surface. The material will either be eliminated through pumping or participate in the plasma chemistry. If the product is not volatile, it will contribute to the formation of a thin film at the surface. Ions, electrons, and photons also interact with the surface. In plasma processing, the sample is, in general, negatively biased with respect to the plasma by means of the external power supply, so positive ions play a very important role in plasma surface interactions. Ions bring energy to the surface and, thus, can assist chemical reactions and desorption of weakly volatile species, or induce direct sputtering.

Microwave plasmas have also been finding applications in large-area plasma processing. With a short wavelength at microwave frequencies, the two general techniques that have been used to obtain large-area uniform plasmas are slot antenna arrays, where the array pattern is designed to obtain uniform radiation through the antenna and traveling wave linear microwave sources. Physical vapor deposition, that is, DC, pulsed DC, or RF sputtering, remains the dominant plasma technology for metal deposition. Since it works well with dc sources, it does not suffer from the electromagnetic nonuniformities. However, target erosion is usually nonuniform due to the nonuniform magnetic field. Traveling wave inductively coupled plasmas and microwave sources appear more promising regarding scaling to larger dimensions.

As familiar, plasmas are created via ionization, which can occur in several ways (collisions of fast particles with atoms, photoionization, or electrical breakdown in strong electric fields), and microscale plasmas (microplasmas) are a special class of electrical discharges formed in geometries where at least one dimension is reduced to submillimeter length scales. This makes microplasma a likely medium to introduce some new phenomena, due to the increase in the surface-to-volume ratio. As a result of their product pD (p stands for pressure and D for the smallest physical dimension) scaling, microplasmas have been found to be characterized by high-pressure stability, nonequilibrium thermodynamics, non-Maxwellian electron energy distribution functions, and high electron densities [51]. It might be expected that confining plasmas to small dimensions will result in new physical behavior. At some critical size, a transition from a plasma with bulk properties to a new phase with size-dependent properties must occur. Two parameters that may determine this transition are the surface-to-volume ratio and the electrode spacing. The surface-to-volume ratio, which increases as the size of the plasma decreases, will alter the overall energy balance, and could lead to plasma stability (or instability) within different regions of the operating parameter space. Decreasing the electrode spacing will change the electric field distribution and, thus, impact the charge distribution, Debye length and global plasma neutrality. These effects should have strong implications on the energy distribution of the different species (electrons, ions, radicals, neutrals) and on the physical structure of the plasma.

Whereas low-pressure plasmas entail a large cost due to the equipment and maintenance of the reactors, operation near or at atmospheric pressure is cheaper, easier to implement, and highly desirable for industrial applications. These properties make microplasmas suitable for a wide range of materials applications, including the synthesis of nanomaterials. Research has shown that vapor-phase precursors can be injected into a microplasma to homogeneously nucleate nanoparticles in the gas phase [51]. Alternatively, microplasmas have been used to evaporate solid electrodes and form metal or metal-oxide nanostructures of various composition and morphology. Because of their small size, microplasmas are capable of locally etching or sputtering films and substrates to directly create microscale patterns without the need for photolithography. Microplasmas have also been coupled with liquids to directly reduce aqueous metal salts and produce colloidal dispersions of nanoparticles [51]. There is also the potential for small-scale applications using microfluidics such as lab-on-a-chip analysis and flow systems for fine synthesis.

The existence of significant populations of energetic electrons (i.e., 10 eV or higher) allows efficient, nonthermal dissociation of molecular gases or other vapor precursors to produce high concentrations of reactive radical species. One important aspect that must be considered for nanomaterials synthesis is how to control the energy flow which will determine the transition from nonequilibrium (i.e., low temperature) to equilibrium (i.e., thermal) plasma. Nonequilibrium plasmas are more desirable for materials synthesis because of the possibility of opening chemical pathways that may not be possible by thermal means. In addition, low-temperature processes allow temperature-sensitive materials to be used. The mechanism for nucleation follows the introduction of a precursor in the microplasma (e.g., SiH_4) that forms reactive radical species (e.g., SiH_3), by electron impact dissociation. These radicals can collide, react, and nucleate small clusters at the appropriate process conditions. The clusters will then grow, by additional radical or vapor deposition on the particle surface, or agglomerate, through collisions with other particles. Finally, the particles will exit the microplasma volume as an aerosol flow.

To enhance the electron multiplication rate at a shorter distance and prevent the wall loss, microplasmas are mostly operated in higher pressure (or density) ranges. Therefore, even if the ionization degree is low or moderate, one can make the electron density n_e larger than 10^{13} cm^{-3}. For a density of about 10^{16} cm^{-3}, the corresponding electron plasma frequency $\omega_{pe}/2\pi$ becomes about 1 THz. This encourages the use of microplasmas also as conductive/dielectric media for electromagnetic waves.

Micro-discharges are among the growing topics in nanotechnology making use of nonequilibrium plasmas. The integrated circuit, MEMS, hard disk drive magnetic recording industry, and flat panel display industries may benefit from the understanding of the breakdown mechanisms in micro gaps under atmospheric pressure and some of these processes may limit the deposition of energy and some applications of nano and microstructures. Generally, nonequilibrium plasmas are produced easily only at low pressures. At atmospheric pressure, the rapid growth of ionization and strong coupling between ions and electrons leads to the tendency to produce thermal plasma. One of the ways to avoid thermalization of plasma at atmospheric pressure is to operate at very small gaps. Standard dimension discharges (~1 cm) at atmospheric pressures operate far from the minimum of the Paschen curve (breakdown voltages vs. pressure × gap (pd)). Thanks to the scaling laws, miniaturization of the discharge gives a possibility to achieve nonequilibrium conditions (near the Paschen minimum, i.e., the lowest and thereby the optimum breakdown potential) even at atmospheric pressure. Micro-discharge sources of various geometry and modes of operation have been developed for use in specific devices, such as dielectric barrier discharges for gas analysis, capacitively coupled and inductively coupled RF plasma sources. The best way to establish a good understanding of the operating conditions in micro-discharges is to start from the low-pressure discharges and employ scaling laws [52]. Standard scaling parameters in gas discharges are electric field to gas number density ratio, pressure times electrode gap, current density normalized by the electrode gap to the square, and frequency times gas number density for RF discharges. Most of the recent studies deal with micro-hollow cathode geometries, which lead to complex behavior that does not resemble conditions in the standard hollow cathodes. Based on modeling results, one can expect a breakdown of scaling at gaps smaller than 10 μm [53], where field emission becomes a dominant mechanism of the breakdown and the discharge maintenance.

Applications of plasmas for film deposition began with the discovery of sputtering [54], and after one century passed, the technique of chemical vapor deposition (CVD) [55] came to make it possible to produce amorphous (a-) carbon films and led to groundbreaking research on plasma polymerization techniques in the 1980s. Even today, CVD is an active area of investigation for producing several types of thin films with increasing interest in nanotribology. In plasma deposition processes the goal is atomic- or molecular-level control during device fabrication. To achieve this, a key approach is the fusion of top-down and bottom-up processes based on self-assembly reactions. For achieving atomic or molecular-level control during thin-film deposition, atomic layer deposition (ALD) [56], and self-assembly, bottom-up processes represent approaches to precise control of the films in realizing atomic-scale devices. Development of bottom-up CVD techniques for organic materials such as carbon nanotubes (CNTs) and graphene sheets is an area of intensive research, especially focused on PECVD as a strong contender since low-temperature growth can be achieved. The self-assembly of CNTs free of catalysts [57] and the crystallographic control of the chirality of CNTs are two relevant challenges to pursue [58]. In fact, among the gas-phase techniques for synthesizing nanostructures where self-assembly plays a role, plasma-enhanced chemical vapor deposition (PECVD) takes a prominent place, and the discovery of carbon nanoclusters was an acceleration factor, as indeed it was for all nanotechnology. Originally derived from high-energy thermal arc plasmas, the synthesis routes have expanded to include laser processing and plasma chemical vapor deposition techniques, in order to meet demanding specifications for both individual nanotube quality and, for example, positional growth. High-temperature arc and laser plasmas are under continued investigation to produce bulk quantities of nanotubes but suffer the inclusion of high levels of metal and amorphous carbon impurities. Post-purification by chemical means is possible but can adversely impact the nanotube quality. The alternative of using PECVD sources for carbon nanotube growth offers some benefits provided that certain requirements are met. These include the identification of the primary precursors and the role of other particles such as atomic hydrogen in the growth, the role of ions in the growth, the mechanism of alignment and how it is influenced by the electric field, how to optimize the substrate bias, and how to control the chirality and fibers number of walls [59, 60]. Individual nanotubes are grown as single-wall (SWNT) or multiple concentric tubes (multi-wall MWNT) and quality parameters to be controlled are diameter, chirality, length, and defects as well as incorporated impurities. They are grown vertically with density ranging from sparse to dense forests and here the quality of the vertical alignment and spacing between the individual nanotube and nanotube bundles may need to be controlled [57].

Carbon nanotubes have captured the interest of the scientific community by the unique structure that provides superior physical and chemical properties. Carbon can form a large variety of structures from the same number of carbon atoms just by changing the bonding configuration with itself. The number of such structures increases nonlinearly with the number of carbon atoms. However, the huge variety of possible carbon nanotube structure geometries (diameters, lengths, and number of walls) poses a challenge of how to increase the production selectivity for the development of carbon nanotube applications. Many of the unique nanoscale properties of CNTs do not have bulk equivalents. The maintenance of these properties in macroscale applications depends on the uniformity of the CNT properties. One can divide the applications of CNTs into two categories: (i) those that use CNTs in a bulk form or in combination with other materials to enhance their bulk thermal, electrical, and mechanical properties; (ii) those that exploit the size-dependent and quantum properties of individual CNTs and molecular scale devices capable of interacting with single molecules.

The chemistry of carbon plays a critical role in the synthesis and processing of CNTs. The reason for the great variety in the carbon structures is in the electronic structure of carbon atoms that can exist in three different electronic states determined by the hybridization of the carbon atomic orbitals. The spatial orientation of the hybridized orbitals provides an illustrative classification scheme for nanocarbon structures and materials (Figure 6.5). Apart from graphite (the thermodynamically stable form of carbon at room temperature), all other structures are metastable [61].

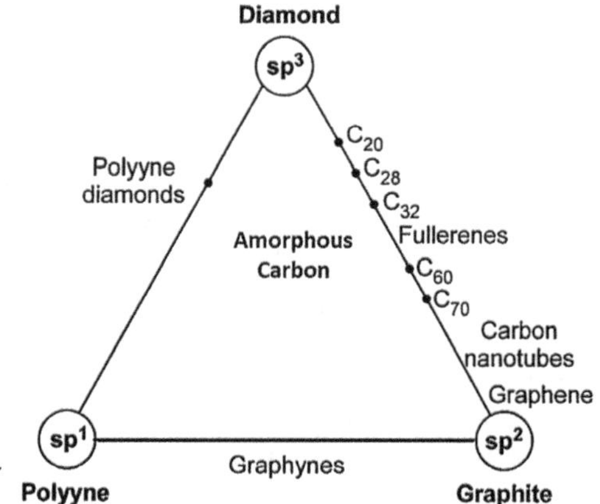

FIGURE 6.5
Ternary phase diagram with known allotropes and intermediate forms of carbon.

Examples of intermediate hybridization structures are fullerenes and CNTs with $sp^{2+\eta}$, where $\eta > 0$. The value of η depends on the degree of curvature of the sp^2 network. The existence of these possible intermediate structures increases the complexity of the product distribution in the synthesis of nanocarbons, such as CNTs. The metastability of the structures implies that the product distribution is controlled by the kinetics, and in the possible reactions a variety of carbon structures are formed, and not only CNTs. The width of the product distribution will depend on the energetic content of the reactants. Consequently, different synthesis methods (arc, laser evaporation, CVD) produce CNTs with different quality, quantity, and purity.

For simplicity, if one imagines building a carbon nanotube by wrapping a single sheet of graphite (graphene) into a cylinder, then we will realize that they are dominated by the properties of the sp^2 carbon that makes up the graphene sheets. In addition, small-diameter CNTs exhibit quantum confinement effects characteristic of 1D periodicity, where the band structure and the electrical properties of CNTs are governed by the rotational and helical symmetry of the CNTs. The properties of CNTs are dictated by the symmetry of their atomic structure. Its honeycomb lattice can be described by making use of a chiral vector concept; this is a vector connecting two equivalent points of the CNT honeycomb lattice and is given as a linear combination of the two basis vectors designated by a pair of integers (n, m). Let $\boldsymbol{u}$ and $\boldsymbol{v}$ be two linearly independent vectors that connect the graphene atom *A1* to two of its nearest atoms with the same bond directions. That is, if one number of consecutive carbons is around a graphene cell with C1 to C6, then $\boldsymbol{u}$ can be the vector from C1 to C3, and $\boldsymbol{v}$ be the vector from C1 to C5. Then, for any other atom *A2* with same class as *A1*, the vector from *A1* to *A2* can be written as a linear combination $n\,\boldsymbol{u} + m\,\boldsymbol{v}$, where n and m are integers; and conversely, each pair of integers (n,m) defines a possible position for *A2*. Given n and m, one can reverse this theoretical operation by drawing the vector $\boldsymbol{w}$ on the graphene lattice, cutting a strip of the latter along lines perpendicular to $\boldsymbol{w}$ through its endpoints *A1* and *A2*, and rolling the strip into a cylinder to bring those two points together. If this construction is applied to a pair (k,0), the result is a zigzag nanotube, with closed zigzag paths of $2k$ atoms. If it is applied to a pair (k,k), one obtains an armchair tube, with closed armchair paths of $4k$ atoms. Then the angle α between $\boldsymbol{u}$ and $\boldsymbol{w}$, which may range from 0 to 30 degrees (inclusive of both), is called the chiral angle of the nanotube. This way, a specific CNT is described by the diameter of the tube and the angle of the chiral vector that specifies how a graphene sheet is rolled up into a CNT [62]. From n and m one can also compute in picometers the length of the vector $\boldsymbol{w}$, which is the circumference c:

$$c = |\boldsymbol{u}|\sqrt{(n^2 + nm + m^2)} \tag{6.30}$$

and the diameter of the nanotube (also in picometers) is thus equal to c/π. To synthesize the study of the geometry of nanotubes, one can define a "zigzag" on a graphene-like lattice as a path that turns 60 degrees, alternating left and right, after stepping through each bond. It is also conventional to define an armchair path as one that makes two left turns of 60 degrees followed by two right turns every four steps. On some carbon nanotubes, there is a closed zigzag path that goes around the tube, and so, one says that the tube is a zigzag nanotube. If the tube is instead encircled by a closed armchair path, it is said to be an armchair nanotube. A nanotube is said chiral if it has type (n,m), with $m > 0$ and $m \neq n$, and so its enantiomer (mirror image) has type (m,n).

The properties of the CNTs have been classified according to the so-called *mod 3 rule*, based on the results of band structures computed from the known tight-binding approximation. All armchair ($n = m$) CNTs are metals at room temperature. CNTs with $n - m = 3j$ ($j = 1, 2, 3, \ldots$) are semimetals. All other CNTs with $n - m = 3j + 1$ and $n - m = 3j + 2$ ($j = 0, 1, 2, \ldots$) are semiconductors with a bandgap that is inversely proportional to the diameter [63].

The three main synthesis methods include arc discharge, laser vaporization, and chemical vapor deposition (CVD). In principle, the carbon reaction mechanisms in these three methods are similar, but in practice the processes are different due to the degree of the initial excitation that starts the nanotube growth process and the intermediate stages that the reactions undergo. As a result of the high temperatures in arc discharge and laser vaporization that are required to evaporate solid carbon, a variety of other carbon structures in the final product along with single-wall CNTs (SWCNTs) are produced. Moreover, the addition of a small amount of transition metal particles (Fe, Co, Ni) was necessary to obtain SWCNTs. The typical by-products of these methods include a large amount of fullerenes, and a variety of graphitic structures and amorphous carbon particles. Usually, a purification process must be performed to extract the SWCNTs.

CVD method has been the easiest to adapt for a larger scale production. The CNT growth in CVD processes occurs by catalytic decomposition of hydrocarbon molecules, which occurs at much lower temperatures than arc discharge and laser evaporation.

The so-called mechanism of carbon fiber growth can be used as the starting point for understanding the CVD growth of CNTs where it is also known as the diffusion/precipitation model [63]. It is based on the concept of the vapor–liquid–solid growth of whiskers, where the fiber growth occurs by precipitation of carbon from a molten metal particle. The diffusion of carbon through the particle may be considered the rate-limiting step based on the similar value of the activation energy for fiber growth and carbon diffusion in the molten metal. In the model, it is assumed that the particle size determines the diameter of the CNTs, and eventually the chirality of the CNT. However, the main assumptions of the diffusion/precipitation model have never been directly proven in experiments and remain a source of controversy in CNT growth. The emergence of observations and data on vertically aligned forest growth of CNTs enables more controlled growth studies [64]. In fact, to reverse the main obstacle of nonreproducibility synthesis methods of pure SWCNT material, a new growth method that uses molecular beams of precursor gases that impinge on a heated substrate coated with a catalyst thin film has been developed. Moreover, the results of these experiments cannot be explained by the diffusion/precipitation model [64]. The main weakness of the diffusion/precipitation model is that it grossly oversimplifies the process by ignoring the role of carbon chemical bond formation.

Regarding carbon evaporation methods, the most remarkable feature is that the carbon structures nucleate before the metal particles are formed. Real-time spectroscopic and optical scattering studies have been used to measure the time scales of CNT formation in expanding carbon laser plumes. These studies revealed that the self-assembly of carbon structures precedes the formation of metal particles in the expanding and cooling laser plume [65]. This implies that the carbon structures must condense on the metal particles, a step that prevents premature closing of the growing carbon network and enables the addition of carbon that is necessary for increasing the length of the CNTs. This fact indicates that carbon bond formation is a driving force and the key to the self-assembly of carbon from highly reactive carbon species [65].

Within the CVD's ability to grow arrays of aligned CNTs, these are grown directly on the substrates of interest and the alignment can be controlled vertically or horizontally with respect to the substrate surface. In certain circumstances, this eliminates the need for purification and processing of CNTs. Vertically aligned nanotube arrays (VANTA) growth is performed on a substrate coated with a catalyst film or a layer of catalyst particles. Upon heating, the film breaks up into particles that are the right size for nucleation and growth of CNTs [66]. VANTA can be grown by both thermal and plasma-induced CVD, containing pure single-wall, pure multiwall, or mixed CNTs. The VANTA method reduces the level of catalyst metal since growth occurs by the base mode in which the catalyst particles are anchored to the substrate and VANTA can be lifted off from the substrate to avoid residual metal contamination. Preventing catalyst poisoning is also an important direction to pursue in CVD growth of CNTs. VANTA growth is governed by the carbon bonding chemistry rather than by the dissolution and precipitation of carbon from metal particles [67]. Carbon self-assembly from reactive carbon species in reactions that are initiated by the action of the catalyst is a spontaneous result of energy minimization by formation of new carbon bonds. For instance, acetylene precursor first forms high-temperature intermediates that have a short lifetime and a chain-like structure attached to the particle surface; these intermediates cross-link to form small graphene sheets that coalesce into incomplete tubular structures that grow longer by addition of acetylene at the base of the growing CNTs. This model requires no metal to dissolve or precipitate carbon and the addition of carbon occurs at active sites in the carbon network. The nanotubes stop growing when there are no more radical type carbon bonds available for further addition of carbon [67].

Controlling the chirality of the CNTs is also an important goal to pursue in the synthesis as there are no synthesis methods that can directly produce CNTs with single predetermined chirality. Currently, polydisperse CNTs are produced, roughly according to the 1:2 metal to semiconductor ratio dictated by the number of ways that a given number of carbon atoms can arrange into a CNT. The most widely used approach is post-growth purification and separation of the CNTs into metal and semiconductor fractions [68]. These methods rely primarily on the slight difference in the side wall chemistry of the metallic and semiconducting CNTs to generate selectivity. A different approach analogous to epitaxial growth has been explored and it is based on using CNT segments with specific chirality as seeds for growing long CNTs. Nanometer-sized metal catalysts are docked to the SWCNTs open ends and subsequently activated to restart growth; SWCNTs thus grow and inherit the diameters and chirality from the seeded SWCNTs [69].

After growth and purification, nanocarbons can be further treated to enhance functionality [70–73]. This procedure is called functionalization, and it is a step that allows some selectivity of the CNTs' diameters, which is often critical in device fabrication of, for example, biosensor electrodes or fuel cell elements. Nitrogen functionalization (i.e., attachment or substitution of nitrogen) has been achieved in low-pressure RF plasmas and at atmospheric pressure as has oxidative functionalization, while more complex tailoring has also been achieved on nanotubes in a range of forms, from powder to vertically aligned forests [33].

Since the invisible individual nanotubes cannot be produced as a continuous material, methods are needed to form macroscale and bulk materials from them. The process of twisting nanotubes together to form a fine thread is called spinning. There are different diameters and lengths and numbers of walls in nanotubes produced under different process conditions. There are also different approaches for producing thread and yarn from nanotubes, and the structure and properties of the thread such as the orientation of nanotubes and strength of the thread can be different for the different types of thread and yarn produced. Nanotubes grown on a substrate self-assemble to form an array or forest with billions of nanotubes per square centimeter [74]. A thin ribbon or sheet of nanotubes may be drawn from the forest. Twisting the ribbon is one way of forming a thread or yarn. The nanotubes when drawn from the forest tend to stick together in parallel in bundles and the bundles overlap each other at the ends of the ribbon. The bundles may be of different diameters and are also called strands. The nanotubes in ribbon and thread are mostly aligned along the longitudinal direction of the ribbon or thread when produced from nanotube forests. Nanotube thread may also be produced directly from a furnace in a simple continuous high-rate-of-production process, but the nanotubes may not be as well aligned along the thread axis [74].

Spinnable carbon nanotubes are useful formats for studying the physical properties of CNT fiber assemblies in their pristine states. They are free of catalyst, uniform in length, with a comparatively narrow diameter distribution, and their assembly into thread does not require additional chemicals or solvents. Good quality drawable CNT arrays can be readily assembled into uniform diameter threads with great control over the number of CNTs incorporated into the thread assembly [57]. This uniformity allows studying the physical properties that result from changes that occur during thread formation, as for instance trends of electrical resistivity and mechanical strength that resulted from alterations in their manufacturing parameters, allowing to change intrinsic physical properties of a material such as electrical resistivity. Correlations between the electrical resistivity and mechanical strength as a function of diameter, density, and turns/meter have been found [57]. Understanding the effects of dry-spinning parameters will allow a better design of the physical properties of CNT threads for specific applications, such as strain or electrochemical sensors.

Plasmas-liquid interaction is also a matter of interest regarding the synthesis of nanoparticles. In fact, discharges in and in contact with water are a rich source of radicals, such as OH, O, and H_2O_2, and UV radiation which constitutes an attractive means for producing nanoparticles at the liquid–plasma interface and also as an advanced oxidation technology to break down organic and inorganic substances in water. Electrical breakdown and ionization in liquids have been investigated for several years and it is assumed that breakdown in water occurs through a bubble mechanism, or due to the presence of voids or even ionization in pure water without phase change [75]. However, details are not well understood and as there is often limited access by laser diagnostics, mainly optical emission spectroscopy is used for diagnostics. Additionally, the electrical properties of the liquid can depend strongly on the electrical field, the local density, and the frequency components of the applied field. Another main challenge is the understanding of the physical and chemical processes occurring at the plasma–liquid interface (Figure 6.6).

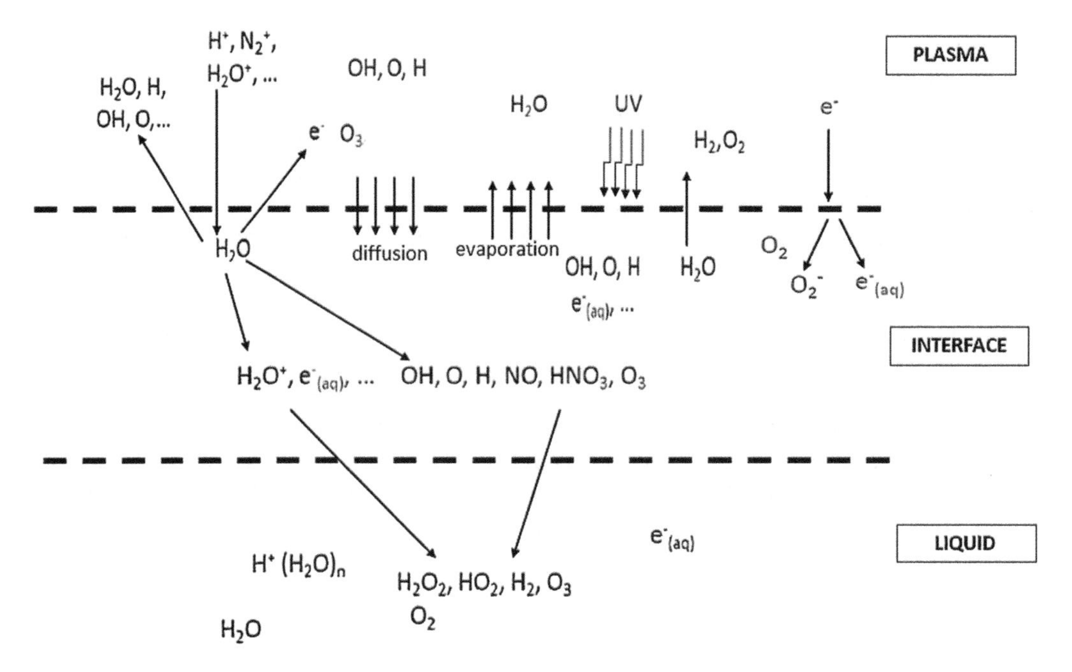

FIGURE 6.6
Some important transfer processes at the plasma-liquid interface.

Apart from the thermal energy transfer at the interface many open questions are still present on how charged species, neutrals, and radicals are transferred from the plasma to the liquid phase and vice versa. Possibly secondary electron emission plays a role too. Since water is electronegative and shows a tendency to cause clustering of the ionic species, this adds more complexity in understanding these plasmas. The time-averaged chemistry in a liquid takes place in seconds while the plasma chemistry happens on microsecond time scales. The development of modeling by including more detailed physical and chemical processes at the plasma–liquid interface could yield considerable insight into plasma properties that are not easily accessible by diagnostics efforts [76].

The steady state of a plasma can be disturbed by the introduction of a solid surface in a way that can play a catalytic action leading to a higher degree of chemical reactivity of the system under study. An early example of that was the decomposition of ammonia ($2NH_3 \rightarrow N_2 + 3H_2$). Other authors investigated the synthesis of methanol from CH_4 and CO_2 in a dielectric barrier discharge with a catalyst and found that the process was enhanced with the introduction of a catalyst [77]. Nowadays, plasma-assisted catalysis has several applications in environmental clean-up removing common pollutants and in the directed synthesis of added value products such as in the reforming of hydrocarbons into fuels [78]. It is even interesting to note that often using atmospheric pressure nonthermal plasmas to activate a catalyst can give significantly reduced operating compared with conventional thermal catalysis. This can reduce commonly occurring problems of catalyst stability such as sintering at high temperatures or poisoning by species such as sulfur. Certain advantages such as low-temperature operation, high selectivity, and improved energy efficiency are priceless. It is timely to remember that energy efficiency means using less energy to get the same job done – and in the process, cutting energy bills, reducing pollution, and combating climate change. This way, energy efficiency reduces energy waste and lower energy costs.

As is well known, catalysis is defined as the process in which the rate of a reaction is increased without changing the overall standard Gibbs energy change, and this rate increase is brought about using a catalyst, which itself is regenerated after each reaction cycle. The overall catalyst activity is measured by the conversion, which is the number of moles of reactant converted into products divided by the number of moles of reactant fed. The optimum activity is obtained for an optimum interaction energy between the adsorbates and the catalyst. When the reactants reach the surface, they need to adsorb and bind sufficiently strongly to the catalyst such that they do not immediately desorb into the gas phase again. Their probability to react is related to their lifetime. After all required elementary reactions have taken place, the formed products need to desorb from the surface. Thus, these products may not be bound too strongly to the surface, as this would prevent them from desorbing, and the optimum reactant adsorption is reflected by the typical volcano plots. In general, the bond formation and bond breaking may be understood similarly to the description of adsorption and desorption. Let us now remind that there are two types of surface reaction barriers corresponding to the so-called concepts of early-barrier reactions and late-barrier reactions; in early-barrier reactions, the transition state resembles the initial configuration of the reactants, whereas in a late-barrier reaction, the transition state resembles the final configuration, resembling the products. A typical example of an early-barrier reaction is the dissociative adsorption of H_2 on transition metal surfaces. Early-barrier reactions are usually enhanced by translational energy instead of vibrational excitation. In this case, the barrier is encountered before the bond has noticeably elongated. Typical examples of late-barrier reactions are the dissociative adsorption of H_2 on noble metals or the catalytic dissociation of CH_4 on transition metal surfaces. The latter process is of interest in plasma-catalytic dry reforming. In this case, one of the hydrogen atoms nearly completely separates from the methane molecule at the transition state, which resembles the final products in which both the hydrogen atom and the CH_3 radical are bonded to the metal surface.

The mechanism of plasma catalysis is complex and far from completely understood. In thermal catalysis, heat activates the catalyst but with plasma activation, the electrical discharge supplies the energy. Electron–gas collisions create ions, reactive atoms, radicals, excited species (electronic and vibrational), and photons. In a nonthermal plasma, there is nonequilibrium with high-energy electrons but little heating of the gas. Many plasma-created species are short-lived particularly at atmospheric pressure where quenching, recombination, and neutralization are rapid. Vibrationally excited species interacting with catalytic surfaces may also play a role. The interactions in plasma catalysis are either from the plasma with the catalyst or the catalyst affecting the discharge. On the one hand, examples of the effect of the plasma on catalyst are the formation of radicals and excited states, reduce active metal and modify the properties of the catalyst; on the other hand, effects of the catalyst in the plasma are adsorption on the catalyst surface, influence on plasma generation, and packed-bed effect. All these synergistic effects converge to enhance the energy efficiency, improve selectivity, increase the concentration of the active species, improve catalyst activity and durability. As well as creating reactive species above the catalyst surface, plasma can change the surface properties by ion, electron, or photon interactions. Packing catalytic materials into the discharge may modify its electrical properties through changing dielectric effects or by altering its nature, for example, from filamentary micro-discharges to surface discharges. Frequently, plasma catalysis offers advantages over a continuous thermal system as the cold plasma is only used intermittently for a small percentage of the time required to saturate the adsorbent catalytic material, giving significant energy savings.

Traditionally, plasma processing makes use of reduced pressures, allowing easy plasma ignition and the formation of a homogeneous discharge. Atmospheric pressure setups, however, do not require expensive vacuum pumps, which is advantageous in an industrial setting. Moreover, ion bombardment at reduced pressure may damage growing nanostructures as in the case of carbon nanotube growth. In contrast, because of the much higher collision rates, ion-induced damage is strongly reduced in atmospheric pressure growth systems.

During the processing of a NiO–Al_2O_3 catalyst with atmospheric-pressure methane plasma [79], NiO is reduced to Ni by the low-temperature plasma ($4NiO + CH_4 \rightarrow 4Ni + CO_2 + 2H_2O$); this is complete when no further CO_2 evolves. Thermally, reduction takes place at temperatures > 400°C but in this case is achieved at lower temperatures. Hydrogen is then produced with high selectivity by the Ni-catalyzed reaction, $CH_4 \rightarrow C + 2H_2$ via the fragmentation of adsorbed CH_4 on active sites of the catalyst surface to form active adsorbed carbon and hydrogen. It happens that carbon appears as nanofibers; a Ni-catalyzed process is normally achievable at temperatures > 600°C: showing increased energy efficiency for low-temperature plasma catalysis over conventional thermal processing and demonstrating a synergistic effect for CH_4 decomposition, where both plasma and catalyst are crucial.

Most nanocatalysts are in the form of nanoparticles and when their diameters are smaller than about 5 nm, they can become powerful catalysts, as in the case of gold nanoparticles for the oxidation of carbon monoxide. The reactivity of the nanoparticles depends not only on their size but also on the material on which they are supported. However, it has been found that the dominant effect is that of the gold nanoparticle size, with the nature of the support playing a secondary role [80]. Figure 6.7 shows the common trend found in the activity of metal nanoparticles on various supports as a function of their size and shows how this is so relevant. Since catalysis can only occur at the surface layer of atoms, the dominant size effect is the proportion of metal atoms that are at the surface. In fact, the most important atoms for catalysis are those at the corners between different facets. These low coordinated atoms are where the reacting molecules preferentially bond to during the reaction. The fraction of this type of atom (highlighted in black in Figure 6.7) is proportional to $1/D^3$, where D is the particle diameter. The black line in Figure 6.7 usually represents a reasonable fit to the experimental data using this law, demonstrating that the dominant size effect is indeed the proportion of corner atoms at the surface.

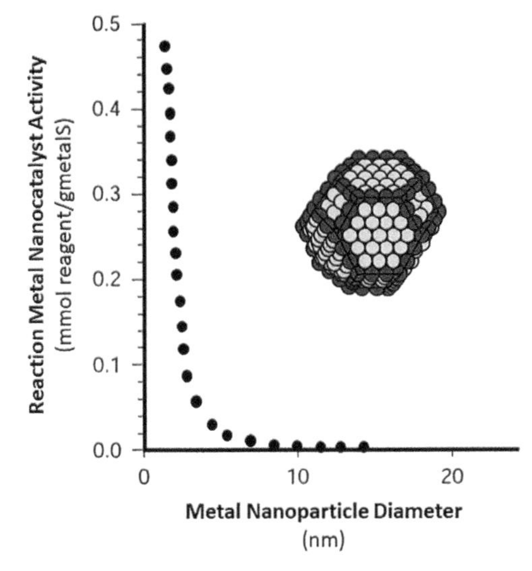

FIGURE 6.7
Reactivity of metal nanoparticles. The activity of most metal nanocatalysts supported on different substrates commonly shows the trend on display as a function of nanoparticle size (D) that follows a $1/D^3$ decay. This variation reflects that the dominant size effect is the proportion of metal atoms that are located at a corner between facets at the surface; such atoms are highlighted in black for an example of a gold nanoparticle shown here.

In the general macroscopic bulk shape, less noble metals interact stronger with the atoms than with undissociated molecules (which leads to molecular dissociation), whereas more noble metals, show the opposite trend, which preserves the molecules intact. In the case of an atomic adsorbate, the (vertically oriented) p_z orbital of the carbon (or oxygen or nitrogen) atom interacts with symmetric surface orbitals like the d_z^2 orbitals or symmetric combinations of s orbitals, typically leading to high coordination. The coupling of the adsorbate to the broad metal s/p states lead to a shift and broadening of the adsorbate states, which is essentially the same for all transition metals. Furthermore, the asymmetric p_x and p_y orbitals may interact with asymmetric surface orbitals like the d_{xz} or asymmetric combinations of s orbitals. This coupling is strongly responsible for the distinct behavior among the different transition metals. The coupling between the adsorbate states and the metal d states typically gives rise to bonding and antibonding states. In more noble metals such as Cu, the antibonding states are below the Fermi level and thus are occupied. This decreases the overall adsorbate–metal bonding interaction. In contrast, in less noble metals such as Ni or Fe, these antibonding states are above the Fermi level and so are unoccupied. Consequently, bonding of adsorbates is stronger at nonnoble metals than it is at noble metals. Therefore, the metal–adsorbate bond becomes stronger when moving to the left in the periodic table.

Moving now the attention to molecular adsorbates, one can say that there are usually more adsorbate valence states to consider than in the case of atomic adsorbates, although the overall picture remains the same. In the case of CO binding to a transition metal, the

CO valence states are the filled 5σ and the empty double degenerate $2\pi^*$ states. The interaction with the metallic s states merely results in a downshift and broadening of both the 5σ and $2\pi^*$ states, as was the case for atomic adsorption. But interaction with the d states leads to the formation of bonding and antibonding states below and above the original states. Therefore, the more noble metals, which show the weakest interactions, show a stronger bonding to molecules than to the corresponding atoms (i.e., CO will bind stronger than the separate C and O atoms, such that molecular dissociation is not favored); but when moving to the left in the periodic table, the increase in metal–adsorbate bond strength is greater in the case of atomic adsorbates than in the case of molecular adsorbates, such that lesser noble metals show stronger bonding to atoms than to the corresponding molecule (i.e., separate C and O atoms will bind more strongly to the metal than the CO molecule, such that molecular dissociation is favored).

The difference in reactivity from one metal to another can be understood by taking into account the position of the corresponding reaction transition state. As it was described, the higher the d-band center (i.e., moving toward the left in the periodic table), the more bonding orbitals will be below the Fermi level, the stronger the bonds will be between adsorbate and the metal, and the more prone the adsorbate will be to dissociation. Hence, the lesser noble metals show a higher reactivity and, correspondingly, a lower activation barrier. This determines the reactivity in early-barrier reactions. In late-barrier reactions, the transition state resembles the products. The differences in transition state energies thus also closely follow the differences in adsorbed products.

As it was earlier explained, due to quantum mechanical effects, the electronic structure of nanoparticles is not composed of complete bands as in the case of macroscale solids but is intermediate between atoms and molecules on the one hand and solids on the other hand. Thus, when the de Broglie wavelength of the electrons becomes comparable to the system size, the particles electronically behave as zero-dimensional. As an example, the de Broglie wavelength of a 1 eV electron (rest mass energy = 0.511 MeV) is 1.23 nm, which is a typical nanocatalyst particle size. The nanofeatures of catalysts typically show a wide variety of surface structures, possibly containing various crystal facets, amorphous local surfaces, and defects. This influences both the interaction energy and the reaction mechanism which are determined by the coordination number of the atoms involved. In general, lower-coordinated surface atoms (kinks, corners, edges, terraces) are more reactive than densely packed surfaces, and that is why catalysts usually come in the form of nanoparticles.

The melting point of nanoparticles is another well-known size-dependent property of nanocatalysts, affecting their catalytic behavior. Melting of nanoparticles is described by the Gibbs–Thomson equation [81],

$$\Delta T = \frac{2T_M \sigma_{SL}}{\rho_S \Delta H r} \tag{6.31}$$

expressing the depression of the melting temperature of small particles relative to the bulk material, where T_M is the bulk melting temperature, σ_{SL} is the solid–liquid interfacial energy, ρ_S is the density of the solid, ΔH is the melting enthalpy, and r is the radius of the particle. Therefore, for processes taking place close to the size-dependent melting, the nanocatalyst may exist in either the solid or the liquid state, and consequently, a small change in temperature may induce a change in process mechanism. This has been observed in metal-catalyzed CNT growth, where carbon transport may occur either by surface diffusion at low temperature (solid nanocatalyst) or by bulk diffusion at higher temperature (liquid nanocatalyst).

Fullerene hybrids have successfully been used as catalysts in some hydrogen transfer reactions. Due to their poor solubility in polar solvents, these hybrids behave as homogeneous/heterogeneous catalysts that can be mechanically separated and reused several times while the final products do not need chromatographic separation [82]. Also, metal-organic frameworks (MOFs) have shown great promise for optimal guest binding, resulting in desirable adsorption and catalysis. MOFs are crystalline porous materials constructed by connecting metal ions/clusters with organic linkers. The highly dispersed and uniformly distributed metal sites can be tailored in MOFs, endowing them the great yet versatile capacity for various applications in catalysis. Since MOFs have poor thermal and hydrolytic stability, this has hindered their applications in conventional catalysis involving thermal treatment and/or the use of water. MOFs containing open metal sites (i.e., coordinatively unsaturated metal sites), have shown a good guest binding, resulting in desirable adsorption and catalysis. However, the most challenging feature of using MOFs with open metal sites in heterogeneous catalysis lies in the role of these, which must provide a balance between activity and stability during the reaction under heating or the presence of water or both.

The behavior of an MOF during the water-gas shift reaction (which normally only occurs at high temperatures) assisted by nonthermal plasma (with Argon as the discharge gas) showed excellent catalytic activity in these conditions [82]. The plasma promoted the dissociation of H_2O supplying the intermediate OH to enable the reaction, as well as preventing the water-induced decomposition of the organic linker, resulting in an enhanced stability. The stability enhancement of MOFs induced by plasma treatment seems promising, as evidenced by several MOF plasma-activated experiments proving the synergy of MOFs-catalysis and nonthermal plasma activation [82]. Thus, the stability of hollow nano-cages like metal-fullerenes or MOFs can be sustained under nonthermal plasma activation and in the presence of water.

The chemical and physical effects (difficult to distinguish) that may cause the synergy between plasma and catalyst in a dielectric barrier discharge (DBD) reactor, are not only dependent on the material of the packing/catalyst but can also be based on changes in the plasma characteristics due to the presence of the packing. This indicates that correlating results from the literature obtained in different setups is not straightforward and should be considered with care. The physical effect of inserting a packing may be attributed to the contact between different beads inside the plasma reactor and between the beads and the dielectric barrier. In fact, the applied potential difference between the electrodes of the DBD reactor causes polarization of the dielectric beads, creating a strong electric field near the contact points. This enhanced electric field strength may be, at least partly, responsible for a higher conversion and energy efficiency. The following three effects may play a role: (a) the positive contribution of the packing, due to electric field enhancement at the contact points, (b) a negative contribution due to the lower residence time in the presence of a packing, and (c) the influence of the voids between particles, with a positive or negative effect depending on the material inserted. Depending on the bead size and material, one or the other effect will dominate, which may explain their eventual different behavior in conversion and energy efficiency.

Therefore, plasma technology can and should play a significant role in solving challenging environmental problems. The oldest and the largest industrial plasma process targeted for environmental purposes is ozone production. Plasma also destroys microorganisms in different environments resulting in plasma sterilization technologies. Variations of pH of deionized water after dielectric barrier discharge (DBD) plasma treatment in different gases is a paradigmatic example [83]. Several environmental tasks already have commercially viable plasma solutions: water sterilization and removal of organic pollutants, waste destruction, dust and chemical fog separation in electrostatic filters and carbon sequestration [84]. Plasma CO_2 dissociation can be rather efficient from the standpoint of conversion of electrical energy to chemical energy, but because of the very low chemical energy of CO_2, this process is attractive only under very specific conditions, for instance involving renewable electrical energy or synthesis of added value products.

To reduce this concentration and the effect of CO_2 on global warming, several technologies and reactions can be considered, like (dry) reforming of methane (DRM) for syngas production, CO_2 hydrogenation for the synthesis of methanol, methane, formaldehyde, dimethyl ether, etc. Anyway, CO_2 can also be directly split into CO and O_2, where CO can be used as a chemical feedstock to produce value-added chemicals. Since CO_2 has a high thermodynamic stability, a large amount of energy is needed to activate the CO_2 gas. For a thermal activation, typically reaching 1,500 K, a lot of energy is lost in heating the entire gas. Because thermal-catalytic CO_2 splitting is also very energy consuming, the only practical way of thermal CO_2 conversion is dry reforming of methane (DRM), that is, the simultaneous conversion of CO_2 with CH_4, yielding the production of syngas ($CH_4 + CO_2 \rightarrow 2CO + 2H_2$). Following Le Châtelier's principle, increasing the pressure of the reactants will drive the reaction backward, and hence the reaction is commonly carried out at atmospheric pressure. But just like direct CO_2 splitting, this is also an endothermic reaction, with a standard reaction enthalpy of 2.56 eV per converted molecule. Therefore, it needs to be carried out at high temperatures (600–900°C), by means of a catalyst (Ni, Co, noble metals, Mo_2C). This inefficiency conversion can be avoided by using nonthermal plasmas, which in addition are an attractive alternative to the conventional (catalytic) thermal route. In fact, this allows to implement chemical reactions close to room temperature in nonequilibrium, low-energy-consuming reactor. Since in a no-thermal plasma the overall gas kinetic temperature is low, while the electrons are accelerated by the applied electric field at energies (1–10 eV), enough to break most chemical bonds (notice that standard reaction enthalpy for CO_2 dissociation is 2.9 eV).

Whether humanity consumes most fossil fuel resources in combination with carbon capture, or progressively uses alternative energies, the need for fuels and transport materials (currently obtained from oil and natural gas) will remain. With the increase in population, carbon-based products (plastics, medicines) will also be needed in increasing amounts. To meet the demand for carbon-based products, carbon dioxide will have to be recycled and converted into new materials and fuels using an alternative clean energy source. Utilization of the CO_2 waste and converting it into a new feedstock using renewable energy and green processes is a sustainable way of closing the carbon loop. Carbon dioxide conversion into value-added chemicals and fuels is a present relevant challenge and due to the limitations of the traditional thermal approaches, several novel technologies have been developed. Plasma-based CO_2 conversion is gaining increasing interest in the aim of present energy sustainability concerns and carbon neutrality or Mars terraforming challenges as well. Gas activation by energetic electrons instead of heat allows thermodynamically difficult reactions to occur with reasonable energy cost: CO_2 splitting into CO and O_2, as well as the reaction with other gases, like CH_4 (dry reforming of methane), H_2 (hydrogenation of CO_2) or H_2O (artificial photosynthesis), aiming for the production of synthetic fuels (syngas, methanol, formaldehyde, and formic acid).

As the reader can infer, improving the energy efficiency of such possible conversion routes is the main goal to pursue. Plasma is certainly a promising medium for energy-efficient CO_2

conversion because it creates energetic electrons that can activate the gas molecules by electron impact. Such collision events lead to dissociation, ionization, and excitation, generating reactive species (i.e., radicals, ions, excited species) that can easily form new products. Consequently, the carbon dioxide gas does not have to be heated as a whole, to allow strongly endothermic reactions to occur (CO_2 splitting and the dry reforming of methane, i.e., simultaneous conversion of $CO_2 + CH_4 \rightarrow 2CO + 2H_2$), at mild reaction conditions of temperature and pressure. To improve the energy efficiency of CO_2 plasma conversion it is desirable to understand the detailed kinetics involved in the process, which implies theoretical modeling of the mechanisms, describing in detail the behavior of the various plasma species and all relevant elastic and inelastic processes, including reactive collisions. Thus, the degree of accuracy of the modeling strongly depends on reliable collision cross-section data, in particular electron impact dissociation. This process is believed to proceed through electron impact excitation, but it is important to specify which excitation channels effectively lead to dissociation. Thus, some modeling efforts have been devoted to discussing the effect of different electron impact dissociation cross-sections reported in the literature on the calculated CO_2 conversion. In this respect, a comparison to experimental data obtained from dielectric barrier discharge (DBD) and microwave (MW) plasmas could elucidate which cross-section might be the most realistic. Some calculations indicate that the choice of the electron impact dissociation cross-section is crucial for the DBD, but in the MW plasma, it is only significant at pressures up to 100 mbar, while it is of minor importance for higher pressures, when dissociation proceeds mainly through collisions of CO_2 with heavy particles [85].

The nature of thermal plasmas makes them unsuitable for the efficient conversion of CO_2.

Since ionization and chemical processes in thermal plasmas are determined by the temperature, the maximum energy efficiency is limited to the thermodynamic equilibrium efficiency and corresponding conversions of 47% and 80% at 3,500 K, respectively. This contrasts with nonthermal plasmas, where lab-scale efficiencies of up to 90% have already been reported [86]. The electrons receive their energy from the electric field in nonthermal plasmas, and subsequently, through collisions, this energy is distributed between elastic energy losses and different channels of excitation, ionization, and dissociation. It is then possible to estimate the fraction of the energy transferred to the different channels of the excitation, ionization, and dissociation of CO_2, as a function of the reduced electric field [86]. The reduced electric field is the ratio of the electric field in the plasma over the neutral gas density and has distinctive values for different plasma types. For example, a dielectric barrier discharge has a typical reduced electric field in the range above 2×10^{-19} V m^2, whereas microwave (MW) and gliding arc (GA) discharges (which belong to the category of warm plasmas) typically operate about 5×10^{-20} V m^2. The reduced electric field has wide implications on the distribution of the electron energy among the different channels of CO_2 nonthermal plasma [86]. In the region above 2×10^{-19} V m^2, 70%–80% of the electron energy goes into electronic excitation, about 5% is transferred to dissociation, 5% is used for ionization, while only 10% goes into vibrational excitation; around 5×10^{-20} V m^2, however, only 10% goes into electronic excitation and 90% of the energy goes into vibrational excitations. Furthermore, the addition of different gases (e.g., Ar, He, N_2, H_2O, H_2, CH_4) has an influence on the distribution of these channels, and so even during the pure decomposition of CO_2 into CO and O_2, there will be an effect on this distribution. The distribution of energy into different modes, and especially the fraction going into vibrational excitation, is very important, since it is known that the vibrational levels of CO_2 play an important role in its efficient dissociation. To achieve direct electron-impact dissociation, an electron needs to have an energy of 47 eV to excite CO_2 into a dissociative (i.e., repulsive) electronic state, which will lead to its dissociation into CO and O [86]. The energy spent is much higher than the theoretical value necessary for CO-O bond breaking (5.5 eV). Due to the special nature of the CO_2 molecule, a more efficient dissociation pathway is based on its vibrational excitation. This pathway starts with electron impact – vibrational-excitation of the lowest vibrational levels, followed by vibrational–vibrational (VV) collisions. This so-called ladder-climbing gradually populates the higher vibrational levels, which eventually leads to dissociation of the CO_2 molecule. In this way, it is possible to dissociate CO_2 more efficiently, since only the minimum amount of 5.5 eV for bond breaking is needed, compared to the overshoot in the case of electronic excitation–dissociation. [86].

Many types of plasma reactors have been investigated for CO_2 conversion and energy efficiency of CO_2 dissociation, including corona discharges, glow discharges, microwave discharges, radio frequency discharges, gliding arc discharges, and dielectric barrier discharges. A dielectric barrier discharge (DBD) reactor operates at ambient temperature and atmospheric pressure, can easily accommodate packing materials, and be scaled up to industrial conditions. Research ongoing towards an energy-efficient conversion of CO_2 in empty DBD reactors have shown modest values obtained for both conversions, and energy efficiencies (at typical plasma parameters of 40W power, frequency of 23.5 kHz, and discharge gap of 1.8 mm). It is possible to change the conditions of the experiments (for instance lowering the specific energy input), to look for improvements, but it has been found that a slight increase in the energy efficiency occurs at the expense of a drop in the conversion. The solution is to implement plasma catalysis in the DBD reactor, and it has been demonstrated for pure CO_2 dissociation that the introduction of a dielectric packing can double both the CO_2 conversion and energy efficiency, in comparison with an unpacked DBD reactor.

The CO adsorption on metal surfaces (typically Co, Ni, Cu) is a critical step in plasma dry reforming of methane (DRM). The largest heat of adsorption for CO is found on Co and of those three metals, Cu has the lowest d-band center, and Co has the highest. Thus, from what was earlier explained, Co is most reactive and shows the strongest binding, both toward the molecule and toward the separate atoms. As the molecular binding energy of CO on Cu is less than the CO bond energy, CO will not dissociate on Cu. The heat of adsorption of CO on Ni is 95 kJ/mol, and on Cu is 80 kJ/mol. This small difference is magnified when the molecule dissociates because after dissociation the valence electrons of the CO molecule are no longer tied to molecular orbitals but are freely available. Consequently, the dissociation process is exothermic on Co, thermoneutral on Ni, and endothermic on Cu. This can explain the difference in selectivity between these metals for the conversion of synthesis gas: while Cu is an active component for methanol synthesis catalysts, Co is a Fischer–Tropsch catalyst producing higher hydrocarbons. The Sabatier principle and the typical volcano curves are a direct result of the above. Indeed, moving toward the left in the periodic table gives a lower activation barrier and higher reactivity but also stronger bonding to the surface and, thus, less free active surface and less desorption. Volcano plots are also dependent on the catalyst size, as adsorbate–catalyst interactions are size dependent. The relationship between size and reactivity is not always monotonic. A clear example is the size dependence of adsorption energies on nanoparticles. Density functional theory (DFT) calculations have shown that the CO adsorption on Pd-nanoclusters is weakest (lowest adsorption energy) for clusters containing 30–50 atoms; below and above this size range, the CO adsorption is stronger. Below this size range, the interaction is stronger because of the decreasing energy gap between the 2π* lowest unoccupied molecular orbital (LUMO) of the CO molecule and the energies of the d levels of the metal. Above this size, the interaction energy increases due to a decrease in lattice contraction with increasing particle size. As a result of these size-dependent interactions, volcano plots are also size dependent.

Polanyi's rules state that translational energy is more efficient than vibrational energy in activating early activation barriers whereas the reverse is true for late activation barriers. It happens that vibrational excitation is particularly important in some plasmas, such as microwave discharges and gliding arcs. In these plasmas, the amount of energy that is deposited in vibrational excitations can be tuned to some extent. This has two important consequences in the kinetics of plasma–surface reactions. First, some of the vibrational energy may be used to decrease the energy barrier to be surmounted, by increasing the reactant energy by an amount E_{vib}, as shown in Figure 6.8a [87]. In this case, the barrier for the back reaction remains unaltered. Note that not all vibrational energy contributes to this, as the vibrational coordinate in general does not fully map onto the reaction coordinate. Second, vibrationally excited species may also experience a lower activation barrier than ground-state species, as ground-state species may not have access to the portion of phase space containing the lowest transition barrier. This is shown in Figure 6.8b [87], where the amount of lowering of the activation barrier due to this second effect is denoted by ΔE^*. In this case, the activation barrier for the back reaction is decreased as well. The resulting energy barrier due to the combination of both effects is E_a^{v3}, which is lower by an amount $E_{vib} + \Delta E^*$ relative to the energy barrier $E_a^{v=0}$ experienced by ground state species. The interplay between catalyst and vibrationally excited species is very complex and not completely understood.

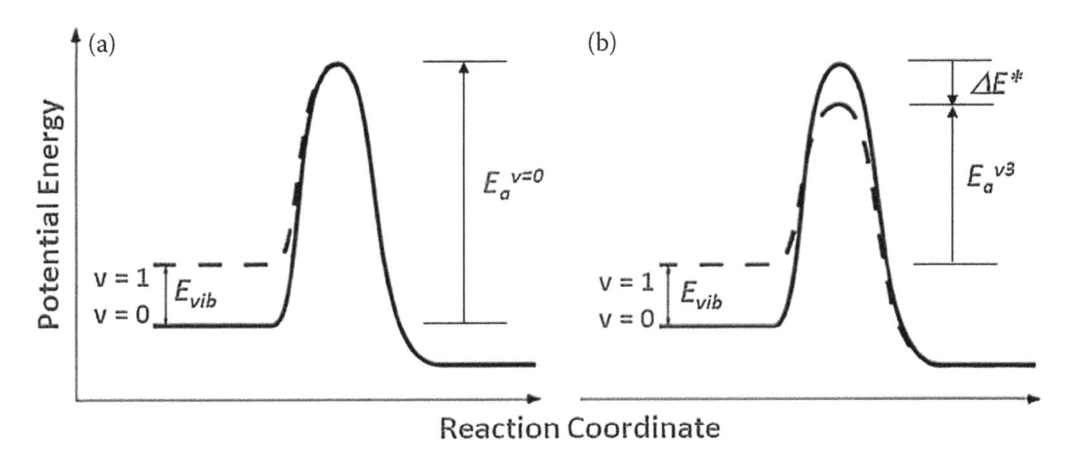

FIGURE 6.8
Double-effect of vibrational excitation of methane on the activation barrier for its adsorption. (a) Decrease of the effective activation barrier by increasing the energy of the reactants by an amount E_{vib}. (b) Also by allowing access to an otherwise inaccessible part of phase space, thus lowering the activation barrier further by an amount ΔE^*.

CO_2 conversion into value-added chemicals is a paradigmatic example of operating synergies in plasma catalysis. This is performed in DBD plasma reactors that can operate at atmospheric pressures. In general, the enhanced performance of plasma catalysis can in part be attributed to vibrational excitation of CO_2 (and the co-reactants CH_4, H_2) in the plasma, which enables easier dissociation at low temperature on the catalyst. Additionally, the plasma electrons can also properly affect the catalyst properties (chemical composition or catalytic structure). One should not also exclude that short-lived plasma species (e.g., excited O atoms) may be formed inside the catalyst pores. Thus, a catalyst with high dielectric constant should be desirable, as this gives more pronounced polarization of the pellets and thus higher electric fields, favouring the dissociation of species inside the catalyst.

Another example where synergy in plasma catalysis assumes relevance is the growth of one-dimensional (1D) carbon nanostructures (SWCNTs, MWCNTs, and carbon nanofibers) on solid surfaces exposed to either a hydrocarbon source gas or a low-temperature nonequilibrium plasma. Carbon nanotubes are most often grown by thermal metal-catalyzed CVD, where the growth process is commonly accepted to follow the vapor–liquid–solid (VLS) mechanism. This occurs in the following four subsequent steps: adsorption of gas-phase hydrocarbon precursor species on the nanocatalyst particle, catalytic decomposition into carbon atoms and dissolution in the liquid bulk, surface carbon segregation with the formation of a solid precipitate, and formation of a solid crystalline structure. The source of carbon atoms for growth is delivered by simple hydrocarbon precursor gases (methane, acetylene) and a hydrogen-based etching carrier gas (H_2 or NH_3), can be added to remove amorphous carbon deposits. The precursors need to be dissociated to release carbon atoms that can then nucleate and give rise to the initial nuclei that eventually develop into the 1D nanostructures. Apart from heating the precursor gas flow in the reaction chamber, the common means of supplying energy required for the nucleation and growth is by heating the deposition substrate. The typical thickness of SWCNTs is of the order of 1–2 nm, and for MWCNTs and carbon nanofibers is about tens of nm. The thickness of these structures is usually limited by the catalyst nanoparticles or catalyst-surface features, which support nucleation and growth of the nanostructures. However, in the case of SWCNTs there is no unique correlation between the catalyst nanoparticle diameter and the tube diameter. The two most common ways to form those catalyst-surface features is by either depositing (or forming otherwise) nanoparticles or surface texturing. The nucleation sites are random and so, poorly controllable on smooth surfaces. This concludes that nucleation may not even be possible on them.

Compared to thermal growth of 1D nanocarbons, in plasma-catalytic growth, a large variety of active species may contribute to the growth process, including atoms, molecules, excited metastable species and radicals, ions and electrons, and photons. In a plasma-catalytic process, the nonequilibrium reactive plasma processes lead to the effective precursor conversion in the gas phase, and to an accelerated carbon species production at the catalyst surface. Moreover, these carbon species can be delivered to the growth surface faster in the plasma, for example, through the ion fluxes. The contributions of these fluxes can be relevant in comparison to the contributions of neutral radical species due to their typically high sticking coefficients. This will lead to larger amounts of carbon atoms on the surface or dissolved within catalyst nanoparticles, which may on the one hand lead to lower diffusion barriers and higher bond switching rates. On the other hand, catalyst nanoparticle supersaturation may occur faster, thereby shortening the lead time before the nucleation (incubation time). However, increased C-fluxes may also cause catalyst nanoparticle poisoning or even complete encapsulation of the catalyst by amorphous carbon.

Ion bombardment may play an important role in plasma-enhanced catalytic CNT growth. During the SWCNT cap nucleation in particular, ion bombardment leads to the top-down energy transfer from the plasma to the C-network. This makes it possible first to improve the structural quality of the C-network. Furthermore, the ion bombardment may also provide the additional bending energy to lift the initial as-nucleated graphene monolayer off the catalyst nanoparticle surface to form an SWCNT. However, ion bombardment may be detrimental to CNT growth, leading to etching and sputtering. The local temperature of the catalyst nanoparticle can also increase because of ions and other plasma species, and this effect may lead to nanostructure nucleation using reduced substrate heating. In this way, the nanostructure growth rates can be substantially increased.

Another relevant synergistic effect of the plasma and catalyst nanoparticles is the vertical alignment along the direction of the applied electric field in the plasma sheath near the substrate surface. Whereas thermal CVD processes usually produce tangled, twisted carbon nanotubes with random orientations and growth directions unless dense crowding effects guide CNTs growth vertically, PECVD produces vertically aligned, free-standing SWCNTs [88]. The electric field increases the degree of directionality of carbon atom movement along the catalyst nanoparticle surfaces as opposed to the completely random motions in all directions without the electric field. Because the direction of the electric field in the plasma is normal to the surface, it guides carbon atoms upward, toward the top of the catalyst nanoparticles, eventually giving rise to the nucleation of SWCNT caps at the summit of the catalyst nanoparticle. This is very different from common thermal CVD cases where nucleation of CNT caps occurs randomly, leading to disordered, misaligned growth, which is often referred to as a "spaghetti-type" growth [89].

Besides plasma-enhanced nanoparticle catalyzed CNT growth and CO_2 conversion, there are other processes displaying synergistic effects when the plasma and catalytic effects are combined, as for instance natural gas reforming (steam reforming), catalytic synthesis of inorganic nanowires, and synthesis of large-area graphene films. Hydrocarbon reforming of natural gas has been receiving a particular attention because it is a major source of hydrogen gas, which in turn is the current focus of the world due to its potential to be a clean energy vector for replacing fossil fuels and implementing a sustainable hydrogen-based economy. In fact, there are unambiguously warnings that current world energy trends are not sustainable, and hydrogen is a product that has the potential of emission-free alternative fuel. Currently, almost 90% of the H_2 is indeed produced via high-temperature steam reforming of natural gas or the light oil fraction. However, this process where water vapor reacts with methane still suffers from some inconveniences: (1) moderate thermodynamic efficiency; (2) fouling/coking on the catalyst surface and sulfur impurities with eventual deactivation, high-temperature requirements of the reforming reaction (a condition that reduces the energy efficiency of the process); (3) need of a further separation of the hydrogen from the syngas product or follow a water shift gas reaction that results in carbon dioxide. The process is endothermic (165 kJ/mol) making high energy requirements and so the wet reforming of methane is a less favorable route. Thus, plasma reforming may offer a number of advantages and it was found that the effect of steam addition increased the CO_2 yield, despite a decrease in methane conversion [90]. Therefore, plasma-assisted catalytic reforming of natural gas in a DBD reactor came to introduce important benefits, like operation at reduced temperature, selectivity toward the formation of value-added products, and an extremely fast switch-on time. Operation at reduced temperature leads to reduced coke formation and catalyst poisoning and, therefore, to increased catalyst lifetime. This is a critical parameter in any industrial setting, in part offsetting the capital costs. The plasma-catalyst synergy effect may be attributed to the presence of vibrationally excited CH_4 molecules, produced in the plasma [91]. This way, plasma-catalytic reforming of natural gas may allow the selective production of syngas, and possibly also other value-added fuels such as methanol or formaldehyde.

The fast plasma switch-on time finally may allow to partially cover the imbalance between supply and demand of energy and, hence, the use of intermittent excess energy, for example, from renewable energy sources, to store this excess electrical energy in the form of liquid fuels (or hydrogen storage). Clearly, large-scale gas conversion requires significant energy input, and hence, the energy efficiency of the process is critical. Thus, plasma-catalytic reforming is prone to drastically reduce the energy costs as much as possible.

Presently, the development of a sustainable energy system is one of the most important industrial challenges and electrocatalytic hydrogen evolution together with suitable safe storage methods are important ways to develop a sustainable energy system based on hydrogen technologies. Hydrogen energy is considered the most powerful candidate to alternate fossil energy due to its clean, renewable, environmentally friendly properties and high energy density. The association of plasma with nanotechnology offers *per se* the potential to significantly increase the energy efficiency of processes required to mitigate pollution and greenhouse gas emissions, whether using renewable energies as power input or not. The desirable use of renewable energies in powering the plasma reactors helps to ensure a full energetic, economic, and environmental sustainability.

Although hydrogen is not widely used today as a transportation fuel, researchers are working toward the goal of clean, economical, and safe hydrogen production as well as fuel-cell electricity for mobility and as a means of mitigating energy delivery fluctuations of current renewables. Finding a safe and not costly technology that can spend as less as possible energy for hydrogen storage and delivery is an issue that presently involved many research laboratories around the world. Though compressed hydrogen and liquefied hydrogen are mature technologies for industrial applications, they require energy consumption and appropriate measures to deal with the issues at high pressure (up to 100 MPa) and low temperature (around 20 K). In turn, storing hydrogen in suitable solids can realize a more compact and much safer approach that does not require high hydrogen pressure and cryogenic temperature. Many studies have been devoted to the use of hydrogen storage solids as active materials and for application in fuel cells. A large number of high-surface nanoporous materials have been investigated, mainly divided into carbon- and MOFs-based groups [92]. Activated carbons are ineffective in hydrogen storage systems because only a small fraction of pores in their pore-size distributions are small enough to interact strongly with hydrogen molecules at the gas phase. For small-sized systems, surface- or interface-related sites assume crucial importance and can change the overall solubility of hydrogen. Thus, for improving the kinetic characteristics of hydrogen storage in materials, great attention has been paid to thermal desorption studies and calibration. Thermal desorption spectrometry assures a good accuracy in monitoring hydrogen absorbed in solid materials, and it is suitable for monitoring hydrogen evolution following the application of a thermal ramp, in contrast to the determination of static sorption isotherms [93]. The use of mass spectrometry allows evolved gas species to be identified, and desorption temperatures give information on the binding energy of adsorbed molecules which varies with the nature of the adsorbate/surface materials. An additional goal consists of looking for low-weight and high-performance hydrogen active storage media made from carbon nanomaterial assemblies. Several carbon nanomaterial assembly-based electrodes have been chosen for comparing their relative performance. For this purpose, the recent molecular-beam thermal desorption spectrometry (MB-TDS) technique has been applied [94, 95]. This technique was developed to detect the hydrogen release by lowermost amounts of solid samples, with several advantages [94]. MB-TDS is a novel variant of thermal desorption based on an effusive molecular beam and represents a significant improvement in the accurate determination of hydrogen mass absorbed on a solid sample. The enhancement in the signal-to-noise ratio for trace hydrogen is on the order of 20%, and no previous calibration with a chemical standard is required.

Light trapping in solar-energy devices is also a relevant subject regarding the efforts that since the last 50 years have been made to put solar radiation harvesting as a player among the family of renewable energies that contribute for sustainability. Metallic nanostructures can contribute to light trapping by scattering and increasing the path length of light, by generating a strong electromagnetic field in the active layer, or by multiple reflections/absorptions. The metallic nanostructures can either be used in reducing the material thickness and device cost or in improving light absorbance and thereby improving conversion efficiency [96]. Light trapping describes the process of confining light to a given space, usually with the intent of converting it to other forms of energy. Some materials (e.g., TiO_2 wafer, because this material has weak visible light absorption), can be designed to slow light, bending it away from its given path, so that it spends more time in the material. A silicon solar cell harvests the energy of the sun as light travels down through light-absorbent silicon. But in order to reduce weight and costs, solar cells must be thin enough, and although silicon absorbs visible light well, it captures less than half of the light in the near-infrared spectrum, which makes up one-third of the sun's energy. As the depth of the material limits absorption, one solution is to channel light horizontally so that silicon could absorb its energy along the width of the cell rather than its depth. Another approach is increasing the optical path-length with the use of cubic 900-nanometer titanium dioxide photonic crystal for enhancing light trapping. This results in an absorption enhanced by almost two orders of magnitude greater than a reference film of the same material [97]. Also, the efficiencies of perovskite solar cells have been further enhanced by the light trapping effect caused by textured fluorine-doped SnO_2 with the incorporation of silver nanoparticles. Aside from enhanced light absorption, the charge transport characteristics of perovskite solar cells can be improved by optimizing the Ag nanoparticles loading levels, which is due to the localized surface plasmon resonance (LSPR) from the incorporated particles [98]. Metallic nanoparticles sustaining localized surface plasmon resonances can enhance light trapping in thin film photovoltaics. A strong resonant response from nanoparticle ensembles can be obtained if the particles have monodispersed physical properties. It can be explored the correlation between the structural and optical properties of self-assembled silver nanostructures fabricated by a solid-state dewetting process on various substrates relevant for silicon photovoltaics and later integrated into plasmonic back reflectors [98].

To finalize let us deliver some words about the production of energy through micromachines, where it is crucial the reduction of friction phenomena at the nanoscale to increase energy efficiency since ultimately one will face the van der Waals and Casimir forces. At this point, the nanotribological considerations and the nature of Casimir forces already addressed with their respective methods of study are not only indispensable but also these forces have the potential to transmit force at short distances without physical contact [99]. A special emphasis should be addressed to the distinction between the Casimir and van der Waals forces since they both have origin in the same source: the zero-point electromagnetic field. Their theoretical descriptions are different because in general the van der Waals force is expressed in terms of surface charges on the two bodies whereas the Casimir force is described in terms of the zero-point electromagnetic field. The surface charges, however, are a result of fluctuations in the zero-point field; hence fundamentally the two forces arise from the same source. As two surfaces are brought together, the power law describing how the force varies with distance changes, and this is the crossover between the two regimes. This crossover can be thought of as the minimum distance between two components before they stick together, which is about 10 nanometers. Also, as the size of these devices has decreased, they have become full of boundaries with submicron gaps where the Casimir force becomes dominant. It is a significant problem because while one can take measures to prevent capillary and electrostatic forces, there is nothing that can be done to prevent the Casimir force as it arises from the fundamental properties of the vacuum. It thus generates a fundamental and ever-present stickiness of components in micromachines and nanomachines.

One reason for difficulties in extracting large amounts of energy via the Casimir effect from the vacuum is the fact that real metals become nearly transparent for wavelengths below 100 nm. Real metals are only effective mirrors down to 100 nm wavelength [100]. The shorter wavelength and the higher frequencies give the highest contributions. Since real metals become nearly transparent for UV, X-ray, and shorter wavelengths, these radiation parts do not contribute to the Casimir force between real metals. If the vacuum could be engineered its energy could be eventually used because in principle energy and heat can be extracted from the vacuum [101]. A proposal for a cyclic engine has been reported [102] by introducing the idea of varying effective distance or cavity size by light-induced conductivity changes in semiconductors. With this change in conductivity, the Casimir force can be modulated, and an engine cycle can be designed. Another approach is to reverse the Casimir force from attraction to repulsion by using surfaces with small cavities. These cavities can change the direction of the Casimir force [103].

The Casimir pressure between any pair of conducting plates (or zero-point pressure of electromagnetic waves) separated by a distance a, is given by $\pi hc/(480a^4)$. This result was expanded by Lifshitz to include forces between dielectrics, and the pressure between two dielectric plates with a dielectric constant ε is as follows:

$$P = \frac{F}{A} = -\frac{\pi}{480}\frac{hc}{a^4}\frac{(\varepsilon - 1)^2}{(\varepsilon + 1)^2}f(\varepsilon) \tag{6.32}$$

where $f(\varepsilon)$ is a function that tends to one when the dielectric constant is very large and 0.35 when $\varepsilon < 4$.

It is worthy of notice that it is possible to obtain the zero-temperature Casimir energy based on the classical (high temperature) Kirchoff law [104]. The basic thermodynamics analysis of Kirchhoff's classical theorem says that the energy density of the radiation in equilibrium is a function of temperature only. Kirchoff's law in the high-temperature limit assures us that the energy density of two Casimir subsystems is the same. This seems to be not true in quantum physics and Casimir effect appears as a violation of this theorem. And so, comes out the necessity of positive zero-point energy. However, the theorem should hold at the high-temperature limit where classical physics is applicable. In fact, it could be shown that the reason for the need of zero-point energy is to comply with Kirchoff's theorem at high temperatures [104]. The intimate connection between the zero-point energy and the thermal Bose occupancy was invoked to account for the results.

Other authors further analyzed the high temperature (or classical) limit of the Casimir effect by introducing the concept of relative Casimir energy [105]. This is a useful quantity that is defined for a configuration of disjoint conducting boundaries of arbitrary shapes, as the difference of Casimir energies between the given configuration and a configuration with the same boundaries infinitely far apart. Using path integration techniques, the authors could show that the relative Casimir energy vanishes exponentially fast in temperature. This is consistent with the simple physical argument based on Kirchhoff's law. Additionally, since they also defined the relative Casimir entropy in an analogous manner, this tends in the classical limit, to a finite asymptotic value that depends only on the geometry of the boundaries. Therefore, the authors could conclude that the Casimir force between disjoint pieces of the boundary, in the classical limit, is entropy driven and is governed by a dimensionless number characterizing the geometry of the cavity [105]. For a particle of mass m localized within a box measuring Δx on the side, the uncertainty of momentum would be of the order of $\hbar/\Delta x$. The associated "zero-point" kinetic energy would be about $(\hbar/\Delta x)^2/2m$. Thus, for $\Delta x \sim 0.1$ nm, the "zero-point" energy for an electron is about 3 eV. This will exert a pressure on the walls of the cube of several million times atmospheric pressure. The lighter the particle, the greater is the zero-point energy. It is precisely this zero-point energy that prevents the lightest inert gases from solidifying even at the absolute zero of temperature; helium remains liquid down to the lowest temperatures known, and although hydrogen solidifies at low enough temperatures, the reason is that the attractive potential between the two hydrogen atoms is sufficiently strong to 'contain' this zero-point motion. For the inert atoms like those of helium, the attraction is relatively much too weak. Again, it is this zero-point energy that prevents the electron in a hydrogen atom from collapsing onto the proton and staying stuck there. It is worth also to note that the uncertainty principle holds for other pairs of dynamical variables too, for example, energy and the lifetime, or angle and the angular momentum.

Despite separations of a few nanometers (or less) between small particles dielectric van der Waals interactions are always present, at larger distances retardation effects associated with the speed of light come into play. This causes the force to fall more rapidly with distance than in the short-range van der Waals limit. These are called Casimir–Lifshitz forces, and it has been shown that their sign can be changed from attractive to repulsive by suitable choice of interacting materials immersed in a fluid [106]. Repulsive forces between macroscopic bodies are dependent on their material dielectric response functions. The interaction of a material 1 with a material 2 across a medium 3 goes as a summation of terms with differences in material permittivities, $-(\varepsilon_1-\varepsilon_3)(\varepsilon_2-\varepsilon_3)$, over frequencies that span the entire spectrum. Between two like materials, $\varepsilon_1 = \varepsilon_2$, these terms are negative and correspond to attraction. However, when the dielectric response ε_3 of the medium is between ε_1 and ε_2, it happens that $\varepsilon_1 > \varepsilon_3 > \varepsilon_2$, and so $-(\varepsilon_1-\varepsilon_3)(\varepsilon_2-\varepsilon_3)$ terms are positive and the force is repulsive. An easy-to-see limit for this repulsion is the case where region 2 is air or vacuum and the polarizability of medium 3 is less than that of a substrate. As a result, substance 3, rather than forming a droplet, spread out to achieve maximum proximity to substance 1. One set of materials, solid-liquid-solid in such conditions over a large frequency range is gold, bromobenzene, and silica. This way, quantum levitation of gold surface above silica surface in a fluid has become measurable [106].

The Hamaker constant, which captures the polarizability and other materials properties of importance, varies by only a single order of magnitude over practical materials. Therefore, there is little opportunity to adjust or manipulate the van der Waals force and reduce the adhesion-induced failure of devices. The Casimir force arises from the interaction of the surfaces with the surrounding electromagnetic spectrum and includes a complex dependence on the full dielectric function of both surfaces and the region between them. The complexity of the Casimir force leads to significantly greater possibility for manipulation through materials, geometries, and other phenomena. The significantly greater complexity of the Casimir force potentially allows greater opportunity for neutralization or for use of Casimir forces to partially cancel van der Waals forces. Once researchers can measure the Casimir force in nanophysics experiments and perform computer calculations of Casimir force between objects made of nanostructured materials comprising nanoparticles [107], one can think about how it could be exploited in technology. It is possible for instance, to consider the effects from changing the parameters of the nano-structured materials that constitute the particles; this last type of study leads to a form of control of the Casimir force and an insight into possible technological applications. Several finite 3D objects consisting of ordered arrangements of spherical particles with dipolar interaction reflect a system that could be interpreted as a model of objects made of metamaterials (where the distance between particles is shorter than the wavelength of incident light). A system of N spherical particles, made of a homogeneous isotropic material characterized by a dielectric function, immersed in the vacuum and under influence of an electric field induces a dipole moment in each sphere because of its polarizability. It is thus possible to control the magnitude and direction of Casimir forces through the modification of the internal structure. Furthermore, it can be used to study the effects of shape, size, temperature, and materials on the Casimir forces between compact 3D objects. Due to the complexity of the theoretical models available, this has not been possible until now.

An interesting basic idea is that if one could exploit the fact that a vacuum is an energy reservoir, thanks to zero-point energy, future space travelers would have access to a limitless energy source. The only thing they need, of course, is a propulsion system that harvests the required energy from the vacuum. A "vacuum-fluctuation battery" can be constructed by using the Casimir force to do work on a stack of charging conducting plates. By applying a charge of the same polarity to each conducting plate, a repulsive electrostatic force will be produced that opposes the Casimir force. If the applied electrostatic force is adjusted to be always slightly less than the Casimir force, the plates will move toward each other, and the Casimir force will add energy to the electric field between the plates. The battery can be recharged by making the electrical forces slightly stronger than the Casimir force to re-expand the foliated conductor.

References

1. G. Binnig, H. Rohrer, Ch. Gerber, and E. Weibel, "7 × 7 Reconstruction on Si(111) resolved in real space," *Physical Review Letters*, vol. 50, no. 2, pp. 120–123, Jan. 1983, doi: 10.1103/PhysRevLett.50.120
2. G. Binnig, C. Gerber, E. Stoll, T. R. Albrecht, and C. F. Quate, "Atomic resolution with atomic force microscope," *Europhysics Letters (EPL)*, vol. 3, no. 12, pp. 1281–1286, Jun. 1987, doi: 10.1209/0295-5075/3/12/006
3. F. J. Giessibl, "High-speed force sensor for force microscopy and profilometry utilizing a quartz tuning fork," *Applied Physics Letters*, vol. 73, no. 26, pp. 3956–3958, Dec. 1998, doi: 10.1063/1.122948
4. D. Rugar, O. Züger, S. Hoen, C. S. Yannoni, H.-M. Vieth, and R. D. Kendrick, "Force detection of nuclear magnetic resonance," *Science (1979)*, vol. 264, no. 5165, pp. 1560–1563, Jun. 1994, doi: 10.1126/science.264.5165.1560
5. D. W. Pohl, W. Denk, and M. Lanz, "Optical stethoscopy: image recording with resolution λ/20," *Applied Physics Letters*, vol. 44, no. 7, pp. 651–653, Apr. 1984, doi: 10.1063/1.94865
6. E. Wolf and M. Nieto-Vesperinas, "Analyticity of the angular spectrum amplitude of scattered fields and some of its consequences," *Journal of the Optical Society of America A*, vol. 2, no. 6, p. 886, Jun. 1985, doi: 10.1364/JOSAA.2.000886

7. R. F. M. Lobo, M. S. Costa, J. H. F. Ribeiro, C. A. C. Sequeira, and P. Pereira, "Fluorescence in nanostructured fulleride films," *Applied Physics Letters*, vol. 89, no. 20, p. 203102, Nov. 2006, doi: 10.1063/1.2388245
8. A. Bouhelier, et al., "Plasmon optics of structured silver films," *Physical Review B*, vol. 63, no. 15, p. 155404, Mar. 2001, doi: 10.1103/PhysRevB.63.155404
9. C. Fokas and V. Deckert, "Towards *in situ* Raman microscopy of single catalytic sites," *Applied Spectroscopy*, vol. 56, no. 2, pp. 192–199, Feb. 2002, doi: 10.1366/0003702021954665
10. J. Kohoutek, et al., "Opto-mechanical force measurement of deep sub-wavelength plasmonic modes," Proceedings of SPIE NanoScience + Engineering, vol. 8097, p. 80971T, San Diego, California, United States, 2011, doi: 10.1117/12.891888
11. M. Li, H. X. Tang, and M. L. Roukes, "Ultra-sensitive NEMS-based cantilevers for sensing, scanned probe and very high-frequency applications," *Nature Nanotechnology*, vol. 2, no. 2, pp. 114–120, Feb. 2007, doi: 10.1038/nnano.2006.208
12. A. Schirmeisen, B. Anczykowski, and H. Fuchs, "Dynamic force microscopy," in *Nanotribology and Nanomechanics*, Berlin/Heidelberg: Springer-Verlag, pp. 243–281. doi: 10.1007/3-540-28248-3_6
13. G. B. M. Fiege, V. Feige, J. C. H. Phang, M. Maywald, S. Gorlich, and L. J. Balk, "Failure analysis of integrated devices by scanning thermal microscopy (SThM)," *Microelectronics Reliability*, vol. 38, no. 6–8, pp. 957–961, Jun. 1998, doi: 10.1016/S0026-2714(98)00086-9
14. J. Leong and C. C. Williams, "Shear force microscopy with capacitance detection for near-field scanning optical microscopy," *Applied Physics Letters*, vol. 66, no. 11, pp. 1432–1434, Mar. 1995, doi: 10.1063/1.113269
15. R. C. Davis, C. C. Williams, and P. Neuzil, "Micromachined submicrometer photodiode for scanning probe microscopy," *Applied Physics Letters*, vol. 66, no. 18, pp. 2309–2311, May 1995, doi: 10.1063/1.114223
16. U. Weierstall and J. C. H. Spence, "Atomic species identification in STM using an imaging atom-probe technique," *Surface Science*, vol. 398, no. 1–2, pp. 267–279, Feb. 1998, doi: 10.1016/S0039-6028(98)80030-7
17. G. M. McClelland and C. T. Rettner, "Scanning aperture photoemission microscopy for magnetic imaging," *Applied Physics Letters*, vol. 77, no. 10, pp. 1511–1513, Sep. 2000, doi: 10.1063/1.1290721
18. B. C. Stipe, M. A. Rezaei, and W. Ho, "Single-molecule vibrational spectroscopy and microscopy," *Science (1979)*, vol. 280, no. 5370, pp. 1732–1735, Jun. 1998, doi: 10.1126/science.280.5370.1732
19. Y. J. L. C. M. Kim, "Nanotube nanotweezers," *Science (1979)*, vol. 286, p. 2148, 1999.
20. E. T. Foley, A. F. Kam, J. W. Lyding, and Ph. Avouris, "Cryogenic UHV-STM study of hydrogen and deuterium desorption from Si(100)," *Physical Review Letters*, vol. 80, no. 6, pp. 1336–1339, Feb. 1998, doi: 10.1103/PhysRevLett.80.1336
21. G. P. Lopinski, D. D. M. Wayner, and R. A. Wolkow, "Self-directed growth of molecular nanostructures on silicon," *Nature*, vol. 406, no. 6791, pp. 48–51, Jul. 2000, doi: 10.1038/35017519
22. A. Aviram and M. A. Ratner, "Molecular rectifiers," *Chemical Physics Letters*, vol. 29, no. 2, pp. 277–283, Nov. 1974, doi: 10.1016/0009-2614(74)85031-1
23. Y. Wada, T. Uda, M. Lutwyche, S. Kondo, and S. Heike, "A proposal of nanoscale devices based on atom/molecule switching," *Journal of Applied Physics*, vol. 74, no. 12, pp. 7321–7328, Dec. 1993, doi: 10.1063/1.354999
24. J. Park, et al., "Coulomb blockade and the Kondo effect in single-atom transistors," *Nature*, vol. 417, no. 6890, pp. 722–725, Jun. 2002, doi: 10.1038/nature00791
25. S. J. Tans, et al., "Individual single-wall carbon nanotubes as quantum wires," *Nature*, vol. 386, no. 6624, pp. 474–477, Apr. 1997, doi: 10.1038/386474a0
26. R. Zhang, et al., "Chemical mapping of a single molecule by plasmon-enhanced Raman scattering," *Nature*, vol. 498, no. 7452, pp. 82–86, Jun. 2013, doi: 10.1038/nature12151
27. R. F. M. Lobo, *Nanophysics for Energy Efficiency*. Cham: Springer International Publishing, 2015. doi: 10.1007/978-3-319-17007-7

28. C. M. Mate, G. M. McClelland, R. Erlandsson, and S. Chiang, "Atomic-scale friction of a tungsten tip on a graphite surface," *Physical Review Letters*, vol. 59, no. 17, pp. 1942–1945, Oct. 1987, doi: 10.1103/PhysRevLett.59.1942
29. W. A. Ducker, T. J. Senden, and R. M. Pashley, "Direct measurement of colloidal forces using an atomic force microscope," *Nature*, vol. 353, no. 6341, pp. 239–241, Sep. 1991, doi: 10.1038/353239a0
30. C. Ritter, M. Heyde, B. Stegemann, K. Rademann, and U. D. Schwarz, "Contact-area dependence of frictional forces: Moving adsorbed antimony nanoparticles," *Physical Review B*, vol. 71, no. 8, p. 085405, Feb. 2005, doi: 10.1103/PhysRevB.71.085405
31. R. Gomer and L. W. Swanson, "Theory of field desorption," *The Journal of Chemical Physics*, vol. 38, no. 7, pp. 1613–1629, Apr. 1963, doi: 10.1063/1.1776932
32. F. London, "Zur Theorie und Systematik der Molekularkräfte," *Zeitschrift fur Physik*, vol. 63, no. 3–4, pp. 245–279, Mar. 1930, doi: 10.1007/BF01421741
33. B. Bhushan, Ed., *Springer Handbook of Nanotechnology*. Berlin, Heidelberg: Springer Berlin Heidelberg, 2017. doi: 10.1007/978-3-662-54357-3
34. B. Bhushan, "Nanotribology, nanomechanics and materials characterization," 2017, pp. 869–934. doi: 10.1007/978-3-662-54357-3_27
35. D. Guo, G. Xie, and J. Luo, "Mechanical properties of nanoparticles: basics and applications," *Journal of Physics D: Applied Physics*, vol. 47, no. 1, p. 013001, Jan. 2014, doi: 10.1088/0022-3727/47/1/013001
36. D. Maharaj and B. Bhushan, "Friction, wear and mechanical behavior of nano-objects on the nanoscale," *Materials Science and Engineering: R: Reports*, vol. 95, pp. 1–43, Sep. 2015, doi: 10.1016/j.mser.2015.07.001
37. E. Gnecco, et al., "Velocity dependence of atomic friction," *Physical Review Letters*, vol. 84, no. 6, pp. 1172–1175, Feb. 2000, doi: 10.1103/PhysRevLett.84.1172
38. H. P. Lang, et al., "A chemical sensor based on a micromechanical cantilever array for the identification of gases and vapors," *Applied Physics A: Materials Science & Processing*, vol. 66, no. 7, pp. S61–S64, Mar. 1998, doi: 10.1007/s003390051100
39. A. Fridman, *Plasma chemistry*. New York: Cambridge University Press, 2008.
40. R. Goldston, *Introduction to Plasma Physics*. Philadelphia: Institute of Physics Publishing, 1997.
41. E. Bormashenko and I. Legchenkova, "Negative effective mass in plasmonic systems," *Materials*, vol. 13, no. 8, p. 1890, Apr. 2020, doi: 10.3390/ma13081890
42. E. Bormashenko, I. Legchenkova, and M. Frenkel, "Negative effective mass in plasmonic systems. II: Elucidating the optical and acoustical branches of vibrations and the possibility of anti-resonance propagation," *Materials*, vol. 13, no. 16, p. 3512, Aug. 2020, doi: 10.3390/ma13163512
43. V. N. Tsytovich, G. E. Morfill, V. E. Fortov, N. G. Gusein-Zade, B. A. Klumov, and S. V. Vladimirov, "From plasma crystals and helical structures towards inorganic living matter," *New Journal of Physics*, vol. 9, no. 8, pp. 263–263, Aug. 2007, doi: 10.1088/1367-2630/9/8/263
44. Y. Watanabe, "Formation and behaviour of nano/micro-particles in low pressure plasmas," *Journal of Physics D: Applied Physics*, vol. 39, no. 19, pp. R329–R361, Oct. 2006, doi: 10.1088/0022-3727/39/19/R01
45. H. Vach and Q. Brulin, "Controlled growth of silicon nanocrystals in a plasma reactor," *Physical Review Letters*, vol. 95, no. 16, p. 165502, Oct. 2005, doi: 10.1103/PhysRevLett.95.165502
46. T. Matsoukas and M. Russell, "Particle charging in low-pressure plasmas," *Journal of Applied Physics*, vol. 77, no. 9, pp. 4285–4292, May 1995, doi: 10.1063/1.359451
47. J. E. Daugherty and D. B. Graves, "Particulate temperature in radio frequency glow discharges," *Journal of Vacuum Science & Technology A: Vacuum, Surfaces, and Films*, vol. 11, no. 4, pp. 1126–1131, Jul. 1993, doi: 10.1116/1.578452
48. T. M. Boyd and J. J. Sanderson, *The Physics of Plasmas*. Cambridge: Cambridge University Press, 2008.

49. K. P. Giapis, T. A. Moore, and T. K. Minton, "Hyperthermal neutral beam etching," *Journal of Vacuum Science & Technology A: Vacuum, Surfaces, and Films*, vol. 13, no. 3, pp. 959–965, May 1995, doi: 10.1116/1.579658
50. S. Samukawa, "Ultimate top-down etching processes for future nanoscale devices: advanced neutral-beam etching," *Japanese Journal of Applied Physics*, vol. 45, no. 4A, pp. 2395–2407, Apr. 2006, doi: 10.1143/JJAP.45.2395
51. D. Mariotti and R. M. Sankaran, "Microplasmas for nanomaterials synthesis," *Journal of Physics D: Applied Physics*, vol. 43, no. 32, p. 323001, Aug. 2010, doi: 10.1088/0022-3727/43/32/323001
52. D. Mari, P. Hartmann, G. Malovi, Z. Donkó, and Z. L. Petrovi, "Measurements and modelling of axial emission profiles in abnormal glow discharges in argon: heavy-particle processes," *Journal of Physics D: Applied Physics*, vol. 36, no. 21, pp. 2639–2648, Nov. 2003, doi: 10.1088/0022-3727/36/21/007
53. M. Radmilović-Radjenović and B. Radjenović, "Theoretical study of the electron field emission phenomena in the generation of a micrometer scale discharge," *Plasma Sources Science and Technology*, vol. 17, no. 2, p. 024005, May 2008, doi: 10.1088/0963-0252/17/2/024005
54. W. R. Grove, "VII. On the electro-chemical polarity of gases," *Philosophical Transactions of the Royal Society of London*, vol. 142, pp. 87–101, Dec. 1852, doi: 10.1098/rstl.1852.0008
55. H. Schmellenmeier, "Solid lubricants: a review", *Experimentelle Technik der Physik*, vol. 1, p. 49, 1953.
56. T. Suntola, "Atomic layer epitaxy," *Materials Science Reports*, vol. 4, no. 5, pp. 261–312, 1989, doi: 10.1016/S0920-2307(89)80006-4
57. N. T. Alvarez, P. Miller, M. R. Haase, R. Lobo, R. Malik, and V. Shanov, "Tailoring physical properties of carbon nanotube threads during assembly," *Carbon N Y*, vol. 144, pp. 55–62, Apr. 2019, doi: 10.1016/j.carbon.2018.11.036
58. B. Liu, F. Wu, H. Gui, M. Zheng, and C. Zhou, "Chirality-controlled synthesis and applications of single-wall carbon nanotubes," *ACS Nano*, vol. 11, no. 1, pp. 31–53, Jan. 2017, doi: 10.1021/acsnano.6b06900
59. M. Meyyappan, "A review of plasma enhanced chemical vapour deposition of carbon nanotubes," *Journal of Physics D: Applied Physics*, vol. 42, no. 21, p. 213001, Nov. 2009, doi: 10.1088/0022-3727/42/21/213001
60. I. Levchenko, K. Ostrikov, M. Keidar, and U. Cvelbar, "Modes of nanotube growth in plasmas and reasons for single-walled structure," *Journal of Physics D: Applied Physics*, vol. 41, no. 13, p. 132004, Jul. 2008, doi: 10.1088/0022-3727/41/13/132004
61. J. P. Harris, *Carbon Nanotube Science: Synthesis, Properties and Applications*. Cambridge: Cambridge University Press, 2009.
62. N. Kumar, H. C. Lam, and H. K. Quang, *Contemporary Physics*, Second ed. Wiley, 2004.
63. F. Léonard, *The Physics of Carbon Nanotube Devices*. Norwich: William Andrew, 2009.
64. G. Eres, A. A. Kinkhabwala, H. Cui, D. B. Geohegan, A. A. Puretzky, and D. H. Lowndes, "Molecular beam-controlled nucleation and growth of vertically aligned single-wall carbon nanotube arrays," *The Journal of Physical Chemistry B*, vol. 109, no. 35, pp. 16684–16694, Sep. 2005, doi: 10.1021/jp051531i
65. A. A. Puretzky, H. Schittenhelm, X. Fan, M. J. Lance, L. F. Allard, and D. B. Geohegan, "Investigations of single-wall carbon nanotube growth by time-restricted laser vaporization," *Physical Review B*, vol. 65, no. 24, p. 245425, Jun. 2002, doi: 10.1103/PhysRevB.65.245425
66. A. A. Puretzky, D. B. Geohegan, S. Jesse, I. N. Ivanov, and G. Eres, "In situ measurements and modeling of carbon nanotube array growth kinetics during chemical vapor deposition," *Applied Physics A*, vol. 81, no. 2, pp. 223–240, Jul. 2005, doi: 10.1007/s00339-005-3256-7
67. G. Eres, C. M. Rouleau, M. Yoon, A. A. Puretzky, J. J. Jackson, and D. B. Geohegan, "Model for self-assembly of carbon nanotubes from acetylene based on real-time studies of

vertically aligned growth kinetics," *The Journal of Physical Chemistry C*, vol. 113, no. 35, pp. 15484–15491, Sep. 2009, doi: 10.1021/jp9001127
68. M. C. Hersam, "Progress towards monodisperse single-walled carbon nanotubes," *Nature Nanotechnology*, vol. 3, no. 7, pp. 387–394, Jul. 2008, doi: 10.1038/nnano.2008.135
69. Y. Wang, et al., "Continued growth of single-walled carbon nanotubes," *Nano Letters*, vol. 5, no. 6, pp. 997–1002, Jun. 2005, doi: 10.1021/nl047851f
70. L. Zhang, et al., "Sidewall functionalization of single-walled carbon nanotubes with hydroxyl group-terminated moieties," *Chemistry of Materials*, vol. 16, no. 11, pp. 2055–2061, Jun. 2004, doi: 10.1021/cm035349a
71. O. V. Kuznetsov, M. X. Pulikkathara, R. F. M. Lobo, and V. N. Khabasheskua, "Solubilization of carbon nanoparticles, nanotubes, nano-onions, and nanodiamonds through covalent functionalization with sucrose," *Russian Chemical Bulletin*, vol. 59, no. 8, pp. 1495–1505, Aug. 2010, doi: 10.1007/s11172-010-0269-y
72. V. N. Khabashesku, M. X. Pulikkathara, and R. Lobo, "Synthesis of carbon nanotube—nanodiamond hierarchical nanostructures and their polyurea nanocomposites," *Russian Chemical Bulletin*, vol. 62, no. 11, pp. 2322–2326, Nov. 2013, doi: 10.1007/s11172-013-0337-1
73. J. Zhang, L. Zhang, V. N. Khabashesku, A. R. Barron, and K. F. Kelly, "Self-assembly of sidewall functionalized single-walled carbon nanotubes investigated by scanning tunneling microscopy," *The Journal of Physical Chemistry C*, vol. 112, no. 32, pp. 12321–12325, Aug. 2008, doi: 10.1021/jp711658b
74. M. Chitranshi, et al., "Carbon nanotube sheet-synthesis and applications," *Nanomaterials*, vol. 10, no. 10, p. 2023, Oct. 2020, doi: 10.3390/nano10102023
75. P. Bruggeman and C. Leys, "Non-thermal plasmas in and in contact with liquids," *Journal of Physics D: Applied Physics*, vol. 42, no. 5, p. 053001, Mar. 2009, doi: 10.1088/0022-3727/42/5/053001
76. N. Y. Babaeva and M. J. Kushner, "Structure of positive streamers inside gaseous bubbles immersed in liquids," *Journal of Physics D: Applied Physics*, vol. 42, no. 13, p. 132003, Jul. 2009, doi: 10.1088/0022-3727/42/13/132003
77. A. Mizuno, A. Chakrabarti, and K. Okazaki, *Application of Corona Technology in the reduction of greenhouse gases and other gaseous pollutants Non-Thermal Plasma Techniques for Pollution Control. ed B M Penetrante and S E Schultheis*. Berlin: Springer, 1993.
78. H. Chen, H. Lee, S. Chen, Y. Chao, and M. Chang, "Review of plasma catalysis on hydrocarbon reforming for hydrogen production—Interaction, integration, and prospects," *Applied Catalysis B: Environmental*, vol. 85, no. 1–2, pp. 1–9, Dec. 2008, doi: 10.1016/j.apcatb.2008.06.021
79. H. J. Gallon, X. Tu, M. V. Twigg, and J. C. Whitehead, "Plasma-assisted methane reduction of a NiO catalyst—low temperature activation of methane and formation of carbon nanofibres," *Applied Catalysis B: Environmental*, vol. 106, no. 3–4, pp. 616–620, Aug. 2011, doi: 10.1016/j.apcatb.2011.06.023
80. N. Lopez, "On the origin of the catalytic activity of gold nanoparticles for low-temperature CO oxidation," *Journal of Catalysis*, vol. 223, no. 1, pp. 232–235, Apr. 2004, doi: 10.1016/j.jcat.2004.01.001
81. N. Wu, X. Lu, R. An, and X. Ji, "Thermodynamic analysis and modification of Gibbs–Thomson equation for melting point depression of metal nanoparticles," *Chinese Journal of Chemical Engineering*, vol. 31, pp. 198–205, Mar. 2021, doi: 10.1016/j.cjche.2020.11.035
82. S. Vidal, J. Marco-Martínez, S. Filippone, and N. Martín, "Fullerenes for catalysis: metallofullerenes in hydrogen transfer reactions," *Chemical Communications*, vol. 53, no. 35, pp. 4842–4844, 2017, doi: 10.1039/C7CC01267E
83. N. Shainsky, et al., "Retraction: plasma acid: water treated by dielectric barrier discharge," *Plasma Processes and Polymers*, vol. 9, no. 6, Jun. 2012, doi: 10.1002/ppap.201100084

84. A. Fridman, A. Gutsol, A. Dolgopolsky, and E. Shtessel, "CO_2-free energy and hydrogen production from hydrocarbons," *Energy & Fuels*, vol. 20, no. 3, pp. 1242–1249, May 2006, doi: 10.1021/ef050247n
85. A. Bogaerts, W. Wang, A. Berthelot, and V. Guerra, "Modeling plasma-based CO_2 conversion: crucial role of the dissociation cross section," *Plasma Sources Science and Technology*, vol. 25, no. 5, p. 055016, Aug. 2016, doi: 10.1088/0963-0252/25/5/055016
86. A. Bogaerts, T. Kozák, K. van Laer, and R. Snoeckx, "Plasma-based conversion of CO_2: current status and future challenges," *Faraday Discussions*, vol. 183, pp. 217–232, 2015, doi: 10.1039/C5FD00053J
87. R. R. Smith, D. R. Killelea, D. F. DelSesto, and A. L. Utz, "Preference for vibrational over translational energy in a gas-surface reaction," *Science (1979)*, vol. 304, no. 5673, pp. 992–995, May 2004, doi: 10.1126/science.1096309
88. T. Kato and R. Hatakeyama, "Growth of single-walled carbon nanotubes by plasma CVD," *Journal of Nanotechnology*, vol. 2010, pp. 1–11, 2010, doi: 10.1155/2010/256906
89. M. Xu, D. N. Futaba, M. Yumura, and K. Hata, "Alignment control of carbon nanotube forest from random to nearly perfectly aligned by utilizing the crowding effect," *ACS Nano*, vol. 6, no. 7, pp. 5837–5844, Jul. 2012, doi: 10.1021/nn300142j
90. M. Sugasawa, T. Terasawa, and S. Futamura, "Effects of initial water content on steam reforming of aliphatic hydrocarbons with nonthermal plasma," *Journal of Electrostatics*, vol. 68, no. 3, pp. 212–217, Jun. 2010, doi: 10.1016/j.elstat.2009.12.008
91. T. Nozaki and K. Okazaki, "Non-thermal plasma catalysis of methane: principles, energy efficiency, and applications," *Catalysis Today*, vol. 211, pp. 29–38, Aug. 2013, doi: 10.1016/j.cattod.2013.04.002
92. D. P. Broom, C. J. Webb, G. S. Fanourgakis, G. E. Froudakis, P. N. Trikalitis, and M. Hirscher, "Concepts for improving hydrogen storage in nanoporous materials," *International Journal of Hydrogen Energy*, vol. 44, no. 15, pp. 7768–7779, Mar. 2019, doi: 10.1016/j.ijhydene.2019.01.224
93. Y. S. Nechaev, et al., "On the problem of 'super' storage of hydrogen in graphite nanofibers," *C (Basel)*, vol. 8, no. 2, p. 23, Mar. 2022, doi: 10.3390/c8020023.
94. R. F. M. Lobo, D. M. F. Santos, C. A. C. Sequeira, and J. H. F. Ribeiro, "Molecular beam-thermal desorption spectrometry (MB-TDS) monitoring of hydrogen desorbed from storage fuel cell anodes," *Materials*, vol. 5, no. 12, pp. 248–257, Feb. 2012, doi: 10.3390/ma5020248
95. R. F. M. Lobo, F. M. V. Berardo, and J. H. F. Ribeiro, "Molecular beam-thermal hydrogen desorption from palladium," *Review of Scientific Instruments*, vol. 81, no. 4, p. 043103, Apr. 2010, doi: 10.1063/1.3385686
96. C. Fei Guo, T. Sun, F. Cao, Q. Liu, and Z. Ren, "Metallic nanostructures for light trapping in energy-harvesting devices," *Light: Science & Applications*, vol. 3, no. 4, pp. e161–e161, Apr. 2014, doi: 10.1038/lsa.2014.42
97. B. J. Frey, P. Kuang, M.-L. Hsieh, J.-H. Jiang, S. John, and S.-Y. Lin, "Effectively infinite optical path-length created using a simple cubic photonic crystal for extreme light trapping," *Scientific Reports*, vol. 7, no. 1, p. 4171, Dec. 2017, doi: 10.1038/s41598-017-03800-y
98. J. Hao, et al., "Light trapping effect in perovskite solar cells by the addition of Ag nanoparticles, using textured substrates," *Nanomaterials*, vol. 8, no. 10, p. 815, Oct. 2018, doi: 10.3390/nano8100815
99. H. B. Chan, V. A. Aksyuk, R. N. Kleiman, D. J. Bishop, and F. Capasso, "Nonlinear micromechanical Casimir oscillator," *Physical Review Letters*, vol. 87, no. 21, p. 211801, Oct. 2001, doi: 10.1103/PhysRevLett.87.211801
100. C. Genet, A. Lambrecht, and S. Reynaud, "Temperature dependence of the Casimir effect between metallic mirrors," *Physical Review A*, vol. 62, no. 1, p. 012110, Jun. 2000, doi: 10.1103/PhysRevA.62.012110
101. D. C. Cole and H. E. Puthoff, "Extracting energy and heat from the vacuum," *Physical Review E*, vol. 48, no. 2, pp. 1562–1565, Aug. 1993, doi: 10.1103/PhysRevE.48.1562

102. F. Pinto, "Engine cycle of an optically controlled vacuum energy transducer," *Physical Review B*, vol. 60, no. 21, pp. 14740–14755, Dec. 1999, doi: 10.1103/PhysRevB.60.14740
103. T. Ludwig, "Quantum field energy sensor based on the Casimir effect," *Physics Procedia*, vol. 38, pp. 54–65, 2012, doi: 10.1016/j.phpro.2012.08.011
104. M. Revzen, R. Opher, M. Opher, and A. Mann, "Kirchhoff's theorem and the Casimir effect," *Europhysics Letters (EPL)*, vol. 38, no. 4, pp. 245–248, May 1997, doi: 10.1209/epl/i1997-00233-9
105. J. Feinberg, A. Mann, and M. Revzen, "Casimir effect: the classical limit," *Annals of Physics*, vol. 288, no. 1, pp. 103–136, Feb. 2001, doi: 10.1006/aphy.2000.6118
106. J. N. Munday, F. Capasso, and V. A. Parsegian, "Measured long-range repulsive Casimir–Lifshitz forces," *Nature*, vol. 457, no. 7226, pp. 170–173, Jan. 2009, doi: 10.1038/nature07610
107. C. R. Velazquez, C. Noguez, R. Villarreal, and E. Sirvent, "Casimir forces between nanoparticles and substrates," *MRS Proceedings*, vol. 738, pp. G7.35.1–G7.35.6, 2003.

Index

Page numbers in italics indicate a figure on the corresponding page.